Artificial Intelligence in Manufacturing

John Soldatos
Editor

Artificial Intelligence in Manufacturing

Enabling Intelligent, Flexible and
Cost-Effective Production Through AI

 Springer

Editor
John Soldatos
Netcompany-Intrasoft S.A
Märel Luxembourg, Luxembourg

ISBN 978-3-031-46451-5   ISBN 978-3-031-46452-2   (eBook)
https://doi.org/10.1007/978-3-031-46452-2

This work was supported by Netcompany-Intrasoft.

This Springer imprint is published by the registered company Springer Nature Switzerland AG
The registered company address is: Gewerbestrasse 11, 6330 Cham, Switzerland

Paper in this product is recyclable.

Preface

For over a decade, Artificial Intelligence (AI) technologies and applications are proliferating in a rapid pace. The rise of AI is driven by a variety of factors including the unprecedented improvements in hardware and software, and the explosion in the amount of generated data. These advances enable the development of sophisticated AI models (e.g., deep learning models, deep reinforcement learning models, large language models), as well as their deployment and execution in realistic settings. This is also the reason why modern manufacturers are undertaking significant investments in AI solutions as part of their digital transformation journey. As a result, AI is rapidly transforming the manufacturing industry, through enabling tangible improvements in the efficiency, quality, and productivity of industrial organizations.

A variety of AI-based use cases are nowadays deployed in Industry 4.0 production lines. Some of the most prominent examples of such AI-enabled use cases can be found in the areas of predictive maintenance, quality control, supply chain optimization, production planning, process automation, and safety monitoring. For instance, a variety of machine learning models are nowadays used to make quality control more practical and more intelligent, by automating product quality inspection, enabling timely detection of defects, and identifying production configurations that could lead into production problems. As another example, deep learning algorithms are commonly used to predict and anticipate machine failures before they occur, based on predictive and accurate estimations of the Remaining Useful Life (RUL) of the machinery. Likewise, there are AI systems that enable the timely detection of anomalies in products and production processes.

These use cases are some of the most disruptive solutions of the Industry 4.0 era, which is transforming manufacturing enterprises by means of Cyber Physical Production Systems (CPPS). In this direction, most AI use cases for Industry 4.0 emphasize the training, development, and deployment of accurate and effective machine learning systems. The latter are integrated with Industrial Internet of Things (IIoT) systems in the scope of scalable and secure cloud/edge environments. To this end, industrial solution integrators leverage structuring principles and blueprints specified in standards-based reference architectures for Industry 4.0

systems. Nevertheless, the AI's potential for manufacturing is still largely underexploited. State of the art systems are usually limited to the extraction of data-driven AI-based insights for improving production processes and related decision making. These insights are based on quite simple models about the production processes and hardly combine capabilities of multiple AI systems and algorithms. To alleviate these limitations, there are research initiatives that explore the integration and collaboration of multiple AI systems in the scope of production processes. In this direction, there is on-going research on:

- *Multi-agent systems* that foster enhanced collaborative intelligence based on the interaction and the development of synergies across different autonomous AI agents.
- *Solutions for AI interoperability* across diverse systems. These solutions leverage advanced knowledge models (e.g., Semantic Knowledge Graphs (SKGs) and embeddings that capture the relationships between different entities) to enable the development of sophisticated AI systems that span entire multi-stage production processes beyond simple ML-based state machines.

During the last couple of years, Industry 4.0 is evolving to a direction where AI serves manufacturing workers, while at the same time interacting closely with them in a variety of human-in-the-loop scenarios such as human-robot collaboration (HRC) scenarios. At the same time, AI use cases are increasingly aiming at increasing production sustainability to ensure that the manufacturing sector contributes to strategic targets such as the European Green Deal (EGD) of the European Union (EU). Sustainability and human-centricity are driving the transition of Industry 4.0 digital manufacturing systems to the Industry 5.0 era, which emphasizes human-centricity and environmental performance.

The advent of Industry 5.0 systems is increasing the functional sophistication and integration complexity of AI systems in manufacturing. It also asks for an evolution of AI in a human-centered dimension, where AI systems operate in trustworthy and reliable manner. Specifically, the evolution of AI systems toward the Industry 5.0 era asks for:

- *Novel AI architectures for Industry 5.0*: Industry 5.0 system comprises multiple AI components (e.g., robots, machine learning models, Natural Language Processing (NLP)) that must safely and effectively interact with humans in industrial environments. The development and deployment of such systems requires novel architectures and structuring principles, beyond classical architectures of Big Data, AI, and Industrial Internet of Things (IIoT) platforms.
- *Knowledge Modelling and Representation*: HRC use cases are usually deployed in highly dynamic environments involving humans, robots, and AI systems that interact with each other. The implementation of advanced and automated reasoning in such an environment asks for novel ways for representing processes in ways that capture the complex interrelationships between the different actors.
- *Models and Learning paradigms for Human-Robot Collaboration*: Industry 5.0 introduces a need for deploying novel learning paradigms that foster the interplay

between humans and AI actors. Such paradigms include, for example, active learning and intelligent Multi-Agent Systems (MAS). They enable outcomes that combine the speed of AI systems with the credibility of human judgment. Likewise, Industry 4.0 solutions like Digital Twins are currently transformed to account for the context of the human workers, i.e., they are evolving toward human-centric digital twins.

- *Explainability, transparency, and regulatory compliance*: Industry 5.0 systems pose their own unique transparency and safety requirements. They involve humans in the loop that must be able to understand the decisions and operation of AI system. Hence, AI use cases cannot be developed based on black-box AI models. Rather, AI systems should be transparent, explainable, trusted, and understandable to humans. Manufacturers must also ensure that their AI systems adhere to the mandates of emerging AI regulations such as the AI Act in Europe.

The aim of this book is to shed light on the limitations of existing solutions for AI in manufacturing and to introduce novel solutions that:

- Improve the functional capabilities and technical performance of state-of-the-art AI systems for manufacturing in a variety of production processes like production scheduling and quality control
- Enhance the human centricity, the trustworthiness, and the overall social performance of AI systems in line with the requirements and concepts of the Industry 5.0 era

The book comprises 27 chapters that present innovative AI systems and solutions spanning both state-of-the-art Industry 4.0 use cases and emerging, human-centric Industry 5.0 use cases. The chapters are contributed by Y EU-funded projects, which are closely collaborated in the context of the AI4Manufacturing Cluster of European projects, as well as in the scope of the activities of the European Factories of the Future Research Association (EFFRA). The contributing projects focus on the development, deployment, and operation of AI systems for production lines. Each of the project addresses a set of unique challenges of AI in Industry 4.0 and/or Industry 5.0 use cases, such as the development and deployment of effective MAS systems, the development of trusted and explainable AI systems, the specification and implementation of knowledge models and semantics for Industry 5.0 applications, as well as the development of novel forms of digital twin systems and applications (e.g., human-centric digital twins).

Specifically, the book is structured in the following three parts:

Part I: Architectures and Knowledge Modelling for AI

This part presents architectures for AI-based Industry 5.0 systems and solutions, ranging from high-level reference architecture models to architecture of specific AI platforms and solutions' marketplaces. The presented architectures illustrate the structure of both Industry 4.0 and Industry 5.0 use cases with emphasis on the structuring principles that drive the integration of AI and ML models with industrial systems. Moreover, this part of the book includes several chapters that illustrate semantic modelling techniques for AI applications in manufacturing, including techniques based on semantic knowledge graphs and embeddings.

Part II: Multi-agent Systems and AI-based Digital Twins for Manufacturing Applications

This part of the book presents multi-agent systems and digital twin solutions for Industry 5.0. The digital twins' solutions can identify the users' context toward modeling and simulating AI-based processes with the human in the loop. In terms of multi-agent systems, the chapter presents human-AI interaction approaches based on the intelligent agents, which empower decentralized collaborative intelligence paradigms for AI in manufacturing.

Part III: Trusted, Explainable, and Human-Centered AI Systems

This part of the book introduces novel approaches to implementing human-centered, trusted, and explainable AI systems for digital manufacturing applications. Most of the presented solutions target human-in-the-loop scenarios such as human-robot interactions and emphasize not only the technical performance but also the social performance of AI systems. Therefore, they are suitable for applications of the Industry 5.0 era.

Overall, the book provides a comprehensive overview of AI technologies and applications in manufacturing covering both Industry 4.0 and Industry 5.0 environments. The book is provided as an open access publication, which ensures that researchers and practitioners will have unlimited access to it. In essence, it is a contribution of the AI4Manufacturing cluster of projects and of various other EU programs to the Industry 4.0 and Industry 5.0 communities. I hope that researchers, practitioners, and providers of industrial automation solutions for manufacturing will find it interesting.

Märel Luxembourg, Luxembourg   John Soldatos
July 2023

Acknowledgments

This book has received funding from the European Union's Horizon 2020 research and innovation program under grant agreements No. 956573 (STAR), No. 957204 (MAS4AI), No. 957362 (XMANAI), No. 101000165 (ASSISTANT), No. 957331 (knowlEdge), and No. 957402 (Teaming. AI), which are part of the AI4Manufacturing Cluster of projects. Several contributions of the book have been also supported financially from other EU projects (e.g., No. 952119 (KITT4SME) No. 870092 (DIMOFAC), No. 869963 (MERGING)) and various national projects as indicated in the acknowledgement sections of each chapter.

The editor and the chapter co-authors acknowledge valuable support from partners of the above-listed EU projects.

Disclaimer: The contents of the book reflect only the contributors' and co-authors' view. The European Commission is not responsible for any use that may be made of the information it contains.

Contents

Editor and Contributors

About the Editor

John Soldatos (http://gr.linkedin.com/in/johnsoldatos) holds a PhD in Electrical and Computer Engineering from the National Technical University of Athens (2000) and is currently Honorary Research Fellow at the University of Glasgow, UK (2014–present). He was Associate Professor and Head of the Internet of Things (IoT) Group at the Athens Information Technology (AIT), Greece (2006–2019), and Adjunct Professor at the Carnegie Mellon University, Pittsburgh, PA (2007–2010). He has significant experience in working closely with large multi-national industries (IBM Hellas, INTRACOM S.A., INTRASOFT International S.A., Netcompany-Intrasoft S.A., and Netcompany S.A.) as R&D consultant and delivery specialist, while being scientific advisor to high-tech startup enterprises. Dr. Soldatos is an expert in Internet-of-Things (IoT) and Artificial Intelligence (AI) technologies and applications, including applications in smart cities, finance (Finance 4.0), and industry (Industry 4.0). Dr. Soldatos has played a leading role in the successful delivery of more than 70 (commercial-industrial, research, and business consulting) projects, for both private and public sector organizations, including complex integrated projects. He is co-founder of the open-source platform OpenIoT (https://github.com/OpenIotOrg/openiot). He has published more than 200 articles in international journals, books, and conference proceedings. He has also significant academic teaching experience, along with experience in executive education and corporate training. Dr. Soldatos is a regular contributor in various international magazines and blogs, on topics related to Artificial Intelligence, IoT, Industry 4.0, and cybersecurity. Moreover, he has received national and international recognition through appointments in standardization working groups, expert groups, and various boards. He has recently coedited and coauthored eight edited volumes (books) on Artificial Intelligence, BigData, and Internet of Things-related themes.

Contributors

Enrico Alberti Nextworks SRL, Pisa, Italy

Patricio Alemany Rovimática SL, Córdoba, Spain

Rubén Alonso R2M Solution s.r.l., Pavia, Italy
Programa de Doctorado, Centro de Automática y Robótica, Universidad Politécnica de Madrid-CSIC, Madrid, Spain

Sergio Alvarez-Napagao High Performance and Artificial Intelligence, Barcelona Supercomputing Center, Barcelona, Spain

Victor Anaya Information Catalyst SL, Xativa, Spain

Veronica Antonello TXT e-solutions SpA, Milan, Italy
TXT e-tech, Milan, Italy

E. Bakopoulos LMS, Laboratory for Manufacturing Systems, Patras, Greece

Marta Barroso High Performance and Artificial Intelligence, Barcelona Supercomputing Center, Barcelona, Spain

Zeki Mert Barut Department of Innovative Technologies, University of Applied Science of Southern Switzerland, Manno, Switzerland

Christian Beecks FernUniversity of Hagen, Hagen, Germany

Alexis T. Bernhard Deutsches Forschungszentrum für Künstliche Intelligenz GmbH (DFKI), Kaiserslautern, Germany

Andrea Bettoni Department of Innovative Technologies, University of Applied Science of Southern Switzerland, Manno, Switzerland

Evmorfia Biliri Suite5 Data Intelligence Solutions, Limassol, Cyprus

Enrica Bosani Whirlpool Management EMEA, Milan, Italy

Cornelis Bouter Data Science, Netherlands Organisation for Applied Scientific Research (TNO), Den Haag, The Netherlands

Mattia Calabresi TXT e-solutions SpA, Milan, Italy
TXT e-tech, Milan, Italy

Andrea Capaccioli Deep Blue, Rome, Italy

Gabriel Gonzalez Castañé The Insight SFI Research Centre of Data Analytics, University College Cork, Cork, Ireland

Fernando Castaño Centro de Automática y Robótica (CSIC-Universidad Politécnica de Madrid), Madrid, Spain

Sara Cavallaro CNH Industrial, Modena, Italy

Fernando Cebrian Fersa Bearings, Zaragoza, Spain

Christos Chadoulos AIDEAS OU, Tallinn, Estonia

Sisay Adugna Chala Department of Data Science and AI, Fraunhofer Institute for Applied Information Technology (FIT), Sankt Augustin, Germany

Sergio Clavijo Fundacion TECNALIA R&I, Madrid, Spain

Javier Colomer-Barbera Ford, Valencia, Spain

Yarens J. Cruz Centro de Automática y Robótica (CSIC-Universidad Politécnica de Madrid), Madrid, Spain

Vincenzo Cutrona University of Applied Science of Southern Switzerland, Switzerland

Theodore Dalamagas Athena Research Center, Marousi, Greece

Silvia Rodríguez Del Rey Asociación de Empresas Tecnológicas Innovalia, Calle Rodríguez Arias, Bilbao, Spain

Angela-Maria Despotopoulou Netcompany-Intrasoft S.A, Luxembourg, Luxembourg

Danilo Dessí Knowledge Technologies for the Social Sciences Department, GESIS – Leibniz Institute for the Social Sciences, Cologne, Germany

Vasileios Dimitriadis Centre for Research and Technology Hellas, Information Technologies Institute (CERTH/ITI), Thessaloniki, Greece

Krzysztof Ejsmont Faculty of Mechanical and Industrial Engineering, Warsaw University of Technology, Warsaw, Poland
Institute of Production Systems Organization, Faculty of Mechanical and Industrial Engineering, Warsaw University of Technology, Warsaw, Poland

Dimitris Eleftheriou CORE Innovation, Athens, Greece

Emmanouil Bakopoulos Laboratory for Manufacturing Systems & Automation (LMS), Department of Mechanical Engineering & Aeronautics, University of Patras, Rio-Patras, Greece

Christos Emmanouilidis University of Groningen, Groningen, The Netherlands

Hector Diego Estrada-Lugo Technological University Dublin, School of Environmental Health, Dublin, Ireland

Andrea Falconi Martel Innovate, Zurich, Switzerland

Álvaro Flores Rovimática SL, Córdoba, Spain

Blaž Fortuna Qlector d.o.o., Ljubljana, Slovenia

Thanassis Giannetsos Ubitech Ltd., Athens, Greece

Victor Gimenez-Abalos High Performance and Artificial Intelligence, Barcelona Supercomputing Center, Barcelona, Spain

Bartlomiej Gladysz Faculty of Mechanical and Industrial Engineering, Warsaw University of Technology, Warsaw, Poland
Institute of Production Systems Organization, Faculty of Mechanical and Industrial Engineering, Warsaw University of Technology, Warsaw, Poland

Marta Gonzalez-Mallo High Performance and Artificial Intelligence, Barcelona Supercomputing Center, Barcelona, Spain

Carlos González-Val AIMEN Technology Centre, Smart Systems and Smart Manufacturing Group, Pontevedra, Spain

Fabio Grandi Università di Modena e Reggio Emilia, Modena, Italy

Alexander Graß Department of Data Science and AI, Fraunhofer Institute for Applied Information Technology (FIT), Sankt Augustin, Germany

David Guillén Fersa Bearings, Zaragoza, Spain

Rodolfo Haber Centre for Automation and Robotics (CAR), Spanish National Research Council-Technical University of Madrid (CSIC-UPM), Madrid, Spain

Rodolfo E. Haber Centro de Automática y Robótica (CSIC-Universidad Politécnica de Madrid), Madrid, Spain

Bernhard Heinzl Software Competence Center Hagenberg GmbH, Hagenberg, Austria

Daniel Hinjos High Performance and Artificial Intelligence, Barcelona Supercomputing Center, Barcelona, Spain

Thomas Hoch Software Competence Center Hagenberg GmbH, Hagenberg, Austria

Maike Holtkemper FernUniversity of Hagen, Hagen, Germany

Erifili Ichtiaroglou Suite5 Data Intelligence Solutions, Limassol, Cyprus

Echeverría Imanol Fundacion TECNALIA R&I, Madrid, Spain

Dimosthenis Ioannidis Centre for Research and Technology Hellas, Information Technologies Institute (CERTH/ITI), Thessaloniki, Greece

Babis Ipektsidis Netcompany-Intrasoft, Brussels, Belgium

Simon Jungbluth Technologie-Initiative SmartFactory KL e.V., Kaiserslautern, Germany

Ali Karnoub Technologie-Initiative SmartFactory KL e.V., Kaiserslautern, Germany

Nikos Kefalakis Netcompany-Intrasoft S.A, Luxembourg, Luxembourg

Elmar Kiesling WU, Institute for Data, Process and Knowledge Management, Vienna, Austria

Timotej Klemenčič University of Ljubljana, Ljubljana, Slovenia

Kosmas Alexopoulos Laboratory for Manufacturing Systems & Automation (LMS), Department of Mechanical Engineering & Aeronautics, University of Patras, Rio-Patras, Greece

Franz Krause University of Mannheim, Data and Web Science Group, Mannheim, Germany

Kabul Kurniawan WU, Institute for Data, Process and Knowledge Management, Vienna, Austria
Austrian Center for Digital Production (CDP), Vienna, Austria

Fenareti Lampathaki Suite5 Data Intelligence Solutions, Limassol, Cyprus

Giuseppe Landolfi University of Applied Science of Southern Switzerland, Switzerland

Eleni Lavasa Athena Research Center, Marousi, Greece

Maria Chiara Leva Technological University Dublin, School of Environmental Health, Dublin, Ireland

Ainhoa Etxabarri Llana UNIMETRIK S.A., Legutiano,Álava, Spain

Afra Maria Petrusa Llopis AIMEN Technology Centre, Smart Systems and Smart Manufacturing Group, Pontevedra, Spain

Alberto Botana López AIMEN Technology Centre, Smart Systems and Smart Manufacturing Group, Pontevedra, Spain

Pedro Lopez Fundacion TECNALIA R&I, Madrid, Spain

Andreas Louca Suite5 Data Intelligence Solutions, Limassol, Cyprus

Paul Lukowicz German Research Center for Artificial Intelligence (DFKI), Kaiserslautern, Germany
Department of Computer Science, RPTU Kaiserslautern-Landau, Kaiserslautern, Germany

Daniel Gordo Martín AIMEN Technology Centre, Smart Systems and Smart Manufacturing Group, Pontevedra, Spain

Andrea Fernández Martínez AIMEN Technology Centre, Smart Systems and Smart Manufacturing Group, Pontevedra, Spain

Jorge Martinez-Gil Software Competence Center Hagenberg GmbH, Hagenberg, Austria

Pablo A. Martin High Performance and Artificial Intelligence, Barcelona Supercomputing Center, Barcelona, Spain

Jawad Masood AIMEN Technology Centre, Smart Systems and Smart Manufacturing Group, Pontevedra, Spain

Antonello Meloni Mathematics and Computer Science Department, University of Cagliari, Cagliari, Italy

Markku Mikkola VTT Technical Research Centre of Finland Ltd., Espoo, Finland

Elena Minisci CRIT S.R.L., Vignola, Italy

Dunja Mladenić Jožef Stefan Institute, Ljubljana, Slovenia

Elias Montini University of Applied Science of Southern Switzerland, Switzerland
Politecnico di Milano, Milan, Italy

David Monzo Tyris AI, Valencia, Spain

Bernhard Moser Software Competence Center Hagenberg GmbH, Hagenberg, Austria

William Motsch Deutsches Forschungszentrum für Künstliche Intelligenz GmbH (DFKI), Kaiserslautern, Germany

Serafeim Moustakidis AIDEAS OU, Tallinn, Estonia

Santiago Muiños-Landin AIMEN Technology Centre, Smart Systems and Smart Manufacturing Group, Pontevedra, Spain

Marco Murgia Mathematics and Computer Science Department, University of Cagliari, Cagliari, Italy

Jose Angel Segura Muros AIMEN Technology Centre, Smart Systems and Smart Manufacturing Group, Pontevedra, Spain

Linda Napoletano Deep Blue, Rome, Italy

Nikolaos Nikolakis Laboratory for Manufacturing Systems & Automation (LMS), Department of Mechanical Engineering & Aeronautics, University of Patras, Rio-Patras, Greece

Nikoletta Nikolova Data Science, Netherlands Organisation for Applied Scientific Research (TNO), Den Haag, The Netherlands

Alexandros Nizamis Centre for Research and Technology Hellas, Information Technologies Institute (CERTH/ITI), Thessaloniki, Greece

Parsha Pahlevannejad German Research Center for Artificial Intelligence (DFKI), Kaiserslautern, Germany
Technologie-Initiative SmartFactory, Kaiserslautern, Germany

Panagiotis Mavrothalassitis Laboratory for Manufacturing Systems & Automation (LMS), Department of Mechanical Engineering & Aeronautics, University of Patras, Rio-Patras, Greece

Dimitrios Papamartzivanos Ubitech Ltd., Athens, Greece

Christos Patsonakis Centre for Research and Technology Hellas, Information Technologies Institute (CERTH/ITI), Thessaloniki, Greece

Sergi Perez-Castanos Tyris AI, Valencia, Spain

Mario Pichler Software Competence Center Hagenberg GmbH, Hagenberg, Austria

Christiane Plociennik German Research Center for Artificial Intelligence (DFKI), Kaiserslautern, Germany
Technologie-Initiative SmartFactory, Kaiserslautern, Germany

Ausias Prieto-Roig Tyris AI, Valencia, Spain

Reforgiato Recupero Diego Mathematics and Computer Science Department, University of Cagliari, Cagliari, Italy
R2M Solution s.r.l., Pavia, Italy

Vitor Fortes Rey German Research Center for Artificial Intelligence (DFKI), Kaiserslautern, Germany
Department of Computer Science, RPTU Kaiserslautern-Landau, Kaiserslautern, Germany

Natalia Roczon Faculty of Mechanical and Industrial Engineering, Warsaw University of Technology, Warsaw, Poland

Sjoerd Rongen Data Ecosystems, Netherlands Organisation for Applied Scientific Research (TNO), Den Haag, The Netherlands

Jože M. Rožanec Jožef Stefan Institute, Ljubljana, Slovenia
Qlector d.o.o., Ljubljana, Slovenia

Martin Ruskowski German Research Center for Artificial Intelligence (DFKI), Kaiserslautern, Germany
Technologie-Initiative SmartFactory KL e.V., Kaiserslautern, Germany

Georg Schlake FernUniversity of Hagen, Hagen, Germany

Jona Scholz FernUniversity of Hagen, Hagen, Germany

Gabriele Scivoletto Nextworks SRL, Pisa, Italy

Michele Sesana TXT e-solutions SpA, Milan, Italy
TXT e-tech, Milan, Italy

Georgios Siachamis Centre for Research and Technology Hellas, Information Technologies Institute (CERTH/ITI), Thessaloniki, Greece

Aleksandr Sidorenko Deutsches Forschungszentrum für Künstliche Intelligenz GmbH (DFKI), Kaiserslautern, Germany

Agastya Silvina Software Competence Center Hagenberg GmbH, Hagenberg, Austria

Athanasios Siouras AIDEAS OU, Tallinn, Estonia

John Soldatos Netcompany-Intrasoft S.A, Luxembourg, Luxembourg

Sungho Suh German Research Center for Artificial Intelligence (DFKI), Kaiserslautern, Germany
Department of Computer Science, RPTU Kaiserslautern-Landau, Kaiserslautern, Germany

Dimitris Syrrafos Suite5 Data Intelligence Solutions, Limassol, Cyprus

Hooman Tavakoli German Research Center for Artificial Intelligence (DFKI), Kaiserslautern, Germany
Technologie-Initiative SmartFactory, Kaiserslautern, Germany

Dimitrios Tzovaras Centre for Research and Technology Hellas, Information Technologies Institute (CERTH/ITI), Thessaloniki, Greece

Luis Usatorre Fundacion TECNALIA R&I, Madrid, Spain

Michael van Bekkum Data Science, Netherlands Organisation for Applied Scientific Research (TNO), Den Haag, The Netherlands

Vasilis Siatras Laboratory for Manufacturing Systems & Automation (LMS), Department of Mechanical Engineering & Aeronautics, University of Patras, Rio-Patras, Greece

Entso Veliou University of West Attica, Aigaleo, Greece

Alberto Villalonga Centro de Automática y Robótica (CSIC-Universidad Politécnica de Madrid), Madrid, Spain

Konstantinos Votis Centre for Research and Technology Hellas, Information Technologies Institute (CERTH/ITI), Thessaloniki, Greece

Eduardo Vyhmeister The Insight SFI Research Centre of Data Analytics, University College Cork, Cork, Ireland

Achim Wagner Deutsches Forschungszentrum für Künstliche Intelligenz GmbH (DFKI), Kaiserslautern, Germany

Stefan Walter VTT Technical Research Centre of Finland Ltd., Espoo, Finland

Snehal Walunj German Research Center for Artificial Intelligence (DFKI), Kaiserslautern, Germany
Technologie-Initiative SmartFactory, Kaiserslautern, Germany

Robert Wilterdink Advanced Computing Engineering, Netherlands Organisation for Applied Scientific Research (TNO), Den Haag, The Netherlands

Abbreviations

ACDS	AI Cyber-Defence Strategies
ACL	Agent Communication Language
ADAS	Advanced Driver Assistance System
AHP	Analytic Hierarchy Process
AI	Artificial Intelligence
AIaaS	AI as a Service
AL	Active Learning
ALE	Accumulated Local Efforts
AMG	AI Model Generation
AMR	Automatic Mobile Robots
ANOVA	Analysis of Variance
APaaS	Application Platform as a Service
API	Application Programming Interface
AR	Augmented Reality
AutoML	Automated Machine Learning
BM	Business Model
BOM	Bill Of Materials
CAD	Computer-Aided Design
CD	Continuous Delivery
CEAP	Circular Economy Action Plan
CI	Continuous Integration
CM	Community Management
CMOS	Complementary Metal Oxide Semiconductor
CNC	Computer Numerical Control
CNN	Convolutional Neural Network
CoAP	Constrained Application Protocol
CPPM	Cyber-Physical Production Modules
CPPS	Cyber-Physical Production Systems
CPS	Cyber-Physical Systems
CPT	Capabilities Periodic Table
CPU	Central Processing Unit

CRUD	Create Update Delete
CSV	Comma Separated Values
CV	Computer Vision
DAIRO	Data, AI, and Robotics
DBFSP	Distributed Blocking Flowshop Scheduling Problem
DCP	Data Collection Platform
DES	Discrete Event Simulation
DL	Deep Learning
DLT	Distributed Ledger Technologies
DnDF	Non-discrimination, and Fairness
DNN	Deep Neural Network
DoS	Denial of Service
DP	Data Type Probing
DPO	Data Protection Officer
DRP	Deep Reinforcement Learning
DS	Digital System
DS	Data Scientist
DSS	Decision Support System
DT	Digital Twin
DTF	Digital Twin Framework
DTI	Decentralised Technical Intelligence
D2C	Direct sales to final Customers
EC	European Commission
EGD	European Green Deal
ERP	Enterprise Resource Planning
ESCO	European Skills, Competences, and Occupations
EU	European Union
EFFRA	European Factories of the Future Research Association
FMEA	Failure Mode and Effects Analysis
FMS	Flexible Manufacturing System
FGSM	Fast Gradient Sign Method
FT	Fungible Token
GAT	Graph ATtention
GCN	Graph Convolutional Network
GDPR	General Data Protection Regulation
GE	Generic Enabler
GNN	Graph Neural Network
GPR	Gaussian Process Regression
GraphML	Graph Machine Learning
GRPN	Global Risk Priority Number
GUI	Graphical User Interface
GW	Griding Wheel
HAR	Human Activity Recognition
HCAI	Human-Centered Artificial Intelligence
HDT	Human Digital Twin

HITL	Human In The Loop
HLF	Hyperledger Fabric
HMI	Human-Machine Interface
HMS	Holonic Manufacturing System
HOTL	Human On The Loop
HR	Human Resources
HRC	Human-Robot Collaboration
HRI	Human-Robot Interface
HTTP	HyperText Transport Protocol
HTTPS	Secure HyperText Transport Protocol
ICT	Information and Communications Technology
IDTA	Industrial Digital Twin Association
IEEE	Institute of Electrical and Electronics Engineers
IEEE-SA	IEEE Standards Association
IDM	IDentity Management and Access Control
IFS	Innovative Factory Systems
IIAF	Industrial Internet Architecture Framework
IIC	Industrial Internet Consortium
IIoT	Industrial Internet of Things
IIRA	Industrial Internet Reference Architecture
IISF	Industrial Internet Security Framework
ILO	International Labour Organization
IMU	Inertial Measurement Unit
IoT	Internet of Things
IoU	Intersection of Union
ISCO	International Standard Classification of Occupations
ISO	International Organization for Standardization
IT	Information Technology
I4.0L	I4.0 Language
JSON	JavaScript Object Notation
JSSP	Job Shop Scheduling Problem
KG	Knowledge Graph
KPI	Key Performance Indicator
KQML	Knowledge Query and Manipulation Language
LIME	Local Interpretable Model-Agnostic Explanations
LLM	Large Language Models
LPG	Labeled Property Graph
LSTM	Long Short-Term Memory
MAP	Mean Average Precision
MAPE	Mean Absolute Percentage Error
MAS	Multi-Agent System
MC	Malicious Control
MDP	Markov Decision Process
MIP	Mixed Integer Programming
ML	Machine Learning

MLP	Multilayer Perceptron
MMD	Maximum Mean Discrepancy
MO	Malicious Operation
MO	Machine Operator
MPMS	Manufacturing Process Management System
MP&L	Material Planning & Logistics
MQTT	Message Queue Telemetry Transport
MSE	Mean Squared Error
MSP	Multi-Sided Platforms
MVP	Minimum Viable Platform
NAICS	North American Industry Classification System
NFT	Non-Fungible Token
NLP	Natural Language Processing
NOC	National Occupation Classification
OD	Object Detection
OECD	Organisation for Economic Cooperation and Development
OEM	Original Equipment Manufacturer
ONNX	Open Neural Network Exchange
OPC-UA	Open Platform Communications United Architecture
OT	Operational Technology
OWL	Web Ontology Language
O*NET	Occupational Information Network
PBT	Population-Based Training
PCA	Principal Component Analysis
PCPSP	Precedence Constrained Production Scheduling Problem
PDT	Platform Design Toolkit
PGD	Projected Gradient Descent
PIAAC	Programme for the International Assessment of Adult Competencies
PLM	Product Lifecycle Management
PLM	Production Line Manager
PMML	Predictive Model Markup Language
POV	Point of View
PPKB	Production Processes Knowledge Base
PRM	Process of Risk Management
PS	Physical System
QFD	Quality Function Deployment
RA	Reference Architecture
RAME	Risk Assessment and Mitigation Engine
RAMI	Reference Architecture Model Industry 4.0
RAMP	Robotics and Automation Marketplace
RASP	Risk Architecture, Strategy, Protocols
RCA	Route Cause Analysis
RDF	Resource Description Framework
REST	Representation State Transfer
RGAN	Relational Graph Attention Network

RGCN	Relational Graph Convolutional Network
RL	Reinforcement Learning
RMS	Reconfigurable Manufacturing System
RMSLE	Root Mean Squared Log Error
ROS	Robot Operation System
RPN	Risk Priority Number
RQ	Research Question
RZSG	Robust Zero-Sum Game
SaaS	Software as a Service
SC	Smart Contract
SHACL	Shapes Constraint Language
SHAP	SHapley Additive exPlanations
SQL	Structured Query Language
SME	Small Medium Enterprise
SNE	Stochastic Neighbor Embedding
SOTA	State of the Art
SPM	Security Policies Manager
SPR	Security Policies Repository
SS	Software Scientist
SSO	Single Sign On
STEP	Skills Towards Employment and Productivity
SVM	Support Vector Machine
SVR	Support Vector Regression
TAI	Trustworthy AI
TEER	Training, Education, Experience, and Responsibilities
TLS	Transport Level Security
UI	User Interface
WS	Wrong Setup
XAI	eXplainable Artificial Intelligence
YOLO	You Only Look Once

Part I
Architectures and Knowledge Modelling for AI in Manufacturing

Reference Architecture for AI-Based Industry 5.0 Applications

John Soldatos ⓘ, **Babis Ipektsidis** ⓘ, **Nikos Kefalakis** ⓘ,
and **Angela-Maria Despotopoulou** ⓘ

1 Introduction

For over a decade, manufacturing enterprises are heavily investing in their digital transformation based on Cyber-Physical Production Systems (CPPS) that enable the digitization of production processing such as production scheduling, products' assembly, physical assets' maintenance, and quality control. The deployment and operation of CPPS in the manufacturing shopfloor is the main enabler of the fourth industrial revolution (Industry 4.0) [1], which boosts automation and efficiency toward improving production speed and quality [2], while lowering production costs and enabling novel production models such as lot-size-one manufacturing and mass customization.

Industry 4.0 applications are usually developed based on advanced digital technologies such as Big Data, Internet of Things (IoT), and Artificial Intelligence (AI), which are integrated with CPPS systems in the manufacturing shopfloor and across the manufacturing value chain. In cases of nontrivial Industry 4.0 systems, this integration can be challenging, given the number and the complexity of the systems and technology involved. For instance, sophisticated Industry 4.0 use cases are likely to comprise multiple sensors and automation devices, along with various data analytics and AI modules that are integrated in digital twins (DTs) systems and applications. To facilitate such challenging integration tasks, industrial automation solution providers are nowadays offered with access to various

J. Soldatos (✉) · N. Kefalakis · A.-M. Despotopoulou
Netcompany-Intrasoft S.A, Luxembourg, Luxembourg
e-mail: john.soldatos@netcompany.com; nikos.kefalakis@netcompany.com;
angelamaria.despotopoulou@netcompany.com

B. Ipektsidis
Netcompany-Intrasoft, Brussels, Belgium
e-mail: babis.ipektsidis@netcompany.com

J. Soldatos (ed.), *Artificial Intelligence in Manufacturing*,
https://doi.org/10.1007/978-3-031-46452-2_1

reference architecture models for Industry 4.0 applications. These models illustrate the functionalities and technological building blocks of Industry 4.0 applications, while at the same time documenting structuring principles that facilitate their integration and deployment in complete systems and applications. Some of these reference architecture models focus on specific aspects of Industry 4.0 such as data collection, data processing, and analytics, while others take a more holistic view that addresses multiple industrial functionalities. Moreover, several architecture models address nonfunctional requirements as well, such as the cybersecurity and safety of industrial systems.

During the last couple of years, there is a surge of interest on Industry 4.0 applications that emphasize human-centered industrial processes, i.e., processes with the human in the loop, as well as the achievement of ambitious sustainability and resilience objectives. The latter are at the very top of the policy agenda of the European Union, as reflected in the European Green Deal (EGD) and Europe's Circular Economy Action Plan (CEAP). This has led to the introduction of the term Industry 5.0, which evolves Industry 4.0 in a direction that complements efficiency and productivity goals with societal targets, notably contributions to sustainability and the workers' well-being [3]. Hence, Industry 5.0 targets a sustainable, human-centric, and resilient industry [4]. In this direction, Industry 4.0 systems must be enhanced with human-centric technologies that put the worker at the center of the production process, while at the same time fostering security, safety, transparency, and trustworthiness. For instance, the shift from Industry 4.0 to Industry 5.0 asks for the deployment and use of transparent, interoperable, and explainable AI systems [5], beyond black-box systems (e.g., deep neural networks) that are typically used in Industry 4.0 deployments. As another example, Industry 5.0 applications comprise technological paradigms that foster the collaboration between humans and industrial systems (e.g., co-bots), rather than systems that aim at replacing the human toward hyper-automation (e.g., fully autonomous industrial robots). Likewise, the scope of digital twins in Industry 5.0 tends to differ from Industry 4.0, as simulations and what-if analysis account for human parameters (e.g., physical characteristics, emotional status, skills) as well. Also, Industry 5.0 pays greater emphasis on nonfunctional requirements such as data protection, security, and safety when compared to Industry 4.0 that prioritizes industrial performance and accuracy.

Despite these technological differences between Industry 5.0 and Industry 4.0 systems, there is still a lack of standards, formal guidelines, and blueprints for developing, deploying, and operating Industry 5.0 systems. In most cases, manufacturers and providers of industrial automation solutions make use of conventional Industry 4.0 and blueprints, which they enhance with the required features and functionalities of their Industry 5.0 use cases at hand. We argue that this is a considerable misstep in the process of designing and implementing Industry 5.0 solutions, as it deprives architects and developers of industrial systems of the opportunity to consider Industry 5.0 functionalities and features from the early stages of the industrial systems' development. State-of-the-art approaches to developing Industry 5.0 systems start from Industry 4.0 reference architectures and address human-centricity, sustainability, and resilience as secondary, complementary concerns rather than as indispensable requirements that must be prioritized.

Motivated by the general lack of reference architecture models for Industry 5.0 systems, this chapter introduces a reference architecture for human-centric, resilient, and sustainable industrial systems, notably digital manufacturing systems that are developed based on cutting-edge technologies such as Artificial Intelligence (AI). The reference architecture highlights the main functional and nonfunctional aspects of Industry 5.0 systems and introduces technological building blocks that can support their implementation. In this direction, the present model specifies technologies that foster human-centricity and trustworthy interactions between humans and industrial systems, such as human-centered digital twins, explainable and interpretable AI, active learning, neurosymbolic learning, and more. Emphasis is also put on technological building blocks that boost cybersecurity and safety, such as technologies for ensuring the trustworthiness of data and machine learning algorithms. Along with the specification of building blocks and structuring principles for the integration in end-to-end Industry 5.0 solutions, the chapter delineates various blueprints that can facilitate the development of Industry 5.0 use cases. The presented blueprints include guidelines for regulatory compliance in the European Union (EU), notably compliance to the European AI regulation proposal (i.e., the "AI Act") [6].

The remainder of the chapter is structured as follows:

- Section 2 presents related work on reference architectures (RAs) for Industry 4.0. Various RAs are briefly reviewed and their concepts that are relevant to Industry 5.0 solutions are highlighted.
- Section 3 introduces a reference architecture for Industry 5.0 systems. A high-level model of functionalities for human-centric and resilient manufacturing is first introduced, followed by a more detailed logical architecture. Moreover, the section presents a set of technological building blocks that can support the development of real-life systems based on the presented architecture.
- Section 4 presents a set of blueprints for developing Industry 5.0 solutions based on the presented architecture. The blueprints include guidelines and best practices for building solutions compliant to the European Regulation for AI systems.
- Section 5 is the final and concluding section of the chapter.

2   Relevant Work

A considerable number of reference architecture models have been recently introduced to facilitate the development, integration, and deployment of Industry 4.0 applications. These include architectural models specified by standards development organizations and research initiatives. As a prominent example, the Industrial Internet Reference Architecture (IIRA) prescribes a standards-based architecture for developing, deploying, and operating Industrial Internet of Things (IIoT) systems [7]. It is destined to boost interoperability across different IoT systems and to provide a mapping on how different technologies can be exploited toward developing IIoT systems. The IIRA is described at a high level of abstraction, as it strives to have broad applicability. Its specification has been driven by the analysis of a rich collection of industrial use cases, notably use cases prescribed in the scope of the activities of the Industrial Internet Consortium (IIC). The IIRA is described based on the ISO/IEC/IEEE 42010:2011 [8] standard, which has been adopted by IIC to define its Industrial Internet Architecture Framework (IIAF). The IIRA is defined in terms of four complementary viewpoints: (1) the "business viewpoint" presents the functional modules that are destined to support the business goals of different stakeholders; (2) the "usage viewpoint" presents the way systems compliant to IIRA are used. It includes various sequences of activities involving human or logical actors, which deliver the functionality prescribed by the architecture; (3) the "functional viewpoint" presents the functional components of an IIoT system, including their structure and interrelation, as well as the interfaces and interactions between them; and (4) the "implementation viewpoint" is devoted to technologies that are used to implement the various functional components and their interactions. In the scope of the IIRA, there are also cross-cutting elements, i.e., elements and functions that are applicable to all viewpoints, including connectivity, industrial analytics, distributed data management, as well as intelligent and resilient control.

The functional viewpoints of the IIRA specify five sets of functionalities, including: (1) the "control domain," which comprises functions conducted by industrial control and automation; (2) the "operations domain," which deals with the management and operation of the control domain. It comprises functions for the provisioning, management, monitoring, and optimization of control domain systems and functions; (3) the "information domain," which focuses on managing and processing data from other domains, notably from the control domain; (4) the "application domain," which provides the application logic required to implement the various business functions; and (5) the "business domain," which implements business logic that supports business processes and procedural activities in the scope of an IIoT system.

Overall, the IIRA provides a taxonomy of the main functional areas of industrial systems, which are relevant to Industry 5.0 systems and applications as well. The IIRA introduces a clustering of functionalities into different domains, which we will leverage in Sect. 3 toward introducing our Industry 5.0 architecture. Also, the IIRA illustrates how specific functions such as asset management and cybersecurity functions can be integrated with IIoT systems and provides insights about how to best structure the logical and implementation views of industrial architectures. The implementation view of the IIRA is based on a cloud/edge computing approach, which is also the suggested implementation approach for Industry 5.0 systems. Nevertheless, the IIRA does not include any specific provisions for human-centric industrial systems such as systems that collect and analyze information about the context of the human user toward customizing the industrial functionalities accordingly.

The Industrial Internet Security Framework (IISF) complements the IIRA with a security viewpoint for industrial systems [9]. One of the main objectives of the IISF is to prescribe the functions needed for the development, deployment, and operation of trusted IIoT. These functions are also essential for ensuring the trustworthiness of Industry 5.0 systems and their AI components in industrial environments. Thus, the structure and functions of IISF provided inspiration about how to support AI trustworthiness for industrial use cases. The IISF specifies functionalities that secure all the different elements of an industrial system such as the various communication endpoints of the system. Most of these functions can be used to boost the security and trustworthiness of Industry 5.0 systems as well, as they safeguard the operation of the networks, the data, and the data processing functions of Industry 5.0 systems. Specifically, the IISF is concerned with the five main characteristics that affect the trustworthiness of IIoT deployments, i.e., security, safety, reliability, resilience, and privacy. The framework specifies a functional viewpoint that is destined to secure IIoT systems compliant to the IIRA. To this end, the functional viewpoint specifies six interacting and complementary building blocks, which are organized in a layered fashion. The top layer comprises four security functions, namely endpoint protection, communications and connectivity protection, security monitoring and analysis, and security configuration management. Likewise, a data protection layer and a system-wide security model and policy layer are specified. Each one of the functional building blocks of the IISF can be further analyzed in more fine-grained functions such as monitoring functionalities, data analytics functionalities, and actuation functionalities. Each of these three types of functionalities include security-related functions.

One more reference architecture for industrial systems, notably for fog computing systems, was introduced by the OpenFog Consortium prior to its incorporation within the Industrial Internet Consortium in 2019 [10]. The OpenFog RA specifies the structure of large-scale fog computing system with emphasis on how fog nodes are connected to enhance the intelligence and to boost the efficiency of Industrial IoT systems. The OpenFog RA specifies some cross-cutting functionalities, which are characterized as "perspectives." One of these perspectives deals with the security functionalities, which implies that security is applicable to all layers and use scenarios from the hardware device to the higher software layers of the architecture. As already outlined, such security functions are key to the development and deployment of trusted industrial systems of the Industry 5.0 era.

The Big Data Value Association (BDVA) has specified the structure of big data systems based on the introduction of a reference model for big data systems [11]. The model illustrates a set of modules that are commonly used in big data systems along with structuring principles that drive their integration. The BDVA reference model consists of the following layers:

- *Horizontal layers* that illustrate the modules and the structure of data processing chains. The modules of data processing chains support functions such as data collection, data ingestion, data analytics, and data visualization. The horizontal layers do not map to a layered architecture, where all layers must coexist in the scope of a system. For instance, it is possible to have a data processing chain that leverages data collection and visualization collection functions without necessarily using data ingestion and data analytics functionalities. Hence, the BDVA horizontal layers can be used as building blocks to construct data pipelines for AI systems.

- *Vertical layers* that deal with cross-cutting issues such as cybersecurity and trust. The latter are applicable to all functionalities of the horizontal layers. Vertical layers can be also used to specify and address nontechnical aspects such as the ever important legal and regulatory aspects of AI systems.

The horizontal and vertical layers of the reference model are used to produce concrete architectures for big data systems. There are clearly many commonalities between big data and AI systems as many AI systems (e.g., deep learning systems) are data-intensive and process large amounts of data. The BDVA RA does not, however, address functionalities that foster the development of Industry 5.0 systems, such as data quality and AI model explainability functionalities. As such it is mostly appropriate for architecting AI systems without special provisions for their trustworthiness and human centricity.

Standards-based functionalities for AI systems are also specified by the ISO/IEC JTC 1/SC 42 technical committee on Artificial Intelligence [12]. The committee has produced several standards that cover different aspects of AI systems, such as data quality for analytics and machine learning (ML) (i.e., ISO/IEC DIS 5259-1), transparency taxonomy of AI systems (i.e., ISO/IEC AWI 12792), a reference architecture of knowledge engineering (i.e., ISO/IEC DIS 5392), functional safety and AI systems (i.e., ISO/IEC CD TR 5469), as well as objectives and approaches for explainability of ML models and AI systems (i.e., ISO/IEC AWI TS 6254). As evident from the above-listed descriptive titles, the ISO/IEC JTC 1/SC 42 technical committee addresses human centricity (e.g., safety, explainability) and trustworthiness (e.g., data quality, explainability) issues for Industry 5.0 systems. Nevertheless, most of the relevant standards are under development and not yet available for practical applications and use.

In recent years, the IEEE (Institute of Electrical and Electronics Engineers) SA (Standards Association) is developing a suite of standards (i.e., standards of the IEEE 7000 family) that deal with the ethical aspects of AI systems. For instance, the IEEE 7000-2021 standard titled "IEEE Standard Model Process for Addressing Ethical Concerns during System Design" [13] specifies a process that organizations can follow to ensure that their AI systems adhere to ethical values and integrate ethical AI concepts within their systems' development lifecycle. These standards can facilitate the development of Industry 5.0. However, they are mostly focused on development, deployment, and operational processes rather on how to structure Industry 5.0 systems.

Overall, there is a still a lack of standards-based architectures and blueprints for the development of Industry 5.0 systems. Hence, AI developers, deployers, and engineers have no easy ways to structure, design, and build nontrivial trustworthy human-centric AI systems [14].

3 Architecture for AI-Based Industry 5.0 Systems (STAR-RA)

3.1 High-Level Reference Model for AI-Based Industry 5.0 Systems

3.1.1 Overview

A high-level reference model for AI-based Industry 5.0 systems is illustrated in Fig. 1. The model specifies a set of functionalities that foster trustworthiness and safety of AI systems in-line with the mandates of Industry 5.0. It clusters Industry 5.0 functionalities in three main categories or functional domains according to the terminology of the Industrial Internet Reference Architecture (IIRA). The three domains are as follows:

- *Cybersecurity domain*: This domain includes functionalities that boost the cybersecurity and cyber resilience of AI systems in industrial settings. These functionalities ensure the reliability and security of industrial data, as well as of the AI algorithms that are trained and executed based on these data. The functionalities of this domain support and reinforce the trustworthiness of the project's functions in the other two domains.
- *Human–robot collaboration (HRC) domain*: This domain provides functionalities for the trusted collaboration between human and robots. It leverages cybersecurity functionalities, while reinforcing functionalities in the safety domain. The specified functionalities aim at boosting trusted interactions between humans and AI systems in ways that yield better performance than humans or AI systems alone. In this direction, this domain leverages AI models that foster the collaboration between humans and AI systems such as active learning [15] and neurosymbolic learning (e.g., [16, 17]).
- *Safety domain*: This domain comprises functionalities that ensures the safety of industrial operations, including operations that involve workers and/or automation systems. For instance, functionalities in this domain reinforce worker safety, while boosting the safe operation of AMRs (automatic mobile robots).

For each one of the functional domains, Fig. 1 presents a set of functionalities, notably functionalities that are actually implemented and validated in the context of the H2020 STAR project. These functionalities are illustrated in the following subparagraphs. Note, however, that they are by no means exhaustive, as it is possible to specify additional functionalities that boost the cybersecurity, HRC and safety of AI-based functionalities in AI systems.

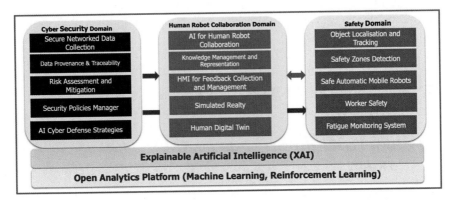

Fig. 1 STAR high-level reference model for Industry 5.0 systems

Explainable AI (XAI) has a prominent role in the high-level reference model of Fig. 1. This is because XAI functionalities provide support to various functionalities of the different domains. For instance, XAI is an integral element of AI models that foster HRC such as neurosymbolic learning. Similarly, XAI can be used to identify potential attempts to tamper AI systems by altering their operations. As such XAI is positioned as a cross-cutting functionality of XAI systems. As shown in Fig. 1, machine learning platforms are among the main pillars of AI-based Industry 5.0 systems as most of the functionalities are deployed and executed over such platforms. This is, for example, the case for AI systems that implement cyber-defense strategies in the cybersecurity domain and reinforcement learning systems that detect safety zones in the safety domain.

3.1.2 Cybersecurity Domain Functionalities

Cybersecurity functionalities are key to ensuring the trustworthiness of AI systems. This is because they ensure the trustworthiness and reliability of industrial data and the AI models that are trained and developed based on these data. Moreover, they protect these systems from cybersecurity attacks that could compromise their operation and break the trust on their operation. A set of indicative functionalities that have been implemented in the STAR project include:

- *Secure networked data collection*: This refers to the implementation of secure networked protocols for accessing industrial data stemming from Cyber-Physical Systems (CPS) and enterprise applications (e.g., ERP (Enterprise Resource Planning) of an industrial site. For instance, they entail the implementation of data collection probes based on secure networked protocols such as TLS (Transport Level Security) and HTTS (Secure HyperText Transport Protocol).
- *Data provenance and traceability*: Data provenance and traceability of industrial data is key toward ensuring industrial data reliability. Specifically, provenance

and traceability functionalities enable AI system developers to continually access properties and characteristics of the source industrial data (e.g., hash codes, statistical properties) in order to implement tamper-proof functionalities [18]. Such functionalities can be implemented for both source data and AI models built on them to help preventing their tampering by malicious actors.

- *Risk assessment and mitigation*: Risk assessment and mitigation functionalities are integral to most cybersecurity systems. In the case of AI-based industrial systems, they provide the means to identify, assess, and install mitigation measures against potential cybersecurity risks. Some of these risks (e.g., data falsification) can be mitigated based on other functionalities of the cybersecurity domain (e.g., provenance and traceability).
- *Security Policies Manager*: This functionality makes provision for the specification and implementation of security policies for AI systems. It is a placeholder for a wide range of policies such as authenticated and authorized access to industrial data and AI functions.
- *AI cyber-defense strategies*: AI systems are subject to additional vulnerabilities and cyber-resilience risks. For example, hackers can launch data poisoning attacks that alter the data that are used for training machine learning systems toward compromising their operation (e.g., [19]). Likewise, AI systems must be robust against evasion attacks that manipulate input data toward producing errors in the operation of machine learning systems (e.g., [20]). AI cyber-defense functionality aim at mitigating and confronting such attacks based on techniques such as auditing of the training data and formal verification of the input data.

3.1.3 HRC Domain Functionalities

The list of HRC functionalities of the high-level reference model includes:

- *AI for human–robot collaboration*: Beyond classical machine learning models and paradigms, there are machine learning approaches that foster the interplay between humans and AI systems to enable effective HRC. This is, for example, the case with active learning systems, where robots and ML systems can consult a human expert to deal with uncertainty during data labeling processes. Active learning approaches accelerate knowledge acquisition by AI systems, facilitate HRC, and improve the overall performance and trust of AI deployments.
- *Knowledge management and representation*: The knowledge of proper representation and management of HRC processes is key to the implementation of effective human–AI interactions. In this direction, semantic modeling techniques (e.g., semantic knowledge graphs) are employed to facilitate AI systems to understand and reason over the context of the HRC process [21].
- *HMI for feedback collection and management*: HRC systems involve interactions between humans and AI systems, including the provision of feedback from the human to the AI system. This is, for example, the case in the above-mentioned active learning systems where humans provide data labeling feedback to AI

systems. To implement such feedback collection and management functionalities, the HRC domain includes placeholders for proper HMI (human machine interfaces) such as NLP (natural language processing) interfaces.

- *Simulated reality*: Simulated reality systems enable the training of HRC systems in the scope of virtual simulated environments. They are usually linked to reinforcement learning systems that are trained in virtual rather than in real environments. Their inclusion in the reference model signals the importance of training humans for HRC functionalities in a safe environment. As such they are also enablers of functionalities in the safety domain of the AI-based Industry 5.0 reference model, such as functionalities for safety zones detection.
- *Human digital twin (HDT)*: This is a placeholder for human-centered digital twins, which are digital twin systems that comprise information about the characteristics and the context of the human workers [22]. HDT systems are ideal for modeling, developing, and implementing HRC systems that blend AI systems with human actors, while modeling and simulating their interactions.

3.1.4   Safety Domain Functionalities

The safety domain of the reference model outlines functionalities that are key to ensuring the safe operation of AI systems (e.g., robots) in the scope of Industry 5.0 scenarios. It identifies the following indicative but important functionalities:

- *Object localization and tracking*: This functionality aims at identifying the location of objects within industrial environments, notably of moving objects such as mobile robots. The localization and tracking of such functionalities are integral elements of applications that safeguard the safe operation of robotics and AI systems in industrial environments.
- *Safety zones detection*: Automation systems that move within an industrial environment (e.g., shopfloor, plant floor) must follow safe trajectories that minimize the risk of collisions between automation systems, humans, and other stationary elements of the workplace. In this direction, the detection of safety zones based on AI technologies (e.g., reinforcement learning [23]) can increase the safety of the AI systems and minimize related risks.
- *Safe automatic mobile robots*: This functionality is a placeholder for systems that ensure the safe movement of automatic mobile robots. The implementation of this functionality can benefit from other functionalities of this domain such as the detection of safety zones.
- *Worker safety*: Apart from ensuring the safe operation and movement of robotic systems, it is important to ensure the safety of the workers. Workers' safety is at the heart of Industry 5.0 system that emphasize human centricity. The respective functionalities ensure that workers act within safe environments and that the emotional and physical context of the human is properly considered in the design, development, deployment, and operation of AI systems.

- *Fatigue monitoring system*: Fatigue monitoring is a very prominent example of a human-centered functionality that can boost both the performance and the safety of industrial systems. It collects and analyzes information about the fatigue of the worker, which can then be used to drive the adaption of AI and Industry 5.0 systems toward a worker-centric direction.

3.2 Logical Architecture for AI-Based Industry 5.0 Systems

3.2.1 Driving Principles

Figure 2 illustrates a specific instantiation of the reference architecture model, which has been implemented in the scope of the STAR project [25] and is conveniently called STAR-RA in the scope of this chapter. The architecture is presented in the form of a logical view, which comprises functional modules along with their structure and their interactions with other systems. It can serve as a basis for implementing, deploying, and operating AI-based Industry 5.0 systems. Systems compliant to the STAR-RA aim at securing existing AI-based CPPS systems in manufacturing production lines based on a holistic approach that pursues the following principles that are fully in-line with the earlier presented high-level architecture:

- *Secure and reliable data*: The STAR AI systems must operate over reliable industrial data, i.e., the architecture makes provisions for alleviating the inherent unreliability of industrial data.

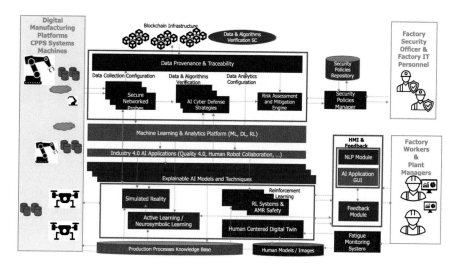

Fig. 2 STAR reference architecture for AI-based Industry 5.0 systems

- *Secure and trusted AI algorithms*: Systems compliant to the STAR-RA enhance the secure operation of the AI systems and algorithms that they comprise. In this direction, they make provisions for implementing cyber-defense strategies that protect and defend AI systems from malicious security attacks. STAR-RA compliant systems focus on defenses against cybersecurity attacks. Physical security attacks may be applicable to some industrial systems (e.g., robotics systems), yet they are not address by the STAR-RA.
- *Trusted human AI interactions*: The presented architectures focus on the implementation of trusted interactions between humans and AI systems. On the one hand, it ensures that AI systems are transparent and explainable to humans toward boosting their acceptance and adoption. On the other hand, it also focuses on safe and trusted interactions between humans and AI systems in HRC scenarios.
- *Safe AI systems*: The architecture boosts the safety of autonomous AI systems such as mobile robots. For example, it focuses on the secure placement and movement of autonomous mobile robots (AMRs) within industrial plants.

The above listed functionalities and the logical modules that implement them can not only work independently but also in synergy with each other. For instance, reinforcement learning (RL) algorithms can be used to ensure the safe operation of AMRs, which contributes to the trusted operation of AI systems. Such RL algorithms can operate independently from other modules. However, they can also be integrated with the industrial data reliability systems of the cybersecurity domain toward ensuring that they operate over trusted and reliable industrial data. This boosts and reinforces their trustworthiness. Moreover, they can be integrated with AI-based cyber-defense strategies to ensure that they cannot be tampered or compromised by malicious parties. This is yet another measure to strengthening the trustworthiness of AI systems for safe AMR operations. Overall, when integrating and combining multiple modules of the architecture, manufacturers and system integrators can gain a multiplicative trustworthiness benefit, as one system can reinforce the other.

This architectural proposition provides the structuring principles for integrating the AI-based industrial systems of the Industry 5.0 era. However, the presented functional modules do not represent an "all-or-nothing" value proposition. Manufacturers and AI systems integrators can adopt and implement parts of the STAR-RA, i.e., specific modules of the logical architecture.

As illustrated in Fig. 2, the architecture enables the development of AI systems that receive data from the shopfloor (i.e., digital manufacturing platforms and other AI-based CPPS systems) and provide different types of AI-based services to factory (cyber)security teams and to other factory stakeholders (e.g., industrial engineers, plant managers, factory workers). The alignment of the different modules to the high-level architecture of the previous subsection is illustrated by means of their color, i.e., blue-colored modules belong to the cybersecurity domain, red-colored modules to the safety domain and green-colored to the HRC domain.

3.2.2 Logical Modules

The main modules and building blocks of the architecture are illustrated in the following subparagraphs.

Digital Manufacturing Platforms and CPPS Systems

The architecture enables the development of secure, safe, and trusted AI systems in production lines. To this end, systems compliant to the STAR-RA collect and process data from AI-based systems in the shopfloor, including machines, robotic cells, AMRs, and other digital manufacturing platforms. Industry 5.0 systems comprise various CPPS systems and digital manufacturing platforms that serve as data sources for other logical modules. The latter may also consume data from other data sources in the shopfloor such as business information systems (e.g., ERP (Enterprise Resource Planning)) and manufacturing databases (e.g., historian systems).

Industry 5.0 Applications

This building block represents different types of AI-based industrial applications such as machine learning (ML) and robotics applications. They leverage information and data sources from the shopfloor. In some cases, they are integrated with the digital manufacturing platforms as well. Other modules of the architecture collect data from them and analyze their behavior toward boosting the security and trustworthiness of their operation. AI-based Industry 5.0 applications can also be data sources that provide data to other logical modules and data-driven systems of the architecture.

Secure Networked Probes (SNP)

This building block provides a secure data collection solution that offers a real time data collection, transformation, filtering, and management services to facilitate data consumers (e.g., the AI cyber-defense module and the Security Policies Manager) in accessing the required data. For example, it can be used to collect security-related data (e.g., network, system, and solution proprietary) from monitored IoT systems and store them to detect patterns of abnormal behavior by applying simple (i.e., filtering) or more elaborate (i.e., deep learning) data processing mechanisms. The solution features specialized probes that may be deployed within the monitored IoT/CPPS system or poling services for acquiring data from shopfloor sources such as CPPS systems and digital manufacturing platforms. The module belongs to the cybersecurity domain of the high-level reference architecture.

Data Provenance and Traceability (DPT)

The DPT module belongs to the cybersecurity domain and provides the means for tracking and tracing industrial data. It interfaces to the data probes to acquire information about the industrial data of the shopfloor such as information about data types, volumes, and timestamps. Accordingly, it records this information (i.e., the metadata) about the acquired data to facilitate the detection of data abuse and data tampering attempts. Specifically, data ingested in the DPT can be queried by other modules to facilitate the validation of datasets and to ensure that the data they consume have not been falsified. In this way, the DPT module reinforces the reliability and the security of the industrial data that flow through the system.

Blockchain – Distributed Ledger Infrastructure

There are different ways for implementing a DPT infrastructure for industrial data. The STAR-RA promotes a decentralized approach, which leverages the benefits of distributed ledger technologies, i.e., a blockchain protocol. Distributed ledger infrastructures offer some advantages for industrial data provenance, such as immutable and tamper-resistant records. They also provide a comprehensive and auditable trail that records the history of data transactions, including creation, modification, and transfer events. In addition, blockchains enable the implementation of Smart Contracts (SC) over the distributed ledger infrastructure, notably SCs that are used to validate the metadata of the industrial datasets that are recorded in the blockchain. SCs enable decentralized applications that provide information about the metadata to interested modules such as the cyber-defense strategies module.

AI Cyber-Defense Strategies (ACDS)

This module implements cyber-defense strategies for AI systems, i.e., strategies that protect AI systems against adversarial attacks. These strategies operate based on access to industrial data from:

- The AI systems (including ML systems) that must be protected from cybersecurity attacks.
- The CPPS and digital manufacturing platforms that act as data sources.
- The metadata of the industrial data that are managed by the DPT module and its blockchain implementation.
- The explainable AI (XAI) module, which implements explainable AI models that illustrate and interpret the operation of various AI systems and algorithms.

The module materializes different strategies in response to attacks against AI systems. For instance, it implements cyber-defense strategies for poisoning and evasion attacks. Nevertheless, additional cyber-defense strategies can be implemented and integrated with the rest of the modules (i.e., secure networked data probes, DPT).

A data integration infrastructure (e.g., based on a data bus pattern) can be used as a data exchange middleware infrastructure to facilitate data transfer and data sharing across different modules involved in the detection of a cybersecurity attacks, i.e., the DPT, the ACDS, and the SNP.

Risk Assessment and Mitigation Engine (RAME)

This module implements the security risk assessment and mitigation service of the STAR-RA in-line with the cybersecurity domain of the high-level architecture. The module assesses risk for assets associated with AI-based systems in manufacturing lines. In this direction, it interacts with the AI cyber-defense strategies modules as follows: (1) the defense strategies communicate to the RAME information about identified risks for AI assets; and (2) the RAME consumes information from the DPT to assess risks. It also offers mitigation actions for the identified risks such as changes to the configuration of a probe via the SNP module.

Security Policies Manager (SPM) – Security Policies Repository (SPR)

This module defines and configures security policies that govern the operation of the DPT, AI cyber-defense, and the RAME modules. Specifically, the module specifies security policies that provide information about the probes and data sources to be integrated, the configurations of the probes, as well as the cyber-defense strategies to be deployed. By changing the applicable policies, the SPM changes the configuration and the operation of other modules of the cybersecurity domain (e.g., DPT, RAME, ACDS). The operation of the SPM is supported by a Security Policies Repository (SPR), where policy files are persisted. Furthermore, the SPM offers a GUI (graphic user interface) to the security officers of the factory (e.g., members of CERT (computer emergency response teams)).

Machine Learning and Analytics Platform

Several modules of the architecture are based on machine learning algorithms, including deep learning and reinforcement learning. This is, for example, the case of the ACDS module, which implements AI-based defense strategies among others. Another prominent example is the XAI module, which produces explainable ML models. The machine learning and analytics platform supports the operation of these AI systems. It enables developers and users of the STAR-RA modules (i.e., data scientists, domain experts, ML engineers) not only to specify and execute ML models but also to access their metadata and outcomes. The platform interacts with modules that provide datasets for training and executing AI algorithms such as the SDP module.

Explainable Artificial Intelligence (XAI)

This module provides and executes XAI models and algorithms. It provides the means for executing different types of XAI algorithms such as algorithms for explaining deep neural networks and general-purpose algorithms (e.g., LIME – Local Interpretable Model-Agnostic Explanations) [24] that explain the outcomes of AI classifiers. As such the module is a placeholder of XAI techniques. The XAI module provides its services to several other modules that leverage explainable algorithms, such as the AI cyber-defense strategies module and the simulated reality (SR) modules.

Simulated Reality (SR)

This module simulates production settings in a virtual world with a twofold objective: (1) producing data to be used by AI algorithms, especially in cases where real-world data are not available at adequate quantities; and (2) utilizing reinforcement learning techniques in artificial settings (i.e., simulated environments) toward accelerating their convergence. SR leverages services from the XAI module, which facilitate humans to assess the appropriateness and correctness of the simulated data that are generated by the SR.

Active Learning (AL) and Neurosymbolic Learning

This module provides a placeholder for machine learning paradigms that foster HRC, i.e., modules of the HRC domain of the high-level reference architecture. Such paradigms include active learning and neurosymbolic learning, which help robots and AI systems to benefit from human expertise in the context of human in the loop industrial processes. These machine learning techniques for HRC fall in the scope of the HRC domain of the high-level architecture.

Production Processes Knowledge Base (PPKB)

This module consolidates domain knowledge about the production processes of the manufacturing environment. It is used for inferencing by the other modules such as the AL and neurosymbolic learning modules. The latter modules can interact and update the module with knowledge acquired by the humans in the scope of human-in-the-loop processes. Therefore, it also falls in the scope of the HRC domain.

AMR Safety

This module comprises RL techniques that boost the safety of AMRs in industrial environments such as manufacturing shopfloors. It provides insights on the safe placement of robots in a manufacturing environment. To this end, it incorporates functionalities such as objective localization and safety zones detection of the safety domain of the high-level architecture of Fig. 1.

Human-Centered Digital Twin

This module implements a digital twin that factors human-centered parameters (e.g., fatigue, emotional status of the worker). It is a placeholder for digital twins of human-centered processes, including AI-based processes that have the human in the loop. It interacts with the analytics platforms, the workers, and the humans' digital models.

The HDT offers a centralized access point to exploit a wide set of workers' related data. It leverages a digital representation of the workers, which is seamlessly integrated with production system DTs. The latter can be exploited by AI-based modules to compute complex features that, enriching the HDT, enable better decisions, and dynamically adapt automation systems behavior toward improving production performance, workers' safety, and well-being.

Human Models – Human Digital Images

This module persists and manages data about the human worker toward supporting the construction, deployment, and operation of HDTs. They provide the means for creating and using digital representations of the workers.

Graphical User Interface (GUI) – Human Machine Interface (HMI)

This module provides a GUI interaction modality between factory workers and AI systems. It comprises visualization elements (e.g., dashboards), while enabling users to interact with the AI-based modules (e.g., provide form-based input).

Natural Language Processing (NLP)

This module enables NLP interactions between the factory users and relevant AI modules (e.g., AL modules). It is a placeholder for different NLP implementations and interfaces.

Feedback Module

This module coordinates the provision of feedback from the human worker to the AI system. It is particularly important for the implementation of human–AI systems interactions (e.g., HRC scenarios). The feedback module interfaces to some interaction module (e.g., GUI or NLP) that enables the transferring of user data to the feedback module and vice versa.

Fatigue Monitoring System

This module leverages sensors and IoT devices (e.g., electroencephalography (EEG) sensors) to collect information about the worker's fatigue. The collected information is transferred to other modules such as the human models and the HDT.

4   Solution Blueprints for Industry 5.0 Applications

4.1   The Industry 5.0 Blueprints Concept

The STAR-RA for Industry 5.0 applications can be used to support the implementation of popular secure and trustworthy data-driven use cases in industrial environments. In this direction, selected functional modules of the STAR-RA can be deployed and operated. The specification of the modules and the information flows that can support specific HRC, cybersecurity, and safety solutions in Industry 5.0 context can be defined as blueprints over the introduced architecture. Each blueprint provides a proven way to implement trusted data processing and AI functionalities for industrial applications. Rather than having to read, browse and understand the entire STAR-RA and its low-level technical details, interested parties (e.g., solution integrators, manufacturers, researchers in industrial automation, and digital manufacturing) could consult blueprints as practical ways for enhancing the trustworthiness and regulatory compliance of their work. Following paragraphs illustrate blueprints for popular technical solutions and for the adherence of Industry 5.0 to the AI regulation proposal of the European Parliament and the Council of Europe.

4.2   Technological Solutions Blueprints

Several blueprints can be defined based on the AI cyber-defense strategies (ACDS) to support use cases for defending cybersecurity attack against AI systems. As a prominent example, Table 1 illustrates the blueprint against defending a poisoning

Table 1 Poisoning attack defense

Blueprint title	Poisoning attack defence
Scope and purpose	Detect with high accuracy a poisoning attack against an AI/ML system, i.e., cases where an attacker compromises the learning process based on adversarial examples, in ways that compromise the AI systems ability to produce correct/credible results.
STAR-RA components involved	Analytics platform, STAR blockchain (distributed ledger infrastructure), DPT (data provenance and traceability), risk assessment and mitigation engine, XAI module

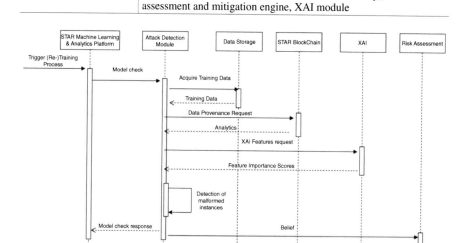

Fig. 3 STAR implementation of AI cyber-defence information flow

attack, where an adversary attempts to contaminate the data used to train an AI system.

Figure 3 presents the information flow between the various components of this blueprint, as implemented in the STAR project.

As another example, Table 2 presents a blueprint for solutions that validate the integrity of industrial data. This is crucial in the scope of Industry 5.0 applications as it is a foundation for ensuring industrial data reliability. Different types of industrial data can be protected based on this blueprint, such as CPPS data and analytics results (including AI outcomes). Figure 4 illustrates the information flow between different blockchain components that implement data integrity validation blueprint.

4.3 Regulatory Compliance Blueprints

Regulatory compliance blueprints illustrate how the STAR-RA and its component could be leveraged to boost the adherence of AI solutions to the AI regulation

Table 2 Validating the integrity of industrial data

Blueprint title	Validating the integrity of industrial data
Scope and purpose	Retrieve persisted critical measurements (e.g., analytics results) from the blockchain to be validated/compared with existing data to verify their authenticity based on their metadata properties
STAR-RA components involved	Data models, blockchain (distributed ledger infrastructure), DPT (data provenance and traceability)

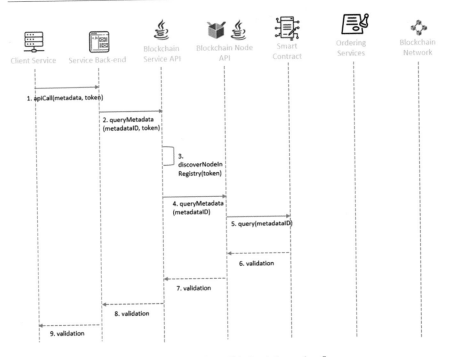

Fig. 4 STAR implementation of data integrity validation information flow

proposal of the European Commission. In April 2021, the European Parliament and the Council of Europe presented an initial proposal for the regulation of AI systems [6]. This proposal is the first organized and structured effort to regulate AI systems worldwide. Its importance for systems deployed within Europe is particularly high, given that it lays a basis for future laws within the various EU member states. The proposal establishes a technology-neutral definition of AI systems in EU law, while presenting a risk-based classification of AI systems. The classification proposes to categorize AI systems in four general classes, ranging from unacceptable risk to no risk (i.e., risk free) systems. It also outlines the requirements and obligations associated with the deployment of systems from each one of the envisaged risk levels. For instance, "high-risk" AI systems can be authorized if and only if they meet requirements spanning the areas of transparency, explainability, data quality,

Table 3 Supporting the deployment of AI systems of minimal risk

Blueprint title	Supporting the deployment of AI systems of minimal risk
Scope and purpose	Support adherence to codes of conduct that mandate transparency of AI system
STAR-RA components involved	(Optional) XAI (for transparency); (optional) security policies manager (for increased cybersecurity); (optional) AI cyber-defence strategies (ACDS) (for AI cybersecurity)

Table 4 Supporting the deployment of AI systems of limited risk

Blueprint title	Supporting the deployment of AI systems of limited risk
Scope and purpose	Support the mandatory transparency of AI system; provide optional support for increasing the security and safety of limited risk AI systems
STAR-RA components involved	(Mandatory) XAI (for transparency); (optional) blockchain (distributed ledger infrastructure) and DPT (data provenance and traceability) for data quality; (optional) security policies manager (for increased cybersecurity); (optional) AI cyber-defence strategies (ACDS) (for AI cybersecurity)

and more. These obligations are significantly lower for medium- and low-risk systems.

STAR-RA includes technical components that can help AI deployers and operators to meet regulatory requirements and obligations. Different components can be used to support systems belonging to the different risk classes of the AI Act. For example, the AI Act specifies that minimal-risk systems (e.g., ML-based calculations and visualization of information about physical assets) can be deployed without essential restrictions. There are no mandatory obligations for minimal risk systems. Compliance to AI code of conduct for them is recommended, yet optional. Deployers may therefore choose to deploy one or more STAR components from the different domains of the platform (cybersecurity, human–robot collaboration, safety), as well as explainable AI components as illustrated in the relevant regulatory blueprint of Table 3.

On the other hand, when deploying a limited risk system, AI deployers must ensure that they are meeting transparency obligations. In this direction, humans must be notified of the existence of an AI system component in the loop of the industrial process. This concerns industrial processes with the human in the loop, where AI systems and human interact. It is, for example, the case of some HDT applications where industrial systems collect information about the status of the worker and adapt their operations to it. The STAR-RA includes XAI components that can help deployers meet the requirements of limited risk deployments. Deployers can optionally use other STAR components to increase the safety, security, and overall trustworthiness of the AI system (see Table 4).

Many AI systems in manufacturing and other industrial environments can be classified as being high risk. This is, for example, the case with systems involving AMRs and other types of industrial robots. In the case of high-risk systems,

Table 5 Supporting the deployment of AI systems of high risk

Blueprint title	Supporting the deployment of AI systems of high risk
Scope and purpose	Support the mandatory transparency, security, data quality, and safety of high-risk AI systems
STAR-RA components involved	(Mandatory) XAI (for transparency); blockchain (distributed ledger infrastructure) and DPT (data provenance and traceability) for data quality; security policies manager (for increased cybersecurity); (optional) AI cyber-defence strategies (ACDS) (for AI cybersecurity)

deployers and operators must comply with a longer list of requirements, including more stringent specifications regarding explainability, transparency, data quality, and more. To support the qualification, deployment and use of such high-risk systems, STAR-RA offers many relevant components that support data reliability, AI algorithms' reliability, increased cybersecurity, safe human–robot collaboration, and more. The use of these systems in a high-risk AI context becomes mandatory rather than optional. This is illustrated in Table 5.

5 Conclusions

Despite the rising interest in trustworthy and human-centered Artificial Intelligence systems for Industry 5.0 deployments, existing reference architectures and blueprints for AI applications do not adequately address the development and deployment of trusted AI solutions. This chapter has introduced an architecture model that can boost the design and development of trustworthy and human-centered AI applications. It has also described few indicative blueprints for the development of technical solutions and regulatory-compliant systems in-line with the architecture.

At a high level, the presented architecture clusters functionalities in three complementary domains, including cybersecurity, human–robot collaboration, and safety. The functionalities in each one of the domains reinforce functionalities in the other domains. Moreover, the XAI components of the project are used to support functionalities in all three domains. The presented architecture and its functional modules do not present an "all or nothing" value proposition. Rather manufacturers and integrators of AI solutions in industrial environments can select subsets of components of the reference architecture to meet different sets of industrial requirements. However, the presented views of the architecture do not go down to implementation detail, but rather provide higher level structuring principles and blueprints for the implementation of trustworthy AI systems.

Some of the presented blueprints provide guidance on how to use STAR-RA components to implement systems that adhere to the mandates of the AI regulation proposal of the European Parliament and the Council of Europe. This guidance is important for companies that seek to comply with the AI Act and to demonstrate regulatory readiness. The inclusion of this chapter in this open access book aspires to raise awareness about both the technical and the regulatory aspects of trustworthy and human-centered solutions for manufacturing.

Acknowledgments This work has been carried out in the H2020 STAR project, which has received funding from the European Union's Horizon 2020 research and innovation program under grant agreement No. 956573. The authors acknowledge valuable support from all partners of the project.

References

1. Soldatos, J., Lazaro, O., Cavadini, F. (eds.): The Digital Shopfloor: Industrial Automation in the Industry 4.0 Era River Publishers Series in Automation, Control and Robotics. River Publishers, Gistrup (2019) ISBN: 9788770220415, e-ISBN: 9788770220408
2. Christou, I., Kefalakis, N., Soldatos, J., Despotopoulou, A.: End-to-end industrial IoT platform for Quality 4.0 applications. Comput. Ind. **137**, 103591. ISSN 0166-3615 (2022). https://doi.org/10.1016/j.compind.2021.103591
3. European Commission.: Industry 5.0 – what this approach is focused on, how it will be achieved and how it is already being implemented. European Commission. https://research-and-innovation.ec.europa.eu/research-area/industrial-research-and-innovation/industry-50_en (2023). Last accessed 26 June 2023
4. Nahavandi, S.: Industry 5.0—a human-centric solution. Sustainability. **11**, 4371 (2019). https://doi.org/10.3390/su11164371
5. Dwivedi R., Dave D., Naik H., Singhal S., Omer R., Patel P., Qian B., Wen Z., Shah T., Morgan G., Ranjan R.: Explainable AI (XAI): Core ideas, techniques, and solutions. ACM Comput. Surv. **55**, 9 (2023), Article 194 (September 2023), 33 pages. https://doi.org/10.1145/3561048
6. Proposal for a Regulation of The European Parliament and of the Council laying down harmonized rules on Artificial Intelligence (Artificial Intelligence Act) and Amending Certain Union Legislative Acts, COM/2021/206 final. https://eur-lex.europa.eu/legal-content/EN/TXT/?uri=celex%3A52021PC0206. Last accessed 26 June 2023
7. Industrial Internet Consortium.: The Industrial Internet Reference Architecture v 1.9. Available at: https://www.iiconsortium.org/IIRA.htm. Last accessed 26 June 2023
8. ISO/IEC/IEEE.: ISO/IEC/IEEE 42010:2011 Systems and software engineering – Architecture description. http://www.iso.org/iso/catalogue_detail.htm?csnumber=50508 (2011). Last accessed 26 June 2023
9. Industrial Internet Consortium.: The Industrial Internet Security Framework. Available at: https://www.iiconsortium.org/IISF.htm. Last accessed 26 June 2023
10. IEEE Standard for Adoption of OpenFog Reference Architecture for Fog Computing. IEEE Std 1934-2018, 1–176, 2 Aug 2018. https://doi.org/10.1109/IEEESTD.2018.8423800
11. European Big Data Value Strategic Research and Innovation Agenda, Version 4.0, October 2017. https://bdva.eu/sites/default/files/BDVA_SRIA_v4_Ed1.1.pdf. Last accessed 26 June 2023
12. ISO/IEC JTC 1/SC 42 Artificial intelligence. https://www.iso.org/committee/6794475.html. Last accessed 26 June 2023
13. IEEE Standard Model Process for Addressing Ethical Concerns during System Design. IEEE Std 7000-2021, 1–82, 15 Sept 2021. https://doi.org/10.1109/IEEESTD.2021.9536679
14. Rožanec, J., Novalija, I., Zajec, P., Kenda, K., Tavakoli, H., Suh, S., Veliou, E., Papamartzivanos, D., Giannetsos, T., Menesidou, S., Alonso, R., Cauli, N., Meloni, A., Reforgiato, R.D., Kyriazis, D., Sofianidis, G., Theodoropoulos, S., Fortuna, B., Mladenić, D., Soldatos,

J.: Human-centric artificial intelligence architecture for industry 5.0 applications. Int. J. Prod. Res. (2022). https://doi.org/10.1080/00207543.2022.2138611

15. Liu, Z., Wang, J., Gong, S., Tao, D., Lu, H.: Deep reinforcement active learning for human-in-the-loop person re-identification. In: 2019 IEEE/CVF International Conference on Computer Vision (ICCV), Seoul, Korea (South), pp. 6121–6130 (2019). https://doi.org/10.1109/ICCV.2019.00622

16. Siyaev, A., Valiev, D., Jo, G.-S.: Interaction with industrial digital twin using neuro-symbolic reasoning. Sensors. **23**, 1729 (2023). https://doi.org/10.3390/s23031729

17. Díaz-Rodríguez, N., Lamas, A., Sanchez, J., Franchi, G., Donadello, I., Tabik, S., Filliat, D., Cruz, P., Montes, R., Herrera, F.: EXplainable Neural-Symbolic Learning (X-NeSyL) methodology to fuse deep learning representations with expert knowledge graphs: the MonuMAI cultural heritage use case. Inf. Fusion. **79**, 58–83., ISSN 1566-2535 (2022). https://doi.org/10.1016/j.inffus.2021.09.022

18. Soldatos, J., Despotopoulou, A., Kefalakis, N., Ipektsidis, B.: Blockchain based data provenance for trusted artificial intelligence. In: Soldatos, J., Kyriazis, D. (eds.) Trusted Artificial Intelligence in Manufacturing: A Review of the Emerging Wave of Ethical and Human Centric AI Technologies for Smart Production, pp. 1–29. Now Publishers, Norwell (2021). https://doi.org/10.1561/9781680838770.ch1

19. Khurana, N., Mittal, S., Piplai, A., Joshi, A.: Preventing poisoning attacks on AI based threat intelligence systems. In: In IEEE 29th International Workshop on Machine Learning for Signal Processing (MLSP), pp. 1–6 (2019). https://doi.org/10.1109/MLSP.2019.8918803

20. Khorshidpour, Z., Hashemi, S., Hamzeh, A.: Learning a secure classifier against evasion attack. In: IEEE 16th International Conference on Data Mining Workshops (ICDMW), pp. 295–302 (2016). https://doi.org/10.1109/ICDMW.2016.0049

21. Umbrico, A., Orlandini, A., Cesta, A.: An ontology for human-robot collaboration. Procedia CIRP. **93**, 1097–1102 (2020)

22. Montini, E., Cutrona, V., Bonomi, N., Landolfi, G., Bettoni, A., Rocco, P., Carpanzano, E.: An IIoT platform for human-aware factory digital twins. Procedia CIRP. **107**, 661–667., ISSN 2212-8271 (2022). https://doi.org/10.1016/j.procir.2022.05.042

23. Andersen, P., Goodwin, M., Granmo, O.: Towards safe reinforcement-learning in industrial grid-warehousing. Inf. Sci. **537**, 467–484., ISSN 0020-0255 (2020). https://doi.org/10.1016/j.ins.2020.06.010

24. Ribeiro, M., Singh, S., Guestrin, C.: "Why should I trust you?": explaining the predictions of any classifier. In: Proceedings of the 22nd ACM SIGKDD International Conference on Knowledge Discovery and Data Mining (KDD '16), pp. 1135–1144. Association for Computing Machinery, New York (2016). https://doi.org/10.1145/2939672.2939778

25. Soldatos, J., Kyriazis, K. (eds.): Trusted Artificial Intelligence in Manufacturing: A Review of the Emerging Wave of Ethical and Human Centric AI Technologies for Smart Production. Now Publishers, Boston-Delft (2021). https://doi.org/10.1561/9781680838770

Designing a Marketplace to Exchange AI Models for Industry 5.0

**Alexandros Nizamis, Georg Schlake, Georgios Siachamis,
Vasileios Dimitriadis, Christos Patsonakis, Christian Beecks,
Dimosthenis Ioannidis, Konstantinos Votis, and Dimitrios Tzovaras**

1 Introduction

Online marketplaces, locations on Internet where people can purchase and sale services and goods, have highly increased in the last couple of decades. Recently, various marketplaces for exchanging AI models have been introduced [1]. In these marketplaces the AI models and machine learning (ML) algorithms have been monetized and offered as products.

AWS Marketplace[1] by Amazon enables its customers to find a large variety of pre-built models and algorithms covering a wide range of use cases and domains related to Business Analytics, Computer Vision, Healthcare, and Text and Language Processing. A subscription-based model with charging per hour, day, etc. is primarily adopted. A monthly based subscription model is also offered by Akira.AI.[2] This marketplace offers AI models especially for solutions related to Text Analysis and Computer Vision along with access to processing, storage, and network resources for enabling the execution of AI models. Pretrained models for a

---

[1] https://aws.amazon.com/marketplace/solutions/machine-learning.

[2] https://www.akira.ai.

A. Nizamis (✉) · G. Siachamis · V. Dimitriadis · C. Patsonakis · D. Ioannidis · K. Votis · D. Tzovaras
Centre for Research and Technology Hellas, Information Technologies Institute (CERTH/ITI), Thessaloniki, Greece
e-mail: alnizami@iti.gr; giosiach@iti.gr; dimvasdim@iti.gr; cpatsonakis@iti.gr; djoannid@iti.gr; kvotis@iti.gr; Dimitrios.Tzovaras@iti.gr

G. Schlake · C. Beecks
FernUniversity of Hagen, Hagen, Germany
e-mail: georg.schlake@fernuni-hagen.de; christian.beecks@fernuni-hagen.de

© The Author(s) 2024
J. Soldatos (ed.), *Artificial Intelligence in Manufacturing*,
https://doi.org/10.1007/978-3-031-46452-2_2

wide variety of sectors are also available at the Gravity AI[3] and Modelplace AI[4] (specialized in Computer Vision) marketplaces. The latter enables the real-time execution of models through web–browser interfaces. Other marketplaces[5] move a step further from live execution to shared building of models among developers by providing software development kits (SDKs).

Some specific AI marketplaces for healthcare domain are also available such as the Imaging AI Marketplace[6] by IBM. It is a centralized marketplace that enables healthcare providers to discover, purchase, and manage applications that provide the latest AI-powered tools. In this marketplace, researchers and developers can reach a large community of customers for their specific AI applications for eHealth domain and take advantage of the provided infrastructure and deployment processes. To this direction, Nuance Communications[7] has introduced its marketplace for Healthcare AI solutions as well providing similar functionalities as the IBM one.

In addition to word-leaders' approaches and market-ready solutions, EC-funded research projects have presented various marketplaces for listing or even trading solutions including AI/ML algorithms. AI4EUROPE or AI on Demand [2] offers a trustworthy open-source platform for the development, training, and sharing of AI/ML models. However it is considered more as an open code repository as it lacks business logic in comparison with commercial marketplaces. MARKET4.0 project [3] develops a multi-sided business platform for plug and produce industrial product service systems. The European Factory Foundation and EFPF project [4] offers a Portal/Marketplace as part of its interoperable Data Spine that includes solutions coming from previous EU projects and third-parties' initiatives. However the solutions are a mix of software solutions, products, and services. Other marketplaces related to manufacturing domain and Industry 4.0 were provided by projects like v-fos [5] (which offers an application marketplace with an embedded SDK) and NIMBLE [6] that has introduced a federated interoperable eco-system for B2B connectivity. Some other approaches [7] coming from research introduced AI as enablers for marketplaces by combining virtual agents with semantics [8, 9] for automated negotiations in manufacturing marketplaces [10]. In Boost 4.0 project a common European Data Space for manufacturing was introduced instead of a Marketplace. However, it contains AI services connected to available data sources based on IDSA[8] architecture. The latter supports also the establishment of AI Marketplace[9] that is a meeting place for AI providers and consumers. Recently,

---

[3] https://www.gravity-ai.com.

[4] https://modelplace.ai/.

[5] https://c3.ai/c3-ai-application-platform/c3-ai-integrated-development-studio/c3-ai-marketplace.

[6] https://www.ibm.com/downloads/cas/6BWYDLDO.

[7] https://www.nuance.com/healthcare.html.

[8] https://internationaldataspaces.org/.

[9] https://ki-marktplatz.com/en/.

PoP-Machina project[10] proposed a collaboration platform for makers that provides a marketplace based on a blockchain network infrastructure. However, it is focused more on collaborative design and not on AI model exchange. A blockchain approach for building marketplaces for AI was also introduced by IBM [11]. It was a back-end implementation regarding trusted transactions and not a full operational marketplace.

As it is perceived there are market-ready solutions in the field of AI marketplaces, however they are focused on cases related to health, text recognition, computer vision, etc. and not to manufacturing and Industry 4.0/5.0 domains. On the contrary, there are marketplaces coming from research field that are related to Industry 4.0/5.0 domain, but either they lack some business logic or they collect heterogeneous solutions and even physical products, so they cannot be considered as marketplaces for exchanging of AI/ML models. In the current work, we are introducing the knowlEdge project's [12, 13] Marketplace that aims to deliver a marketplace for exchanging AI models for smart manufacturing using blockchain services and smart contracts as the core of its business logic. The introduced marketplace can act as an enabler for intelligent production as it collects and offers AI solutions related to manufacturing domain able to solve various kinds of problems in factories.

Following this introductory section, the next section presents the knowlEdge AI Marketplace's main functionalities and its high-level architecture. Section 3 presents the core technical parts and interfaces of the knowlEdge Marketplace. The conclusions are drawn in Sect. 4.

2 Functionalities and Proposed System Architecture of knowlEdge Marketplace

The introduced marketplace for AI models regarding smart manufacturing offers a series of functionalities common to normal marketplaces and stores for services and products. In particular, it offers:

- A user-friendly web-based interface to enable trading of AI algorithms and models
- Trusted trades among the stakeholders, protection of Intellectual Property Rights (IPR) and security
- Profiles and role management functionalities
- Search functionalities based on various features
- Reviews and ratings regarding users and AI models

To support the abovementioned functionalities, the knowlEdge Marketplace incorporates a series of technologies and components in its architecture as it is depicted in Fig. 1. They are distinguished in three main categories: the back-end part related to AI models description and management, the back-end part related

---

[10] https://pop-machina.eu.

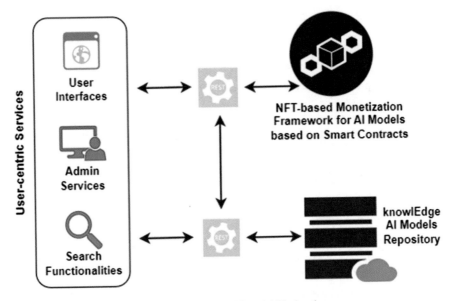

Fig. 1 High-level architecture of the knowlEdge AI model Marketplace

to business logic and transactions, and the front-end part related to user-centric services such as interfaces.

In particular, the *User-Centric Services* module provides a series of functionalities related to user experience such as UIs, search functionalities, user profile management, etc.

NFT-Based Monetization Framework for AI Models based on Smart Contracts provides all the functionalities related to business logic based on blockchain. It offers smart contracts (or chaincode) for fungible and non-fungible Tokens (FTs and NFTs), marketplace, and mint notary. Furthermore, secure access services are also part of this module.

AI Model Repository is responsible for the modeling of AI/ML algorithms, their storage, and the provision of management services such as CRUD (Create, Read, Update, and Delete) operations.

Various *APIs* have also been developed to enable the different module communication based on HTTP protocol.

All the core modules of the introduced marketplace are presented in the following section.

3 Implementation of knowlEdge Marketplace for Exchanging AI Models in Industry 5.0

In this section, the core modules of the knowlEdge Marketplace are described. We start with the way, the data is stored in the knowlEdge AI Model Repository

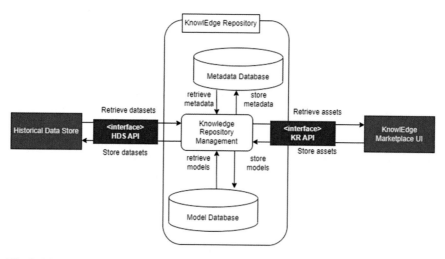

Fig. 2 The internal architecture of the knowlEdge AI Model Repository

(Sect. 3.1), before we discuss the Monetization Framework (Sect. 3.2) and conclude it with the User Interfaces (Sect. 3.3).

3.1 knowlEdge AI Model Repository

The knowlEdge AI Model Repository is a central cloud-hosted component, which manages a database of AI models and their corresponding meta-information. For these purposes, it consists of four main components: the *knowlEdge Repository Management*, the *Model Database*, the *Metadata Database*, and the *Historical Data Store* (see Fig. 2). In general, the AI Model Repository provides all the necessary functionality to Marketplace regarding the modeling of the AI and the management of data and metadata in the Marketplace that are related to AI models. So, besides the four aforementioned components, an Ontology that enables the modeling of Marketplace's metadata is a core part of the Repository as well.

3.1.1 Overview of Key Components

The *knowlEdge Repository Management* is the central component connecting the other components and offering services over a REST API, which follows the OpenAPI [14] specification and is the single interface of the knowlEdge Repository. It offers a feature-rich possibility to query for AI models and datasets to identify similarities between different problems and solutions.

The *Metadata Database* is a MongoDB [15], which stores the metadata of the knowlEdge Repository. This metadata follows the knowlEdge Ontology (see Sect. 3.1.2) to ensure a high level of usable information for all datasets and models present in the repository.

The *Model Database* is used to store the actual model specification files in a Hadoop [16] Distributed File System. The model files can be presented in ONNX[11] or PMML [17] format to make sure that as many different models as possible can be described for the knowlEdge Repository.

The *Historical Data Store* is used to store the datasets a model is trained on. With these datasets also present, it is possible to benchmark new models on the same datasets and directly compare the performance of models.

3.1.2 Ontology

The knowlEdge Ontology has been developed to ensure that a wide variety of metadata are available for the knowlEdge repository. The Ontology consists of 12 different types of entities (see Fig. 3). In this entity–relationship diagram, the most

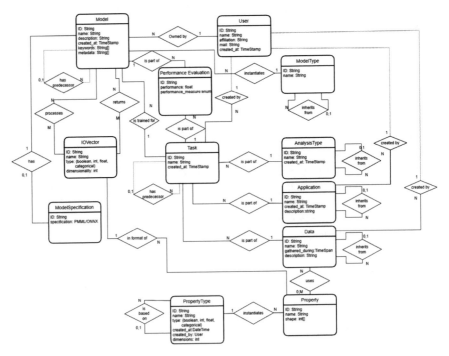

Fig. 3 The technical structure of the knowlEdge Ontology

---

[11] https://onnx.ai.

important entities and relationships of the knowlEdge Ontology are given and show the possibilities of the hierarchical structure of this Ontology. It shows the technical ways the knowlEdge Repository stores the data according to it. These entities can be split into user-related (User), model-related (Model, IO Vector, Model Specification and Model Type), data-related (Task, Analysis Type, Application, Data, Property and Property Type), and performance-related (Performance Evaluation) Entities. The main entities/classes are as follows:

- **User**: A User is identified by its unique name and email address. Furthermore, the timestamp of the time it was created is stored. A User can be owner of several Models and creator of several Application, Data, and Property Type entities.
- **Model**: A Model contains its name, a description, and the timestamp of its creation as well as a link to a Model Specification and the Model Type it instantiates. It can have hierarchical children and a parent as well as multiple IO Vectors as input and output.
- **IO Vector**: An IO Vector contains its name, dimensionality, and data type, which can be of type integer, float, Boolean, or categorical. It can be input and output to several Models.
- **Model Specification**: A Model Specification contains the actual model file, i.e., an ONNX or PMML file describing the model.
- **Model Type**: A Model Type has its name. It can be child to another Model Type and parent to several Model Types. It can be instantiated by several Models.
- **Task**: A Task consists of its name, the timestamp of its creation, its Analysis Type, Application, and Data. It is created by a single User, can have multiple Tasks as children, and can have several Models trained for it. It may be child of another Task.
- **Analysis Type**: An Analysis Type has a name and the timestamp of its creation. It is part of a Task and can be child to one and parent to many Analysis Types.
- **Application**: An Application has a name, a description, and the timestamp of its creation. It is part of a Task, created by a User, and can be child to one and parent to many Applications.
- **Data**: A Data entity consists of its name and description, the Task it is part of, and the timespan it was gathered during. It is created by a User, consists of several Properties, and may inherit from one and be parent for several Data entities.
- **Property**: A Property consist of its name, the Property Type it instantiates, and the Data entity it belongs to.
- **Property Type**: A Property Type consists of its name, creation time, and type, which may be Boolean, integer, float, or categorical. It is created by a User and may be instantiated by several Properties. A Property Type can be based on one Property Type, and there can be multiple Property Types based on one Property Type.
- **Performance Evaluation**: A Performance Evaluation represents the actual performance of an AI Model on a Task. It is linked to both these entities and contains the performance measurement as well as the information which measure this represents.

The model specifications in the Model Database will be stored in the ONNX or PMML Format. These two formats offer compatibility with a wide range of different machine learning techniques and frameworks to boost interoperability regarding AI models in the proposed marketplace. While PMML focuses on traditional machine learning methods, ONNX is specialized for exchanging deep neural networks. The combination of both formats makes it possible to store a wide range of machine learning models in an easy to use and deploy way.

3.2   NFT-Based Monetization Framework for AI Models

Besides the Ontology for describing the Marketplace metadata and the repository services for the management of AI models, a set of services regarding the monetization-, security-, and business-related concepts was required for the delivery of the proposed AI Model Marketplace.

Therefore, in the context of the knowlEdge Marketplace, a number of blockchain-based services have been developed and deployed. An end-to-end decentralized AI Model Marketplace that enables the monetization of AI models while ensuring security, auditability, and verifiability has been implemented based on blockchain technology. To guarantee ownership of AI models, each model is treated as a unique asset on the distributed ledger, represented as non-fungible tokens (NFTs). The use of NFTs provides additional functionalities, including ownership transfer. The marketplace is based on the presupposition that participants share a common value system, and fungible tokens are used as the equivalent of real-world fiat currency.

In Table 1, we provide the various actor roles and the means under which they engage with the marketplace platform, i.e., their capabilities.

Note that the terms "AI Model Producer," "AI Model Researcher," and "AI Model Developer" are used interchangeably to refer to the same actor. Similarly, the terms "AI Model Consumer" and "Marketplace Customer" refer to the same actor. Lastly, note that the same real-world entity can potentially enact in all of the aforementioned roles, e.g., an AI model producer can also act as a consumer, or marketplace customer for AI models produced by others.

The following functionalities are available in the DLT-based AI Model Marketplace, building on the core set of functionalities that were previously outlined. The functionalities are as follows:

- AI model producers, represented as NFT owners, can advertise their willingness to sell access to individual AI model binary files at a price of their choice.
- Each AI model producer can query the AI models they have advertised on the marketplace.
- Each AI model producer can retract advertised AI models at any time.
- Any entity can query all AI models advertised on the marketplace.
- Interested AI model consumers can purchase access to any advertised AI model, provided they have sufficient coin balance.

Table 1 Roles and capabilities of users in the system

Role	Capabilities
Admin	Entities that have access to privileged functionalities, such as:
	• Anything related to Identity and access management (IAM)
	• Acting as principals of notarized functionalities wherever and if deemed necessary. For instance, approving the minting of fungible tokens (coins)
AI Model Producer	Entities that produce or develop new AI models. Their engagement with the platform is as follows:
	• Querying the platform for a complete list of all the AI models that they own
	• Uploading AI models to the repository
	• Minting AI models (NFTs) on the DLT (Distributed Ledger Technology) and storing the corresponding metadata file on the off-chain metadata store
	• Publishing their willingness to sell access to AI models (NFTs) that they own on the marketplace
	• Retracting from the marketplace selling access to AI models (NFTs)
	• Creating (bank) accounts, transferring coins to accounts of other users, and so on
AI Model Consumer	These are the main clients of the marketplace, i.e., the ones that purchase access to AI models. Their engagement with the platform is as follows:
	• Retrieve a list of all (or even individual) AI models for which they can buy access to.
	• Query the NFT metadata store to obtain additional information of an AI model, such as the URL of the corresponding repository on which it is stored.
	• Create (bank) accounts, transferring coins to accounts of other users, and so on.
	• Purchase access to AI models that are advertised by specifying the fungible token account that will be used for the payment. Obviously, the specified account must have a sufficient coin balance.
	• Once a purchase has been completed, these entities retain indefinitely their right to access the AI model, regardless of whether the corresponding producer has retracted it.
	• Query the platform for a complete list of all the purchases that they have performed in the marketplace.

- AI model consumers retain access rights to purchased AI model binary files, even if the models are later retracted from the marketplace.
- Each consumer can query the marketplace for a list of all successful purchases.
- External entities, such as an AI Model repository, can securely verify that an actor requesting access to an AI model binary file is a legitimate consumer who has previously performed a successful purchase.

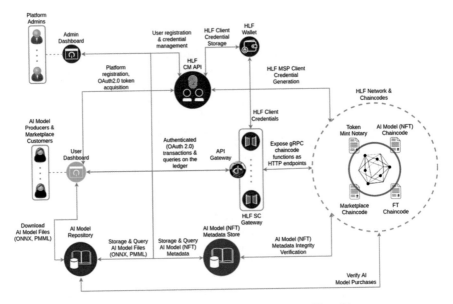

Fig. 4 High-level overview of the AI model monetization framework's architecture

A detailed diagram of the architecture can be found in Fig. 4. The diagram depicts the involved users, the components of the blockchain infrastructure, and intuitive descriptions of their interactions.

There are also several off-chain components that play different roles. For example, we need to account for the physical storage of AI model files and provide an identity and access management infrastructure for the actors and the services they consume. Additionally, we need to include various integration and deployment-related components such as API gateways and dashboards (user interfaces) for the involved actors. The list of components along with a succinct description of their function can be found below:

- Hyperledger Fabric (henceforth, HLF) Community Management (CM) API: A federated identity authorization service that, apart from supporting all standard OAuth 2.0 and OpenID Connect flows, encompasses, as part of the user registration (onboarding) process, private key and X.509 digital certificate generation, which are subsequently stored in an HLF-compliant secure wallet store.
- HLF Wallet: Implementation of HLF's Wallet interface employing MongoDB as a storage medium for Hyperledger Fabric's Go SDK. This is employed internally by the HLF SC Gateway component (see below).
- HLF Smart Contract (SC) Gateway: a configurable microservice that exposes functions of any smart contract deployed on any arbitrary Hyperledger Fabric channel as HTTP endpoints

- NFT Metadata Store: a REST API that exposes endpoints for storing and retrieving metadata files associated with NFTs
- NFT Chaincode: the smart contract that implements the entire non-fungible token-related functionality
- FT Chaincode: the smart contract that implements the entire fungible token-related functionality

3.3 User Interfaces and Functionalities

Besides the core back-end services and the corresponding modules that were presented in the previous section, a module focused on the delivery of front-end-related services is also included in knowlEdge Marketplace. This module does not only include interfaces but also supports some user-related functionalities such as search capabilities and user's profile management. They are considered in the same building block as they are strictly connected to a front-end theme that was used and their functionality is derived from them.

Regarding the interfaces design, best practices were used. The design pillars that were followed were the esthetic and minimalistic design, the use of common and consistent UI elements, the adoption of widely used input controls and navigational elements, the error prevention and good error messaging, etc. The UIs were implemented as web-based interfaces using technologies such as Angular, Bootstrap, and Nebular. The ngx-admin template was used to enable faster implementation as it is a popular admin dashboard based on Angular and it is free and open source. It is efficient as it is packed with a huge number of UI components and it is highly customizable.

The UIs enable the exploration of various available AI models in different views (see Fig. 5) based on the user's preferences (grid view and list view). The search functionalities provide various filters such as AI models owner, category of the algorithm, price range, rating, etc. Text-based search is also supported, so the user can type text related to the model's name, keywords, and other metadata.

By selecting a model, the user is able to read details (see Fig. 6) such as description, specifications of the model, and metadata such as rating, price, and owner. Furthermore, any datasets connected to a model are also visible. All the data available to UI are dynamically retrieved from Repository and Monetization modules. The user can also select to add to cart a model in order to purchase it based on the NFT monetization module.

Besides exploring and purchasing AI models, a user can act as a provider and deploy his/her own AI model by using corresponding interfaces (Fig. 7) that are available in a kind of a step wizard form. First, the user adds the dataset details that were used for training a model. Then the general details regarding the task/application that the model is related to (e.g., predictive maintenance) are added. After that, the user adds AI model details such as the type, input and output, model

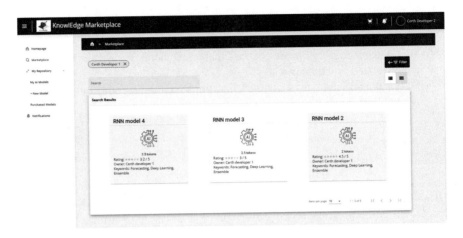

Fig. 5 List of AI models

Fig. 6 AI model details

format, and connections with other models and deploys the model itself (e.g., an ONNX file, etc.).

4 Conclusions

The design and the implementation of a marketplace for exchanging AI models related to Industry 5.0 and smart manufacturing are introduced. The knowlEdge Marketplace highlights the main components that a marketplace for AI models should include. A component to enable the modeling of AI models/algorithms

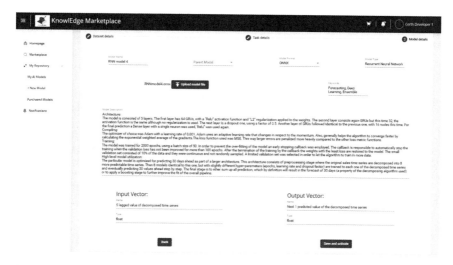

Fig. 7 Uploading an AI model

and their metadata based on standards has been defined as a necessity. Moreover, AI developers should be able to provide their models based on widely used formats and standards in such marketplaces. Furthermore, services to enable trusted transactions and sharing, along with security and protection of ownership in AI marketplaces, have been found as core concepts that should be covered as well. The use of blockchain technology for this kind of services has been proved as an ideal candidate as it provides all the necessary concepts regarding monetization and secure and trusted transactions. Moreover, user-friendly and easy-to-use interfaces were another important factor that should be considered as in the end, as any other marketplace, it is focused to end-users.

Regarding the next steps, the knowlEdge Marketplace focuses on further testing and evaluation by domain experts targeting to final improvements so to be considered as a completely market-ready solution. The plan is for this marketplace to be one of the core AI marketplaces in Europe related to AI models exchanged in Industry 5.0 era.

Acknowledgments This work has received funding from the European Union's Horizon 2020 research and innovation program under Grant Agreement No. 957331—KNOWLEDGE.

References

1. Kumar, A., Finley, B., Braud, T., Tarkoma, S., Hui, P.: Sketching an ai marketplace: Tech, economic, and regulatory aspects. IEEE Access **9**, 13761–13774 (2021)
2. Cortés, U., Cortés, A., Barrué, C.: Trustworthy AI. The AI4EU approach. In: Proceedings of Science (2019)

3. Mourtzis, D., Angelopoulos, J., Panopoulos, N.: A survey of digital B2B platforms and marketplaces for purchasing industrial product service systems: a conceptual framework. Procedia CIRP **97**, 331–336 (2021)
4. Deshmukh, R.A., Jayakody, D., Schneider, A., Damjanovic-Behrendt, V.: Data spine: a federated interoperability enabler for heterogeneous IoT platform ecosystems. Sensors **21**(12), 4010 (2021)
5. Anaya, V., Moalla, N., Stellingwerff, L., Flores, J.L., Fraile, F.: vf-OS IO Toolkit. Enterprise Interoperability: Smart Services and Business Impact of Enterprise Interoperability, pp. 91–97 (2018)
6. Gönül, S., Çavdaroğlu, D., Kabak, Y., Glachs, D., Gigante, F., Deng, Q.: A B2B marketplace eCommerce platform approach integrating purchasing and transport processes. In: International Conference on Interoperability for Enterprise Systems and Applications, pp. 105–121. Springer, Cham (2023)
7. Bonino, D., Vergori, P.: Agent marketplaces and deep learning in enterprises: the composition project. In: 2017 IEEE 41st Annual Computer Software and Applications Conference (COMPSAC), vol. 1, pp. 749–754. IEEE, New York (2017)
8. Nizamis, A.G., Ioannidis, D.K., Kaklanis, N.T., Tzovaras, D.K.: A semantic framework for agent-based collaborative manufacturing eco-systems. IFAC-PapersOnLine **51**(11), 382–387 (2018)
9. Nizamis, A., Vergori, P., Ioannidis, D., Tzovaras, D.: Semantic framework and deep learning toolkit collaboration for the enhancement of the decision making in agent-based marketplaces. In: 2018 5th International Conference on Mathematics and Computers in Sciences and Industry (MCSI), pp. 135–140. IEEE, New York (2018)
10. Mertens, C., Alonso, J., Lázaro, O., Palansuriya, C., Böge, G., ..., Poulakidas, A.: A framework for big data sovereignty: the European industrial data space (EIDS). In: Data Spaces: Design, Deployment and Future Directions. pp. 201–226. Springer International Publishing, Cham (2022)
11. Sarpatwar, K., Sitaramagiridharganesh Ganapavarapu, V., Shanmugam, K., Rahman, A., Vaculin, R.: Blockchain enabled AI marketplace: the price you pay for trust. In: Proceedings of the IEEE/CVF Conference on Computer Vision and Pattern Recognition Workshops (2019)
12. Alvarez-Napagao, S., Ashmore, B., Barroso, M., Barrué, C., Beecks, C., Berns, F., Bosi, I., Chala, S.A., Ciulli, N., Garcia-Gasulla, M., Grass, A., Ioannidis, D., Jakubiak, N., Köpke, K., Lämsä, V., Megias, P., Nizamis, A., Pastrone, C., Rossini, R., Sànchez-Marrè, M., Ziliotti, L.: knowlEdge Project–Concept, methodology and innovations for artificial intelligence in industry 4.0. In: 2021 IEEE 19th International Conference on Industrial Informatics (INDIN), pp. 1–7. IEEE, New York (2021)
13. Wajid, U., Nizamis, A., Anaya, V.: Towards Industry 5.0–A Trustworthy AI Framework for Digital Manufacturing with Humans in Control. Proceedings https://ceur-ws.org ISSN, 1613, 0073 (2022)
14. Tzavaras, A., Mainas, N., Petrakis, E.G.M.: OpenAPI framework for the Web of Things. Internet of Things **21**, 100675 (2023)
15. Banker, K., Garrett, D., Bakkum, P., Verch, S.: MongoDB in action: covers MongoDB version 3.0. Simon and Schuster, New York (2016)
16. White, T.: Hadoop: the definitive guide. O'Reilly Media Inc., California (2012)
17. Guazzelli, A., Zeller M., Lin, W., Williams, G., et al.: PMML: an open standard for sharing models. R J. **1**(1), 60 (2009)

Human-AI Interaction for Semantic Knowledge Enrichment of AI Model Output

Sisay Adugna Chala and Alexander Graß

1 Introduction

Modern day agile manufacturing [6] requires developing a framework of AI solutions that capture and process data from various sources including from human-AI collaboration [1]. Enhancing a manufacturing process by (semi-)automatized AI solutions that can support different stages in a production process that involves inter-company data infrastructure is one of the challenges in data-intensive AI for manufacturing. This challenge is exacerbated by the lack of contextual information and nontransparent AI models. In this chapter, we describe the concept of domain knowledge fusion in human-AI collaboration for manufacturing. Human interaction with AI is enabled in such a way that the domain expert not only inspects the output of the AI model but also injects engineered knowledge in order to retrain for iterative improvement of the AI model. It discusses domain knowledge fusion, the process to augment learned knowledge of AI models with knowledge from multiple domains or sources to produce a more complete solution. More specifically, a domain expert can interact with AI systems to observe and decide the accuracy of learned knowledge and correct it if needed.

The purpose of a domain ontology is to serve as a repository for domain-specific knowledge. Ontology enrichment system, as a part of human-AI collaboration, enables domain experts to contribute their expertise with the goal of enhancing the knowledge learned by the AI models from the patterns in the data. This enables the integration of domain-specific knowledge to enrich the data for further improvement of the models through retraining.

S. A. Chala (✉) · A. Graß
Department of Data Science and AI, Fraunhofer Institute for Applied Information Technology (FIT), Sankt Augustin, Germany
e-mail: sisay.adugna.chala@fit.fraunhofer.de; alexander.grass@fit.fraunhofer.de

© The Author(s) 2024
J. Soldatos (ed.), *Artificial Intelligence in Manufacturing*,
https://doi.org/10.1007/978-3-031-46452-2_3

Domain knowledge fusion is a technique that involves combining knowledge from multiple domains or sources to produce a more complete solution by augmenting learned knowledge of AI models. It is used to improve the accuracy of predictive models, e.g., to guide feature selection in a machine learning model, resulting in better predictive performance [16]. Domain knowledge fusion also helps improve effectiveness of predictive models by supporting efficient dimension reduction techniques that are able to capture semantic relationships between concepts [17].

After reviewing prior research, we describe our concept domain knowledge fusion in agile manufacturing use case scenarios for human-AI interaction. We identify two kinds of knowledge: (i) learned knowledge, i.e., the knowledge generated by the AI model and (ii) engineered knowledge, i.e., the knowledge provided by the domain expert. We identify three aspects of domain expert interaction with our AI systems to observe and (a) reject if the learned knowledge is incorrect, (b) accept if the learned knowledge is correct, (c) adapt if the learned knowledge is correct but needs modification. We demonstrate these concepts for researchers and practitioners to apply human-AI interaction in agile manufacturing.

The rest of this chapter is organized as follows: in the Related Works section, we examine related works in order to identify research gaps of human-AI interaction in agile manufacturing. In the Human Feedback into AI Model section, we discuss the methodology (sub-components and interfaces) developed in the human-AI collaboration to enhance agile manufacturing. The Interaction for Model Selection and Parameter Optimization section covers the implementation of the proposed system and presents the preliminary results. Finally, in the Conclusion and Future Works section, we summarize the results and outline the future research works of the chapter.

2   Related Works

Fact-checking, a task to evaluating the accuracy of AI models, is a crucial, pressing, and difficult task. Despite the emergence of numerous automated fact-checking solutions, the human aspect of this collaboration has received minimal attention [13], though some advancement is being observed in conversational AI [8]. Specifically, it remains unclear how end users can comprehend, engage with, and build confidence in AI-powered agile manufacturing systems. In other words, enabling interaction of domain experts to AI model outputs, in order that they inspect the output and provide their feedback, helps fix errors that could lead to undesirable outcomes in production process.

Existing studies on human-AI collaboration predominantly focus on user interface (UI) and user experience (UX) aspects, i.e., whether (and how) the AI systems provide an intuitive user interface. A number of them assessed human-AI collaboration with respect to human-AI interaction guidelines as opposed to features that enable human actor to provide feedback to the AI model [4, 9]. Regarding its effect on decision-making of users has been studied using different eXplainable AI (XAI) interface designs [12].

Apart from data fact-checking and UI/UX, human-AI interaction can be done for data labeling. For example, data such as time series measurements are not intuitive for users to understand, and AI is used to generate descriptive labels [10] for a given data. Expert knowledge can augment the result of AI models by inspecting the output and complementing it. Ontology enrichment is being studied in areas of knowledge management [7], natural language processing [14], medical [3], and energy [5]. Although manufacturing with human in the loop is recently getting traction, studies on ontology enrichment have minimal attention for manufacturing.

There are several existing human-AI collaboration solutions that aim to leverage the strengths of both humans and AI systems through human-AI collaboration. Advanced applications like virtual assistants like Google Assistant [11] and Apple Siri [2] are common examples of human-AI collaboration systems. These AI-powered voice-activated assistants interact with humans to perform tasks, answer questions, and control connected devices. Most of these advancements concentrate in natural language processing, healthcare, and energy. However, the role of AI in manufacturing is mainly focused on automation and control.

This chapter focuses on data analytics and insights generation through AI models that had minimal attention despite the fact that manufacturing domain generates massive amount of sensor data. Processing and analyzing the large volumes of data can help identify patterns, trends, and anomalies, providing valuable insights to support decision-making. The chapter develops a tool that enables humans to collaborate with AI systems through intuitive interfaces that help domain experts in interpreting insights, validating the findings, and applying domain knowledge to gain a deeper understanding of the data.

3 Human Feedback into AI Model

The purpose of human feedback into AI model is enabling domain experts to inject their knowledge via predefined interfaces allowing for collaboration with the system in order to connotate previous knowledge with semantics, as for instance with a description of a specific process or data. It helps better understanding of the data, as it also provides the possibility for a better evaluation of the whole AI pipeline. In other words, human-AI collaboration is a component that offers interfaces between domain expert and AI system. The functionalities offered by the human-AI collaboration are to enable human feedback for domain experts, i.e., machine operators and managers without the need to understand the intricacies of AI models.

As shown in Fig. 1, the human-AI collaboration is composed of multiple sub-components and interfaces that enable communication with external systems such as data sources, model repositories, machine configurations, and decision support systems. The main sub-components described below are interface abstraction,

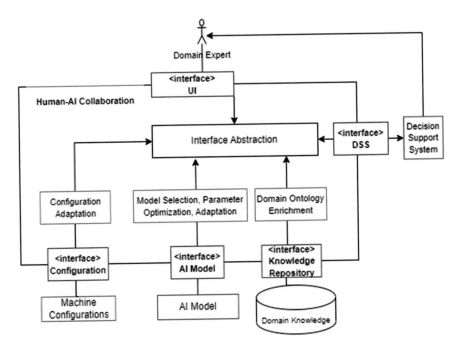

Fig. 1 Human-AI collaboration components and interfaces

model and data selection, parameter optimization, configuration adaptation, domain knowledge enrichment, and domain knowledge repository.

3.1 Interface Abstraction

Interface abstraction component is a container for configuration adaptation, model/data selection, parameter optimization and adaptation, and domain ontology enrichment components. It provides an interface to the domain expert through the decision support system through intuitive and user-friendly interfaces that enable effective communication and cooperation between humans and AI. Interface abstraction is beneficial for human-AI collaboration by playing a crucial role in enabling seamless cooperation and enhancing the productivity and usability of AI technologies. The goal of interface abstraction is to bridge the gap between the capabilities of AI systems and the understanding and expertise of human users. It allows users to interact with complex AI technologies without requiring them to have in-depth knowledge of the underlying algorithms of AI models. In effect, it empowers users to leverage the capabilities of AI systems while focusing on their own areas of expertise. By abstracting the complexities of AI algorithms and technologies, interface abstraction facilitates effective communication and collaboration between humans and AI.

3.2 Model and Data Selection

The human-AI collaboration system offers features for data and model selection. Operators select models and data from the available list of options in order to execute them for a specific scenario. Model and data selection are critical factors in human-AI collaboration. Because the choice of model and data significantly influences the performance, accuracy, and overall effectiveness of the AI system. When considering human-AI collaboration, several key considerations come into play. One aspect is determining the specific requirements of the task at hand, understanding the problem domain, the type of input data, and the desired output. This knowledge helps guide the selection of an appropriate model for the dataset. Another aspect is understanding capabilities of the AI model because different AI models and algorithms are suitable for the task. Considering factors such as the model's architecture, complexity, interpretability, and scalability affect the choice of a model that aligns with the task requirements.

3.3 Parameter Optimization

Parameter optimization is an important step in human-AI collaboration to ensure optimal performance and effective interaction between humans and AI systems. Operators and managers perform the optimization of parameters that offer the best outcome for the given scenario. The system provides them with an interface where the operators can select the parameters, try various values, and observe the results.

It involves continuous evaluation of the performance of the system and collecting feedback from the human collaborator. This feedback can be used to identify areas for improvement and guide the parameter optimization process to iteratively refine and fine-tune the parameters based on the evaluation results and feedback. The parameter optimization is necessary for the domain expert to deal with trade-offs between different performance metrics or constraints that need to be satisfied. For example, optimizing for accuracy may lead to longer response times, which may impact the user experience.

The first step in parameter optimization is to identify the metrics or criteria that will be used to measure success. This will help guide the parameter optimization process. Once the parameters are identified, the next step is to determine the metrics that evaluate the performance of AI model, such as efficiency and accuracy. In this chapter, an example implementation of parameter optimization is shown in Fig. 4.

3.4 Configuration Adaptation

Configuration adaptation is the process of adjusting or fine-tuning the configuration settings of AI systems to better align with the needs, preferences, and context of human users. It involves customizing the equipment, parameters, or policies of AI models to optimize their performance. Feedback of domain expert plays a vital role in configuration adaptation as it provides valuable insights into the effectiveness and suitability of the AI model's behavior in that the AI system can learn and adjust its configuration settings to improve its performance and align more closely with the user's requirements in response to incorporating domain expert feedback. Moreover, when a model offers a need for specific configurations of machines that need to be modified, operators/managers can adapt the configurations of machines so that it suits to the model under consideration, for example, if new machines need to be added to the human-AI collaboration system, their configuration should be extracted and stored in such a way that they are accessible and usable to the modules.

3.5 Domain Knowledge Enrichment

Enriching learned knowledge with engineered knowledge describes the scenario where the AI model analyzes the given data for a task (e.g., outlier detection) and produces its result (e.g., that a given data point is an outlier), the domain expert realizes that the output of the model is not right (e.g., that the data point is not an outlier), and the information provided by the domain expert (i.e., the data point is not an outlier) is stored in the repository of ground truth and sent back to the AI model for retraining. It is used by operators and managers to enrich the knowledge repository with new entries obtained from the execution of the system using diverse setting of models, parameters, and configurations.

The key challenge of this approach is that it relies on the availability of domain experts. Scarcity of domain experts (that most of them spend their time on machine monitoring, operation, and management), limited availability of domain expertise, rapidly evolving AI landscape, and its demand for interdisciplinary skills make this challenge difficult to handle. Developing AI models often requires deep domain expertise in specific fields, e.g., manufacturing in this case, and experts who possess both domain expertise and a solid understanding of AI techniques are difficult to find. Moreover, effective collaboration between domain experts and AI practitioners often necessitates interdisciplinary skills. Domain experts need to understand AI concepts and methodologies, while AI practitioners need to comprehend the nuances and complexities of the specific domain. The scarcity of individuals with expertise in both areas makes the task of domain knowledge enrichment challenging.

In this chapter, the limitation of AI model development knowledge of domain expert is taken into account. Having listed these challenges, this research assumes that the involvement of humans in enriching knowledge will potentially reduce over time. As such, an initial set of AI models are trained and made available for the domain expert to experiment with them before trying to perform parameter optimization and feedback provision. At the beginning of a collaboration between humans and AI, there will be a significant effort to optimize parameters and transfer human knowledge and expertise to the AI models by providing more data, defining rules, and setting up the initial knowledge base. However, through retraining, the AI model learns and accumulates more data that it can gradually require less feedback from the domain expert.

Organizations can benefit from this system despite these challenges by defining feasible objectives of the human-AI collaboration in the manufacturing setting whereby they identify the specific areas where AI can enhance manufacturing processes, such as quality control, predictive maintenance, or supply chain optimization and establish key performance indicators (KPIs) to measure success. For example, companies can utilize this approach to perform what-if analysis in order to explore the potential implications of different scenarios and make more informed decisions by combining the analytical capabilities of AI models with human judgment, expertise, and contextual understanding. Domain experts can modify the input parameters, adjust variables, or introduce new constraints to observe the potential changes in the outcomes. The AI system then performs the simulations and presents the results to the human collaborator.

3.6   Domain Knowledge Repository

Domain Knowledge is the repository of knowledge (both learned knowledge generated by the AI model and engineered knowledge curated by the domain expert. Machines and production processes are undergoing a rapid digital transformation, opening up a wide range of possibilities. This digitalization enables various opportunities, including early detection of faults and pricing models based on actual usage. By leveraging sensor data analytics, it becomes possible to monitor machine operations in real time, providing valuable insights and applications. This is better achieved if the domain experts assist in improving the quality of AI model output by providing domain knowledge, for which this component is responsible to store.

4   Interaction for Model Selection and Parameter Optimization

Improving the effectiveness of AI model requires a comprehensive understanding of the model's design and implementation and it can be achieved in a number of ways: (i) reviewing the input data, including the quality, completeness, and relevance, to determine if it can be modified to improve the output, (ii) analyzing the output data of the interaction model can help identify patterns and trends that can be used to modify the model's output and identify areas for improvement or optimization, and (iii) modifying the algorithms used in the interaction model can help improve the output. In this chapter, the second method is used, i.e., the domain expert provides feedback on the output of the AI model.

An example scenario that shows the procedure for the human-AI interaction is shown below:

- Fetch predicted labels form the output of automatic label detection of models.
- Present the data with predicted label to the domain expert.
- Present the domain expert with choices to (i) accept the predicted label and (a) confirm the predicted label or (b) offer an alternative label or (ii) reject the predicted label and offer the correct label.
  - If the domain expert accepts and confirms the label, the process ends.
  - If the domain expert accepts the predicted label and offers an alternative label or a refinement of the label or rejects the predicted label altogether and offers the correct label, the domain expert's input will be sent as input to retrain the model.
- Visualization of behavior of the model with/without the domain expert's input will be shown for comparison of the effect of the domain fusion.
- Human-AI collaboration system will expose an API of the visualization to the DSS component through which the user will inspect the outputs of the model.

Figure 3 shows the process of data/model selection and parameter optimization including data flow and UI mockup for model selection and parameter optimization user interface through which the domain expert selects the model and parameter and optimizes the parameter values. The UI presents visualization of processing results for the selected model, parameter, and values. Once the domain expert determines the model, parameter, and values, the UI then enables the domain expert to export the result which will then be consumed by the Decision Support System (DSS).

The domain expert selects a section of the visualization and provides engineered knowledge, i.e., manual labeling of data points. This helps the user to visually inspect the dataset and enrich it with domain knowledge to boost the quality of the data to be used as training dataset for better performance of the ML model. For example, for an AI model built for anomaly detection, this is achieved by enabling the user to select the data point on the visualization plot in order to display and review (and when applicable, modify) the data that are marked by the system as anomalies. This is implemented by providing point, box, or lasso [15] selection where the user can select a single (or multiple data points on the graphs) and get the corresponding data points back, to provide the domain knowledge.

As depicted in Figs. 2 and 3, the domain expert will load data and models from the model repository, run the models on the data, observe the visualization, and adjust parameters in order to achieve the desired behavior of the AI model. Once the domain expert obtains satisfactory output from the model, she/he can then provide feedback. The algorithm shown in Fig. 4 shows the detailed operations during the domain knowledge enrichment.

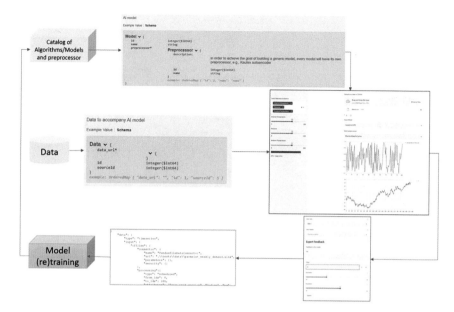

Fig. 2 Human-AI collaboration flow diagram

5 Conclusion and Future Works

This chapter discusses the concept of a framework of human-AI collaboration in manufacturing for injecting domain knowledge provided by human experts into AI models, as provided by machine learning processes in order to iteratively improve AI models. It explores the importance of human feedback in enhancing the effectiveness of AI models and improving usefulness of their outputs through incorporating human feedback. It describes a use case scenario to showcase the implementation of human-AI interaction where human feedback is utilized to enrich learned knowledge.

There are a number of future works in this chapter. An implementation of a full-fledged human-AI software prototype should be implemented and deployed so as to measure its effectiveness and also to experiment on the varied use cases to measure its actual usability in the real world. For this, the questions such as how the introduction of human-AI interaction affects the performance and effectiveness of the AI model, and for the given test, how much of the output of the AI model is rejected, accepted, and modified need to be answered.

Another aspect of the future research is to analyze whether AI model produces erroneous results even after retraining using expert feedback, i.e., whether the accuracy of the retrained AI model shows any improvement.

Yet another aspect is a study on scaling human-AI collaboration that takes into account the scarcity of domain experts who have understanding of intricacies of AI modeling or AI developers who have sufficient knowledge of the domain of

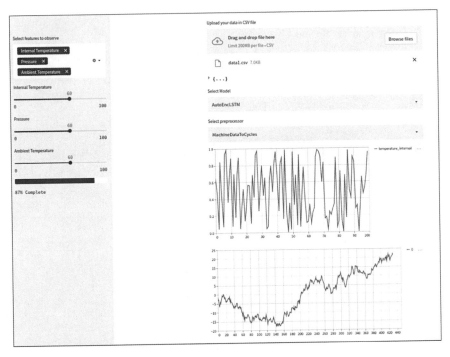

Fig. 3 Human-AI collaboration main interface

ALGORITHM1: PSEUDOCODE FOR DOMAIN KNOWLEDGE ENRICHMENT

Input: data, url of model repository
Output: data enriched by domain expert feedback

1	*data ← load data // obtain data*
2	*model_url ← url of model repository // obtain url of model repository*
3	*models ← load_models (model_url) // load models from the model repository*
4	*parameters ← selected parameters // user adjusts parameters for the model*
5	**While** *(model in models)* **do**
6	*accuracy[i] ← run_models (model, data, parameters) // obtain accuracy of models*
7	*best_model = argmax{accuracy[i]}*
8	*visualize output of best model*
9	*item_selected ← index (user_selection) // obtain index of the selected item from visualization*
10	*item_class_update ← index (user input) // update class or label of item with expert input*
11	*End*

Fig. 4 Pseudocode showing high-level process of human-AI collaboration

---

manufacturing. Long-term studies of how to scale human-AI collaboration are of paramount importance because scaling up the human-AI collaboration approach to large-scale manufacturing settings presents challenges in maintaining consistent collaboration and effectively incorporating human feedback across various use cases, domains, and data volumes. Therefore, it is important to further research scaling up human-AI collaboration in large-scale manufacturing settings through a well-planned approach that involves aligning technology, data, technical/domain expertise, and processes to create a seamless integration of human and AI capabilities, ultimately enhancing productivity, quality, and efficiency in manufacturing operations.

Acknowledgments The research leading to these results has received funding from the Horizon 2020 Program of the European Commission under Grant Agreement No. 957331.

References

1. Arinez, J.F., Chang, Q., Gao, R.X., Xu, C., Zhang, J.: Artificial intelligence in advanced manufacturing: current status and future outlook. J. Manuf. Sci. Eng. **142**(11), 110804 (2020)
2. Aron, J.: How Innovative is Apple's New Voice Assistant, Siri? (2011)
3. Baghernezhad-Tabasi, S., Druette, L., Jouanot, F., Meurger, C., Rousset, M.-C.: IOPE: interactive ontology population and enrichment. In: Workshop on Ontology-Driven Conceptual Modelling of Digital Twins co-located with Semantics 2021 (2021)
4. Fan, M., Yang, X., Yu, T., Liao, Q.V., Zhao, J.: Human-ai collaboration for UX evaluation: effects of explanation and synchronization. Proc. ACM Hum.-Comput. Interact. **6**(CSCW1), 1–32 (2022)
5. Fernández-Izquierdo, A., Cimmino, A., Patsonakis, C., Tsolakis, A.C., García-Castro, R., Ioannidis, D., Tzovaras, D.: OpenADR ontology: semantic enrichment of demand response strategies in smart grids. In: International Conference on Smart Energy Systems and Technologies (SEST), pp. 1–6. IEEE, New York (2020)
6. Gunasekaran, A., Yusuf, Y.Y., Adeleye, E.O., Papadopoulos, T., Kovvuri, D., Geyi, D.G.: Agile manufacturing: an evolutionary review of practices. Int. J. Prod. Res. **57**(15–16), 5154–5174 (2019)
7. Idoudi, R., Ettabaa, K.S., Solaiman, B., Hamrouni, K.: Ontology knowledge mining for ontology conceptual enrichment. Knowl. Manag. Res. Pract. **17**(2), 151–160 (2019)
8. Khadpe, P., Krishna, R., Fei-Fei, L., Hancock, J.T., Bernstein, M.S.: Conceptual metaphors impact perceptions of human-ai collaboration. Proc. ACM Hum.-Comput. Interact. **4**(CSCW2), 1–26 (2020)
9. Li, T., Vorvoreanu, M., DeBellis, D., Amershi, S.: Assessing human-AI interaction early through factorial surveys: a study on the guidelines for human-AI interaction. ACM Trans. Comput.-Hum. Interact. **30**(5), 45 (2023). https://doi.org/10.1145/3511605
10. LLC, C.T.: Time Series Data Labeling: A Complete Know-how for Efficient AI Implementation (2022). https://www.cogitotech.com/blog/time-series-data-labeling-a-complete-know-how-for-efficient-ai-implementation/. Last accessed: November 16, 2023
11. López, G., Quesada, L., Guerrero, L.A.: Alexa vs. Siri vs. Cortana vs. Google assistant: a comparison of speech-based natural user interfaces. In: Advances in Human Factors and Systems Interaction: Proceedings of the AHFE 2017 International Conference on Human Factors and Systems Interaction, July 17–21, 2017, The Westin Bonaventure Hotel, Los Angeles, California, USA 8, pp. 241–250. Springer, Berlin (2018)
12. Mucha, H., Robert, S., Breitschwerdt, R., Fellmann, M.: Interfaces for explanations in human-ai interaction: proposing a design evaluation approach. In: Extended Abstracts of the 2021 CHI Conference on Human Factors in Computing Systems, CHI EA '21, New York, NY, USA. Association for Computing Machinery, New York (2021)
13. Nguyen, A.T., Kharosekar, A., Krishnan, S., Krishnan, S., Tate, E., Wallace, B.C., Lease, M.: Believe it or not: designing a human-ai partnership for mixed-initiative fact-checking. In: Proceedings of the 31st Annual ACM Symposium on User Interface Software and Technology, UIST '18, pp. 189–199. Association for Computing Machinery, New York (2018)

14. Stratogiannis, G., Kouris, P., Alexandridis, G., Siolas, G., Stamou, G., Stafylopatis, A.: Semantic enrichment of documents: a classification perspective for ontology-based imbalanced semantic descriptions. Knowl. Inf. Syst. **63**(11), 3001–3039 (2021)
15. Tsang, M., Enouen, J., Liu, Y.: Interpretable artificial intelligence through the lens of feature interaction. arXiv preprint arXiv:2103.03103 (2021)
16. Urbanowicz, R.J., Moore, J.H.: Exstracs 2.0: description and evaluation of a scalable learning classifier system. Evol. Intell. **8**, 89–116 (2015)
17. Wang, W., Barnaghi, P.M., Bargiela, A.: Probabilistic topic models for learning terminological ontologies. IEEE Trans. Knowl. Data Eng. **22**(7), 1028–1040 (2009)

Examining the Adoption of Knowledge Graphs in the Manufacturing Industry: A Comprehensive Review

Jorge Martinez-Gil, Thomas Hoch, Mario Pichler, Bernhard Heinzl, Bernhard Moser, Kabul Kurniawan, Elmar Kiesling, and Franz Krause

1   Introduction

Advancements in Artificial Intelligence (AI) have enabled automation, prediction, and problem-solving, leading to increased productivity, adaptability, and efficiency in both the service and industrial sectors. In the latter, the fourth industrial revolution, commonly referred to as Industry 4.0 [19], represents a significant shift in industrial production. Driven by the integration of digital technology, industrial advancements increasingly hinge on data, which has opened up new possibilities beyond traditional applications.

A key goal toward Industry 5.0 is to combine human adaptability with machine scalability. Knowledge Graphs (KGs) provide a foundation for developing frameworks that enable such integration because they facilitate to dynamically integrate human decision-making with AI-generated recommendations and decisions [34]. KGs represent knowledge in a graph-based structure which connect entities and

J. Martinez-Gil (✉) · T. Hoch · M. Pichler · B. Heinzl · B. Moser
Software Competence Center Hagenberg GmbH, Hagenberg, Austria
e-mail: Jorge.Martinez-Gil@scch.at; thomas.hoch@scch.at; mario.pichler@scch.at; bernhard.heinzl@scch.at; bernhard.moser@scch.at

K. Kurniawan
WU, Institute for Data, Process and Knowledge Management, Vienna, Austria

Austrian Center for Digital Production (CDP), Vienna, Austria
e-mail: kabul.kurniawan@wu.ac.at

E. Kiesling
WU, Institute for Data, Process and Knowledge Management, Vienna, Austria
e-mail: elmar.kiesling@ai.wu.ac.at

F. Krause
University of Mannheim, Data and Web Science Group, Mannheim, Germany
e-mail: franz.krause@uni-mannheim.de

J. Soldatos (ed.), *Artificial Intelligence in Manufacturing*,
https://doi.org/10.1007/978-3-031-46452-2_4

their relationships. In the context of hybrid human-AI intelligence, KGs can represent shared conceptualizations between humans and AI components, providing a foundation that facilitates their collaboration in dynamically integrating human decision-making.

Furthermore, KGs provide a critical abstraction for organizing semi-structured domain information. By utilizing KGs, decision-making can be improved, knowledge management can be enhanced, personalized interactions can be enabled, predictive maintenance can be supported, and supply chain operations can be optimized. Therefore, KGs serve as a foundation for creating a shared knowledge space for AI components and humans, an environment for the representation of policies that govern the interactions between agents, and a view of real-world physical processes on the shop floor by extracting and integrating relevant events.

Thus, KGs hold enormous potential for facilitating collaboration in this field, making production lines more efficient and flexible while producing higher quality products. As such, companies seeking to achieve the goals of Industry 5.0 find that KGs can realize their vision [20]. However, research in this area is still in its early stages, and further studies are required to analyze how KGs might be implemented. This overview aims to provide a review of the current state of research in this field, as well as the challenges that remain open.

The rest of this work is structured in the following way: Sect. 2 reviews the current usage scenarios of KGs in industrial settings, Sect. 3 outlines the research questions and search strategy, Sect. 4 presents the major findings that can be extracted when analyzing the previously mentioned research questions, and Sect. 5 discusses the lessons learned and future research directions.

2 Antecedents and Motivation

The industrial landscape has been revolutionized by the emergence of collaborative paradigms between human–machine systems, such as the Internet of Things (IoT), Internet of Services (IoS), and Cyber-Physical Systems (CPS), resulting in the so-called Industry 5.0. This has led to a shift in focus toward enhancing collaboration between humans and machines [13]. The creation and connection of newly available devices generate enormous data with significant potential value, which can be used to extend the product's life cycle, on-demand manufacturing, resource optimization, machine maintenance, and other logistical arrangements [8].

KGs have recently gained much attention due to their potential to boost productivity across various sectors. KG can primarily empower industrial goods and services and their development process in two areas. First, they may save time and labor costs while enhancing accuracy and efficiency in domain information retrieval for requirements gathering, design and implementation, and service and maintenance management, by offering a semantic-based and in-depth knowledge management approach. Second, the development of KG has made it possible to build a data architecture suitable for corporate use by combining intelligent discovery and knowledge storage, utilizing KG embedding techniques to gather more information from KGs. KGs are most employed to model a particular and frequently complex domain semantically, explicitly modeling domain knowledge used to support and increase the accuracy of tasks performed further down the pipeline. Furthermore, advanced methodologies founded on KGs have become essential for knowledge representation and business process modeling.

In recent years, there has been a rise in interest in analyzing KGs using Machine Learning (ML) techniques, such as predicting missing edges or classifying nodes. To input feature vectors into most ML models, much research has been devoted to developing methods to build embeddings from KG. The transformation of nodes and, depending on the technique, edges into a numerical representation via KG embedding enables direct input into an ML model [2].

Furthermore, KG is widely assumed to be a tool for optimizing supply chain operations, reducing costs, and improving overall efficiency. Manufacturers can model their supply chain using KGs to fully understand how their suppliers, customers, and operations depend on one another, enabling them to make real-time data-based decisions.

In summary, many KGs have been built, both open to the public and closed for internal company use. Enterprise KGs are closed applications that can only be used by authorized personnel, while Open KGs are usually academic or open-source projects available for use by anyone on the Web. By using modeling and ML organizations can gain insights and make data-based decisions thanks to the creation of these KGs. This research work describes the current state of Open KGs.

3 Research Questions and Search Strategy

KGs serve as semantic representations of various aspects involved in the manufacturing process. These aspects include all phases of system engineering, such as the phases of development (e.g., layouts), organizational development (e.g., collaboration and worker roles), and operational development (e.g., user stories). These KGs can improve the processes by considering the data coming from the industrial monitoring and the human work themselves and additional contextual data and knowledge sources. Some examples include technical documentation about the process, questionnaires about maintenance cases, constraints, and rules for representing standards and policies for safety or ethical issues, protocols about teaming workflows, logging about process states, and user feedback. This chapter investigates the present state of KGs in manufacturing. In this chapter, potential areas are identified, and chances for future works are highlighted. The following are the primary research questions that guided this study.

3.1 Research Questions

We propose some research questions to provide specific insights into how KGs are used in manufacturing. These Research Questions (RQs) consider that the two most popular KG types are Resource Description Framework (RDFs) and Labeled Property Graph (LPG). Our RQs are designed to cover the essential aspects of bibliometric facts and application scenarios.

RQ1: Which areas within manufacturing are most interested in KGs?

The purpose of RQ1 is to demonstrate the significance and relevance of the topic by providing an overview of bibliometric facts from previously published studies on the applications of KGs in manufacturing.

RQ2: Which manufacturing domains commonly employ KGs?

RQ2 investigates KG application scenarios within manufacturing. Specifically, we will examine the manufacturing domains in which KGs have been used, the specific use cases, and the types of systems developed.

RQ3: What is the popularity of RDF and LPG as KG types?

RQ3 aims to evaluate the degree to which KG applications have matured by investigating some research aspects such as the format and standards used.

RQ4: How are industrial KGs currently used?

RQ4 discusses which building, exploitation, and maintenance procedures are commonly followed in manufacturing-related KGs. This provides insight into the structure of KGs that is vital for researchers and practitioners.

3.2 Dataset

To address these RQs, we analyze a significant sample of literature published in recent years. The search scope considers gray literature such as professional forums and publications as well as academic publications published in journals or academic conferences or in books that have been peer-reviewed. In total, we have identified 40 items of publication using KG published between 2016 and 2022. The authors of these items come from a diverse range of academic disciplines and represent institutions from various parts of the world. Overall, the sample of publications provides a comprehensive and diverse set of perspectives on the research questions at hand. The following is an analysis of the main characteristics of these sources.

3.3 Subject Area

KG in manufacturing is an emerging field that has drawn considerable attention from both industry and academic communities. The current body of research primarily originates from Computer Science. Conversely, there is a significant gap in research output from the fields of Engineering and Business, which are the other two most represented areas of knowledge. The scope of research in other areas, such as Chemistry, Physics, and Astronomy, as well as Materials Science, remains limited, with only a marginal number of proposals. Figure 1 presents a comprehensive categorization of the research venues considered in this chapter. The classification scheme is based on the self-description provided by each venue where the research works have been published.

3.4 Manufacturing Domain

The presented findings illustrate the prevailing domains explored in the literature on applying KGs in manufacturing, as summarized in Fig. 2. To determine whether a given paper pertains to the manufacturing domain, the North American Industry Classification System (NAICS[1]) was employed. However, most of the examined literature does not specify any particular application domain.

Machinery is identified as the second most frequently represented domain, after which Materials, Chemistry, and Automotive follow. Furthermore, Additive Manufacturing, Aerospace, Mining, Operations, and Textile, albeit less frequently

---

[1] https://www.census.gov/naics/.

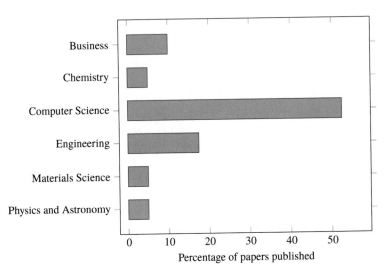

Fig. 1 Research communities that have published the most research on KGs in manufacturing

investigated, are also observed in the literature. The identified domains highlight the diverse industries that benefit from leveraging KGs.

Most reviewed works employ knowledge fusion techniques in general scenarios where KGs combine data from multiple sources. Additionally, KGs are applied to automate the merging of isolated production processes, generate digital twins based on KGs, and utilize them for automated source code development. These findings demonstrate the versatility of KGs in manufacturing and their potential to revolutionize various aspects of industrial production.

3.5 Kinds of KGs

A KG may be modeled as either an RDF graph or an LPG, depending on the data requirements. As shown in Fig. 3, RDF-based solutions currently dominate the field. However, a considerable proportion of solutions are also represented by LPGs.

RDF is a recommended standard from the World Wide Web Consortium[2] that provides a language for defining resources on the Web. The representation of resources is accomplished using triples that consist of a subject, predicate, and object. RDF Schema, commonly referred to as RDFS, defines the vocabulary used in RDF descriptions. The RDF data model is specifically designed for knowledge representation and is used to encode a graph as a set of statements. By standardizing data publication and sharing on the Web, RDF seeks to ensure

---

[2] https://www.w3.org/.

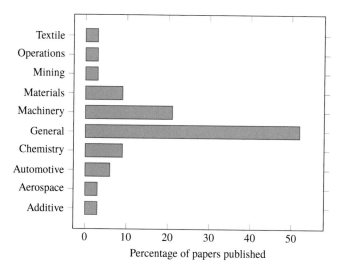

Fig. 2 Manufacturing domains that have carried out the most research work around KGs

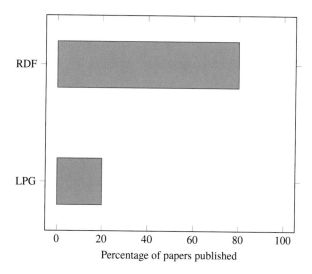

Fig. 3 Knowledge Graph adoption in the manufacturing industry by representation paradigm

semantic interoperability. The semantic layer of the available statements, along with the reasoning applied to it, forms the foundation of intelligent systems in the RDF domain.

On the other hand, LPG representation primarily emphasizes the graph's structure, properties, and relationships. This highlights the unique characteristics of graph data, opening new opportunities for data analysis and visualization. It also brings a window of opportunity for developing ML systems that use graphs to infer additional information.

Different approaches to KG have a significant impact on the user experience. When developers and analysts work with RDF data, they use statements and SPARQL query language to make changes. On the other hand, LPG use the Cypher query language, which provides a more intuitive way to interact with nodes, edges, and related properties within the graph structure.

3.6 Different Approaches for KG Creation

Compared to the broader scope of research on KGs, the development of KGs in an industrial context often employs a knowledge-driven approach. Consequently, knowledge-driven KGs are more used in the industry. This trend may stem from the practical advantages of a more closed-world approach, which is better suited to the constraints and contingencies inherent in a production environment. It also suggests that the manufacturing industry remains cautious about adopting the latest advancements in KG embeddings to enhance their analytical capabilities.

Figure 4 depicts the distribution of popularity between the two distinct approaches for building KGs. Currently, the knowledge-driven approach prevails, but recent years have witnessed a significant surge in the number of data-driven solutions. These solutions are better equipped to deal with ML and other computational intelligence techniques.

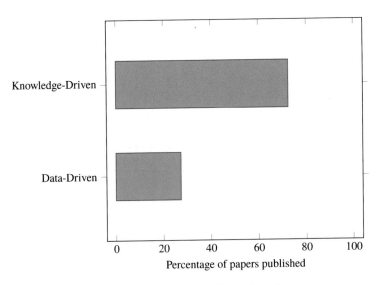

Fig. 4 Manufacturing industry Knowledge Graphs by form of creation

4 Insights

This section summarizes the results obtained from our analysis and highlights potential areas for future research on the use KGs in the manufacturing domain. The findings are structured according to the RQs addressed earlier in the study.

4.1 Answers to the Research Questions

Based on our study, we can deduce the most active research communities in the field of KGs. The answer to RQ1 (AQ1) is as follows:

> **AQ1.** The majority of primary research in the field of KGs is conducted in the discipline of **Computer Science**. Research in KGs is less common in other areas of knowledge.

This could be because computer scientists have been developing new representation models since the beginning. Today, KGs are considered the natural progression of such models to make them more adaptable to new platforms and emerging methods for managing large amounts of data.

Regarding the answer to RQ2 (AQ2), it is unsurprising that the most common case is the preference for proposing generic models that can be easily adapted to various domains.

> **AQ2.** The literature primarily covers the **manufacturing industry** as a general concept. In most of the works examined, no specific application domain was provided.

The domains related to machinery and materials are the next most represented, followed by chemistry and material. Finally, some KGs have also been developed in the aerospace, additive manufacturing, mining, operations, and textile fields.

Regarding the representation of KGs, the two most commonly used data models are RDF and LPG. However, in answering RQ3 (AQ3), we seek to identify the current prevailing choice for representing KGs.

> **AQ3.** In the industrial domain, **RDF is the preferred format for building.** **KGs.** This is due to RDF's ability to represent complex data and relationships in a structured and interoperable manner, which allows for the building of integrated knowledge spaces for both humans and AI components.

RDF is beneficial for industrial applications as it facilitates the integration of diverse sources and a more comprehensive understanding of the data. Moreover, the ability to query across multiple sources makes it easier for people to analyze relevant information for their specific needs.

Regarding the question of the predominant approach to constructing industrial KGs, it has been observed that knowledge-driven approaches are most commonly used, as stated in Answer to RQ4 (AQ4):

> **AQ4. Knowledge-driven approaches** are predominant. However, new developments using data-driven approaches are expected to be increasingly incorporated into the existing body of literature as new solutions are proposed in combination with more mature techniques.

It is worth noting that existing knowledge-driven methods still encounter several general challenges, such as the interoperability and heterogeneity of data, incompleteness, and other specific challenges that arise from the goal of integrating them as active components rather than passive artifacts or mere data stores.

4.2 Additional Lessons Learned

In light of our study, the utilization of KGs within the manufacturing industry has experienced substantial growth in recent years as manufacturers seek to enhance their operational efficiency and decision-making capabilities. The structural design of KGs facilitates a more intuitive and comprehensive representation of data than traditional database models, rendering KGs well suited for the manufacturing industry.

> **Additional Lesson Learned #1.** Although still nascent, the application of KGs within the manufacturing industry has garnered substantial interest from academia and industry.

One of the primary reasons for this keen interest is that by modeling relationships between suppliers, manufacturers, and customers, organizations can better understand the flow of goods, services, and information through their supply chain. This, in turn, can assist them in identifying bottlenecks, optimizing production processes, and ensuring product delivery to customers.

> **Additional Lesson Learned #2.** The majority of the studies examined have been published in conference proceedings. In many instances, this indicates that the subject of investigation is still in the developmental stages. The state of the art is gradually maturing in almost every research area, leading to more journal publications with archival significance.

KGs can aid manufacturers in enhancing their ability to predict and respond to shifts in demand. This can help reduce waste, optimize production processes, and boost efficiency. However, most of the research is a work in progress, and there is still a long way to go to consolidate the results of archival value.

4.3 Open Problems

As a result of our study, we have identified several issues that limit the adoption of KGs in manufacturing and production environments. Some of the most critical issues are described below. The first issue concerns tabular data. This kind of data is frequently represented in values separated by commas. It is typically one of the most common input methods in industrial environments because it enables modeling a wide variety of data associated with temporal aspects (timestamps) and spatial aspects (coordinates). However, more optimal solutions still need to be proposed.

> **Problem 1 (Dealing with Tabular Data)** Most solutions today are created to deal with information that is predominately textual in its presentation. Although this information category is crucial in the sector, it is not dominant in manufacturing settings, which involve working with machinery and equipment that generate numerical data in tabular form.

Another fact that is taken for granted by both researchers and practitioners is that it is possible for KGs to effectively deal with information of varying types that may arrive via a variety of channels and sources. However, our research has not found a large number of papers concerned with the temporal component of processing KGs.

Problem 2 (Real-time and Synchronization) Because many of the processes involved in manufacturing are automated and must have a high degree of synchronization, the manufacturing industry demands solutions that can perform adequately in environments with substantial time constraints and synchronization needs.

Last but not least, according to the results of our investigation, work still needs to be done in compiling the best practices for manufacturing KGs. In this sense, we miss work in the direction of design and proposal of best practices for the sector.

Problem 3 (Lack of Standardized Procedures) A substantial obstacle still exists in identifying reference architectures to build, implement, and use KGs in industrial and production settings. A compilation of best practices can be of genuine benefit in several ways, including high standards of quality results and resource saving while developing new systems or making changes to existing ones.

KGs are suitable for the manufacturing industry because they can provide systems with contextual data to achieve efficient and effective solutions. This contextual data includes human experience, environmental knowledge, technical conventions, etc. Creating such solutions becomes critical when the influence on human life is essential, as in the case of a factory that employs human workers.

5 Conclusion

In this chapter, we have seen how the amount of data generated in the industrial sector at a high velocity is bringing new challenges. For example, this data emanates from multiple sources, each utilizing distinct formats and standards. Consequently, integrating these divergent pieces of information is not only essential but also critical. Contextualizing data elements utilizing relevant relationships is imperative to ensure consistency and high-quality data.

The study also examines KGs as multifaceted knowledge bases that capture interlinked descriptions of entities. KGs facilitate the smooth integration and structuring of information at large scale, even from heterogeneous sources. Unlike other knowledge bases, KGs are not homogeneous and do not require rigid schemas. This makes KGs highly scalable and suitable for integrating and connecting diverse data representations.

Semiautomatic methods, employing available data sources and manual effort, are used to construct manufacturing KGs. However, manual KGs construction is only practical for small-scale KGs, and automated methods are necessary for large-scale KGs. Therefore, automating the construction and maintenance of KGs in the manufacturing domain is essential for successful implementation.

In conclusion, utilizing KGs in the manufacturing industry can offer several advantages, including better decision-making processes and the ability to predict and respond to changes in demand. With the manufacturing industry evolving at an unprecedented rate, KGs will likely play an increasingly critical role in driving operational efficiency and competitiveness.

Acknowledgments We would like to thank the anonymous reviewers for their constructive comments to improve this work. Our study is based on a sample of literature on KGs in manufacturing. The primary sources used in the study are not cited in the text, but are listed in the References section. SCCH co-authors have been partially funded by the Federal Ministry for Climate Action, Environment, Energy, Mobility, Innovation, and Technology (BMK), the Federal Ministry for Digital and Economic Affairs (BMDW), and the State of Upper Austria in the frame of SCCH, a center in the COMET Program managed by Austrian FFG. All co-authors have also received funding from Teaming.AI, a project supported by the European Union's Horizon 2020 program, under Grant Agreement No. 957402.

References

1. Aggour, K.S., Kumar, V.S., Cuddihy, P., Williams, J.W., Gupta, V., Dial, L., Hanlon, T., Gambone, J., Vinciquerra, J.: Federated multimodal big data storage & analytics platform for additive manufacturing. In: 2019 IEEE International Conference on Big Data (Big Data), pp. 1729–1738. IEEE, New York (2019)
2. Alam, M., Fensel, A., Martinez-Gil, J., Moser, B., Recupero, D.R., Sack, H.: Special issue on machine learning and knowledge graphs. Future Gener. Comput. Syst. **129**, 50–53 (2022). https://doi.org/10.1016/j.future.2021.11.022
3. Bachhofner, S., Kiesling, E., Kurniawan, K., Sallinger, E., Waibel, P.: Knowledge graph modularization for cyber-physical production systems. In: Seneviratne, O., Pesquita, C., Sequeda, J., Etcheverry, L. (eds.) Proceedings of the ISWC 2021 Posters, Demos and Industry Tracks: From Novel Ideas to Industrial Practice co-located with 20th International Semantic Web Conference (ISWC 2021), Virtual Conference, October 24–28, 2021. CEUR Workshop Proceedings, vol. 2980. CEUR-WS.org (2021). https://ceur-ws.org/Vol-2980/paper333.pdf
4. Bachhofner, S., Kiesling, E., Revoredo, K., Waibel, P., Polleres, A.: Automated process knowledge graph construction from BPMN models. In: Strauss, C., Cuzzocrea, A., Kotsis, G., Tjoa, A.M., Khalil, I. (eds.) Database and Expert Systems Applications—33rd International Conference, DEXA 2022, Vienna, Austria, August 22–24, 2022, Proceedings, Part I. Lecture Notes in Computer Science, vol. 13426, pp. 32–47. Springer, Berlin (2022). https://doi.org/10.1007/978-3-031-12423-5_3
5. Bachhofner, S., Kurniawan, K., Kiesling, E., Revoredo, K., Bayomie, D.: Knowledge graph supported machine parameterization for the injection moulding industry. In: Villazón-Terrazas, B., Ortiz-Rodríguez, F., Tiwari, S., Sicilia, M., Martín-Moncunill, D. (eds.) Knowledge Graphs and Semantic Web—4th Iberoamerican Conference and Third Indo-American Conference, KGSWC 2022, Madrid, Spain, November 21–23, 2022, Proceedings. Communications in Computer and Information Science, vol. 1686, pp. 106–120. Springer, Berlin (2022). https://doi.org/10.1007/978-3-031-21422-6_8

6. Bader, S.R., Grangel-Gonzalez, I., Nanjappa, P., Vidal, M.E., Maleshkova, M.: A knowledge graph for industry 4.0. In: The Semantic Web: 17th International Conference, ESWC 2020, Heraklion, Crete, Greece, May 31–June 4, 2020, Proceedings 17, pp. 465–480. Springer, Berlin (2020)
7. Banerjee, A., Dalal, R., Mittal, S., Joshi, K.P.: Generating digital twin models using knowledge graphs for industrial production lines. In: Proceedings of the 2017 ACM on Web Science Conference, pp. 425–430 (2017)
8. Buchgeher, G., Gabauer, D., Martinez-Gil, J., Ehrlinger, L.: Knowledge graphs in manufacturing and production: a systematic literature review. IEEE Access 9, 55537–55554 (2021). https://doi.org/10.1109/ACCESS.2021.3070395
9. Chhetri, T.R., Aghaei, S., Fensel, A., Göhner, U., Gül-Ficici, S., Martinez-Gil, J.: Optimising manufacturing process with Bayesian structure learning and knowledge graphs. In: Computer Aided Systems Theory—EUROCAST 2022—18th International Conference, Las Palmas de Gran Canaria, Spain, February 20–25, 2022, Revised Selected Papers. Lecture Notes in Computer Science, vol. 13789, pp. 594–602. Springer, Berlin (2022). https://doi.org/10.1007/978-3-031-25312-6_70
10. Dombrowski, U., Reiswich, A., Imdahl, C.: Knowledge graphs for an automated information provision in the factory planning. In: 2019 IEEE International Conference on Industrial Engineering and Engineering Management (IEEM). pp. 1074–1078. IEEE, New York (2019)
11. Duan, W., Chiang, Y.Y.: Building knowledge graph from public data for predictive analysis: a case study on predicting technology future in space and time. In: Proceedings of the 5th ACM SIGSPATIAL International Workshop on Analytics for Big Geospatial Data, pp. 7–13 (2016)
12. Eibeck, A., Lim, M.Q., Kraft, M.: J-park simulator: an ontology-based platform for cross-domain scenarios in process industry. Comput. Chem. Eng. 131, 106586 (2019)
13. Freudenthaler, B., Martinez-Gil, J., Fensel, A., Höfig, K., Huber, S., Jacob, D.: Ki-net: Ai-based optimization in industrial manufacturing—A project overview. In: Computer Aided Systems Theory—EUROCAST 2022—18th International Conference, Las Palmas de Gran Canaria, Spain, February 20–25, 2022, Revised Selected Papers. Lecture Notes in Computer Science, vol. 13789, pp. 554–561. Springer, Berlin (2022). https://doi.org/10.1007/978-3-031-25312-6_65
14. Garofalo, M., Pellegrino, M.A., Altabba, A., Cochez, M.: Leveraging knowledge graph embedding techniques for industry 4.0 use cases. In: Cyber Defence in Industry 4.0 Systems and Related Logistics and IT Infrastructures, pp. 10–26. IOS Press, New York (2018)
15. Grangel-González, I., Halilaj, L., Vidal, M.E., Lohmann, S., Auer, S., Müller, A.W.: Seamless integration of cyber-physical systems in knowledge graphs. In: Proceedings of the 33rd Annual ACM Symposium on Applied Computing, pp. 2000–2003 (2018)
16. Grangel-González, I., Halilaj, L., Vidal, M.E., Rana, O., Lohmann, S., Auer, S., Müller, A.W.: Knowledge graphs for semantically integrating cyber-physical systems. In: International Conference on Database and Expert Systems Applications, pp. 184–199. Springer, Berlin (2018)
17. Haase, P., Herzig, D.M., Kozlov, A., Nikolov, A., Trame, J.: metaphactory: a platform for knowledge graph management. Semantic Web 10(6), 1109–1125 (2019)
18. He, L., Jiang, P.: Manufacturing knowledge graph: a connectivism to answer production problems query with knowledge reuse. IEEE Access 7, 101231–101244 (2019)
19. Hermann, M., Pentek, T., Otto, B.: Design principles for Industrie 4.0 scenarios. In: 2016 49th Hawaii International Conference on System Sciences (HICSS), pp. 3928–3937. IEEE, New York (2016)
20. Hoch, T., Heinzl, B., Czech, G., Khan, M., Waibel, P., Bachhofner, S., Kiesling, E., Moser, B.: Teaming.ai: enabling human-ai teaming intelligence in manufacturing. In: Zelm, M., Boza, A., León, R.D., Rodríguez-Rodríguez, R. (eds.) Proceedings of Interoperability for Enterprise Systems and Applications Workshops co-located with 11th International Conference on Interoperability for Enterprise Systems and Applications (I-ESA 2022), Valencia, Spain, March 23–25, 2022. CEUR Workshop Proceedings, vol. 3214. CEUR-WS.org (2022). https://ceur-ws.org/Vol-3214/WS5Paper6.pdf

21. Kalaycı, E.G., Grangel González, I., Lösch, F., Xiao, G., ul Mehdi, A., Kharlamov, E., Calvanese, D.: Semantic integration of Bosch manufacturing data using virtual knowledge graphs. In: The Semantic Web–ISWC 2020: 19th International Semantic Web Conference, Athens, Greece, November 2–6, 2020, Proceedings, Part II 19, pp. 464–481. Springer, Berlin (2020)
22. Kattepur, A.: Roboplanner: autonomous robotic action planning via knowledge graph queries. In: Proceedings of the 34th ACM/SIGAPP Symposium on Applied Computing, pp. 953–956 (2019)
23. Ko, H., Witherell, P., Lu, Y., Kim, S., Rosen, D.W.: Machine learning and knowledge graph based design rule construction for additive manufacturing. Addit. Manuf. 37, 101620 (2021)
24. Kumar, A., Bharadwaj, A.G., Starly, B., Lynch, C.: FabKG: a knowledge graph of manufacturing science domain utilizing structured and unconventional unstructured knowledge source. arXiv preprint arXiv:2206.10318 (2022)
25. Leijie, F., Yv, B., Zhenyuan, Z.: Constructing a vertical knowledge graph for non-traditional machining industry. In: 2018 IEEE 15th International Conference on Networking, Sensing and Control (ICNSC), pp. 1–5. IEEE, New York (2018)
26. Li, R., Dai, W., He, S., Chen, X., Yang, G.: A knowledge graph framework for software-defined industrial cyber-physical systems. In: IECON 2019-45th Annual Conference of the IEEE Industrial Electronics Society, vol. 1, pp. 2877–2882. IEEE, New York (2019)
27. Li, X., Chen, C.H., Zheng, P., Wang, Z., Jiang, Z., Jiang, Z.: A knowledge graph-aided concept–knowledge approach for evolutionary smart product–service system development. J. Mech. Des. 142(10), 101403 (2020)
28. Li, X., Zhang, S., Huang, R., Huang, B., Xu, C., Kuang, B.: Structured modeling of heterogeneous cam model based on process knowledge graph. Int. J. Adv. Manuf. Technol. 96(9–12), 4173–4193 (2018)
29. Liebig, T., Maisenbacher, A., Opitz, M., Seyler, J.R., Sudra, G., Wissmann, J.: Building a Knowledge Graph for Products and Solutions in the Automation Industry (2019)
30. Liu, M., Li, X., Li, J., Liu, Y., Zhou, B., Bao, J.: A knowledge graph-based data representation approach for IIoT-enabled cognitive manufacturing. Adv. Eng. Inform. 51, 101515 (2022)
31. Martinez-Gil, J., Buchgeher, G., Gabauer, D., Freudenthaler, B., Filipiak, D., Fensel, A.: Root cause analysis in the industrial domain using knowledge graphs: a case study on power transformers. In: Longo, F., Affenzeller, M., Padovano, A. (eds.) Proceedings of the 3rd International Conference on Industry 4.0 and Smart Manufacturing (ISM 2022), Virtual Event/Upper Austria University of Applied Sciences—Hagenberg Campus—Linz, Austria, 17–19 November 2021. Procedia Computer Science, vol. 200, pp. 944–953. Elsevier, Amsterdam (2021). https://doi.org/10.1016/j.procs.2022.01.292
32. Meckler, S., Steinmüller, H., Harth, A.: Building a knowledge graph with inference for a production machine using the web of things standard. In: Advances and Trends in Artificial Intelligence. From Theory to Practice: 34th International Conference on Industrial, Engineering and Other Applications of Applied Intelligent Systems, IEA/AIE 2021, Kuala Lumpur, Malaysia, July 26–29, 2021, Proceedings, Part II 34, pp. 240–251. Springer, Berlin (2021)
33. Nayak, A., Kesri, V., Dubey, R.K.: Knowledge graph based automated generation of test cases in software engineering. In: Proceedings of the 7th ACM IKDD CoDS and 25th COMAD, pp. 289–295 (2020)
34. Noy, N.F., Gao, Y., Jain, A., Narayanan, A., Patterson, A., Taylor, J.: Industry-scale knowledge graphs: lessons and challenges. Commun. ACM 62(8), 36–43 (2019). https://doi.org/10.1145/3331166
35. Peroni, S., Vitali, F.: Interfacing fast-fashion design industries with semantic web technologies: the case of imperial fashion. J. Web Semant. 44, 37–53 (2017)
36. Ringsquandl, M., Kharlamov, E., Stepanova, D., Lamparter, S., Lepratti, R., Horrocks, I., Kröger, P.: On event-driven knowledge graph completion in digital factories. In: 2017 IEEE International Conference on Big Data (Big Data), pp. 1676–1681. IEEE, New York (2017)

37. Ringsquandl, M., Lamparter, S., Lepratti, R., Kröger, P.: Knowledge fusion of manufacturing operations data using representation learning. In: IFIP International Conference on Advances in Production Management Systems, pp. 302–310. Springer, Berlin (2017)
38. Rožanec, J.M., Zajec, P., Kenda, K., Novalija, I., Fortuna, B., Mladenić, D.: XAI-KG: knowledge graph to support XAI and decision-making in manufacturing. In: Proceedings of the Advanced Information Systems Engineering Workshops: CAiSE 2021 International Workshops, Melbourne, VIC, Australia, June 28–July 2, 2021, pp. 167–172. Springer, Berlin (2021)
39. Tushkanova, O., Samoylov, V.: Knowledge net: model and system for accumulation, representation, and use of knowledge. IFAC-PapersOnLine 52(13), 1150–1155 (2019)
40. Wang, Z., Zhang, B., Gao, D.: A novel knowledge graph development for industry design: a case study on indirect coal liquefaction process. Comput. Ind. 139, 103647 (2022)
41. Yan, H., Yang, J., Wan, J.: KnowIME: a system to construct a knowledge graph for intelligent manufacturing equipment. IEEE Access 8, 41805–41813 (2020)
42. Zhang, X., Liu, X., Li, X., Pan, D.: MMKG: an approach to generate metallic materials knowledge graph based on DBpedia and Wikipedia. Comput. Phys. Commun. 211, 98–112 (2017)
43. Zhao, M., Wang, H., Guo, J., Liu, D., Xie, C., Liu, Q., Cheng, Z.: Construction of an industrial knowledge graph for unstructured Chinese text learning. Appl. Sci. 9(13), 2720 (2019)
44. Zhao, Y., Liu, Q., Xu, W.: Open industrial knowledge graph development for intelligent manufacturing service matchmaking. In: 2017 International Conference on Industrial Informatics-Computing Technology, Intelligent Technology, Industrial Information Integration (ICIICII), pp. 194–198. IEEE, New York (2017)
45. Zhou, B., Bao, J., Li, J., Lu, Y., Liu, T., Zhang, Q.: A novel knowledge graph-based optimization approach for resource allocation in discrete manufacturing workshops. Robot. Comput. Integr. Manuf. 71, 102160 (2021)
46. Zhou, X., Lim, M.Q., Kraft, M.: A Smart Contract-Based Agent Marketplace for the j-park Simulator–a Knowledge Graph for the Process Industry (2020)

Leveraging Semantic Representations via Knowledge Graph Embeddings

Franz Krause, Kabul Kurniawan, Elmar Kiesling, Jorge Martinez-Gil, Thomas Hoch, Mario Pichler, Bernhard Heinzl, and Bernhard Moser

1 Introduction

Knowledge graphs are becoming increasingly recognized as a valuable tool in data-driven domains like healthcare [1], finance [2], and manufacturing [3], where they have gained considerable popularity in recent research. They are commonly employed to represent and integrate both structured and unstructured data, providing a standardized approach to encode domain knowledge [4]. Built on ontologies that conceptualize domain classes, relations, and logical inference rules, KGs represent specific instantiations of ontological models and their inherent semantic characteristics. Typically, KGs are divided into two modules: a terminological TBox containing concepts (such as the class of a manufacturing process) and an assertive ABox containing real-world instances (such as unique executions of a manufacturing process).

F. Krause (✉)
University of Mannheim, Data and Web Science Group, Mannheim, Germany
e-mail: franz.krause@uni-mannheim.de

K. Kurniawan
WU, Institute for Data, Process and Knowledge Management, Vienna, Austria

Austrian Center for Digital Production (CDP), Vienna, Austria
e-mail: kabul.kurniawan@wu.ac.at

E. Kiesling
WU, Institute for Data, Process and Knowledge Management, Vienna, Austria
e-mail: elmar.kiesling@wu.ac.at

J. Martinez-Gil · T. Hoch · M. Pichler · B. Heinzl · B. Moser
Software Competence Center Hagenberg GmbH, Hagenberg, Austria
e-mail: Jorge.Martinez-Gil@scch.at; thomas.hoch@scch.at; Mario.Pichler@scch.at; bernhard.heinzl@scch.at; Bernhard.Moser@scch.at

© The Author(s) 2024
J. Soldatos (ed.), *Artificial Intelligence in Manufacturing*,
https://doi.org/10.1007/978-3-031-46452-2_5

We adopt the notion of a (standard) KG $G = (V, E)$ as described in [5], which is represented by a set of nodes V (also referred to as vertices) and a set of triples $E \subseteq V \times R \times V$ consisting of directed and labeled edges. Here, R denotes the set of valid relation types defined in the underlying ontology. Thus, an edge in the form of a triple $(s, p, o) \in E$ implies an outgoing relation from the subject $s \in V$ to the object $o \in V$ via the predicate $p \in R$. Given such a KG, embedding techniques aim to exploit the topology of the graph to generate latent feature representations

$$\gamma : V \to \Gamma \tag{1}$$

of its nodes V in a latent representation space Γ, e.g., $\Gamma = \mathbb{R}^d$ with $d \in \mathbb{N}$, thereby enabling their utilization in downstream applications, e.g., graph-based machine learning (ML). However, the findings of this work can be applied almost analogously to the most well-known KG extensions, such as labeled property graphs like Neo4j [6].

In addition to the improved applicability of graph-based data in tasks like recommendation systems [7] or question answering [8], embedding formalisms have also proven to be valuable as intrinsic complements to graph-structured data. This is due to their ability to provide an empirical approach for enhancing the expressivity of graph topologies by means of downstream tasks like entity linking [9] and link prediction [10]. Consequently, related areas such as relational ML are receiving significant attention in both literature and applications [11].

In this chapter, we first provide a brief overview of representation learning as the enabler of KG embeddings, addressing state-of-the-art embedding formalisms for generating lean feature representations and describing their functionalities. An analysis of the advantages and drawbacks of employing KG embeddings is provided, along with a discussion of associated open research questions. We focus specifically on potential challenges and risks that may hinder the usage of KG embeddings in the highly dynamic manufacturing domain. Accordingly, we present the methodologies developed within the Teaming. AI project to address those problems. In this context, we describe the applicability and potential benefits of KG embeddings in the human–AI-based manufacturing use cases of the project. Furthermore, we showcase the Navi approach as an enabler of dynamic KG embeddings that allows for real-time and structure-preserving computations of new or updated node representations.

2 Knowledge Graph Embeddings

The generation of KG embeddings as per Eq. (1) denotes a subdiscipline of representation learning. In the context of KGs, representation learning is applied to determine lean feature representations that are able to capture inherent semantic relationships between KG elements. Thus, we first provide a general overview of representation learning to subsequently describe its application in KG embeddings.

3 Representation Learning

Representation learning comprises techniques for the automatic detection of appropriate feature representations that can be employed by downstream models or tasks, such as machine learning models [12]. Thus, the main objective of representation learning is to eliminate the need for preprocessing raw input data. Given a set of observable variables V with semantic representations $\pi : V \rightarrow \Pi$ within an inherent representation space Π (which is not necessarily compatible with the downstream model), these techniques aim to generate an alternative feature mapping $\gamma : V \rightarrow \Gamma$ into a representation space Γ that satisfies the requirements of the desired task.

Representation learning can be performed in a supervised, unsupervised, or self-supervised manner. One example of a supervised approach for learning latent feature representations is the training of deep neural networks on labeled input data. Namely, given an input feature $\pi(v)$ for some $v \in V$, the hidden layer outputs (and also the output layer) obtained from the forward pass of the network can be considered as alternative representations $\gamma(v)$, as illustrated in Fig. 1.

Contrarily, unsupervised representation learning techniques can be utilized for unlabeled representations $\pi(v)$. Methods like principal component analysis or auto-encoders intend to reduce the dimensionality of high-dimensional input features. Accordingly, the goal of these algorithms is to determine alternative, low-dimensional representations without the consideration of any target feature except the input feature $\pi(v)$ itself. For example, auto-encoders feed a representation $\pi(v) \in \mathbb{R}^{d'}$ into a deep neural network and attempt to reconstruct it, i.e., $\pi(v)$ also serves as the output feature. However, the hidden layers are assumed to be low-dimensional to serve as alternative representations $\gamma(v) \in \mathbb{R}^{d}$ of $v \in V$ with $d \ll d'$ as depicted in Fig. 2.

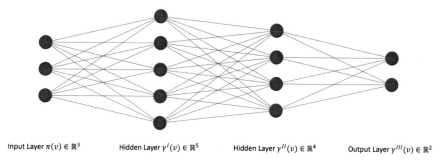

Input Layer $\pi(v) \in \mathbb{R}^3$ Hidden Layer $\gamma^I(v) \in \mathbb{R}^5$ Hidden Layer $\gamma^{II}(v) \in \mathbb{R}^4$ Output Layer $\gamma^{III}(v) \in \mathbb{R}^2$

Fig. 1 Deep neural networks as supervised representation learning formalisms

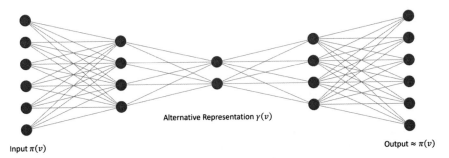

Input $\pi(v)$

Alternative Representation $\gamma(v)$

Output $\approx \pi(v)$

Fig. 2 Auto-encoders as unsupervised representation learning formalisms

Smart manufacturing is arriving. It promises a future of mass-producing highly personalized **PRODUCTS** via responsive autonomous manufacturing operations at a competitive cost. Of utmost importance, smart manufacturing requires end-to-end integration of intra-business and inter-business manufacturing processes and systems. ...

... Subsequently, focusing on meeting changing demands of efficient production of highly personalized **PRODUCTS**, we detail several future-proofing manufacturing automation scenarios via integrating various existing standards. We believe that existing automation standards have provided a solid foundation for developing smart manufacturing solutions.

Fig. 3 Extract from the abstract in [15]. The semantics of the word *products* is encoded within the sentences that contain it

Finally, self-supervised representation learning aims to leverage the underlying structure S_V of unlabeled data that contains the variables $v \in V$ and which allows for deriving meaningful initial representations $\pi(v)$. For example, a word $v \in V$ may appear in a set of sentences $\pi(v)$ within a shared text corpus S_V, as exemplified in Fig. 3. While state-of-the-art NLP models like BERT [13] usually split words into frequently occurring subword tokens via subword segmentation algorithms such as Wordpiece [14], the inherent methods can be applied analogously to sets of complete words. In the course of training such NLP models, numerical embeddings $\gamma(v) \in \mathbb{R}^d$ are assigned to the domain variables $v \in V$ with respect to their original representations $\pi(v)$. These alternative representations are optimized by backpropagating the output of the LLM for at least one element of its initial representation $\pi(v)$.

Analogously, most NLP techniques can be applied to KG structures $G = (V, E)$ by characterizing directed graph walks $(v_1, p_1, v_2, p_2, v_3, \ldots, v_{l-1}, p_{l-1}, v_l)$ of depth $l - 1 \in \mathbb{N}$ as sentences that are composed of edges $(v_i, p_i, v_{i+1}) \in E$. For instance, the sample manufacturing KG depicted in Fig. 4 contains the 4-hop walk

*(**John**, executes, **Task 1**, output, **Product 1**, input, **Task 2**, output, **Product 2**).*

One of these transfer approaches is RDF2Vec [16], which utilizes random graph walks to generate input data for the NLP-based Word2Vec algorithm [17]. By doing so, a mapping $\overline{\gamma} : V \cup R \to \mathbb{R}^d$ is trained and thus, alternative representations of the graph nodes in V, but also for the relation types in R as well. Therefore,

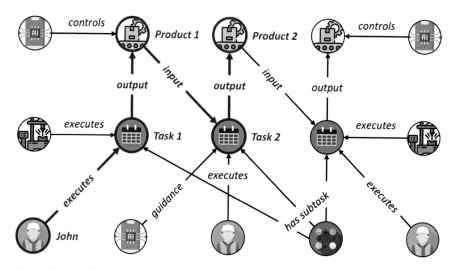

Fig. 4 Sample KG containing process flows within a production process

node embeddings can be derived via $\gamma(v) := \overline{\gamma}(v)$. Besides transfer approaches like RDF2Vec, various embedding algorithms exist, which are specifically tailored toward KG structures. These are further discussed in the following.

3.1 Representation Learning for Knowledge Graphs

KG embedding techniques denote a subdiscipline of representation learning, taking into account KG structures as initial input data. Given a KG $G = (V, E)$, these approaches intend to provide numerical representations $\gamma : V \rightarrow \Gamma$ as per Eq. (1). However, as exemplified by RDF2Vec, KG embeddings may contain alternative representations of graph elements $y \notin V$ as well, such as embeddings of relations, but also edges or subgraphs. Thus, in general, a KG embedding is a mapping $\overline{\gamma} : \Omega \rightarrow \Gamma$, where Ω represents a collection of KG elements pertaining to G. The node embedding of some $v \in V$ is accordingly obtained by restricting $\overline{\gamma}$ to V, i.e., $\gamma(v) = \overline{\gamma}(v)$.

Based on the research conducted in [10], KG embedding methods can be categorized into three model families, namely tensor decomposition models, geometric models, and deep learning models. We adopt this subdivision in the following.

3.1.1 Tensor Decomposition Models

Tensor decomposition models for KG embeddings are based on the concept of tensor decompositions within the area of multilinear algebra [18]. These attempt

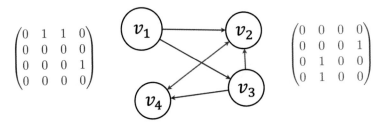

Fig. 5 Sample KG with $n = 4$ nodes and $k = 2$ relations r_1 (blue) and r_2 (red), including their respective adjacency matrices $\mathcal{A}_1$ and $\mathcal{A}_2$

to characterize tensors via sequences of simplified tensor operations. For a KG $\mathcal{G}$, this approach is applied to its unique adjacency tensor $\mathcal{A} \in \{0, 1\}^{k \times n \times n}$, defined as

$$\mathcal{A}_{h,i,j} = 1 \iff (v_i, r_h, v_j) \in E.$$

Here, $k \in \mathbb{N}$ denotes the cardinality of the underlying relation set R and $n \in \mathbb{N}$ is the number of nodes in V. Accordingly, without loss of generality, we may assume labeled sets $R = \{r_1, \ldots, r_k\}$ and $V = \{v_1, \ldots, v_n\}$, as exemplified in Fig. 5.

Accordingly, tensor decomposition-based KG embedding methods intend to approximate $\mathcal{A}$ by a sequence of lower dimensional tensor operations. Among these methods, RESCAL [19] is considered to be the first work to apply this methodology for determining KG embeddings. Regarding $\mathcal{A}$, it proposes a rank-d factorization

$$\mathcal{A}_h \approx X \cdot \mathcal{R}_h \cdot X^T$$

of its h-th slice $\mathcal{A}_h \in \{0, 1\}^{n \times n}$ by means of matrices $X \in \mathbb{R}^{n \times d}$ and $\mathcal{R}_h \in \mathbb{R}^{d \times d}$ with $d \ll n$. Therefore, the i-th row of the matrix X contains an alternative representation $\gamma(v_i) := (X_{i,1}, \ldots, X_{i,d}) \in \mathbb{R}^d$ of $v_i \in V$. The optimization of the matrices X and $(\mathcal{R}_h)_{1 \leq h \leq k}$ is accordingly achieved by solving the minimization problems

$$\min_{X, \mathcal{R}_h} f(X, \mathcal{R}_h) \text{ for } f(X, \mathcal{R}_h) = \frac{1}{2} \left(\sum_{h=1}^{k} \| \mathcal{A}_h - X \cdot \mathcal{R}_h \cdot X^T \|_F^2 \right),$$

with the Frobenius norm $\| \cdot \|_F$ and the corresponding element-wise operations

$$f(h, i, j) = \frac{1}{2} \left(\mathcal{A}_{h,i,j} - \gamma(v_i)^T \cdot \mathcal{R}_h \cdot \gamma(v_j) \right)^2.$$

To reduce the complexity of these optimizations, DistMult proposes to use diagonal matrices $(\mathcal{R}_h)_{1 \leq h \leq k}$ [20]. However, by doing so, DistMult is limited to symmetric relations. ComplEx solves this problem by employing $\mathbb{C}$-valued embedding spaces [21]. In addition to the mentioned models, numerous other tensor decomposition models for KG embeddings exist, including ANALOGY [22], SimplE [23], and HolE [24].

3.1.2 Geometric Models

Geometric KG embedding models represent semantic relations as geometric transformations within a corresponding embedding space. In contrast to tensor decomposition models, embeddings are not determined based on characteristics of the unique adjacency tensor $\mathcal{A}$, but with respect to individual facts $(s, p, o) \in E$.

As outlined in [10], transformations $\tau_p(s) := \tau\left(\gamma(s), \overline{\gamma}(p)\right) \in \Gamma$ are applied for subject nodes $s \in V$ regarding predicates $p \in R$. Accordingly, based on a distance measure $\delta : \Gamma \times \Gamma \to \mathbb{R}_{\geq 0}$, KG embeddings are computed via score functions

$$f(s, p, o) := \delta\left(\tau_p(s), \gamma(o)\right).$$

Among the family of geometric KG embedding methods, TransE [25] constitutes the most famous approach. As a translational model, it approximates object representations $\gamma(o)$ via $\gamma(o) \approx \tau_p(s) = \gamma(s) + \overline{\gamma}(p)$. Various geometric KG embedding models build upon the idea of TransE, improving the representation of nodes and relations by introducing additional components or transformations, such as

- Relationship-specific hyperplanes to capture complex interactions between nodes and relationships more effectively (TransH) [26]
- Relationship-specific node projection matrices to handle entities and relationships with different characteristics more flexibly (TransR) [27]
- Adaptive projection matrices regarding differing node-relation-pairs (TransD) [28]
- Relationship clustering to group similar relations (TransG) [29]

For a comprehensive overview of these methods, we refer to [10]. This work also introduces negative sampling as a common obstacle of KG embedding formalisms. Due to the open-world assumption of KGs, $(s, p, o) \notin E$ does not necessarily imply that the fact is false. Rather, it means that the KG does not contain information about its validity. Thus, negative sampling is applied to create a set of false facts $E_{neg} \subseteq V \times R \times V$ with $E \cap E_{neg} = \emptyset$ to train the embeddings in a supervised way.

3.1.3 Deep Learning Models

Graph-based deep learning (DL) approaches, also referred to as Graph Neural Networks (GNNs), exist for some time already, especially in the context of complex network systems and their underlying undirected graph structures [30]. However, the application of such algorithms on directed and labeled KGs may lead to a loss of relevant information. To address this issue, Graph Convolutional Networks

(GCNs) were first introduced to account for directed edges [31]. Furthermore, to accommodate different relation types, Relational Graph Convolutional Networks (RGCNs) were elaborated as extensions of GCNs [32], which were subsequently extended by means of attention mechanisms [33] in Relational Graph Attention Networks (RGATs) [34].

In contrast to geometric KG embedding models that apply score functions to individual triples and tensor decomposition models that intend to reduce the dimensionality of the adjacency tensor $\mathcal{A}$, DL-based models perform predictions for labeled nodes $v \in \overline{V}$, taking into account itself and its outgoing neighbors

$$N(v) := \{y \in V \mid \exists (s, p, o) \in E : (s = y \wedge o = v) \vee (s = v \wedge o = y)\}.$$

These labels can be derived from the KG itself via node assertions or link assignments, or they can be external, such as numerical or nominal node attributes. Adjacent node representations are meant to be aggregated to receive a composite node representation of v. By backpropagating a suitable loss, initial embeddings of v and its neighbors are optimized. This process is repeated for each labeled training node to generate latent feature representations for all $v \in \overline{V} \cup \{N(v) : v \in \overline{V}\}$. The formalism proposed in [32] subdivides $N(v)$ into relation-specific neighborhoods

$$N_r(v) := \{y \in V \mid \exists (s, p, o) \in E : (s = y \wedge o = v) \vee (s = v \wedge o = y) \wedge p = r\},$$

regarding relation types $r \in R$. Thus, given a matrix of (initial) feature representations $X \in \mathbb{R}^{n \times d}$ (i.e., the i-th row of X is an embedding of $v_i \in V$), embeddings of outgoing neighbors can be incorporated in the forward pass of a GNN via

$$\mathcal{A}_h \cdot X \in \mathbb{R}^{n \times d},$$

where $\mathcal{A}_h$ denotes the h-th slice of $\mathcal{A}$. For instance, in the context of the KG from Fig. 5, the composite representation of v_1 regarding the relation r_1 equals the sum of the initial embeddings of v_2 and v_3. To account for differing impacts of incoming and outgoing edges, R is typically extended via inverse relations r' for each $r \in R$. Some works also consider a self-relation r_0. Accordingly, by taking into account the adjacency matrices $\mathcal{A}_0 = Id$ and $\mathcal{A}_{2h} = \mathcal{A}_h^T$ for $1 \leq h \leq k$, we extend the set R via

$$\widehat{R} := R \cup \{r' \mid r \in R\} \cup \{r_0\} \quad \text{with } r_h' = r_{2h}.$$

By doing so, GNN models capture the semantics of directed and labeled graphs by summing up weighted composite representations to receive a convoluted matrix

$$\sum_{h=0}^{2k} \widehat{\mathcal{A}}_h \cdot X \cdot W_h \in \mathbb{R}^{n \times d'},$$

including relation-specific weight matrices $\mathcal{W}_h \in \mathbb{R}^{d \times d'}$. Moreover, the extended adjacency tensor $\widehat{\mathcal{A}} \in \mathbb{R}^{(2k+1) \times n \times n}$ is not necessarily $\{0, 1\}$-valued. Rather, it is intended to contain normalization constants or attention scores to encode the significance of individual nodes and relations to the forward pass of a GNN. However,

$$\left(v_i, r_h, v_j\right) \notin E \Rightarrow \widehat{\mathcal{A}}_{h,i,j} = 0$$

still holds. If no normalization constants or attention mechanisms are to be implemented, this tensor can be directly derived from $\mathcal{A} \in \{0, 1\}^{k \times n \times n}$ by means of matrix transpositions and the insertion of an additional identity matrix. Finally, by introducing an activation function $\sigma : \mathbb{R} \to \mathbb{R}$ such as ReLu, the generalized forward pass of a GNN layer (including RGCNs and RGATs) can be defined as

$$\sigma \left(\sum_{h=0}^{2k} \widehat{\mathcal{A}}_h \cdot \mathcal{X} \cdot \mathcal{W}_h \right) =: \mathcal{X}' \in \mathbb{R}^{n \times d'}. \tag{2}$$

4 Industrial Applications of Knowledge Graph Embeddings

The lack of use case scenarios poses a significant challenge to the application of KGs and corresponding KG embeddings in the manufacturing domain. Without specific applications, it becomes difficult to identify the relevant data sources, design appropriate KG structures, and create meaningful embeddings that capture the intricate relationships within manufacturing processes. Thus, the absence of concrete use cases hinders the exploration of the full potential of KGs and KG embeddings in improving efficiency, decision-making, and knowledge sharing within this domain.

As a result of the research conducted within the Teaming.AI project, which aims to enhance flexibility in Industry 4.0, while prioritizing human involvement and collaboration in maintaining and advancing AI systems, we identified several application scenarios within the manufacturing domain that can be leveraged by introducing industrial KGs and KG embeddings. These are introduced in the following.

Data Integration and Fusion Manufacturing involves diverse and complex data from various sources, such as sensors, process logs, or maintenance records. While KGs can integrate these heterogeneous data sources, KG embeddings map them into a shared representation space. By representing KG nodes and their relationships in this shared embedding space, it becomes easier to combine and analyze data from different sources, leading to enhanced data fusion capabilities.

Semantic Similarity and Recommendation KG embeddings allow for quantifying the semantic similarity between nodes. In the manufacturing domain, this can be useful for recommending similar products, materials, or processes based on their embeddings. For example, embeddings can help to identify alternative materials with desired properties or characteristics, thereby aiding in material selection.

Supply Chain Management Effective supply chain management is crucial for manufacturing. KGs and corresponding KG embeddings can help model and analyze complex supply chain networks by representing suppliers, products, transportation routes, and inventory levels as graph entities. By considering their semantic relations, embeddings can facilitate supply chain optimization, demand forecasting, and identifying potential risks in the supply chain.

Decision Support Systems KG embeddings and relational ML techniques can serve as a foundation for developing decision support systems in manufacturing. By learning from empirical semantic observations, these systems can provide recommendations, insights, and decision-making support to operators, engineers, and managers. For example, based on the current state of the manufacturing environment, the system can suggest optimal operating conditions or maintenance actions. Moreover, models can be learned to recommend ML models for AI activities, given the current manufacturing environment.

Fault Detection and Diagnosis KG embeddings combined with relational ML techniques can aid in fault detection and diagnosis in manufacturing systems. By analyzing historical data and capturing the relationships between machines, process variables, and failure events, embeddings can be used to build systems that identify faults or failures in advance. This facilitates proactive maintenance, reduces downtime, and improves overall effectiveness.

In conclusion, KGs allow for representing manufacturing concepts and entities (such as processes, machines, and human workers) and their semantic relationships. KG embeddings, on the other hand, capture inherent semantics in lean numerical representations which facilitate (i) the analysis of existing manufacturing knowledge and (ii) the extraction of new manufacturing knowledge based on empirical observations. As a powerful tool for representing domain knowledge in a human- and machine-interpretable way, KGs enable the combination of human comprehensibility with the computational capabilities of machines. This synergy of human and machine intelligence enables effective collaboration, decision-making, and efficient problem solving in the manufacturing domain. Moreover, it represents a step toward optimized human-in-the-loop scenarios [35] and human-centric Industry 5.0 [36].

However, the manufacturing domain is inherently dynamic, with continuous changes in its processes, equipment, materials, and market demands. Therefore, it is crucial to incorporate these dynamics into KG embeddings, which are typically designed for static snapshots of a domain (cf. Sect. 3.1). In the end, KG embeddings should be able to capture the evolving relationships, dependencies, and contextual information, preferably in real time. By incorporating dynamics, the embeddings can adapt to changes in manufacturing operations, such as process modifications, equipment upgrades, or variations in product requirements. This enables the representations to accurately reflect the current state of the manufacturing system and to capture the evolving aspects of runtime observations and data.

5 The Navi Approach: Dynamic Knowledge Graph Embeddings via Local Embedding Reconstructions

Most of the existing works on dynamic graph embeddings do not account for directed and labeled graphs. Rather, they are designed to be applicable to undirected and/or unlabeled graphs [37, 38], or they aim to embed temporally enhanced snapshots of non-dynamic graphs [39, 40]. Moreover, approaches like the one proposed in [41] exist that intend to perform an online training of KG embeddings by focusing on regions of the graph which were actually affected by KG updates. However, the overall embedding structure is still affected, leading to a need for continuous adjustments of embedding-based downstream tasks, such as graph-based ML models. Thus, we require a dynamic KG embedding formalism that (i) can produce real-time embeddings for dynamic KGs and (ii) is able to preserve the original structure of KG embeddings to allow for consistent downstream applications.

We propose to utilize the dynamic Navi approach [42], which is based on the core idea of GNNs as per Eq. (2). Given an initial KG $\mathcal{G}_{t_0} = (V_{t_0}, E_{t_0})$ at timestamp t_0, we assume an embedding $\widetilde{\gamma}_{t_0} : V_{t_0} \rightarrow \mathbb{R}^d$ based on some state-of-the-art KG embedding method from Sect. 3.1. Accordingly, a dynamic KG is defined as a family of stationary snapshots $(\mathcal{G}_t)_{t \in \mathcal{T}}$ with respect to some time set $\mathcal{T}$. Given a future timestamp $t > t_0$, the Navi approach provides a consistent embedding $\gamma_t : V_t \rightarrow \mathbb{R}^d$ so that previously trained downstream models can still be employed.

Since we leverage the idea of GNNs to reconstruct $\widetilde{\gamma}_{t_0}(v)$ through local neighborhoods, these reconstructions are based on the unique adjacency tensors $(\mathcal{A}(t))_{t \in \mathcal{T}}$ with $\mathcal{A}(t) \in \mathbb{R}^{k \times n_t \times n_t}$. Here, $n_t = \left| \bigcup_{\tau \leq t} V_\tau \right|$ denotes the number of nodes that were known to exist since the graph's initialization and thus $n_t \geq n_{t_0}$ holds. Thus, we assume an initial embedding matrix $\widetilde{X}_{t_0} \in \mathbb{R}^{n_{t_0} \times d}$ that contains the initial embeddings as per $\widetilde{\gamma}_{t_0}$. This matrix is then reconstructed based on itself via a single-layer GNN

$$\sigma \left(\widehat{\mathcal{A}}(t_0)_0 \cdot \Theta_{t_0} \cdot W_0 + \sum_{h=1}^{2k} \widehat{\mathcal{A}}(t_0)_h \cdot \widetilde{X}_{t_0} \cdot W_h \right) =: X_{t_0} \approx \widetilde{X}_{t_0}$$

by taking into account the extended adjacency tensor $\widehat{\mathcal{A}}(t_0)$ (cf. Sect. 3.1.3). During the training process, a global embedding $\gamma_{r_0} \in \mathbb{R}^d$ is implemented regarding the self-relation r_0 so that $\Theta_{t_0} \in \mathbb{R}^{n_{t_0} \times d}$ contains n_{t_0} copies of γ_{r_0}. Moreover, instead of zero-value dropouts, overfitting is prevented by randomly replacing node embeddings with γ_{r_0} in the input layer, simulating the semantic impact of nodes that are not known at time t_0. It is also used to represent self-loops, enabling reconstructions that are independent of the (potentially unknown) initial representations. A detailed overview, including training settings and benchmark evaluation results, can be found in [42]. The evaluation implies that, given a timestamp $t > t_0$, this approach allows for high-qualitative and consistent embeddings $\gamma_t : V_t \rightarrow \mathbb{R}^d$ that are computed via

$$\sigma \left(\widehat{\mathcal{A}}(t)_0 \cdot \Theta_t \cdot \mathcal{W}_0 + \sum_{h=1}^{2k} \widehat{\mathcal{A}}(t)_h \cdot \widetilde{X}_t \cdot \mathcal{W}_h \right) =: X_t,$$

i.e., the i-th row of X_t represents the embedding $\gamma_t(v_i)$ of the node $v_i \in V_t$. In the case of new nodes, $\widetilde{X}_t$ and Θ_t are the extensions of $\widetilde{X}_{t_0}$ and Θ_{t_0} by inserting copies of γ_{r_0}, respectively. Moreover, the update of the adjacency tensor can be performed via

$$\mathcal{A}(t)_h = I(t_0, t)^T \cdot \mathcal{A}(t_0)_h \cdot I(t_0, t) + \mathcal{B}(t_0, t)_h.$$

First, the matrix $I(t_0, t) \in \{0, 1\}^{n_{t_0} \times n_t}$ accounts for newly inserted nodes, i.e.,

$$I(t_0, t)_{i,j} = 1 \iff i = j.$$

Second, the update matrices $\mathcal{B}(t_0, t)_h \in \{-1, 0, 1\}^{n_t \times n_t}$ identify KG updates

$$\mathcal{B}(t_0, t)_{i,j} == \begin{cases} 1 & \iff \text{ the edge } (v_i, r_h, v_j) \text{ was inserted between } t_0 \text{ and } t \\ -1 & \iff \text{ the edge } (v_i, r_h, v_j) \text{ was deleted between } t_0 \text{ and } t. \end{cases}$$

After the KG update, a synchronizing assistant is to provide (i) the number of nodes n_t and (ii) the update tensor $\mathcal{B}(t_0, t) \in \{-1, 0, 1\}^{k \times n_t \times n_t}$. For instance, given an Apache Jena Fuseki[1] KG, existing logging tools like rdf-delta[2] can be extended to use them as synchronizing assistants. Moreover, while we focus on a single update at time $t \in \mathcal{T}$, transitions between arbitrary timestamps can be handled as well, i.e.,

$$\mathcal{A}(t')_h = I(t, t')^T \cdot \mathcal{A}(t)_h \cdot I(t, t') + \mathcal{B}(t, t')_h \text{ for } t_0 < t < t'.$$

In conclusion, the late shaping of KG embeddings via Navi reconstructions represents a promising approach for incorporating dynamic KG updates and semantic evolutions into KG embeddings as lean feature representations of domain concepts and instances. Besides the ability to allow for consistent embeddings, the results in [42] even showed that the reconstruction of existing embeddings often leads to an improved performance in downstream tasks like link predictions and entity classifications as key enablers of the industrial use case applications outlined in Sect. 4.

---

[1] Apache Software Foundation, 2021. Apache Jena, Available at https://jena.apache.org/.
[2] https://afs.github.io/rdf-delta/.

6 Conclusions

In this work, we highlighted the increasing importance of representing and exploiting semantics, with a specific emphasis on the manufacturing domain. While industrial KGs are already employed and utilized to integrate and standardize domain knowledge, the generation and application of KG embeddings as lean feature representations of graph elements have been largely overlooked. Existing KGs lack either domain dynamics or contextuality, limiting the applicability of context-dependent embedding algorithms. Thus, we provide an overview of state-of-the-art KG embedding techniques, including their characteristics and prerequisites. In this context, we emphasized the need for dynamic embedding methods and their implementation in concrete manufacturing scenarios, describing potential KG embedding applications in industrial environments, which were identified as a result of the Teaming.AI project. Furthermore, we introduced the concept of Navi reconstructions as a real-time and structure-preserving approach for generating dynamic KG embeddings.

To summarize, KGs and KG embeddings offer significant advantages for the manufacturing domain. The structured representation of complex relationships in KGs enables context-awareness, dynamic analysis, and efficient information retrieval. Furthermore, the utilization of KG embeddings promotes process optimization, leading to improved product quality, reduced errors, and an increased overall productivity.

Acknowledgments This work is part of the TEAMING.AI project which receives funding in the European Commission's Horizon 2020 Research Program under Grant Agreement Number 957402 (www.teamingai-project.eu).

References

1. Mohamed, S.K., Nováček, V., Nounu, A.: Discovering protein drug targets using knowledge graph embeddings. Bioinformatics **36**(2), 603–610 (2020)
2. Fu, X., Ren, X., et al.: Stochastic optimization for market return prediction using financial knowledge graph. In: IEEE International Conference on Big Knowledge, ICBK, pp. 25–32. IEEE Computer Society, New York (2018)
3. Buchgeher, G., Gabauer, D., et al.: Knowledge graphs in manufacturing and production: a systematic literature review. IEEE Access **9**, 55537–55554 (2021)
4. Hogan, A., Blomqvist, E., et al.: Knowledge graphs. ACM Comput. Surv. (CSUR) **54**(4), 1–37 (2021)
5. Krause, F., Weller, T., Paulheim, H.: On a generalized framework for time-aware knowledge graphs. In: Towards a Knowledge-Aware AI—Proceedings of the 18th International Conference on Semantic Systems, vol. 55, pp. 69–74. IOS Press, New York (2022)
6. Neo4j: Neo4j—the world's leading graph database (2012)
7. Palumbo, E., Rizzo, G., et al.: Knowledge graph embeddings with node2vec for item recommendation, In: The Semantic Web: ESWC Satellite Events, pp. 117–120 (2018)

8. Diefenbach, D., Giménez-García, J., et al.: Qanswer KG: designing a portable question answering system over RDF data. In: The Semantic Web: ESWC 2020, pp. 429–445 (2020)
9. Sun, Z., Hu, W., et al.: Bootstrapping entity alignment with knowledge graph embedding. In: Proceedings of the 27th International Joint Conference on Artificial Intelligence, pp. 4396–4402. AAAI Press, New York (2018)
10. Rossi, A., Barbosa, D., et al.: Knowledge graph embedding for link prediction: a comparative analysis. ACM Trans. Knowl. Discov. Data **15**(2), 1–49 (2021)
11. Nickel, M., Murphy, K., et al.: A review of relational machine learning for knowledge graphs. Proc. IEEE **104**(1), 11–33 (2016)
12. Bengio, Y., Courville, A., Vincent, P.: Representation learning: A review and new perspectives. IEEE Trans. Pattern Anal. Mach. Intell. **35**(8), 1798–1828 (2013)
13. Devlin, J., Chang, M.-W., et al.: BERT: Pre-training of deep bidirectional transformers for language understanding. In: Proceedings of the Conference of the North American Chapter of the Association for Computational Linguistics: Human Language Technologies, pp. 4171–4186. Association for Computational Linguistics, Kerrville (2019)
14. Schuster, M., Nakajima, K.: Japanese and Korean voice search. In ICASSP, pp. 5149–5152. IEEE, New York (2012)
15. Lu, Y., Xu, X., Wang, L.: Smart manufacturing process and system automation—a critical review of the standards and envisioned scenarios. J. Manuf. Syst. **56**, 312–325 (2020)
16. Ristoski, P., Rosati, J., et al.: Rdf2vec: RDF graph embeddings and their applications. Semantic Web **10**, 721–752 (2019)
17. Mikolov, T., Sutskever, I., et al.: Distributed representations of words and phrases and their compositionality. In: Advances in Neural Information Processing Systems, vol. 26. Curran Associates, Inc., New York (2013)
18. Kolda, T.G., Bader, B.W.: Tensor decompositions and applications. SIAM Rev. **51**(3), 455–500 (2009)
19. Nickel, M., Tresp, V., Kriegel, H.-P.: A three-way model for collective learning on multi-relational data. In: Proceedings of the 28th International Conference on International Conference on Machine Learning, pp. 809–816. Omnipress, New York (2011)
20. Yang, B., Yih, W.-T., et al.: Embedding entities and relations for learning and inference in knowledge bases. In: 3rd International Conference on Learning Representations, ICLR 2015, San Diego, CA, USA, May 7–9, 2015, Conference Track Proceedings (2015)
21. Trouillon, T., Welbl, J., et al.: Complex embeddings for simple link prediction. In: Proceedings of The 33rd International Conference on Machine Learning, vol. 48, pp. 2071–2080. PMLR, New York (2016)
22. Liu, H., Wu, Y., Yang, Y.: Analogical inference for multi-relational embeddings. In: Proceedings of the 34th International Conference on Machine Learning, vol. 70, pp. 2168–2178 (2017). JMLR.org
23. Kazemi, S.M., Poole, D.: Simple embedding for link prediction in knowledge graphs. In: NeurIPS, pp. 4289–4300. Curran Associates Inc., New York (2018)
24. Nickel, M., Rosasco, L., Poggio, T.: Holographic embeddings of knowledge graphs. In: Proceedings of the Thirtieth AAAI Conference on Artificial Intelligence, pp. 1955–1961. AAAI Press, New York (2016)
25. Bordes, A., Usunier, N., et al.: Translating embeddings for modeling multi-relational data. In: Proceedings of the 26th International Conference on Neural Information Processing Systems, vol. 2, pp. 2787–2795. Curran Associates Inc., New York (2013)
26. Wang, Z., Zhang, J., et al.: Knowledge graph embedding by translating on hyperplanes. In: Proceedings of the AAAI Conference on Artificial Intelligence, vol. 28(1) (2014)
27. Lin, Y., Liu, Z., et al.: Learning entity and relation embeddings for knowledge graph completion. In: Proceedings of the Twenty-Ninth AAAI Conference on Artificial Intelligence, pp. 2181–2187. AAAI Press, New York (2015)
28. Ji, G., He, S., et al.: Knowledge graph embedding via dynamic mapping matrix. In: Proceedings of the 53rd Annual Meeting of the Association for Computational Linguistics, pages 687–696. Association for Computational Linguistics, New York (2015)

29. Xiao, H., Huang, M., Zhu, X.: TransG: a generative model for knowledge graph embedding. In: Proceedings of the 54th Annual Meeting of the Association for Computational Linguistics, pp. 2316–2325. Association for Computational Linguistics, New York (2016)
30. Wu, Z., Pan, S., et al.: A comprehensive survey on graph neural networks. IEEE Trans. Neural Networks Learn. Syst. **32**(1), 4–24 (2021)
31. Kipf, T.N., Welling, M.: Semi-supervised classification with graph convolutional networks. In: ICLR (2017)
32. Schlichtkrull, M., Kipf, T.N., et al.: Modeling relational data with graph convolutional networks. In: The Semantic Web ESWC, pp. 593–607. Springer, Berlin (2018)
33. Vaswani, A., Shazeer, N., et al.: Attention is all you need. In: Advances in Neural Information Processing Systems, pp. 5998–6008 (2017)
34. Busbridge, D., Sherburn, D., et al.: Relational Graph Attention Networks (2019)
35. Schirner, G., Erdogmus, D., et al.: The future of human-in-the-loop cyber-physical systems. Computer **46**(1), 36–45 (2013)
36. Leng, J., Sha, W., et al.: Industry 5.0: prospect and retrospect. J. Manuf. Syst. **65**, 279–295 (2022)
37. Pareja, A., Domeniconi, G., et al.: EvolveGCN: Evolving graph convolutional networks for dynamic graphs. In: The Thirty-Fourth AAAI Conference on Artificial Intelligence AAAI, The Thirty-Second Innovative Applications of Artificial Intelligence Conference IAAI, The Tenth AAAI Symposium on Educational Advances in Artificial Intelligence EAAI, pp. 5363–5370. AAAI Press, New York (2020)
38. Trivedi, R., Farajtabar, M., Biswal, P., Zha, H.: DyRep: learning representations over dynamic graphs. In: International Conference on Learning Representations (2019)
39. Dasgupta, S.S., Ray, S.N., Talukdar, P.: HyTE: hyperplane-based temporally aware knowledge graph embedding. In: Proceedings of the Conference on Empirical Methods in Natural Language Processing, pp. 2001–2011. Association for Computational Linguistics, Belgium (2018)
40. Liao, S., Liang, S., et al.: Learning dynamic embeddings for temporal knowledge graphs. In: Proceedings of the 14th ACM International Conference on Web Search and Data Mining, pp. 535–543. Association for Computing Machinery, New York (2021)
41. Wewer, C., Lemmerich, F., Cochez, M.: Updating embeddings for dynamic knowledge graphs. CoRR, abs/2109.10896 (2021)
42. Krause, F.: Dynamic knowledge graph embeddings via local embedding reconstructions. In: The Semantic Web: ESWC Satellite Events, pp. 215–223. Springer, Berlin (2022)

Architecture of a Software Platform for Affordable Artificial Intelligence in Manufacturing

Vincenzo Cutrona ⓘ, Giuseppe Landolfi ⓘ, Rubén Alonso ⓘ,
Elias Montini ⓘ, Andrea Falconi ⓘ, and Andrea Bettoni ⓘ

1 Introduction

The huge transformation brought by the fourth industrial revolution into the manufacturing world has forced any company to take on the digitalization journey, regardless of its size, sector, or location. In this context, Artificial Intelligence (AI) technologies are ready to take off as a new approach to solve business issues, and, recently, AI tools are proliferating [1]. Forward-thinking results can be obtained by analyzing huge amounts of data from a wide range of sources in the production system and by identifying deviations and trends in real time for making decisions [2]. The greater intelligence brought by AI embedded in production systems can not only bring advantages for large companies but also support Small-Medium Enterprises (SMEs) and mid-caps in achieving better operational performance. Yet

V. Cutrona · G. Landolfi · A. Bettoni
University of Applied Science of Southern Switzerland, Viganello, Switzerland
e-mail: vincenzo.cutrona@supsi.ch; giuseppe.landolfi@supsi.ch; andrea.bettoni@supsi.ch

R. Alonso
R2M Solution s.r.l., Pavia, Italy

Programa de Doctorado, Centro de Automática y Robótica, Universidad Politécnica de Madrid-CSIC, Madrid, Spain
e-mail: ruben.alonso@r2msolution.com

E. Montini (✉)
University of Applied Science of Southern Switzerland, Viganello, Switzerland

Politecnico di Milano, Milan, Italy
e-mail: elias.montini@supsi.ch

A. Falconi
Martel Innovate, Zurich, Switzerland
e-mail: andrea.falconi@martel-innovate.com

J. Soldatos (ed.), *Artificial Intelligence in Manufacturing*,
https://doi.org/10.1007/978-3-031-46452-2_6

several challenges are still preventing them from embracing AI on a large scale. To reduce the barriers, two conditions have to be satisfied: the technology has to be affordable and accessible enough for mass use, and the level of awareness of individuals and companies should be high enough to be able to understand how and where to use it.

The first condition can be tackled by democratizing AI tools: by exploiting the "as-a-service-model," technologies can be made available to SMEs at an affordable price and on-demand, thus reducing the financial gap with large companies and avoiding SMEs getting lost in the hype around AI [3]. This is the best solution since, on the one hand, the adoption of ad hoc solutions for specific company requirements leads to integration problems, long implementation times, and flexibility limits. On the other hand, adopting an all-in-one solution requires big investments for complex systems, which exceed the effective needs and strictly depend on legacy providers.

The second condition is more difficult to be satisfied at the level of single companies since, often, SMEs lack the skills and knowledge needed to turn AI into a major boost for their business, thus lagging behind larger organizations in the uptake level [4]. Successful implementation of AI requires identifying the right tools to be chosen among a plethora of solutions and their harmonization with existing IT systems and processes from both a technical and a strategic point of view so that they can become real enablers for performance improvement. Upskilling workers is essential to both empower a mutually supportive human–machine interaction and lower adoption barriers, but building internal competences requires time. Support is needed now to accompany European SMEs in their digitization journey so that they can keep the pace with their larger counterparts and be key players in their value chains. An innovation ecosystem should be created around SMEs so that they can easily find locally the needed support to draw customized AI adoption plans and be immersed in a vibrant and stimulating environment that makes them progress across the digital innovation flow.

At the European level, initiatives have been launched to promote the development of platforms that could support SMEs in the digital uptake, and the creation of local Digital Innovation Hubs (DIHs) is promoted to create an innovation ecosystem providing services to lower the entry barriers for SMEs. The AI uptake has to pivot on digital platforms that act as one-stop shop for SMEs, showcasing advances brought forward by synergistic efforts of DIHs, research centers, and technologies providers and offering services to smooth the adoption. Being able to offer to SMEs solutions tailored to their specific needs, built on modular kits, at a reasonable cost, easy and fast to implement is a must to strengthen the European economy's competitiveness.

KITT4SME recognizes that SMEs are among the companies that could benefit the most from the opportunities brought by AI solutions while, at the same time, being the ones with the least capabilities and resources to embrace them. KITT4SME specifically targets European SMEs and mid-caps to provide them with scope-tailored and industry-ready hardware, software, and organizational kits, delivered as a modularly customizable digital platform that seamlessly introduces AI in their production systems. Uptake of the resulting packages and of the provided services is strongly supported by the clear characterization and market readiness of the individual components and by the platform grounding on the already established RAMP marketplace. Seamless adoption of the customized kits is made possible by a Powered by FIWARE infrastructure,[1] which flawlessly combine factory systems (such as Manufacturing Execution System (MES) and Enterprise Resource Planning (ERP)), Internet of Things (IoT) sensors and wearable devices, robots, collaborative robots, and other factory data sources with functional modules capable of triggering data-driven value creation.

The rest of the chapter is structured as follows: Sect. 2 examines existing platforms and alternative methods for delivering AI services to manufacturing SMEs; Sect. 3 introduces the concept underlying the proposed platform, its architecture, and the provided functionalities for supporting AI developers; in Sect. 4, a real-world use case illustrating the advantages of the proposed platform for an SME is presented; and Sect. 5 concludes with a discussion of limitations and related future work.

2 Platforms in the AI Ecosystem

The KITT4SME platform aims to assist SMEs in adopting AI-based solutions by offering various services. These services, ranging from analyzing clients' requirements and implementing technical solutions to developing AI applications and training AI algorithms, coexist in an environment with platforms providing AI solutions, technology providers, AI consulting firms, and DIHs.

Platform-based services and aPaas are cloud computing services that allow customers to provide, instantiate, run, and manage modular software solutions comprising a core infrastructure and one or more applications without the complexity of building and maintaining the whole system, typically associated with developing and launching the applications [5]. These solutions allow also developers to create, develop, and package such software bundles. Gartner sees AI Platform as a Service (AI PaaS) as a set of separate AI services. However, it is possible to consider the concept of AI PaaS from the perspective of the classic Platform as a Service (PaaS) model. Such an environment usually includes two main components required for application development: hardware infrastructure (computing power, data storage, networking infrastructure, and virtual machines) and software solutions (tools and services).

The key hurdle to generalize a similar scheme for the AI PaaS architecture is that there is no general model for AI PaaS yet. The market is still forming, and different vendors offer completely different services under the same umbrella term. Yet many elements are common to the majority of today's AI PaaS and AI service

---

[1] https://www.fiware.org/.

platforms: infrastructure, data storage, pre-trained AI models, and Application Program Interfaces (APIs).

AI as a Service (AIaaS) allows individuals and companies to experiment with AI for various purposes without a large initial investment and with lower risk [6]. In this market, different AI providers offer several styles of Machine Learning (ML) and AI. These variations can be more or less suited to an organization's AI needs since organizations must evaluate features and pricing to see what works for them. To date, there are two kinds of platforms, depending on how they offer the service:

- *Platforms to develop code to build AI programs*: comparable to an opensource solution that allows users to create and configure applications through a graphical user interface instead of a traditional hand-coding computer program
- *Platforms providing already developed applications*: similarly to KITT4SME, these platforms allow users to deploy and implement ready-to-use solutions that do not require users to have advanced technology and IT skills.

Since KITT4SME addresses SMEs, and very few of them have in-house competencies (data scientists, analysts, and developers) or a specialized team able to develop AI models and applications [4], the following paragraphs focus on the platforms providing already developed applications.

Acumos AI Acumos AI[2] is an open-source platform that enables the training, integration, and deployment of AI models. It was launched in 2018 by the LF AI Foundation,[3] which supports open-source innovation in AI, ML, and Deep Learning (DL), making these technologies available to developers and data scientists. The platform provides a marketplace for AI solutions that are not tied to any specific infrastructure or cloud service. It aims to create a flexible mechanism for packaging, sharing, licensing, and deploying AI models securely through a distributed catalog among peer systems. Acumos AI aims to make AI and machine learning accessible to a broad audience by creating a marketplace of reusable solutions from various AI toolkits and languages. This way, ordinary developers who are not machine learning experts or data scientists can easily create their applications [7].

Bonseyes Bonseyes[4] was a H2020 project that ended in 2020. It was led by NVISO SA,[5] based in Lausanne, and aimed to create a platform with a Data Marketplace, DL Toolbox, and Developer Reference Platforms. The platform was designed for organizations that wanted to implement AI in low-power IoT devices, embedded computing systems, or data center servers. The platform had an engagement strategy where platform experts published challenges and requests for AI solutions that met specific technical requirements based on real industrial problems faced by companies. Data scientists proposed their own AI applications to be deployed on the

---

[2] https://www.acumos.org/.

[3] https://lfaidata.foundation/.

[4] https://www.bonseyes.eu/.

[5] https://www.nviso.ai.

platform. Companies evaluated and paid the winners after the call ended. Bonseyes used a collaborative AI Marketplace to provide real-world solutions to the industry, supporting scenarios where data must remain on the data provider's premises and online learning with distributed Cyber-Physical Systems (CPSs). The platform allowed continuous feedback from human actors to evaluate model performance and obtain metadata about context and users' perspectives [8, 9].

GRACE AI Grace AI[6] is an AI platform launched by 2021.AI in 2018, with the mission to help customers in realizing their vision of AI by identifying innovative business opportunities in key processes and functions. Grace AI Platform and the AIaaS portfolio are the company's main assets. The Grace platform is built for both organizations at the beginning of their AI and ML journey and organizations that have already established a data science team but are looking for ways to infuse continuous intelligence into their business.

Grace AI aims to provide any organization access to AI implementation, including automated documentation, validation, and certification through data exploration, AI development, deployment, and operation.

PTC Inc. PTC Inc.[7] is a software and services company founded in 1985, based in Boston. It offers a range of products and services that support innovation and Industry 4.0. It is a platform for developing IoT and Augmented Reality (AR) solutions. PTC Marketplace is a digital space where customers and partners can access IoT apps, market-ready solutions, and innovative technologies. PTC has made recent enhancements to its marketplace, making it easier for solution builders to find market-ready solutions and customized accelerators. It also provides a platform for PTC partners to showcase their technologies, solutions, services, and industry expertise to customers and prospects.

The platform offers a rich set of capabilities that enable solutions for design, manufacturing, service, and industrial operations and incorporates modular functionality that simplifies development. These include pre-built applications for the fast, easy implementation of Industrial Internet of Things (IIoT) solutions for common use cases in various industries.

3   KITT4SME: A Platform Delivering AI to SMEs

The KITT4SME project aims to provide AI solutions to SMEs in the manufacturing domain through a five-step workflow. This workflow consists of interconnected activities designed to facilitate the adoption of AI technologies on the shop floor. The activities are detailed as follows:

---

[6] https://2021.ai/offerings/grace-ai-platform/.

[7] https://www.ptc.com/.

- *Diagnose*: In this step, the KITT4SME platform utilizes a smart questionnaire to identify how AI can be beneficial in transitioning the shop floor. The questionnaire helps assess the specific needs and challenges of the SMEs, enabling a better understanding of where AI technologies can be applied effectively.
- *Compose*: The platform recommends a minimal set of AI tools from a marketplace catalog based on the diagnosis obtained in the previous step. It considers the unique requirements and constraints of each SME, aiming to maximize the benefits derived from the AI technologies. The platform provides guidance on the wiring and configuration of these AI tools, ensuring their seamless integration into the existing workflow.
- *Sense*: This activity focuses on establishing the connection between the shop floor and the cloud platform. By enabling this connection, new data become available to AI services. The KITT4SME platform provides a tailor-made kit that can output insights about the status of the shop floor (e.g., to detect and explain anomalies). Additionally, it offers visualization of Key Performance Indicators (KPIs), allowing SMEs to gain valuable insights into their operations.
- *Intervene*: In this step, the platform suggests corrective actions to address ongoing issues and anomalies identified on the shop floor. Leveraging the power of AI, the platform provides recommendations for resolving problems and improving the overall performance of the manufacturing processes.
- *Evolve*: The final step involves analyzing the outcomes and feedback generated from the previous steps. The platform uses this information to continuously improve the *Diagnose* and *Compose* steps. It also provides personalized staff training recommendations to further enhance the adoption and utilization of AI technologies within the SME.

The underlying concepts of the KITT4SME platform revolve around understanding the specific needs of SMEs, recommending tailored AI solutions, establishing seamless connections between the shop floor and the cloud platform, providing real-time analyses and KPI visualization, offering intervention recommendations, and continuously improving the overall workflow based on feedback and outcomes.

By following this five-step workflow, the KITT4SME platform aims to empower SMEs in the manufacturing domain to harness the potential of AI technologies, enhance their operational efficiency, and drive growth and innovation in their businesses. This section reports on the basic concepts underlying the platform and explains its main functionalities.

3.1   High-Level Concept and Architecture

The KITT4SME architecture is designed to address the challenges associated with deploying and utilizing AI models developed by data scientists or AI developers in SMEs. One of the key challenges is the discrepancy between the pace of AI model

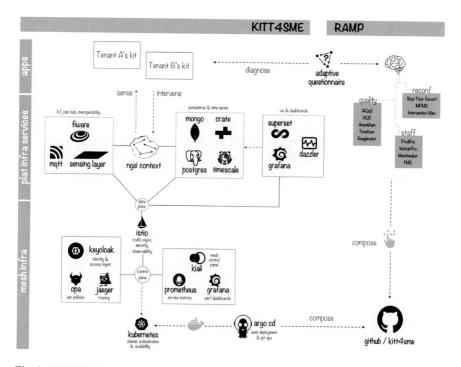

Fig. 1 KITT4SME platform three-tier architecture

development and the capabilities of SMEs' IT systems. This often leads to situations where models are not deployed or where the deployment and update process is time-consuming.

To tackle these challenges, KITT4SME proposes a conceptual pipeline consisting of six steps, which cover the process from data preparation to the practical use of the model. The steps (presented in Fig. 1) are as follows:

1. *Prepare data*: This step involves collecting and preparing the data required for training the AI model. It includes tasks such as data cleaning, transformation, and feature engineering to ensure the data are suitable for model development.
2. *Develop the model*: In this step, AI researchers and developers focus on building and training the AI model using the prepared data. This is where the core value of the AI solution is generated.
3. *Package the model*: Once the model is developed, it needs to be packaged in a way that it can be easily deployed and integrated into the existing systems of the SME. Packaging involves encapsulating the model and its associated dependencies into a deployable form.
4. *Validate the model*: Before deployment, it is crucial to validate the model to ensure its accuracy, reliability, and suitability for the intended use. Validation may involve testing the model's performance on a separate dataset or using techniques like cross-validation.

5. *Deploy the model*: This step focuses on deploying the validated model into the SME's IT infrastructure. It involves integrating the model with the existing systems, ensuring compatibility, and addressing any technical requirements or constraints.

6. *Use the model*: The final step is when the SME can actively utilize the deployed model in its operations. This includes making predictions, generating insights, and incorporating the model's outputs into decision-making processes.

The three intermediate steps, namely packaging, validating, and deploying the model, are often complex and time-consuming. KITT4SME aims to simplify and automate these steps, reducing the overall time and effort required to deploy and update the AI model. By streamlining these processes, the platform enhances the repeatability and efficiency of the entire pipeline, making it easier for SMEs to leverage AI technologies effectively.

The software platform implementing the KITT4SME workflow is based on a service mesh, multi-tenant cloud architecture. It provides a means to assemble various AI components from a marketplace and facilitates their connection to the shop floor while ensuring interoperability, security, and privacy-preserving data exchange. The platform consists of loosely coupled web services running in a cluster environment and relies on a dedicated cluster software infrastructure. Several key concepts and guiding principles underpin the architecture of the KITT4SME platform:

- *Leveraging state-of-the-art technology and standards*: The platform utilizes a dedicated cluster software infrastructure, referred to as *mesh infrastructure*. This infrastructure is built on industry-standard technologies such as Kubernetes[8] and Istio[9]. The platform reuses open communication and data standards as much as possible to foster service interoperability (e.g., REST principles for services interaction and NGSI standard for data exchange).

- *Platform services*: The platform comprises two types of services: *application services*, which are integral to the KITT4SME workflow and provide the functionality required for the platform's core activities, and *infrastructure services*, which consist of a network of intermediaries within the mesh infrastructure. These intermediaries handle essential operational aspects such as routing, security, and monitoring. By separating these concerns, AI developers can focus on implementing service-specific features while relying on the platform for operational support.

- *Multi-tenancy*: The platform is designed to support multiple SMEs sharing the same instance. Each company is associated with a security protection domain, referred to as a *tenant*, which isolates its data and users from other tenants. The platform also allows for explicit sharing policies that enable companies to selectively share data and resources if desired.

---

[8] https://kubernetes.io/.

[9] https://istio.io/.

- *Containerized deployment and orchestration*: The platform adopts a container-based virtualization approach for service deployment and orchestration. Services are packaged and executed within containers, enabling independent development using appropriate technology stacks. This containerization allows for the decoupling of services and facilitates their independent deployment, potentially through automated release processes such as Continuous Integration (CI) and Continuous Delivery (CD).

By adhering to these principles and utilizing modern technologies, the KITT4SME platform ensures efficient and scalable execution of the AI workflow. It promotes service interoperability, simplifies deployment and management, and provides a secure and isolated environment for SMEs to leverage AI capabilities within their manufacturing processes. The KITT4SME high-level architecture provides the ecosystem enabling the streamlined AI packaging, validation, and deployment while also fostering and facilitating the composability and integration of AI solutions.

As depicted in Fig. 1, the architecture is organized into a three-tier structure on top of the hardware layer. Each layer comes with a set of components dealing with certain operational functionalities, as follows:

1. *Mesh Infrastructure Layer*: This layer, depicted as "mesh infra" in Fig. 1, is responsible for managing computational resources, network proxies, and interconnection networks. It utilizes Kubernetes for containerized workloads and services, while Istio acts as a service mesh for traffic management, observability, and security. The tasks performed by the mesh infrastructure layer include:

    - Managing computational resources (e.g., CPU, memory, storage) and allocating them to processes in the upper layers, acting as the *Cluster Orchestration Plane*
    - Handling the network of proxies for transparent routing, load balancing, and securing communication, which represents the *Control Plane*
    - Managing proxies and interconnection networks for capturing and processing application traffic, serving as the *Data Plane* of the mesh infrastructure

2. *Platform Services Infrastructure Layer*: This layer, labeled as "plat infra services" in Fig. 1, comprises processes that support the operation of application services in the upper layer. It includes components such as IoT sensor connectors, context brokers, databases, and software for creating dashboards and visualizations. These components rely on well-known software and IoT middlewares like FIWARE [10]. Each component exposes interfaces for use by higher layers while utilizing the lower layer for interconnection.

3. *Application Layer*: This layer, represented as "apps" in Fig. 1, hosts services and components that provide functionality to the manufacturing SME. Examples include anomaly detection, data augmentation components, and dashboards. The application layer focuses on application-specific concerns while leveraging

the security, traceability, scalability, integration, and communication mechanisms provided by the lower layers.

Additionally, the KITT4SME platform benefits from its connection to an application marketplace. This marketplace, facilitated by discovery solutions like adaptive questionnaires, enables the identification of new applications and components, supporting the *Compose* activity in the KITT4SME workflow. A detailed description of the components and their functionality in each layer is provided in Sect. 3.2.

3.2   Functionalities and Component Description

Pursuing the idea of an open-source platform for the uptake of AI solutions in manufacturing SME, KITT4SME has chosen FIWARE[10] as the underlying open-source platform for its AI solutions in manufacturing SMEs. FIWARE is renowned as a top-quality open-source platform for IoT [10]. By leveraging FIWARE, the KITT4SME platform, branded as "Powered by FIWARE," inherits a range of capabilities that are beneficial for managing context information and data in the manufacturing domain. These capabilities include:

- *Handling and managing context information*: The KITT4SME platform can efficiently handle and manage context information from diverse data sources. This allows for the collection and aggregation of data from various sensors, machines, and other sources in the manufacturing environment.
- *Distributing and streaming data*: The platform is equipped with mechanisms for distributing and streaming data to external components. This enables the seamless transfer of data to external systems for various purposes, such as persistence or AI-based processing.
- *Integration with AI-based processing*: The KITT4SME platform can integrate with AI-based processing components, leveraging the capabilities of FIWARE. This integration facilitates the application of AI algorithms and techniques to analyze and derive insights from manufacturing data. The results obtained from AI processing can be seamlessly integrated back into the platform, enriching the current context and enabling data-driven decision-making.

Overall, by utilizing FIWARE as the foundation, the KITT4SME platform gains powerful tools and features that are instrumental in the management of IoT data and seamless integration of AI-based processing capabilities. Figure 2 depicts the logical architecture of the platform, illustrating the hierarchical layout in which intelligent services and AI applications are placed on top of the FIWARE ecosystem. In the subsequent discussion, we present a comprehensive overview of the platform's functionalities, accentuating the advantages derived from harnessing FIWARE as the bedrock of its technological infrastructure.

---

[10] https://www.fiware.org/.

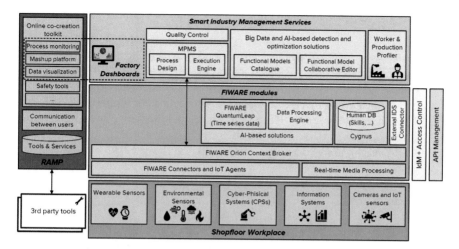

Fig. 2 Powered by FIWARE KITT4SME architecture. Components represented as blue boxes are from the FIWARE reference architecture [10, 11]

Data Gathering The data gathering aspect of the KITT4SME architecture encompasses the collection of data from diverse devices, situated at the lowest layer of the architecture (as depicted in Fig. 2). These devices, deployed within the factory, serve to enrich the system's knowledge base with both raw and preprocessed data. The following categories of devices contribute to the data gathering process:

- *Wearable Sensors*: These sensors are specifically designed to monitor the health and well-being of workers within the factory setting. They provide valuable insights into various physiological parameters and indicators.
- *Environmental Sensors*: Scattered throughout the factory, environmental sensors play a vital role in monitoring and capturing data related to the prevailing environmental conditions. This includes parameters such as air pollution levels, temperature, and humidity.
- *CPSs*: The architecture also incorporates CPSs, with a particular emphasis on those commonly involved in the manufacturing processes, such as machining equipment and collaborative robots. These CPSs facilitate the capture of relevant data pertaining to the operational aspects of the production line.
- *Information Systems*: Information systems represent a valuable source of raw and value-added data, which contribute to update the contextual information of the platform also with aggregated data.
- *Cameras and IoT sensors*: Together with environmental sensors, cameras and IoT sensors are needed to monitor the production, usually requiring a real-time processing to extract valuable knowledge from data streams.

Communication Interfaces In the subsequent layer, the FIWARE framework encompasses a collection of Generic Enablers (GEs) that serve as interfaces between devices, enabling the retrieval of contextual information and the initiation of actuations in response to context updates. Examples of FIWARE GEs available in the catalog[11] include:

- *Connectors and IoT agents*: These modules facilitate the interaction with devices utilizing widely adopted IoT protocols, including LWM2M over CoaP, OneM2M, and OPC-UA. It provides a standardized approach to interface and communicate with diverse IoT devices. Also, connector supporting FAST Real-Time Publish–Subscribe (RTPS) for efficient and real-time processing of data streams is provided, based on the ROS 2 [12] framework.
- *Real-Time Media Processing*: These GEs are designed to support real-time processing and manipulation of media streams (e.g., to transform video cameras into sensor-like devices) to extract valuable information from visual data streams.

Data Broker In the layer above, the FIWARE Orion Context Broker represents the fundamental component of any solution powered by FIWARE. This context broker facilitates the decentralized and scalable management of context information, allowing data to be accessed through a RESTful API. Serving as the authoritative source of information, the Context Broker stores the latest update status of all devices, components, and processes that contribute data to the platform.

However, for the purpose of training and fine-tuning AI tools, it is often necessary to access historical data. To address this requirement, FIWARE offers dedicated GEs called QuantumLeap that automatically generate time series data from the evolving context information, enabling AI tools to leverage the valuable insights gained from historical data analysis.

Smart Industry Management Services The topmost layer of the architecture encompasses analytical services and profilers that leverage the knowledge base within the system. These services include Big Data applications and components utilizing AI-based detection and optimization tools. It is in this layer that AI developers and researchers can greatly benefit from the historical data and up-to-date context information made available by the Powered by FIWARE platform. Additionally, the KITT4SME architecture incorporates utility components in this layer to extract additional knowledge from persistent information and provide insights to human actors regarding the factory's status. These components include:

- *Human Description Database*, which stores a comprehensive representation of factory workers derived from physiological parameters, worker information, machine parameters, and environmental data
- *External IDS Connector*, a component from the IDSA reference architecture[12] that ensures a trustworthy interface between internal data sources and external data consumers. This connector plays a critical role in enabling the integration

---

[11] https://www.fiware.org/catalogue/.

[12] https://docs.internationaldataspaces.org/ids-ram-4/.

of external value-added services, where data exchange is governed by IDS policies

The outputs of analytical models, such as anomaly detection, can be fed back into the FIWARE Context Broker. This triggers decision-making mechanisms, whose logic can be modeled and managed during execution by decision support systems, such as the Manufacturing Process Management System (MPMS). The activation processes of the platform can involve human-in-the-loop interactions, such as collective intelligence, or rely on behavioral updates for groups of involved CPSs. The decisions thus triggered must be identified by IDAS IoT Agents through the FIWARE Context Broker to effectively enable feedback to the CPSs.

Marketplace and Identity Management and Access Control To facilitate the widespread adoption of AI applications and enhance their discoverability, the KITT4SME platform leverages an existing marketplace called Robotics and Automation MarketPlace (RAMP). RAMP enables the Software as a Service (SaaS) provision of these applications, making them easily accessible to users. By incorporating FIWARE-compatible equipment (e.g., robots, machines, sensors) on the production floor, businesses can directly utilize the various tools offered by KITT4SME without the need for complex software deployments and extensive IT expertise. This allows manufacturing SMEs to focus on their core business activities and adds value to their operations.

Furthermore, the distributed nature of the architecture promotes collaborative usage of tools and production data between manufacturing SMEs and technology providers. It facilitates online co-creation and minimizes the necessity for continuous on-site inspections and system installations. Access to platform resources is facilitated by an IDentity Management and Access Control (IDM) GE. This IDM GE provides robust support for secure and private OAuth2-based authentication of users and devices. It also offers features such as user profile management, privacy-preserving handling of personal data, Single Sign-On (SSO), and Identity Federation across multiple administrative domains. These capabilities ensure secure access to the platform's resources while maintaining user privacy and data protection.

4 KITT4SME to Bring AI to an Injection Molding Use Case

The KITT4SME platform has been applied in 4 use cases within the KITT4SME project and 18 external demonstrators made via Open Calls.[13]

In this section, we discuss how the KITT4SME platform has been exploited to create an AI kit supporting one of the internal use cases. This use case is from the injection molding industry, and it aims at facilitating an assembly task mainly composed of screwdriving operations. The assembly task starts with a molding

---

[13] https://kitt4sme.eu/open-call/.

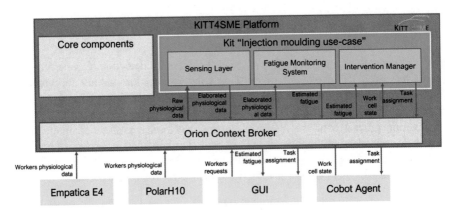

Fig. 3 The KITT4SME solution for the use case in the injection molding sector

press producing a molded piece every 90 seconds. Then, the task foresees a gantry robot that automatically extracts the molded piece from the injection molding machine and places it onto a conveyor belt. Subsequently, a human operator works at a designated workstation to perform the assembly operations while also being responsible for periodic quality checks on the molded pieces or quick maintenance operations on the injection molding machine.

The KITT4SME platform has introduced an AI solution to mitigate workers' physical stress caused by heavy workloads and the injection molding machine's demanding pace during operations. In particular, the use case relies on the concept of human digital twin [13]. A dynamic task assignment between the collaborative robot and the operator is performed by creating a digital representation of the operator and the production system.

The Kit used for this use case, represented with its whole architecture in Fig. 3, includes:

- *Sensing Layer*: This module supports the collection and use of IoT sensor data to be used by data analysis and decision-making modules to take decisions or to be visualized on dashboards. It provides a solution including the interoperability elements (APIs and broker client) for bidirectional data exchange between sensors and the KITT4SME's Orion Context Broker. Data are also preprocessed if needed.
- *Fatigue Monitoring System*: It is an AI model that estimates the perceived fatigue of the workers based on physiological data (e.g., heart rate) from wearable devices and on quasi-static characteristics (e.g., age). The estimation is made using physiological data collected from wearable devices selected by applying an Analytic Hierarchy Process (AHP)-based methodology [14] and operator's characteristics, including age, sex, handedness, exercise/healthy habits, and routines, collected via interviews.
- *Intervention Manager*: It monitors the real-time status of the worker–factory ecosystem, elaborating data from sensors, machines, workers monitoring sys-

tems, and ERP, and it knows what interventions can be applied and which are the rules to decide which is the best one given a particular situation. It applies AI models specifically developed to support decision-making.

The kit has been deployed to the platform to support the task assignment in a screwdriving process in the following process:

1. The operator retrieves two molded parts from a conveyor belt and positions them on the working bench.
2. The operator inserts six nuts into each part, flips one part, and places it on top of the other.
3. The operator positions nine screws on one side of the assembled parts.
4. The Intervention Manager assigns each screw to the operator or the cobot. The operator and a cobot simultaneously perform the screwdriving process. Depending on the number of screws assigned to the operator, they may also engage in other support activities, such as monitoring other machines, conducting short maintenance operations, or removing the pallet.
5. The operator flips the assembled parts and repeats steps 3 and 4.
6. The assembled parts are stacked on a pallet.

The task assignment, performed by the Intervention Manager and confirmed by the operator, consists of the allocation of the screwing operations (9x2 for each assembled part), and it is made considering the following parameters:

- Current perceived fatigue of the operator as estimated by the Fatigue Monitoring System.
- Work In Progress level.
- Cobot state (idle, current operation, and error).

Discussion The above use case exemplifies how the KITT4SME platform can actually ease AI adoption by SMEs, compared to other platforms in the AI ecosystems. Indeed, compared to platforms to develop AI solutions, the SME from the use case did not spend any effort on developing AI, given that they exploited the existing application available on the platform. Also, the platform helped the company compose the best kit to solve a real need, i.e., to facilitate an assembly task mainly composed of screwdriving operations. The proposed kit already included all the components needed to be implemented in the factory, i.e., data acquisition (Sensing Layer), AI solution to derive data-driven knowledge (Fatigue Monitoring System), and a reasoning engine (Intervention Manager), relieving the company from extra development activities needed to connect the shop floor to the platform. Instead, by considering platforms providing already developed applications, a similar use case has been successfully tested in a laboratory environment [15], exploiting a different IIoT platform [16]. This kind of platform enables the handling of third-party applications, with no guarantees about the interoperability of components in terms of application interfaces and data models, which are covered within the KITT4SME platform by the FIWARE components. Also, while this kind of platform comes with a ready-to-use solution, integrating and deploying such solutions is often a burden solely on the developers. Again, the KITT4SME platform offers a distinct advantage since it effortlessly facilitated the integration and deployment of three distinct modules, two of which leverage AI, resulting in a smooth and reliable operation.

5 Conclusion

In this chapter, we discussed the potential of AI solutions in increasing the profitability of SMEs (e.g., by improving product quality or optimizing production line configurations), and we presented a new platform, namely the KITT4SME platform, intended to deliver affordable, tailor-made AI kits to the manufacturing industry. The cloud platform presented in this chapter supports the KITT4SME workflow by relying on widely adopted platforms, e.g., FIWARE, in such a way to ease the development of new AI services, as well as their deployment in real industrial settings.

Specifically, the platform is capable of composing AI components from a marketplace (i.e., the RAMP marketplace) into a tailor-made service offering for a factory, a functionality that is not provided by any of the existing AI platforms. Once the factory shop floor is connected to the AI services, the platform enables data storage and exchange in an interoperable, secure, privacy-preserving, and scalable way. The architecture has been designed by leveraging state-of-the-art technology and standards, reusing open-source software and technologies whenever possible, thus promoting both its adoption by small manufacturing companies on a budget and further extensions by other researchers and practitioners in the reference community. The exploitation of the platform has been demonstrated with real-world use cases, which have been conducted as part of the KITT4SME project thanks to a platform prototype publicly available in the KITT4SME online repository.[14]

Future work will be focused on further increasing the interoperability of platform services, also relying on semantic data interoperability, to better define the composability of different AI components, possibly available in different marketplaces, enabling cross-platform service composition.

Acknowledgments This work has been partly supported by EU H2020 research and innovation program project KITT4SME (grant no. 952119).

References

1. Wu, S.-Y.: Key technology enablers of innovations in the ai and 5g era. In: 2019 IEEE International Electron Devices Meeting (IEDM), pp. 36–3. IEEE (2019)
2. Alexopoulos, K., Nikolakis, N., Chryssolouris, G.: Digital twin-driven supervised machine learning for the development of artificial intelligence applications in manufacturing. Int. J. Comput. Integr. Manuf. **33**(5), 429–439 (2020)

---

[14] https://github.com/c0c0n3/kitt4sme.

3. Elger, P., Shanaghy, E.: AI as a Service: Serverless Machine Learning with AWS. Manning Publications (2020)
4. Bettoni, A., Matteri, D., Montini, E., Gladysz, B., Carpanzano, E.: An ai adoption model for SMEs: a conceptual framework. IFAC-PapersOnLine **54**(1), 702–708 (2021). 17th IFAC Symposium on Information Control Problems in Manufacturing INCOM 2021
5. IBM: What is platform-as-a-service (PaaS)? https://www.ibm.com/topics/paas (2021). Accessed 08 Aug 2023
6. Lins, S., Pandl, K.D., Teigeler, H., Thiebes, S., Bayer, C., Sunyaev, A.: Artificial intelligence as a service. Bus. Inf. Syst. Eng. **63**(4), 441–456 (2021)
7. Zhao, S., Talasila, M., Jacobson, G., Borcea, C., Aftab, S.A., Murray, J.F.: Packaging and sharing machine learning models via the Acumos AI open platform. In: 2018 17th IEEE International Conference on Machine Learning and Applications (ICMLA), pp. 841–846. IEEE (2018)
8. Llewellynn, T., Fernández-Carrobles, M.M., Deniz, O., Fricker, S., Storkey, A., Pazos, N., Velikic, G., Leufgen, K., Dahyot, R., Koller, S., et al.: Bonseyes: platform for open development of systems of artificial intelligence. In: Proceedings of the Computing Frontiers Conference, pp. 299–304, 2017
9. Prado, M.D., Su, J., Saeed, R., Keller, L., Vallez, N., Anderson, A., Gregg, D., Benini, L., Llewellynn, T., Ouerhani, N., et al.: Bonseyes ai pipeline—bringing ai to you: End-to-end integration of data, algorithms, and deployment tools. ACM Trans. Internet Things **1**(4), 1–25 (2020)
10. Ahle, U., Hierro, J.J.: Fiware for data spaces. In: Otto, B., ten Hompel, M., Wrobel, S. (eds.), Designing Data Spaces: The Ecosystem Approach to Competitive Advantage, pp. 395–417. Springer International Publishing (2022)
11. Fiware Smart Industry Reference Architecture: https://www.fiware.org/about-us/smart-industry/ (2022). Access 08 Aug 2023
12. Macenski, S., Foote, T., Gerkey, B., Lalancette, C., Woodall, W.: Robot operating system 2: Design, architecture, and uses in the wild. Sci. Robot. **7**(66), eabm6074 (2022)
13. Montini, E., Bettoni, A., Ciavotta, M., Carpanzano, E., Pedrazzoli, P.: A meta-model for modular composition of tailored human digital twins in production. Procedia CIRP **104**, 689–695 (2021)
14. Montini, E., Cutrona, V., Gladysz, B., Dell'Oca, S., Landolfi, G., Bettoni, A.: A methodology to select wearable devices for industry 5.0 applications. In: 2022 IEEE 27th International Conference on Emerging Technologies and Factory Automation (ETFA), pp. 1–4. IEEE (2022)
15. Montini, E., Cutrona, V., Dell'Oca, S., Landolfi, G., Bettoni, A., Rocco, P., Carpanzano, E.: A framework for human-aware collaborative robotics systems development. Procedia CIRP **120**, 1083–1088 (2023). 56th CIRP Conference on Manufacturing Systems 2023. https://doi.org/10.1016/j.procir.2023.09.129. https://www.sciencedirect.com/science/article/pii/S2212827123008612
16. Montini, E., Cutrona, V., Bonomi, N., Landolfi, G., Bettoni, A., Rocco, P., Carpanzano, E.: An IIoT platform for human-aware factory digital twins. Procedia CIRP **107**, 661–667 (2022)

Multisided Business Model for Platform Offering AI Services

Krzysztof Ejsmont ⓘ**, Bartlomiej Gladysz** ⓘ**, Natalia Roczon,**
Andrea Bettoni ⓘ**, Zeki Mert Barut, Rodolfo Haber** ⓘ**, and Elena Minisci**

1 Introduction

Platform businesses have become one of the latest research topics in various management disciplines [10]. A platform is an interface that facilitates interactions between different parties, usually complementors and customers [9]. In the platform business, the platforms and their complementors have a strong one-way complementarity, where the total value of the platform and its complementors is more than the sum of the two combined [18], and this complementarity requires the interdependencies between the platforms and the complementarities to be managed in an ecosystem level.

There are two basic types of platforms: innovation platforms (as an intermediary for direct exchange or transactions) and transaction platforms (as a technological foundation upon which other firms develop complementary innovations). Some

K. Ejsmont (✉) · B. Gladysz · N. Roczon
Faculty of Mechanical and Industrial Engineering, Warsaw University of Technology, Warsaw, Poland
e-mail: krzysztof.ejsmont@pw.edu.pl; bartlomiej.gladysz@pw.edu.pl

A. Bettoni · Z. M. Barut
Department of Innovative Technologies, University of Applied Science of Southern Switzerland, Manno, Switzerland
e-mail: andrea.bettoni@supsi.ch; zekimert.barut@supsi.ch

R. Haber
Centre for Automation and Robotics (CAR), Spanish National Research Council-Technical University of Madrid (CSIC-UPM), Madrid, Spain
e-mail: rodolfo.haber@car.upm-csic.es

E. Minisci
CRIT S.R.L., Vignola, Italy
e-mail: minisci.e@crit-research.it

© The Author(s) 2024
J. Soldatos (ed.), *Artificial Intelligence in Manufacturing*,
https://doi.org/10.1007/978-3-031-46452-2_7

companies combine the features of the two and create "hybrid platforms" [13]. Multisided platforms (MSPs) allow direct interactions between two or more different entities, where each entity is associated with the platform [17]. Examples of well-known MSPs include Facebook, Uber, PayPal, Airbnb, Alibaba, eBay. The growing interest in MSPs is due to two key factors: their essential role in minimizing the transaction costs between sides [15] and the power of the business models (BM) in the digital economy because of their ability to adapt and cope with complexity, rapid scaling, and value capture [1]. Although many companies are opting for MSP BMs, only a few have been successful. MSPs should strive to attract users and must achieve direct and indirect network effects to be successful. More importantly, they ought to solve the chicken-or-egg problem, which refers to a network effect meaning "one side of the market realizes the value only if another side is fully engaged" [13].

2 Methodologies for MSPs Business Modeling

The pioneering models of MSPs were introduced by Armstrong, Valillaud and Jullien, Parker and Van Alstyne, Rochet and Tirol, as described in more detail in Hagiu and Wright [17]. Allweins et al. [2] proposed a Business Model Canvas [21] to illustrate the MSP businesses. As a result, the cited paper proposed Platform Canvas. The focus of this study was not on the definition of individual entities (having different value propositions) but on the modeling of MSPs' business transactions. For this purpose, only methodologies dedicated to MSPs' business models were considered. The Business Model Kit is proposed by the Board of Innovation.[1] It consists of 16 blocks filled with details on various stakeholders and value propositions, resulting in a marketing tool for communicating the BM to different entities. Leanstack[2] offers a Lean Canvas, adjusted from the Business Model Canvas, with procedures to complete the nine blocks starting with problem definition, modeling customer segments, and finally, the derived unique value proposition. Lean Canvas introduces a phase of finding the solution, identifying channels to reach customer segments, estimating revenue and cost structure, and defining crucial metrics and unfair advantages.

Most papers develop analytical models focusing on a specific characteristic of the MSPs business model, such as pricing structure, network externalities, or competition (i.e., [3, 6, 14, 16]), while a holistic approach to building a business model for MSPs is lacking. Therefore, a methodology that seems to meet the expectations of MSPs in the context of business model development is the Platform Design Toolkit (PDT). This methodology is the first codified platform design method, released in 2013.[3] The PDT was developed by a team led by S. Cicero

---

[1] https://www.boardofinnovation.com/tools/business-model-kit/

[2] https://leanstack.com/lean-canvas

[3] https://www.boundaryless.io/pdt-toolkit/

to provide companies with support in describing the platform's vision, the core and ancillary value propositions, the platform's infrastructure and core components, and the characteristics of the platform ecosystem expressed through transaction dynamics [5, 11]. It was optimized to support the development of multisided, transformative platform strategies to empower ecosystems to create shared value. It is an open-source method adopted worldwide by global Fortune 500 leaders, leading institutions, start-ups, and scale-ups. The PDT covers all stages, from exploration to design, validation, and growth. The core of the PDT methodology in developing a business model is the design stage: an extensive and proven step-by-step process that helps move from contextualizing entities in the ecosystem, their role and relationships, detailing possible transactions between entities, to designing the platform experience.

PDT, in the design stage, contains eight templates (canvases) to be completed, considering as many aspects of the business. The steps are as follows:

1. *Mapping the ecosystem*: entities present in the ecosystem are mapped onto the canvas, allowing us to understand the role they may play and identify possible clusters.

2. *Portraying ecosystem's entities roles*: a coherent and deep picture of the role of each of the entities identified in step 1 is created by defining what their context is, what they want to achieve, with whom and how they want to integrate, what potential they can represent and what kind of experience gains they are looking for, and what the platform shaper can provide them with.

3. *Analyzing the potential to exchange value*: using the so-called "ecosystem's motivation matrix," entities' potential to exchange value flows is analyzed. This is a mapping of what type of value exchange is already being performed (or attempted to be performed) by the entities and what additional value they could exchange if adequately enabled.

4. *Choosing the core relationships you want to focus on*: the platform shaper needs to identify which entities in the ecosystem they want to focus on and which relationships will form the core of the platform design.

5. *Identifying the elementary transactions*: the "transaction board" tool is used to map how the ecosystem currently exchanges value (focusing on the entities and relationships prioritized in step 4) and how the platform's strategy is to help them make value transactions more manageable, faster, and cheaper by providing and curating channels and contexts that increase the likelihood of interactions and transactions.

6. *Designing the learning engine*: through the "learning engine canvas," a step-by-step process has been designed to support/enable services that will support entities to adopt the platform strategy. These services will not only help them evolve and become better producers and consumers but also radically evolve and discover new opportunities and behaviors that were not initially intended.

7. *Assembling the platform experiences*: with the "platform experience canvas," the elements emerged from the transaction board (step 5) and those from the learning engine canvas (step 6) are combined to create an experience persistence

model that summarizes the key value propositions arising from the strategy being developed. This allows consideration of what resources and components need to be put in place and managed to deliver these experiences and derive value from them.

8. *Setting up the minimum viable platform (MVP)*: this allows us to test in the natural environment (market) whether the design assumptions are suitable for the future. By analyzing design products, in particular the compiled "platform experience canvases" (step 7), the riskiest assumptions of the strategy are isolated, as well as experiments and indicators to validate them with the ecosystem are identified.

The resulting business model is then summarized in the platform design canvas, which is the final output of this reference methodology. According to the author's knowledge and experience, by far, the most essential element of business models for MSPs is to identify the value that can be transferred to the different entities through the platform [12]. Taking this into account, it was decided to focus on the first five steps of the PDT methodology.

3 Application of PDT for the Design of AI Platform as a Service Business Model – KITT4SME Case Study

3.1 Introduction to the KITT4SME Project

KITT4SME (platform-enabled KITs of arTificial intelligence FOR an easy uptake by SMEs) is a Horizon 2020 project (GA 952119). It is explicitly aimed at European SMEs and mid-caps to provide them with scope-tailored and industry-ready hardware, software, and organizational bundles, delivered as modularly customizable digital platform that seamlessly introduce AI into their production systems.[4]

Among the main objectives of the KITT4SME project that need to be included in the business model are [20]:

- to provide SMEs with ready-to-use, customized digital packages to harness the capabilities of AI at an affordable price and a proper scale,
- seamlessly combine AI and human problem-solving expertise (know-how) into a single digital platform with unparalleled shop floor orchestration capabilities, and
- expanding the local ecosystem offerings so that entities with different competencies can grow by collaborating on customizable AI kits.

---

[4] https://kitt4sme.eu/

3.2 Needs Elicitation

The process of creating a BM for AI platform as a service was initiated by identifying the main stakeholders (manufacturing SMEs, AI developers, DIHs) and their needs. The needs elicitation process was conducted by adhering to iterative stakeholder engagement based on interviews and workshops, as proposed by Azadegan et al. [4] and confirmed by Bettoni et al. [8]. To identify needs/expectations, 29 interviews were conducted with samples of different types of stakeholders. AI developers (13 respondents) and SMEs (10 respondents) are the most represented entities [7], as they will be the primary and direct users of the KITT4SME platform (supply and demand side). Entities of different sizes, from different EU countries, and with different scopes of activity (from national to global) were involved (for more details about this analysis, see [19]).

The following needs were identified:

- modularity of solutions,
- the possibility of integrating implemented AI solutions with already existing ones,
- increased data transparency and traceability,
- identification of hidden problems to improve processes,
- defining solutions to the identified problems,
- personalizing the platform, allowing to tailor solutions to individual needs,
- matching potential partners,
- access to multilevel knowledge transfer,
- simplified AI implementation algorithms,
- generalization of implementation middleware,
- integration of modules to facilitate deployments,
- ability to integrate with low-digitized infrastructure,
- introduction of preventive maintenance,
- improved analytics and a better understanding of customer behavior and purchase decisions, and
- personalization of actions in real-time.

3.3 KITT4SME Business Model

The first step in developing the KITT4SME platform BM consisted of identifying the crucial entities that will form the platform ecosystem and have a significant impact (direct or indirect) on the functioning of the platform. These entities have been mapped into a unique canvas, as shown in Fig. 1.

The idea behind the canvas is to divide the entities in the ecosystem into three main groups: impact entities (platform owners, external stakeholders) – they are not involved in the continuous interactions happening in the ecosystem; demand entities (peer consumers) – they are interested in "consuming" the value produced

THE KITT4SME ECOSYSTEM CANVAS

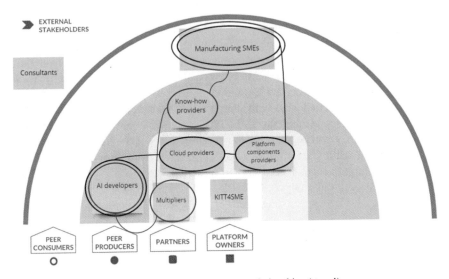

Fig. 1 KITT4SME ecosystem canvas (step 1) + core relationships (step 4)

in the ecosystem; supply entities (partners, peer producers) – they are interested in "producing" the value consumed in the ecosystem.

Considering a single entity, its position in this framework may vary. For example, an AI developer (peer producer) may become a partner after a certain period of time if it provides many AI solutions and takes an active part in the development of the platform. An entity may also have a dual role, as access to the platform may create new opportunities: a company initially interested in offering its products (peer producer) may later be interested in using its belonging to the ecosystem to seek ideas for improving manufacturing processes in SMEs (peer consumer).

In the second step, the aim is to develop a portrait of the leading entities accessing the platform from both the demand and supply sides. It should be noted that this second step aims to map what the entities are currently looking for rather than what the idea behind the platform service is. Thus, it is possibly better to characterize the value from their point of view. In the KITT4SME ecosystem, six different entities have been identified (Fig. 1). Figure 2 shows a portrait of AI developers, as they appear to be the most important in the initial lifecycle of the platform – they will be responsible for delivering AI solutions/services that can be transacted. Similarly, portraits should be taken for all other identified entities.

The ecosystem motivation matrix (step 3) maps the values exchanged between pairs of entities through the KITT4SME platform. Money is undoubtedly exchanged as a consequence of interactions through the platform, but even more important for shaping the KITT4SME BM is the identification of intangible values resulting from the opportunities the platform brings. The matrix shown in Fig. 3 details the central

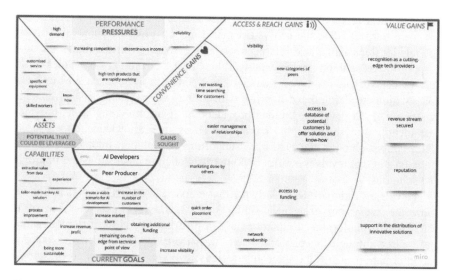

Fig. 2 Portrait of the entity "AI developers" (step 2)

gives to	Manufacturing SMEs	AI developers	Know-how providers	Platform components providers	Cloud providers	Multipliers
Manufacturing SMEs PA PP PC		· real scenarios to develop AI solutions · training data · money · feedback for improve AI solutions · visibility (advertising) · inspiration and confrontation	· money · feedback · help to better understand expectations regarding AI (in which direction the market is going) · inspiration and confrontation	· ideas for new or extended platform functional ties · business contacts	· new customers cross-selling opportunities	· trends of AI · needs in manufacturing sector regarding AI · paying association fee (money)
AI developers PA PP PC	· AI solutions · extract value from data · knowledge · customized services · improved processes · access to product, new competences and technologies		· AI solutions as a basis for training and knowledge expansion	· enhance value of solutions	· new customers (indirectly)	· services to offer · reach new association · satisfy better some needs
Know-how providers PA PP PC	· access to new competences, skills and knowledge · timeout on support	· visibility · dissemination of the solutions offered through training and user support		· ideas for new or extended platform functionalities · feedback · keep up with market trends	· more cloud contents · demostration of cloud usefulness · feedback · keep up with market trends	· training · reputation · money
Platform components providers PA PP PC	· access to AI modules · timeout on support		· customers · increase revenue · keep up with market trends · expertize and knowledge on the specific platform component			· extend the network · knowledge
Cloud providers PA PP PC	· access to cloud infrastructure · innovation support · computing services · servers	· computing services · servers	· customers · increase revenue · keep up with market trends · expertize and knowledge on the specific cloud solutions	· technical competence and experience (development in cloud)		· extend the network · knowledge
Multipliers PA PP PC	· innovation support · new ideas · feedback · networking opportunities · consulting · support in contact with the platform's offer page · networks of contacts	· networking opportunities · consulting · impact on local/regional innovation markets and networks · communication with the demand side · networks of contacts	· extended market · increase revenue and customers (visibility) · impact on local/regional innovation markets and networks · communication with the demand side · networks of contacts	· visibility · business contacts · European dimension	· visibility · European dimension	

Fig. 3 KITT4SME ecosystem motivation matrix (step 3)

values exchanged between peer consumers (PC), peer producers (PP), and partners (Pa) – previously mapped in the ecosystem canvas (Fig. 1). The cells report what the entity in the first column from the left can "give to" the entities on the upper axis.

The goal of the fourth step is to decide which subset of relationships to focus on to ensure that enough attention is paid to defining and implementing the core experience. The value flows identified in the ecosystem motivation matrix (Fig. 3) were transferred to the ecosystem map (Fig. 1). Figure 1 shows the division of relationships into those relating to resource sharing (brown lines) and those supporting AI solution implementation (blue lines). In the first case, entities contact each other to share resources. Manufacturing SMEs in this context seek dedicated AI solutions to develop and improve their production capabilities. The remaining entities, i.e., AI developers, cloud providers and platform components providers, are identified as suppliers and partners, offering their knowledge, expertise, and AI solutions through the platform. Supporting AI solution implementation is a relationship that involves entities seeking to collaborate on creating and improving AI solutions.

The identification of the underlying transactions and channels serves to illustrate how the ecosystem exchanges value (step 5) and highlights the role of the KITT4SME platform as an intermediary in this process. Most of the interactions take place through the platform itself, which creates value from the exchange of information, while the three interactions involving the exchange of software (AI solution/module), AI service (e.g., support to solving problems using AI, implementation AI solution, consultation), and payment are realized outside the platform.

The transaction matrix helps analyze the relationship between the demand side (entity 1) and supply side (entity 2). It helps identify all transactions/interactions and their channels that are already taking place or may take place. In addition, for each transaction/interaction is assigned what is the unit of value. One of the key roles of the platform (owner) is to create channels that can reduce coordination/transaction costs.

The transaction matrix (Table 1) confirms that the KITT4SME platform is the main channel of interaction and, to be successful during each interaction, the exchange of information must add value for the stakeholders. A crucial role of the platform is to participate in the facilitation of the communication process actively and the interaction between stakeholders, thereby reducing transaction costs and facilitating transactions.

3.4 Business Model Design Canvas

The analyses conducted in the previous chapters were finally aggregated into the platform design canvas and structured as follows:

- *Enabling services (platform to partners)*: focused on helping partners generate/create value from their assets and capabilities, access to potential consumers, increasing competitiveness and visibility, and decisively improving as professional entities (reputation). For KITT4SME, these are designed services to facilitate the implementation of technical specifications and core service stan-

Table 1 KITT4SME transaction board for core relationships

Entity 1	Transaction/interaction	Entity 2	Currency/value unit	Channel or context
Mfg. SMEs	Request for support to solve a problem/implement a solution High level of request Request or complete NDA Sending enquiry Communication/interaction Sending offer Reply to the offer	AI developers Cloud providers Platform components providers Multipliers Know-how providers	Information	KITT4SME platform
Mfg. SMEs	Development/tailored of an AI solution Problem-solving/implementation of AI solution	AI developers Cloud providers Platform components providers Know-how providers	AI solution AI module AI service	Companies developing AI solutions Software houses Companies implementing AI solutions AI consulting firms
Mfg. SMEs	Testing/validation of the AI solution	AI developers Cloud providers Platform components providers	Information	KITT4SME platform
Mfg. SMEs	Corrections/elimination of errors	AI developers Cloud providers Platform components providers	AI solution AI module	Companies developing AI solutions Software houses
Mfg. SMEs	Reply to the AI solution Reply to the problem-solving/implementation of the AI solution	AI developers Cloud providers Platform components providers Know-how providers	Information	KITT4SME platform
Mfg. SMEs	Payment	AI developers Cloud providers Platform components providers Know-how providers	Information Money	Independent channel
Mfg. SMEs	Feedback/reputation	AI developers Cloud providers Platform components providers Multipliers Know-how providers	Information	KITT4SME platform

dards for AI developers providing solutions for KITT4SME and disseminating KITT4SME in the AI field.

- *Empowering services (platform to peer producers)*: aimed at helping peer producers start executing transactions, improve their capabilities, improve on the platform, and enter the development stage (growth phase). The KITT4SME platform aims to support the development of EU-compliant applications, modules and services for AI solutions through dedicated consulting, training, success stories, and best practices.
- *Other services (platform to peer consumers)*: there are many cases in which platforms provide more classical industrialized services to users. They are complementary to the value exchanged, experiences provided by the ecosystem, and they provide powerful, robust usability for the individual user. Like empowering services, support and training will also be provided for those consumers who intend to use other AI platforms or switch to solutions offered by other AI vendors.
- *Core value proposition*: stands for the core value that the platform is trying to create for the main purpose of its operation. It usually targets consumers, as they usually represent the broadest market segment of peers and are the customers who buy products or services. Particularly, in dynamic market networks and in more niche contexts, where transaction value is higher and transaction volume is lower, partners or peer producers may be the basic recipients of the core value proposition.
- *Ancillary value propositions*: these are ancillary values offered by the platform. Ancillary value propositions can be aimed at the same market segment as the core value proposition or at others. It is common for MSPs to supplement the core value proposition for the demand side of the platform (manufacturing SMEs) with a proposition aimed at the supply side (AI developers, know-how providers). KITT4SME provides an entire environment (infrastructure) that enables not only real interaction between entities in a multisided ecosystem but also the resources necessary to increase their visibility in the AI field. Ancillary value propositions for the KITT4SME platform could be SME issues assessment, modules combination and kit composition, kit deployment and maintenance, shop floor data acquisition, extraction, synthetization and reporting of data, generation of real-time interventions, workforce assessment and upskilling, best practices, and knowledge creation. Most of them can be assigned dedicated assistance, and these services can be the basis of the membership fee. Advertising services can also be considered as ancillary values. With the development of the platform, the growth of the number of users and increasing platform reach – it will be possible to provide advertising services to interested entities (e.g., advertising AI services), which may be the basis of advertiser fees.

At the beginning of the platform's life, the platform will charge mostly transaction fees for the transfer of AI solutions and apply membership fees only for some services while the rest will be offered for free. In the future, when a critical mass of consumers is reached, the platform will charge mostly membership fees.

- *Infrastructure and core components*: these are assets owned and controlled by the platform owner. They are managed according to the platform's governance rules. Assets can be tangible (e.g., server or venue) or intangible (e.g., common standard – FIWARE). They guarantee the platform's operation and use by the ecosystem. KITT4SME identifies the critical elements of the platform's IT environment as the core components of the platform BM, namely the AI module standards, protocols, the standard enablers (CPS-izers, runtime), codes, and the functionalities and channels that enable its dissemination (such as RAMP).
- *Transactions*: are part of a more complex "experience." They should be understood as a sub-activity during which value is created, delivered, exchanged, or transferred between typically two (or more) platform users. KITT4SME assumes two main types of transactions: the first is intangibles (information), which the platform completes by providing it through the systems typically used in such kinds of platforms; the second is monetary and related to AI services that are exchanged through the platform (AI solutions, applications, modules, services, runtime).
- *Channels and contexts*: enable exchanges within the platform and are the platform's interface with users. Channels are user touch points that play an essential role in the user experience. They are crucial in creating added value: they should be actively created and continuously improved by the platform owner. The marketplace should be considered the principal channel provided by the KITT4SME ecosystem, where AI solutions, applications, modules, and services are purchased, exchanged, transferred, and downloaded, respectively. Channels for exchanging/obtaining information and processing payments are also important.

3.5 Revenue Model for the KITT4SME Platform

After a literature analysis of MSPs' pricing strategies, a review of the monetization strategies of other platforms offering AI services, and an internal workshop of the partners involved in developing the KITT4SME revenue model, it seems possible for the platform to generate revenue through all three main streams [22]:

- subscriptions (membership fee),
- advertising (advertisers fee), and
- transactions (transaction fee).

For the KITT4SME platform, several revenue streams can be combined and different models can be adopted at different stages of the platform lifecycle.

When designing a business model that assumes revenue from all three main streams, a fundamental issue to have in mind is the evolving network effects. These are generated from the interaction of user pairs and strongly influence the level of interest in the platform. Given the resource-sharing scenario of the platform, network effects are generated when the availability of more resource providers

(AI developers, know-how providers) attracts more entities seeking resources (manufacturing SMEs), which in turn causes more providers (peer producers) to join the platform. Finding the right balance at the outset is problematic because if there are not so many providers, there is a risk that the peer consumers may not find what they are looking for and will use a competitor's platform. The same consumer could abandon the KITT4SME platform and not return when it is upgraded with updated versions of its services, such as an advanced matchmaking mechanism or new AI solutions/modules. On the other hand, a provider that does not receive contacts may choose to post its offer in multiple places (e.g., AI platforms) if the cost of staying on the platform is affordable. The first effort should be to build a good peer producers base, while the right message needs to be sent to potential peer consumers.

In order to support the creation of this kind of dynamics while generating revenue for the platform, the following approach can be used, especially in the initial lifecycle stage of the platform:

- A free trial period is offered to each type of entity. This gives access to a primary or all set of services. The KITT4SME platform owner has to decide whether to keep the free access with no time limits forever;
- After the trial period, a peer producer (AI developer, know-how provider) and peer consumer (manufacturing SME) access fee is required;
- A transaction fee is charged and paid by the peer producer, who will set the final price offered to the peer consumer.

For the solution implementation scenario, a different revenue mode should be used. Most likely also, in this case, the initial access will be free of charge for each type of entity. Then a lead fee model is considered more appropriate than the one based on commissions, as the final exchange value may be differently related to creating and improving customized AI solutions.

For both scenarios, some incentives can be offered to active entities of the KITT4SME platform ecosystem. For example, an opportunity can be created for a platform member to invite some of their contacts (e.g., suppliers or customers) to the KITT4SME platform; if onboarding is achieved, the platform member may receive some benefits (e.g., discounts on the transaction fee, extension of the trial period, special rates). It can also be more complex and linked to the actual activity of the invited new members. For example, a platform member may receive the first set of benefits when their contacts are onboard and the second when their contacts start transacting on the platform. This can also be valuable for partners who can use the platform to gain benefits by including their network in the KITT4SME ecosystem.

All the considerations so far have allowed the construction of an initial revenue streams model to determine the pricing strategy for the KITT4SME platform and to assess the financial sustainability of the KITT4SME platform (Fig. 4).

Figure 4 illustrates the different revenue streams of the KITT4SME platform, which include several interdependent groups of entities (manufacturing SMEs, AI developers, know-how providers, cloud providers, platform components providers, consultants, multipliers), the KITT4SME platform owner, and their interactions. For instance, a usage externality exists when peer producers and peer consumers

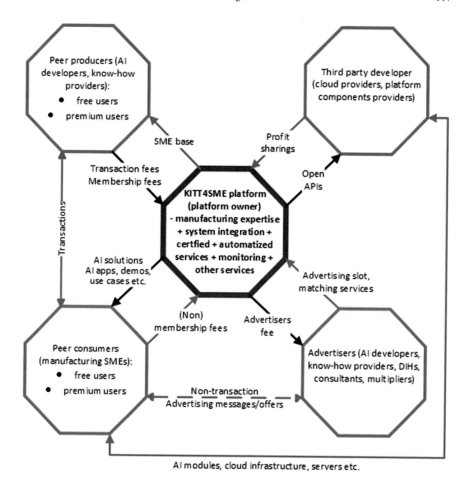

Fig. 4 Revenue streams in the platform within the KITT4SME ecosystem entities defined in PDT

need to work together to generate value using the KITT4SME platform (enhancing the quality of the match). Interactions can also occur between peer consumers and advertisers (very often advertisers will be peer producers, but not only, e.g., consultants). In this case, no transaction is taking place. Furthermore, the KITT4SME platform can enable advertising services or matching offers and charge an advertising fee for this and charge a premium fee for continued access to all KITT4SME services (i.e., a membership fee).

4   Conclusions and Next Steps

In addition to most traditional strategies for defining business models, this study allowed us to understand better the users' needs of the platform offering AI services, to identify the values that can be exchanged through the platform, and to formalize the relationships and partnership mechanisms between entities accessing the MSP. This was done using the platform business model developed for the KITT4SME ecosystem as a case study.

The adoption of the PDT method has shown that this tool provides a relevant methodological approach to define business model scenarios dedicated to MSPs qualitatively. Dividing the development of a business model into a few canvases allows one to focus on the different steps and to go deeper into the details of their design. The first five stages of the PDT have made it possible to define which entities can exchange values through which transaction channels. Although the completion of the canvases still does not allow a quantitative approach to assess the extent to which the elaborated BM can remain sustainable under the dynamic evolution of the boundary conditions.

The following steps should be setting up the MVP and determining the value of the different fees charged for using the platform. The KITT4SME project will be used as a case study for these steps. In this way, the canvases proposed by Cicero [11] will be expanded by developing a methodology that guides the user to quantify the BM elements required for economic feasibility.

Acknowledgments The work presented in this chapter was developed within the project "KITT4SME – platform-enabled KITs of arTificial intelligence FOR an easy uptake by SMEs" and received funding from the European Union's Horizon 2020 research and innovation program under grant agreements No. 952119.

References

1. Abdelkafi, N., Raasch, C., Roth, A., Srinivasan, R.: Multi-sided platforms. Electr. Mark. **29**, 553–559 (2019). https://doi.org/10.1007/s12525-019-00385-4
2. Allweins, M.M., Proesch, M., Ladd, T.: The platform canvas—conceptualization of a design framework for multi-sided platform businesses. Entrepren. Educ. Pedagogy. **4**(3), 455–477 (2021). https://doi.org/10.1177/2515127420959051
3. Armstrong, M.: Competition in two-sided markets. Rand J. Econ. **37**, 668–691 (2006)
4. Azadegan, A., Papamichail, K.N., Sampaio, P.: Applying collaborative process design to user requirements elicitation: a case study. Comput. Ind. **64**(7), 798–812 (2013). https://doi.org/10.1016/j.compind.2013.05.001
5. Barni, A., Montini, E., Menato, S., Sorlini, M., Anaya, V., Poler, R.: Integrating agent based simulation in the design of multi-sided platform business model: a methodological approach. In: 2018 IEEE International Conference on Engineering, Technology and Innovation (ICE/ITMC), Stuttgart, Germany, 17–20 June 2018 (2018) https://doi.org/10.1109/ICE.2018.8436360

6. Belleflamme, P., Peitz, M.: Platform competition and seller investment incentives. Eur. Econ. Rev. **54**, 1059–1076 (2010). https://doi.org/10.1016/j.euroecorev.2010.03.001
7. Bettoni, A., Corti, D., Matteri, D., Montini, E., Fiorello, M., Masiero, S., Barut, Z.M., Gretenkord, S., Gladysz, B., Ejsmont, K.: KITT4SME report 2021: Artificial Intelligence adoption in European Small Medium Enterprises (2021). https://kitt4sme.eu/wp-content/uploads/2021/09/AI-readinessSurvey_resultstorespondentsV8.1.pdf. Accessed 12 July 2023
8. Bettoni, A., Matteri, D., Montini, E., Gładysz, B., Carpanzano, E.: An AI adoption model for SMEs: a conceptual framework. IFAC-Papers OnLine. **54**(1), 702–708 (2021). https://doi.org/10.1016/j.ifacol.2021.08.082
9. Boudreau, K., Hagiu, A.: Platform rules: multi-sided platforms as regulators. SSRN Electron. J. (2008). https://doi.org/10.2139/ssrn.1269966
10. Chen, L., Tong, T.W., Tang, S., Han, N.: Governance and design of digital platforms: a review and future research directions on a meta-organization. J. Manag. **48**(1), 147–184 (2022). https://doi.org/10.1177/01492063211045023
11. Cicero, S.: That's cognitive capitalism (2015). https://medium.com/@meedabyte/that-s-cognitivecapitalism-baby-ee82d1966c72. Accessed 12 July 2023
12. Corti, D., Bettoni, A., Montini, E., Barni, A., Arica, E.: Empirical evidence from the design of a MaaS platform. IFAC Papers OnLine. **54**(1), 73–79 (2021). https://doi.org/10.1016/j.ifacol.2021.08.008
13. Cusumano, M.A., Gawer, A., Yoffie, D.B.: The Business of Platforms: Strategy in the Age of Digital Competition, Innovation, and Power. Harper Business, New York (2019)
14. De Matta, R., Lowe, T.J., Zhang, D.: Competition in the multi-sided platform market channel. Int. J. Prod. Econ. **189**, 40–51 (2017). https://doi.org/10.1016/j.ijpe.2017.03.022
15. Hagiu, A.: Pricing and commitment by two-sided platforms. Rand J. Econ. **37**(3), 720–737 (2006) http://www.jstor.org/stable/25046268
16. Hagiu, A., Wright, J.: Marketplace or reseller? Manag. Sci. **61**, 184–203 (2015). https://doi.org/10.2139/ssrn.2794585
17. Hagiu, A., Wright, J.: Multi-sided platforms. Int. J. Ind. Organ. **43** (2015). https://doi.org/10.2139/ssrn.2794582
18. Holgersson, M., Baldwin, C.Y., Chesbrough, H., Bogers, M.L.A.M.: The forces of ecosystem evolution. Calif. Manag. Rev. **64**(3), 5–23 (2022). https://doi.org/10.1177/00081256221086038
19. KITT4SME D1.1 Stakeholder analysis: (2021) https://kitt4sme.eu/wp-content/uploads/2021/09/kitt4sme-d1.1-stakeholder-analysis.pdf. Accessed 12 July 2023
20. KITT4SME Trifold-brochure: (2021) https://kitt4sme.eu/wp-content/uploads/2021/03/Trifold-brochure-KITT4SME.pdf.pdf. Accessed 12 July 2023
21. Osterwalder, A., Pigneur, Y.: Business Model Generation: a Handbook for Visionaries, Game Changers, and Challengers. Wiley, New Jersey (2010)
22. Wang, Y., Tang, J., Jin, Q., Ma, J.: On studying business models in mobile social networks based on two-sided market (TSM). J. Supercomput. **70**, 1297–1317 (2014). https://doi.org/10.1007/s11227-014-1228-4
23. Zhao, Y., von Delft, S., Morgan-Thomas, A., Buck, T.: The evolution of platform business models: exploring competitive battles in the world of platforms. Long Range Plan. **53**(4), 101892 (2020). https://doi.org/10.1016/j.lrp.2019.101892

Self-Reconfiguration for Smart Manufacturing Based on Artificial Intelligence: A Review and Case Study

Yarens J. Cruz ⓘ, Fernando Castaño ⓘ, Rodolfo E. Haber ⓘ,
Alberto Villalonga ⓘ, Krzysztof Ejsmont ⓘ, Bartlomiej Gladysz ⓘ,
Álvaro Flores ⓘ, and Patricio Alemany

1 Introduction

In the context of manufacturing systems, reconfiguration refers to the practice of changing a production system or process to meet new needs or to improve its performance. This might involve varying the structure of the production process, the order of the steps in which operations are executed, or the manufacturing process itself to make a different product.

Reconfiguration may be necessary for several reasons, including changes in raw material availability or price, changes in consumer demand for a product, the need to boost productivity, save costs, or improve product quality, among others. It is a complex process that requires careful planning and coordination to ensure that production is not disrupted and that the changes result in the desired outcomes. In return, it may offer substantial advantages including enhanced product quality, reduction of waste, and greater productivity, making it a crucial strategy for enterprises trying to maintain their competitiveness in a rapidly changing market.

Y. J. Cruz (✉) · F. Castaño · R. E. Haber · A. Villalonga
Centro de Automática y Robótica (CSIC-Universidad Politécnica de Madrid), Madrid, Spain
e-mail: y.cruz@car.upm-csic.es; fernando.castano@car.upm-csic.es;
rodolfo.haber@car.upm-csic.es; alberto.villalonga@car.upm-csic.es

K. Ejsmont · B. Gladysz
Institute of Production Systems Organization, Faculty of Mechanical and Industrial Engineering, Warsaw University of Technology, Warsaw, Poland
e-mail: krzysztof.ejsmont@pw.edu.pl; bartlomiej.gladysz@pw.edu.pl

Á. Flores · P. Alemany
Rovimática SL, Córdoba, Spain
e-mail: alvaro.flores@rovimatica.eu; patricio.alemany@rovimatica.eu

J. Soldatos (ed.), *Artificial Intelligence in Manufacturing*,
https://doi.org/10.1007/978-3-031-46452-2_8

Self-reconfiguration is the capacity of a manufacturing system to autonomously modify its configuration or structure to respond to dynamic requirements. This concept is frequently linked to the development of modular and adaptive manufacturing systems. These systems exhibit high flexibility, efficiency, and adaptability by allowing the self-reconfiguration of their assets. However, self-reconfiguration is not directly applicable to all manufacturing systems. To implement self-reconfiguration, a particular level of technological maturity is required, including the following requirements [1]:

- *Modularity*: The system is made up of a collection of standalone components.
- *Integrability*: The components have standard interfaces that facilitate their integration into the system.
- *Convertibility*: The structure of the system can be modified by adding, deleting, or replacing individual components.
- *Diagnosability*: The system has a mechanism for identifying the status of the components.
- *Customizability*: The structure of the system can be changed to fit specific requirements.
- *Automatability*: The system operation and modifications can be carried out without human intervention.

Additionally, self-reconfiguration may involve a variety of techniques and technologies, including IT infrastructure, robotic systems, intelligent sensors, and advanced control algorithms. These technologies enable machines to automatically identify and select the appropriate components or configurations needed to complete a given task, without requiring manual intervention or reprogramming. However, in some practical scenarios, human validation is still required before executing the reconfiguration.

Self-reconfiguration in manufacturing typically focuses on process reconfiguration and capacity reconfiguration with success stories in the automotive industry. Process reconfiguration involves changes in the manufacturing process itself, such as changing the sequence of operations or the layout of the production line, as well as modifications to the equipment. On the other hand, capacity reconfiguration involves adjusting the capacity of the manufacturing system to meet changes in demand. This may involve adding or removing production lines, or modifying the parameters of machines. It should be noted that modifying the parameters of existing equipment can increase production throughput without requiring significant capital investment; however, it may also require changes to the production process, such as modifying the material flow or introducing new quality control measures.

2 Reconfiguration in Manufacturing

2.1 Precursors of Reconfigurable Systems: Flexible Manufacturing Systems

Current self-reconfigurable manufacturing systems are the result of the evolution of ideas that emerged more than 50 years ago. During the 1960s and 1970s, the production methods were primarily intended for mass production of a limited range of products [2]. Due to their rigidity, these systems needed a significant investment of time and resources to be reconfigured for a different product. During that period, supported by the rapid advancements and affordability of computer technology, the concept of flexible manufacturing system (FMS) emerged as a solution to address this scenario [3]. FMSs are versatile manufacturing systems, capable of producing a diverse array of products utilizing shared production equipment. These systems are characterized by high levels of automation and computer control, enabling seamless adaptation for manufacturing different goods or products.

FMSs typically consist of a series of integrated workstations, each containing a combination of assets. These workstations are connected by computer-controlled transport systems that can move raw materials, workpieces, and finished products between workstations. When FMSs were introduced, they were primarily focused on achieving reconfigurability through the use of programmable controllers and interchangeable tooling. These systems may be configured to carry out a variety of manufacturing operations such as milling, drilling, turning, and welding. FMSs can also incorporate technologies such as computer-aided design/manufacturing (CAD/CAM) and computer numerical control (CNC) to improve efficiency and quality. This paradigm has been widely adopted in industries such as automotive [4], aerospace [5], and electronics [6] and continues to evolve with advances in technology.

However, despite the adaptability to produce different products, the implementation of FMSs has encountered certain drawbacks such as lower throughput, high equipment cost due to redundant flexibility, and complex design [7]. In addition, they have fixed hardware and fixed (although programmable) software, resulting in limited capabilities for updating, add-ons, customization, and changes in production capacity [3].

2.2 Reconfigurable Manufacturing Systems

Although FMSs can deal with the market exigence for new products or modifications of existing products, they cannot efficiently adjust their production capacity. This means that if a manufacturing system was designed to produce a maximum number of products annually and, after 2 years, the market demand for the product is reduced to half, the factory will be idle 50% of the time, creating a big financial loss.

On the other hand, if the market demand for the product surpasses design capability and the system is unable to handle it, the financial loss can be even greater [8]. To handle such scenarios, during the 1990s, a new type of manufacturing system known as reconfigurable manufacturing system (RMS) was introduced. RMSs adhere to the typical goals of production systems: to produce with high quality and low cost. However, additionally, they also aim to respond quickly to market demand, allowing for changes in production capacity. In other words, they strive to provide the capability and functionality required at the right time [3]. This goal is achieved by enabling the addition or removal of components from production lines on demand.

Design principles such as modularity, integrability, and open architecture control systems started to take more significance with the emergence of RMSs, given the relevance of dynamic equipment interconnection in these systems [9]. Considering their advantages, RMSs have been applied to the manufacturing of medical equipment [10], automobiles [11], food and beverage [12], and so on. Because they require less investments in equipment and infrastructure, these systems often offer a more cost-effective alternative to FMSs.

Although these systems can adapt to changing production requirements, the reconfiguration decisions are usually made or supervised by a human, which means the systems cannot autonomously reconfigure themselves. This gives more control to the plant supervisor or operator, but the downside is that it limits the response speed.

2.3 Evolution Towards Self-Reconfiguration

As technology advanced and the demands of manufacturing increased, production systems began to incorporate more sophisticated sensing, control, and robotics capabilities. This allowed them to monitor and adjust production processes in real time, adapt to changes in the manufacturing environment, and even reconfigure themselves without human intervention. This shift from reconfigurable to self-reconfigurable systems was driven by several technological advancements:

- *Intelligent sensors*: sensors that are capable of not only detecting a particular physical quantity or phenomenon but also processing and analyzing the data collected to provide additional information about the system being monitored [13].
- *Adaptive control*: control systems that can automatically adjust the manufacturing process to handle changes in the production environment while maintaining optimal performance [14].
- *Autonomous robots*: robots that can move and manipulate objects, work collaboratively, and self-reconfigure. These robots can be used to assemble components, perform quality control checks, and generate useful data for reconfiguring production lines [15].

- *Additive manufacturing*: 3D printing and additive manufacturing techniques allow to create complex and customized parts and structures on demand, without the need for extensive changes in the production system. Additionally, this technique is very useful for quick prototyping [16].

Compared to conventional RMSs, self-reconfigurable manufacturing systems enable to carry out modifications to the production process in a faster and more autonomous way [17]. Today, these systems are at the cutting edge of advanced manufacturing, allowing the development of extremely complex, specialized, and efficient production systems that require little to no human involvement. Self-reconfiguration is receiving significant attention in the context of Industry 4.0, where the goal is to create smart factories that can communicate, analyze data, and optimize production processes in real time [18].

3 Current Approaches

Currently, there are several approaches for designing self-reconfiguration solutions including computer simulation, which is one of the most reported in the literature with proof-of-concepts based on simulation results. Other alternative techniques include those based on artificial intelligence (AI), which provide powerful methods and tools to deal with uncertainty, such as fuzzy and neuro-fuzzy approaches, machine learning and reinforcement learning strategies. These approaches are not mutually exclusive and, in many cases, are used in a complementary way.

3.1 Computer Simulation

Computer simulation is a particularly valuable tool for the design and optimization of self-reconfigurable manufacturing systems. In this context, these tools aim to enhance the system's responsiveness to changes in production requirements. The recent increase in computational capacities has enabled the testing of various configurations and scenarios before their actual implementation [19]. Currently, commercial applications such as AutoMod, FlexSim, Arena, Simio, and AnyLogic, among others, allow to create high-fidelity simulations of industrial processes [20], that even include three-dimensional recreations of factories for use in augmented/virtual reality applications. Computer simulation becomes a powerful tool when integrated with the production process it represents. Based on this idea, digital twins have gained significant attention in both industry and academia [21]. Digital twins enable real-time data integration from the production process into the simulation, replicating the actual production environment. By evaluating different options and identifying the optimal configuration for the new scenarios, digital twins provide feedback to the production process, facilitating real-time modifications.

3.2 Fuzzy Systems

Fuzzy logic is a mathematical framework that can be used to model and reason with imprecise or uncertain data. This capability makes fuzzy logic particularly useful in situations where the system may not have access to precise data or where the data may be subject to noise or other sources of uncertainty. In the context of self-reconfiguration, fuzzy systems can be used to model the behavior of the physical processes and make decisions about how to reconfigure them based on imprecise data. For instance, it is often very complex to assign a precise value to indicators such as expected market demand, product quality, or energy consumption [22]. These variables can be assigned to fuzzy membership functions and then, following predefined rules, combined using fuzzy operators to determine how the production system should be optimally reconfigured depending on the available data.

3.3 Data-Driven Methods

Data-driven methods deal with the collection and analysis of data, the creation of models, and their use for decision-making. This approach is extensively applied when historical data of the production process is available. By using data analytics, it is possible to identify bottlenecks or the inefficient use of assets in the production process. Also, data-driven methods make extensive use of machine learning algorithms for modeling the production process behavior [23]. Machine learning methods can be trained with datasets containing a large number of features and samples, learning to identify correlations, patterns, and anomalies that are beyond human perception [24]. Moreover, by collecting new data of the production process, machine learning models can be retrained or fine-tuned to improve their performance over time. Once the machine learning model has been trained with production data, it can be used as an objective function of an optimization algorithm to make decisions about how to reconfigure the manufacturing process to optimize desired indicators.

3.4 Reinforcement Learning

Reinforcement learning is a subfield of machine learning that has shown great capacity in the development of algorithms for autonomous decision-making in dynamic and complex environments. In reinforcement learning, an agent learns to make decisions based on feedback from the environment. The agent performs actions in the environment and receives feedback in the form of rewards or penalties. The goal of the agent is to maximize the cumulative reward over time by learning which actions are most likely to lead to positive outcomes. Self-reconfigurable manufacturing systems present a unique challenge for reinforcement learning algorithms because the environment is constantly changing [25]. The agent should be able to adapt to changes in the production environment, such as changes in demand or changes in the availability of resources. The agent can learn which modules are more effective for specific tasks and reconfigure itself accordingly [18]. Another important benefit of using reinforcement learning is the ability to learn from experience. These algorithms can learn from mistakes and errors and try to avoid repeating them.

4 Lighthouse Demonstrator: GAMHE 5.0 Pilot Line

To evaluate how AI tools can be applied for self-reconfiguration in manufacturing processes and how they can be integrated with one another, an Industry 4.0 pilot line was chosen for demonstration. The selected pilot line was the GAMHE 5.0 laboratory, which simulates the slotting and engraving stages of the production process of thermal insulation panels. Figure 1 illustrates the typical workflow of the process. Initially, a robot picks up a panel and positions it in a machining center to create slots on all four sides. Subsequently, the same robot transfers the panel to a conveyor belt system that transports it to a designated location, where a second robot takes over the handling of the panel. Next, the panel is positioned in a visual inspection area by the robot. If the slotting is deemed correct, the panel is then moved to a second machining center for the engraving process. Finally, the robot transfers the panel to a stack of processed panels.

Occasionally, due to poor positioning in the slotting process, some sides of the panels are not slotted or the depth of the slot is smaller than required. In those cases, the visual inspection system should detect the irregularity and the workflow of the process should be modified to repeat the slotting process. Figure 2 illustrates this

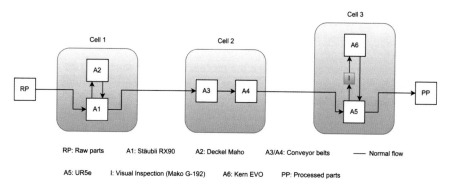

Fig. 1 Normal workflow of the pilot line

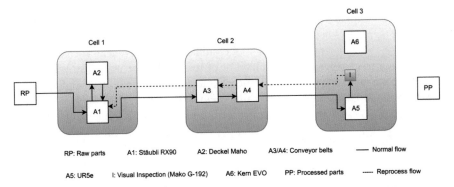

Fig. 2 Workflow for reprocessing in the pilot line

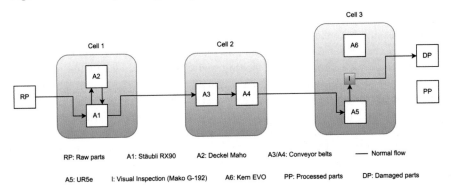

Fig. 3 Workflow for defective panels in the pilot line

situation. Once the slotting irregularities are corrected, the system continues with the normal workflow.

In some cases, the slotting process may cause damage to the panels. This can happen when working with new materials or previously unverified machining configurations. In those cases, the visual inspection system should detect that the panel is damaged and it should be sent directly to a stack of damaged parts. Figure 3 shows this situation.

Making accurate decisions about the process workflow depending on the quality of products, specifically on the result of the slotting process, has a direct impact on the productivity of the pilot line. For instance, in cases where a panel is damaged during slotting, it is crucial to remove it from the production line to prevent unnecessary time and resources from being spent on machining it during the engraving stage. To achieve this, the presence of a reliable visual inspection system becomes essential. Although a deep learning classifier could be used for this task, one drawback is that it is very hard to understand how the decision is made. For this reason, it is proposed a deep learning segmentation model, whose function is to separate the desired areas of the images from the unwanted regions.

The output of a segmentation model provides a pixel-level understanding of objects and their boundaries in an image, enabling a detailed visual interpretation of the model prediction. Then, using the segmentation result, a reasoned decision can be made, making the outcome of the system more interpretable. Section 4.1 deals with this situation.

On the other hand, a common situation is that the pilot line should deal with small batches of panels made of different materials and with different dimensions. Thus, the configuration of the assets for reaching an optimal performance varies frequently. For dealing with this situation a self-reconfiguration approach based on automated machine learning (AutoML) and fuzzy logic is proposed. Although the approach proposed in this work is generalizable to multiple objectives, for the sake of simplicity, the improvement of only one key performance indicator (KPI) will be considered. Sections 4.2 and 4.3 cover this topic.

4.1 Deep Learning-Based Visual Inspection

The segmentation model developed for application in the pilot line intends to separate the side surface of the panel from other elements within an image, allowing for later decisions on the panel quality and modifications of the process workflow. This model is based on a U-net architecture, which consists of an encoder path that gradually downsamples the input image and a corresponding decoder path that upsamples the feature maps to produce a segmentation map of the same size as the input image. This network also includes skip connections between the encoder and decoder paths that allow to retain and fuse both high-level and low-level features, facilitating accurate segmentation and object localization [26].

A dataset containing 490 images with their corresponding masks was prepared for training and evaluating the model. The image dataset was split into three subsets: training (70% of the data), validation (15% of the data), and testing (15% of the data). In this case, the validation subset serves the objective of facilitating early stopping during training. This means that if the model's performance evaluated on the validation subset fails to improve after a predetermined number of epochs, the training process is halted. By employing this technique, overfitting can be effectively mitigated and the training time can be significantly reduced.

A second version of the dataset was prepared by applying data augmentation to the training set while keeping the validation and test sets unchanged. The dataset was augmented using four transformations: horizontal flip, coarse dropout, random brightness, and random contrast. This helps increase the number of training examples, improving the model's prediction capability and making it more robust to noise. Using the two versions of the dataset, two models with the same architecture were trained. Table 1 presents the output of the two models for three examples taken from the test set. As can be observed, the predictions obtained with the model trained on the augmented dataset are significantly better than those obtained with the model

Table 1 Segmentation results for sample images

Sample	Original image	Ground truth	Prediction of the model trained on the original dataset	Prediction of the model trained on the augmented dataset
1				
2				
3				

Table 2 Metric values obtained by segmentation models on the test set

Model	Accuracy	F1 score	Jaccard index	Precision	Recall
Trained on the original dataset	0.892	0.639	0.563	0.997	0.564
Trained on the augmented dataset	0.995	0.992	0.984	0.99	0.993

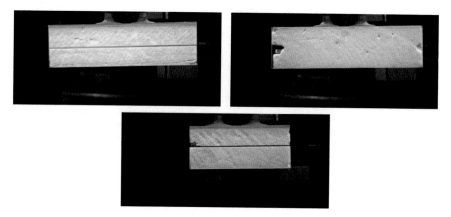

Fig. 4 Squared contours detection for a compliant panel, a panel without slot, and a damaged panel, respectively

trained on the original dataset. This is also confirmed by the values obtained in several metrics, which are shown in Table 2.

After the image is segmented by the deep learning model, a second algorithm is used. Here, a convex hull is adjusted to each separate contour in the segmented image. Then, a polygonal curve is generated for each convex hull with a precision smaller than 1.5% of the perimeter of the segmented contour. Finally, if the polygonal curve has four sides, it is drawn over the original image. After this procedure, if two rectangles were drawn over the image it is assumed that the slotting was correct and the panel did not suffer any significative damage; thus, it can be sent to the next stage of the line. On the other hand, if only one rectangle was drawn, it is assumed that the slotting was not carried out or the panel was damaged during this process. Figure 4 shows the results obtained for illustrative cases of a compliant panel, a panel with missing slots, and a damaged panel, respectively. If only one rectangle was drawn, depending on its size and location, the panel will be sent to the slotting stage again or removed from the line. This method was applied to the test set images and in all the cases the output produced matched the expected output.

4.2 Automating the Machine Learning Workflow

As outlined in the previous sections, the working conditions of the pilot line are subject to rapid variations. To effectively address these variations and generate

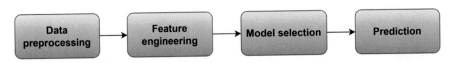

Fig. 5 General machine learning steps

optimal parametrizations for the assets, machine learning emerges as a promising tool.

The usual machine learning workflow is composed of a series of steps that are executed one by one by a team of specialists. However, this workflow can be automated. This research area is known as AutoML and recently has gained considerable attention. AutoML plays a crucial role in streamlining workflows, saving time, and reducing the effort required for repetitive tasks, thereby enabling the creation of solutions even for nonexperts. Noteworthy tools in this domain include Google Cloud AutoML, auto-sklearn, Auto-Keras, and Azure AutoML, among others. Typically, these tools encompass various stages, from data preprocessing to model selection. Moreover, in line with the automation philosophy of these systems, the process optimization step can also be integrated. This way the system would receive a dataset and return the parameter values that make the process work in a desired regime. Considering this idea, an end-to-end AutoML solution has been developed to be applied to GAMHE 5.0 pilot line. The following subsections describe the typical machine learning workflow, as well as the specificities of its different steps and how AutoML can be used for optimizing the production process.

4.2.1 Typical Machine Learning Workflow

Machine learning aims to create accurate and reliable models capable of identifying complex patterns in data. The creation and exploitation of these models is typically achieved through a series of steps that involve preparing the dataset, transforming the data to enhance its quality and relevance, selecting and training an appropriate machine learning model, evaluating the model's performance, and deploying the model in a real-world setting. Figure 5 depicts these steps. By following this workflow, machine learning practitioners can build models that harness the power of data-driven learning, enabling them to effectively derive meaningful insights and make accurate predictions in practical applications.

Data Preprocessing

Data preprocessing is the initial step in the creation of a machine learning system. The data to be used may have a variety of sources and formats, thus it should be prepared before being used by any algorithm. If data are originated from different sources, it must be merged into a single dataset. Furthermore, most methods are not designed to work with missing data, so it is very common to remove samples with missing information. Preprocessing may also include filtering data to remove noise, which can result later in more robust models. In this stage, the data may be transformed to a format that is suitable for analysis, which can include operations such as normalization, bucketizing, and encoding. Finally, one common operation carried out in this stage is splitting. This refers to the partition of the dataset into two subsets, which will be used for training and evaluation purposes. Additionally, a third subset can be created if it is planned to carry out a hyperparameter optimization or neural architecture search over the model.

Feature Engineering

The goal of the feature engineering stage is to convert raw data into relevant features that contain the necessary information to create high-quality models. One of the most interesting techniques that can be used in this stage is feature selection. Feature selection aims to determine which features are the best predictors for a certain output variable. Then, when these features are selected, they can be extracted from the original dataset to build a lower dimensional dataset, allowing to build more compact models with better generalization ability and reduced computational time [27, 28]. Typically, for problems with numerical input and output variables, Pearson's [29] or Spearman's correlation coefficients [30] are used. If the input is numerical but the output is categorical, then the analysis of variance (ANOVA) [31] or Kendall's rank coefficient [32] are employed. Other situations may require the use Chi-squared test or mutual information measure [33].

Other techniques that can be applied in the feature engineering stage include feature creation and dimensionality reduction. Feature creation implies creating new features either by combining the existing ones or by using domain knowledge [34]. On the other hand, dimensionality reduction techniques such as principal component analysis (PCA) or t-distributed stochastic neighbor embedding (t-SNE) algorithms are used to map the current data in lower dimensional space while retaining as much information as possible [35].

Model Selection

The model selection step implies the creation, training, and evaluation of different types of models to, in the end, select the most suitable for the current situation. This practice is carried out since it does not exist a methodology for determining a priori which algorithm is better for solving a problem [36]. Therefore, the most adequate model may vary from one application to another as in the following cases: long short-term memory network (LSTM) [37], multilayer perceptron (MLP) [38], support vector regression (SVR) [39], Gaussian process regression (GPR) [40], convolutional neural network (CNN) [41], gradient boosted trees (GBT) [42]. The number and types of models to explore in this stage will depend on the characteristics of the problem and the available computational resources. The selection of the model is carried out taking into consideration one or more metrics. For regression problems is common to rely on the coefficient of determination (R^2), mean squared error (*MSE*), and mean absolute percentage error (*MAPE*), among other metrics [43]. On the other hand, for classification problems, typical metrics are accuracy, recall, precision, F1-score, and so on.

Optionally, this stage can also include hyperparameter optimization. Hyperparameters determine a model's behavior during training and, in some cases, also how its internal structure is built. They are set before a model is trained and cannot be modified during training. The selection of these values can greatly affect a model's performance. However, finding an optimal or near-optimal combination of hyperparameters is not a trivial task and, usually, it is computationally intensive. The most commonly used techniques for this task include grid search, random search, Bayesian optimization, and so on.

4.2.2 Process Optimization

Once a model has been created for representing a process, it can be used for optimizing it. Assuming the model exhibits robust predictive capabilities and the constraints are accurately defined, various input values can be evaluated in the model to determine how the system would respond, eliminating the need for conducting exhaustive tests on the actual system. In other words, the model created can be embedded as the objective function of an optimization algorithm for finding the input values that would make the production process work in a desired regime. In this context, popular strategies such as particle swarm optimization [44], simulated annealing [45], evolutionary computation [46], and Nelder-Mead [47], among others, are commonly employed.

4.2.3 Application of AutoML to the Pilot Line

To apply an AutoML methodology to the selected pilot line, it is essential to collect operational data from the runtime system under varying asset parametrization. This data should include recorded measurements of variables and KPIs. Since not all the collected data have to be necessarily recorded using the same rate, it is necessary to transform the data to the same time base. This is commonly done by downsampling or averaging the data recorded with a higher rate to match the time base of the data recorded with a lower rate. In this case, averaging was used. Once the historical dataset has been prepared, an AutoML methodology can be applied. While typical AutoML methodologies automate the steps shown in Fig. 5, the proposed methodology also includes the process optimization procedure by embedding the selected model as objective function of an optimization algorithm for automatically finding the assets' configuration as depicted in Fig. 6.

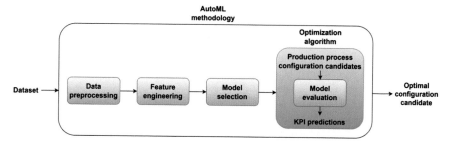

Fig. 6 Overall description of the proposed AutoML methodology

First, in the data preprocessing step, the dataset is inspected searching for missing values. If any are found, the corresponding sample is eliminated. Next, the features' values are standardized and the dataset is divided into training and validation sets. In this case, hyperparameter optimization was not implemented for making the methodology applicable in scenarios with low computational resources. For this reason, a test set is not required. Following that, feature selection is carried out by computing the Pearson's correlation coefficient (r) individually between each feature and the output variable on the training set, using the following equation:

$$r = \frac{\sum_{i=1}^{n} (x_i - \overline{x})(y_i - \overline{y})}{\sqrt{\sum_{i=1}^{n} (x_i - \overline{x})^2} \sqrt{\sum_{i=1}^{n} (y_i - \overline{y})^2}}$$

where n is the number of samples, x_i represents the value of the i-th sample of feature x, y_i represents the value of the i-th sample of the output variable, and $\overline{x}$ and $\overline{y}$ represent the mean of the respective variables.

Pearson's correlation coefficient is a univariate feature selection method commonly used when the inputs and outputs of the dataset to be processed are numerical [48]. By using this method, it is possible to select the features with higher predictive capacity, resulting not only in a reduction of the dimensionality of data but also leads to more compact models with better generalization ability and reduced computational time [27, 28]. In the proposed approach, the features for which $|r| > 0.3$ are selected as relevant predictors, and the rest are discarded from both, the training and validation sets. Typically, a value below the 0.3 threshold is considered an indicator of low correlation [49]. During the application of the AutoML methodology to the data of the pilot line for improving the throughput, the number of features was reduced from 12 to 7. This intermediate result is important to guide the technicians on which parameters they should focus on while looking for a certain outcome.

The next step involves model selection. Among the different models to evaluate in the proposed approach are MLP, SVR, GPR, and CNN, which have been previously used for modeling industrial KPIs [40, 50–52]. Table 3 presents details of these models. Each one of these models is trained on the training set and then

Table 3 Details of the evaluated models

Model	Details
MLP	Architecture: Fully connected layer (128 units, ReLU activation) + Fully connected layer (64 units, ReLU activation) + Fully connected layer (1 unit, linear activation), Optimizer: RMSprop, Learning rate: 0.001, Epochs: 5000
SVR	Kernel: rbf, C: 1.0, Epsilon: 0.2, Tolerance: 0.001
GPR	Kernel: Dot Product + White Kernel, Alpha: 1e-10
CNN	Architecture: 1-D Convolution layer (64 filters, kernel size: 3, strides: 1, padding: same) + 1-D Max pooling layer (pool size: 2, strides: 1, padding: valid) + Flatten layer + Fully connected layer (64 units, ReLU activation) + Dropout layer (dropout rate: 0.1) + Fully connected layer (32 units, ReLU activation) + Fully connected layer (1 unit, linear activation), Optimizer: RMSprop, Learning rate: 0.001, Epochs: 5000

they are evaluated on the validation set. The metric used for comparison was the coefficient of determination (R^2). After this process is finished, the model that produced the best result is selected. The model selected during the application of the methodology to the pilot line was MLP with $R^2 = 0.963$ during validation. The R^2 value for the remaining candidate models was 0.958 for GPR, 0.955 for CNN, and 0.947 for SVR. One of the enablers of these results was the feature selection process, which allowed to retain the relevant predictors.

Finally, an optimization method is applied for determining the most favorable parametrization of the production process to minimize or maximize the desired KPI using the selected model as the objective function. In this case, the goal is to maximize throughput. The optimization is carried out using random search, which is a simple, low-complexity, and straightforward optimization method [53]. This method can be applied to optimizing diverse types of functions, even those that are not continuous or differentiable. It has been proven that random search is asymptotically complete, meaning that it converges to the global minimum/maximum with probability one after indefinitely run-time computation and, for this reason, it has been applied for solving many complex problems [54]. One aspect to consider before executing the optimization is that the feasible range of the parameters must be carefully decided to prevent the result of the optimization from being invalid. In the case analyzed, where the objective is to maximize the throughput of the pilot line, the obvious choice is to make the assets work at the maximum speed within the recommended ranges. To evaluate if the proposed methodology was capable of inferring this parametrization, during the preparation of the dataset the samples where all the assets were parametrized with the maximum speed were intentionally eliminated. As desired, the result of the methodology was a parametrization where all the assets were set to the maximum speed, yielding an expected throughput value of 163.37 panels per hour, which represents an expected improvement of 55.1% with respect to the higher throughput value present in the dataset. It is noticeable that the higher throughput value of the samples that were intentionally eliminated from the dataset is 158.52. The reason why the proposed methodology slightly overestimates this value is that the model is not perfectly accurate.

4.3 Fuzzy Logic-Based Reconfigurator

Once the parametrization of the assets has been determined by the AutoML methodology to meet a desired KPI performance, it is important to ensure that the system will continue to work as desired. Unfortunately, some situations may prevent the system from functioning as intended. For instance, a degradation in one of the assets may result in a slower operation, reducing the productivity of the entire line. For such cases, a fuzzy logic-based reconfigurator is developed. The intuition behind this component is that if the behavior of some assets varies from their expected performance, the reconfigurator can modify the parameters of the assets to make them work in the desired regime again, as long as the modification of the parameters is within a predefined safety range. Additionally, if the deviation from the expected performance is significant, the component should be able to detect it and inform the specialists that a problem needs to be addressed.

The proposed reconfigurator has two inputs and generates three outputs using the Mamdani inference method [55]. These variables are generic, so the reconfigurator can use them without any modification to try to keep each asset's throughput level constant. The first input is the deviation from nominal production time (ΔT) and its safety range was defined as $\pm 50\%$ of the nominal production time. The second input is the change in the trend of the deviation from nominal production time (ΔT^2) and its safety range was defined as $\pm 20\%$ of the nominal production time. There is an instance of these two variables for each asset in the line and they are updated whenever a panel is processed. These values are normalized in the interval $[-1, 1]$ before being used by the reconfigurator. Figure 7 presents the membership functions defined for the two inputs.

On the other hand, the first output is the operation that the reconfigurator must apply to the current asset's working speed ($Reco1$). If the operation is *Increase* or *Decrease*, the values in the interval $[-1, 1]$ are denormalized to a range comprising $\pm 50\%$ of the nominal asset speed. The second output represents the timing when the modifications should be applied ($Reco2$), and the third output represents the operation mode ($Reco3$), which specifies if the previous reconfigurator outputs

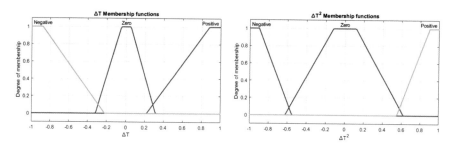

Fig. 7 Membership functions for inputs ΔT and ΔT^2

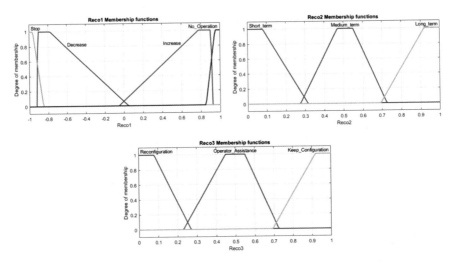

Fig. 8 Membership functions for outputs *Reco1*, *Reco2*, and *Reco3*

should be applied automatically, presented as recommendations for the operators, or ignored. Figure 8 shows the membership functions of the three outputs.

Once the membership functions of the input and output variables were defined, a rule base was created for each output variable. Each rule base is formed by nine If–Then rules that associate a combination of the input membership functions with an output membership function, as in the following example:

If ΔT is *Negative* **And** ΔT^2 is *Negative* **Then** *Reco1* is *Increase*

The defined rule bases allow to obtain the output surfaces illustrated in Fig. 9 for the fuzzy inference systems corresponding to each output variable.

To evaluate the reconfigurator, the nominal speed of each asset was set to 70% of its maximum speed and several disturbances were emulated. The first one was reducing the speed of all assets to 50% of their maximum speed, the second increasing the speed of all assets to their maximum speed, and finally, the speed of all assets was set to 30% of their maximum speed. As expected after the first disturbance the system recommended increasing the speed, after the second it recommended decreasing the speed, and after the third it recommended stopping the production. The results are shown in Table 4.

5   Conclusions

This work has addressed self-reconfigurable manufacturing systems from both theoretical and practical points of view, emphasizing how AI is applied to them. The emergence and evolution until the current state of these systems have been

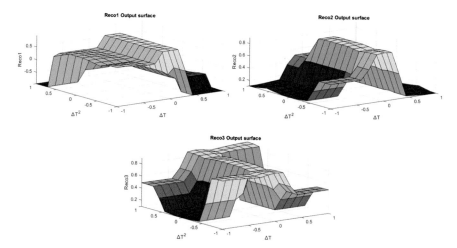

Fig. 9 Output surfaces of *Reco1*, *Reco2*, and *Reco3*

presented. Likewise, their potential benefits such as improved responsiveness, flexibility, and adaptability have been analyzed. Current approaches for implementing self-reconfiguration in manufacturing have also been discussed. Additionally, the application of self-reconfiguration and AI techniques to a pilot line was tested. First, the integration in the pilot line of an AI-based solution for visual inspection was evaluated. This component has a direct relation with the workflow of the pilot line, thus influencing the productivity. Two segmentation models were trained for the visual inspection task and the best one, with an accuracy of 0.995 and a F1 score of 0.992, was deployed in the pilot line, enabling the correct handling of products. Furthermore, an AutoML approach that includes generating the models and optimizing the production process was used for determining the optimal parametrization of the line. This way, a model with $R^2 = 0.963$ was obtained and the expected improvement in throughput with respect to the data seen during training is 55.1%, which matches the values reached in real production at maximum capacity. Then, a fuzzy logic-based reconfigurator was used for dealing with the degradation in performance. This component demonstrated a correct behavior and showed robustness when tested against three different perturbations. The findings of this study suggest that self-reconfiguration is a key area of research and development in the field of advanced manufacturing. Future research will explore additional applications of self-reconfiguration in different manufacturing contexts.

Table 4 Fuzzy logic reconfigurator results for disturbances

Asset	Nominal speed	Disturbance	Reco1	Reco2	Reco3
Stäubli RX90	7.7 m/s	1	Increase to 7.05 m/s	Medium term	Keep configuration
		2	Decrease to 8.94 m/s	Medium term	Keep configuration
		3	Stop	Short term	Operator assistance
Deckel Maho	7200 mm/min	1	Increase to 7027.8 mm/min	Medium term	Keep configuration
		2	Decrease to 8132.68 mm/min	Medium term	Keep configuration
		3	Stop	Short term	Operator assistance
Conveyor belts	5040 rpm	1	Increase to 4013.6 rpm	Medium term	Keep configuration
		2	Decrease to 5904 rpm	Medium term	Keep configuration
		3	Stop	Short term	Operator assistance
UR5e	126°/s	1	Increase to 115.28°/s	Medium term	Keep configuration
		2	Decrease to 146.1°/s	Medium term	Keep configuration
		3	Stop	Short term	Operator assistance
Kern EVO	11,200 mm/min	1	Increase to 10,250.2 mm/min	Medium term	Keep configuration
		2	Decrease to 12,992 mm/min	Medium term	Keep configuration
		3	Stop	Short term	Operator assistance

Acknowledgments This work was partially supported by the H2020 project "platform-enabled KITs of arTificial intelligence FOR an easy uptake by SMEs (KITT4SME)," grant ID 952119. The work is also funded by the project "Self-reconfiguration for Industrial Cyber-Physical Systems based on digital twins and Artificial Intelligence. Methods and application in Industry 4.0 pilot line," Spain, grant ID PID2021-127763OB-100, and supported by MICINN and NextGenerationEU/PRTR. This result is also part of the TED2021-131921A-I00 project, funded by MCIN/AEI/10.13039/501100011033 and by the European Union "NextGenerationEU/PRTR."

References

1. Hees, A., Reinhart, G.: Approach for production planning in reconfigurable manufacturing systems. Proc. CIRP. **33**, 70–75 (2015). https://doi.org/10.1016/j.procir.2015.06.014
2. Yin, Y., Stecke, K.E., Li, D.: The evolution of production systems from Industry 2.0 through Industry 4.0. Int. J. Prod. Res. **56**, 848–861 (2018). https://doi.org/10.1080/00207543.2017.1403664
3. Mehrabi, M.G., Ulsoy, A.G., Koren, Y., Heytler, P.: Trends and perspectives in flexible and reconfigurable manufacturing systems. J. Intell. Manuf. **13**, 135–146 (2002). https://doi.org/10.1023/A:1014536330551
4. Cronin, C., Conway, A., Walsh, J.: Flexible manufacturing systems using IIoT in the automotive sector. Proc. Manuf. **38**, 1652–1659 (2019). https://doi.org/10.1016/j.promfg.2020.01.119
5. Parhi, S., Srivastava, S.C.: Responsiveness of decision-making approaches towards the performance of FMS. In: 2017 International Conference of Electronics, Communication and Aerospace Technology (ICECA), pp. 276–281 (2017)
6. Bennett, D., Forrester, P., Hassard, J.: Market-driven strategies and the design of flexible production systems: evidence from the electronics industry. Int. J. Oper. Prod. Manag. **12**, 25–37 (1992). https://doi.org/10.1108/01443579210009032
7. Singh, R.K., Khilwani, N., Tiwari, M.K.: Justification for the selection of a reconfigurable manufacturing system: a fuzzy analytical hierarchy based approach. Int. J. Prod. Res. **45**, 3165–3190 (2007). https://doi.org/10.1080/00207540600844043
8. Koren, Y.: The emergence of reconfigurable manufacturing systems (RMSs) BT - reconfigurable manufacturing systems: from design to implementation. In: Benyoucef, L. (ed.) , pp. 1–9. Springer International Publishing, Cham (2020)
9. Koren, Y., Shpitalni, M.: Design of reconfigurable manufacturing systems. J. Manuf. Syst. **29**, 130–141 (2010). https://doi.org/10.1016/j.jmsy.2011.01.001
10. Epureanu, B.I., Li, X., Nassehi, A., Koren, Y.: An agile production network enabled by reconfigurable manufacturing systems. CIRP Ann. **70**, 403–406 (2021). https://doi.org/10.1016/j.cirp.2021.04.085
11. Koren, Y., Gu, X., Guo, W.: Reconfigurable manufacturing systems: principles, design, and future trends. Front. Mech. Eng. **13**, 121–136 (2018). https://doi.org/10.1007/s11465-018-0483-0
12. Gould, O., Colwill, J.: A framework for material flow assessment in manufacturing systems. J. Ind. Prod. Eng. **32**, 55–66 (2015). https://doi.org/10.1080/21681015.2014.1000403
13. Shin, K.-Y., Park, H.-C.: Smart manufacturing systems engineering for designing smart product-quality monitoring system in the Industry 4.0. In: 2019 19th International Conference on Control, Automation and Systems (ICCAS), pp. 1693–1698 (2019)
14. Arıcı, M., Kara, T.: Robust adaptive fault tolerant control for a process with actuator faults. J. Process Control. **92**, 169–184 (2020). https://doi.org/10.1016/j.jprocont.2020.05.005
15. Ghofrani, J., Deutschmann, B., Soorati, M.D., et al.: Cognitive production systems: a mapping study. In: 2020 IEEE 18th International Conference on Industrial Informatics (INDIN), pp. 15–22 (2020)

16. Scholz, S., Mueller, T., Plasch, M., et al.: A modular flexible scalable and reconfigurable system for manufacturing of microsystems based on additive manufacturing and e-printing. Robot. Comput. Integr. Manuf. **40**, 14–23 (2016). https://doi.org/10.1016/j.rcim.2015.12.006

17. Cedeno-Campos, V.M., Trodden, P.A., Dodd, T.J., Heley, J.: Highly flexible self-reconfigurable systems for rapid layout formation to offer manufacturing services. In: 2013 IEEE International Conference on Systems, Man, and Cybernetics, pp. 4819–4824 (2013)

18. Lee, S., Ryu, K.: Development of the architecture and reconfiguration methods for the smart, self-reconfigurable manufacturing system. Appl. Sci. **12** (2022). https://doi.org/10.3390/app12105172

19. Mourtzis, D.: Simulation in the design and operation of manufacturing systems: state of the art and new trends. Int. J. Prod. Res. **58**, 1927–1949 (2020). https://doi.org/10.1080/00207543.2019.1636321

20. dos Santos, C.H., Montevechi, J.A.B., de Queiroz, J.A., et al.: Decision support in productive processes through DES and ABS in the Digital Twin era: a systematic literature review. Int. J. Prod. Res. **60**, 2662–2681 (2022). https://doi.org/10.1080/00207543.2021.1898691

21. Guo, H., Zhu, Y., Zhang, Y., et al.: A digital twin-based layout optimization method for discrete manufacturing workshop. Int. J. Adv. Manuf. Technol. **112**, 1307–1318 (2021). https://doi.org/10.1007/s00170-020-06568-0

22. Abdi, M.R., Labib, A.W.: Feasibility study of the tactical design justification for reconfigurable manufacturing systems using the fuzzy analytical hierarchical process. Int. J. Prod. Res. **42**, 3055–3076 (2004). https://doi.org/10.1080/00207540410001696041

23. Lee, S., Kurniadi, K.A., Shin, M., Ryu, K.: Development of goal model mechanism for self-reconfigurable manufacturing systems in the mold industry. Proc. Manuf. **51**, 1275–1282 (2020). https://doi.org/10.1016/j.promfg.2020.10.178

24. Panetto, H., Iung, B., Ivanov, D., et al.: Challenges for the cyber-physical manufacturing enterprises of the future. Annu. Rev. Control. **47**, 200–213 (2019). https://doi.org/10.1016/j.arcontrol.2019.02.002

25. Schwung, D., Modali, M., Schwung, A.: Self-optimization in smart production systems using distributed reinforcement learning. In: 2019 IEEE International Conference on Systems, Man and Cybernetics (SMC), pp. 4063–4068 (2019)

26. Ronneberger, O., Fischer, P., Brox, T.: U-net: convolutional networks for biomedical image segmentation BT - medical image computing and computer-assisted intervention – MICCAI 2015. In: Navab, N., Hornegger, J., Wells, W.M., Frangi, A.F. (eds.) , pp. 234–241. Springer International Publishing, Cham (2015)

27. Al-Tashi, Q., Abdulkadir, S.J., Rais, H.M., et al.: Approaches to multi-objective feature selection: a systematic literature review. IEEE Access. **8**, 125076–125096 (2020). https://doi.org/10.1109/ACCESS.2020.3007291

28. Solorio-Fernández, S., Carrasco-Ochoa, J.A., Martínez-Trinidad, J.F.: A review of unsupervised feature selection methods. Artif. Intell. Rev. **53**, 907–948 (2020). https://doi.org/10.1007/s10462-019-09682-y

29. Jebli, I., Belouadha, F.-Z., Kabbaj, M.I., Tilioua, A.: Prediction of solar energy guided by Pearson correlation using machine learning. Energy. **224**, 120109 (2021). https://doi.org/10.1016/j.energy.2021.120109

30. González, J., Ortega, J., Damas, M., et al.: A new multi-objective wrapper method for feature selection – accuracy and stability analysis for BCI. Neurocomputing. **333**, 407–418 (2019). https://doi.org/10.1016/j.neucom.2019.01.017

31. Alassaf, M., Qamar, A.M.: Improving sentiment analysis of Arabic Tweets by One-way ANOVA. J. King Saud Univ. Comput. Inf. Sci. **34**, 2849–2859 (2022). https://doi.org/10.1016/j.jksuci.2020.10.023

32. Urkullu, A., Pérez, A., Calvo, B.: Statistical model for reproducibility in ranking-based feature selection. Knowl. Inf. Syst. **63**, 379–410 (2021). https://doi.org/10.1007/s10115-020-01519-3

33. Bahassine, S., Madani, A., Al-Sarem, M., Kissi, M.: Feature selection using an improved Chi-square for Arabic text classification. J. King Saud Univ. Comput. Inf. Sci. **32**, 225–231 (2020). https://doi.org/10.1016/j.jksuci.2018.05.010

34. Lu, Z., Si, S., He, K., et al.: Prediction of Mg alloy corrosion based on machine learning models. Adv. Mater. Sci. Eng. **2022**, 9597155 (2022). https://doi.org/10.1155/2022/9597155
35. Anowar, F., Sadaoui, S., Selim, B.: Conceptual and empirical comparison of dimensionality reduction algorithms (PCA, KPCA, LDA, MDS, SVD, LLE, ISOMAP, LE, ICA, t-SNE). Comput. Sci. Rev. **40**, 100378 (2021). https://doi.org/10.1016/j.cosrev.2021.100378
36. Cruz, Y.J., Rivas, M., Quiza, R., et al.: A two-step machine learning approach for dynamic model selection: a case study on a micro milling process. Comput. Ind. **143**, 103764 (2022). https://doi.org/10.1016/j.compind.2022.103764
37. Castano, F., Cruz, Y.J., Villalonga, A., Haber, R.E.: Data-driven insights on time-to-failure of electromechanical manufacturing devices: a procedure and case study. IEEE Trans. Ind. Inform, 1–11 (2022). https://doi.org/10.1109/TII.2022.3216629
38. Mezzogori, D., Romagnoli, G., Zammori, F.: Defining accurate delivery dates in make to order job-shops managed by workload control. Flex. Serv. Manuf. J. **33**, 956–991 (2021). https://doi.org/10.1007/s10696-020-09396-2
39. Luo, J., Hong, T., Gao, Z., Fang, S.-C.: A robust support vector regression model for electric load forecasting. Int. J. Forecast. **39**, 1005–1020 (2023). https://doi.org/10.1016/j.ijforecast.2022.04.001
40. Pai, K.N., Prasad, V., Rajendran, A.: Experimentally validated machine learning frameworks for accelerated prediction of cyclic steady state and optimization of pressure swing adsorption processes. Sep. Purif. Technol. **241**, 116651 (2020). https://doi.org/10.1016/j.seppur.2020.116651
41. Cruz, Y.J., Rivas, M., Quiza, R., et al.: Computer vision system for welding inspection of liquefied petroleum gas pressure vessels based on combined digital image processing and deep learning techniques. Sensors. **20** (2020). https://doi.org/10.3390/s20164505
42. Pan, Y., Chen, S., Qiao, F., et al.: Estimation of real-driving emissions for buses fueled with liquefied natural gas based on gradient boosted regression trees. Sci. Total Environ. **660**, 741–750 (2019). https://doi.org/10.1016/j.scitotenv.2019.01.054
43. Chicco, D., Warrens, M.J., Jurman, G.: The coefficient of determination R-squared is more informative than SMAPE, MAE, MAPE, MSE and RMSE in regression analysis evaluation. PeerJ. Comput. Sci. **7**, e623 (2021). https://doi.org/10.7717/peerj-cs.623
44. Eltamaly, A.M.: A novel strategy for optimal PSO control parameters determination for PV energy systems. Sustainability. **13** (2021). https://doi.org/10.3390/su13021008
45. Karagul, K., Sahin, Y., Aydemir, E., Oral, A.: A simulated annealing algorithm based solution method for a green vehicle routing problem with fuel consumption BT - lean and green supply chain management: optimization models and algorithms. In: Weber, G.-W., Huber, S. (eds.) Paksoy T, pp. 161–187. Springer International Publishing, Cham (2019)
46. Cruz, Y.J., Rivas, M., Quiza, R., et al.: Ensemble of convolutional neural networks based on an evolutionary algorithm applied to an industrial welding process. Comput. Ind. **133**, 103530 (2021). https://doi.org/10.1016/j.compind.2021.103530
47. Yildiz, A.R.: A novel hybrid whale–Nelder–Mead algorithm for optimization of design and manufacturing problems. Int. J. Adv. Manuf. Technol. **105**, 5091–5104 (2019). https://doi.org/10.1007/s00170-019-04532-1
48. Gao, X., Li, X., Zhao, B., et al.: Short-term electricity load forecasting model based on EMD-GRU with feature selection. Energies. **12** (2019). https://doi.org/10.3390/en12061140
49. Mu, C., Xing, Q., Zhai, Y.: Psychometric properties of the Chinese version of the Hypoglycemia Fear SurveyII for patients with type 2 diabetes mellitus in a Chinese metropolis. PLoS One. **15**, e0229562 (2020). https://doi.org/10.1371/journal.pone.0229562
50. Schaefer, J.L., Nara, E.O.B., Siluk, J.C.M., et al.: Competitiveness metrics for small and medium-sized enterprises through multi-criteria decision making methods and neural networks. Int. J. Proc. Manag. Benchmark. **12**, 184–207 (2022). https://doi.org/10.1504/IJPMB.2022.121599
51. Manimuthu, A., Venkatesh, V.G., Shi, Y., et al.: Design and development of automobile assembly model using federated artificial intelligence with smart contract. Int. J. Prod. Res. **60**, 111–135 (2022). https://doi.org/10.1080/00207543.2021.1988750

52. Zagumennov, F., Bystrov, A., Radaykin, A.: In-firm planning and business processes management using deep neural network. GATR J. Bus. Econ. Rev. **6**, 203–211 (2021). https://doi.org/10.35609/jber.2021.6.3(4)

53. Ozbey, N., Yeroglu, C., Alagoz, B.B., et al.: 2DOF multi-objective optimal tuning of disturbance reject fractional order PIDA controllers according to improved consensus oriented random search method. J. Adv. Res. **25**, 159–170 (2020). https://doi.org/10.1016/j.jare.2020.03.008

54. Do, B., Ohsaki, M.: A random search for discrete robust design optimization of linear-elastic steel frames under interval parametric uncertainty. Comput. Struct. **249**, 106506 (2021). https://doi.org/10.1016/j.compstruc.2021.106506

55. Mamdani, E.H.: Application of fuzzy algorithms for control of simple dynamic plant. Proc. Inst. Electr. Eng. **121**, 1585–1588 (1974). https://doi.org/10.1049/piee.1974.0328

Part II
Multi-agent Systems and AI-Based Digital Twins for Manufacturing Applications

Digital-Twin-Enabled Framework for Training and Deploying AI Agents for Production Scheduling

Emmanouil Bakopoulos (iD)**, Vasilis Siatras** (iD)**, Panagiotis Mavrothalassitis** (iD)**,
Nikolaos Nikolakis** (iD)**, and Kosmas Alexopoulos** (iD)

1 Introduction

Production scheduling problems are essential for optimizing manufacturing processes and ensuring effective resource utilization. In other words, scheduling defines where and when production operations will be performed [1]. The production scheduling aims to optimize resource utilization, minimize the makespan, reduce global setup time, and satisfy customer demands [2]. According to Lawler et al. [3], in the majority of the time, scheduling falls under the category of non-deterministically polynomial (NP) time problems. To address the complex nature of production scheduling problem, advanced techniques have been developed. Some of these techniques are the mathematical optimization models, heuristic algorithms, and machine learning (ML) approaches. Important inputs that the methods above take into account is information as setup matrices, processing times, quantities to be produced, material availability, due dates, technical information, production capabilities of the production lines that such a technique is modelled. Additionally, the use of real-time data from the shop floor and the use of artificial intelligence (AI) techniques can improve adaptability in dynamic manufacturing environments. Moreover, using AI techniques in parallel with real-time data from the shop floor arises new challenges, i.e., real-time decision-making.

Digital twin (DT) is a technology that allows real-time monitoring and optimization of the physical environment, prosses, or assets [4]. DT is a technology that enables the user to gain insights on the model's accuracy [5]. Using data from sensors and Internet of Thing (IoT) devices, DT creates a digital replica of the physical environment sharing the same characteristics and interactions of the

E. Bakopoulos · V. Siatras · P. Mavrothalassitis · N. Nikolakis · K. Alexopoulos (✉)
Laboratory for Manufacturing Systems & Automation (LMS), Department of Mechanical
Engineering & Aeronautics, University of Patras, Rio-Patras, Greece
e-mail: alexokos@lms.mech.upatras.gr

J. Soldatos (ed.), *Artificial Intelligence in Manufacturing*,
https://doi.org/10.1007/978-3-031-46452-2_9

physical counterpart [6]. DT have been used in many different domains, including healthcare, urban planning and manufacturing [7]. In a variety of industries, the combination of real-time data and the digital replica of the physical environment promotes the decision-making and allow the industries to continuously improve their performance [8, 9].

Asset Administration Shell (AAS) technology was introduced within the Reference Architecture Model Industry 4.0 (RAMI4. 0) [10] and has become a ground-breaking idea in manufacturing of how assets are managed and used [11]. AAS are standardized models that allow industries to combine the physical assets with their digital counterparts (i.e., machines, production systems, or tools), where AASs provide a framework to control and monitor the physical assets. AI scheduling agents have been used coupled with AAS concept in the literature [9, 12, 13]. Additionally, AI scheduling agents are intelligent autonomous systems that take as an input production system information to plan resource allocation tasks [14]. The AI scheduling agents play an important role due to the fact that they can generate a real-time efficient schedule. Coupled with the AAS and DT technologies, AI scheduling agents can be used for the real-time decision-making or for predictions [14, 15].

Multi-agent system (MAS) are systems that are used to compose many autonomous agents, where these agents interact to each other [15]. MAS provides decentralized and collaborative decision-making, where it allows the collaboration of different agents. Each agent in the MAS has some capabilities, decision-making abilities and takes decisions. MAS is used in order to solve complex problems where one agent is impossible to solve. The idea of dived and conquer is used to divide the problem into subproblems, where each agent solves a subproblem, and provides solutions that are adaptable, robust, and able to handle real-time uncertainties. MAS is also combined with DT, AAS, and AI scheduling agents. In a further analysis, AASs can enable interaction between different agents, where the agents can be models as different assets.

The contributions of this work are the use of the DT in order to accurately simulate and validate the AI agents that have been developed, as well as training some of the agents. Moreover, the use of the AAS technology to exchange data between the DT and AI agents within the MAS and finally the developed AI scheduling agents that were developed and modeled based on the bicycle industry's requirements and challenges.

This chapter is organized in four sections, where the first section introduces the concepts of digital twins (DT), Asset Administration Shells (AAS), scheduling problem, and artificial intelligence (AI) applications. In second section are discussed related works. In the third section is explained the proposed MAS framework and explain the optimization tools that have been developed. In the fourth section is described the case study that the proposed framework is implemented. Finally, the last section is the conclusion of this work, where some future works are discussed.

2   Related Works

The manufacturing sector has experienced an evolution thanks to Artificial Intelligence (AI). AI offers innovative methods for increasing productivity, quality, and efficiency [16]. Due to its capacity to handle complicated issues, evaluate big amount of information, and make precise predictions, AI approaches have become utilized more and more in manufacturing. Numerous manufacturing processes, including quality assurance, preventative maintenance, supply chain management, and production scheduling, have benefited from the application of AI approaches [17].

Two of the most important tasks in the industrial sector are production planning and scheduling. For creating effective and efficient production plan or schedule, a variety of strategies, methods, and technologies have been developed and deployed. Production planning comprises considering what to do and how to do something in advance. Scheduling, on the other hand, entails allocating resources or manufacturing facilities to handle work orders. Effective production scheduling lowers production costs, boosts productivity, and lastly improves customer satisfaction. Due to their capacity to handle complicated scheduling issues and offer precise solutions, artificial intelligence (AI) systems have been gaining prominence in production scheduling. Machine learning (ML) is one of the most frequently used AI approaches [5]. More effective production planning and scheduling algorithms have been created using genetic algorithms, artificial neural networks, and reinforcement learning.

Heuristics is one of the approaches to solve the dynamic flexible job-shop scheduling problem [18]. Another popular technique, genetic algorithms, has been utilized in several research to improve production scheduling by describing the issue as a combinatorial optimization issue [19]. Nevertheless, the rise of Industry 4.0 has made ML techniques an attractive alternative to address manufacturing difficulties, due to the availability of data, powerful processing, and plenty of storage capacity. Neural networks and deep learning have gained more attention in recent years [20]. Additionally, reinforcement learning (RL), which uses experience to improve scheduling policies, has been proposed for production scheduling. The scheduling policy is represented in RL as a function that connects the state of the system at the moment with an action [21]. In conclusion, future research may concentrate on combining several AI techniques to develop production scheduling algorithms that are more potent and effective.

The digital twin (DT) interest is growing from both an academic and an industry perspective. However, the definition of that concept in the scientific literature lacks distinctiveness. DT provides virtual representations of systems along their lifecycle. Then, decisions and optimizations would be based on the same data, which is updated in parallel with the physical system [22]. DT can be briefly described as a framework or concept that combines the physical and real environment with the digital and virtual one, with the use of novel interconnection methods and technological innovations [23]. This physical to virtual connection for addressing real processes and assets to their digital representative ones can be characterized as twinning.

One of the main technologies used, in order to realize most of DT implementation approaches, is simulation [24]. As already mentioned, the idea of DT is to build a virtual version of a real system. This replica can be used to simulate and forecast how the physical system will respond to certain situations. Thus, one of the best methods to construct a virtual representation of the physical system seems to be simulation, which enables engineers to test and improve the system before it is built, lowering costs and increasing efficiency. Digital twin and simulation technology are being used more and more frequently in sectors such as manufacturing and aerospace, exhibiting their ability to completely change how complex systems are created and optimized [25].

Furthermore, digital twin implementation methods can support decision-making related to the scheduling task for a production system with potential uncertainties [26]. A crucial aspect for the development of a digital twin is the achievement of a high level of standardization and interoperability with systems outside the digital environment. The digital twin simulates some of the behaviors of the physical environment, and thus requires some kind of seamless information exchange with the physical entities and the information they provide. OPC UA is a standard that can provide standardization in the data exchange between the digital twin and production hardware, achieving real-time monitoring and control, interconnectivity, security, access control, while also data modelling and semantics [27].

The Asset Administration Shell (AAS) could be also used in order to standardize the description and management of assets. The digital twin technology can exchange information with the asset via the establishment of a common information language [28]. In addition, the AAS and OPC UA are complementary standards that can be both used to define the framework and protocol for that communication [29]; it is worth noticing that AAS is a collection of standards, namely IEC 62875, DIN SPEC 91345, IEC 62541 (OPC UA), and RAMI 4.0. In cases where the digital twin composes a higher level system such as a production line, a station, or a production system, it is usually composed of multiple assets and thus AAS models. From the digital twin side, the AAS can be the middleware for exchanging information with the assets or managing their behavior. It is important, however, to highlight that there is no standard way for describing an asset using the AAS; although the metamodel will always be the same, there is a freedom to the different submodels and submodel elements that will be selected in order to describe any different asset. It is thus usual to exploit additional information modelling standards or framework to define the specific components and information structures within the AAS metamodel – e.g., ISA-88, ISO 22400, and ISA-95.

Digital twin is not only about simulating the environment but also taking decisions over the next actions, which can then be used on the physical environment. Simulation on its own cannot address this issue, and AI agents is a way that this challenge can be solved. Multi-agent systems are preferred over centralized software components in cases where the problem is hard enough to be solved from a monolithic software component. It is a decentralized approach that breaks the problem into subproblems and each agent has access only to the subproblems compatible with its skills. In the case of production scheduling, this is a useful approach as it enables different types of scheduling problems being solved by different AI methods based on which method best satisfies the requirements. AI is a broad term and in scheduling in particular the most common methods are heuristic, metaheuristic, mathematical optimization, machine learning, reinforcement learning, and policy-making.

Mathematical optimization, also referred as mathematical programming, is an optimization model consisted of input sets and parameters, decision variables, constraints/expressions, and the objective function. Based on the constraints and the objectives, the model may be classified as linear, nonlinear, convex, integer, and mixed-integer problem, with different type of algorithms to optimize the objectives. As such, as important as the model, the algorithm that is used in order to find a both feasible and accurate solution is also crucial for the quality of the solution. The algorithms may be exact or heuristic-based, while metaheuristic methods are also popular for various optimization problems.

Heuristics have been deployed to solve various production scheduling optimization problems. A combination of constructive heuristics and iterated greedy algorithm was used to solve the distributed blocking flowshop scheduling problem (DBFSP) and lead to makespan minimization [30]. Montiel et al. (2017) proposed an approach for the stochastic optimization of mine production schedules with the use of heuristics, implementing iterative improvement by swapping periods and destinations of the mining blocks to create the final solution [31]. Heuristics can also be successfully deployed to optimize the scheduling task, aiming at reducing total energy consumption [32]. Jélvez et al. (2020) worked for a new hybrid heuristic algorithm to solve the Precedence Constrained Production Scheduling Problem (PCPSP) for an open-pit mining industry [33].

Heuristic and metaheuristic algorithms focus on an intelligent search along the solution space, which does not ensure the quality of the solution, and in complex optimization problems require flexible time delays. Deep learning methods, on the other hand, do not depend on searching the solution space, but rather predicting the solution based on patterns from historical information. Although in most cases the results are guaranteed to be fast, it is not necessarily of high quality. In reality it depends on the deep learning model that was used, the dataset quality and quantity. In some cases, there is also a dataset shortage which makes the problem even more difficult to solve. In practice, researchers may address this problem via the utilization of a system digital replica which is able to simulate the behaviors of the actual system in a realistic manner. This can support the development of either reinforcement learning methods that use the simulation as a reward retrieval plugin or for extracting artificial dataset that can then be used in supervised learning models to learn and adapt to the actual system implementation. Especially, deep reinforcement learning has showed great potential in recent years in dealing with complex scheduling optimization problems. Researchers have focused on the implementation of deep reinforcement learning techniques for production scheduling-related problems where there is lack of data, and the problem appears high complexity. The Job-Shop Scheduling Problem (JSSP) is one of the most common optimization problems related to production scheduling that the scientific community has tried to solve with the application of deep reinforcement learning. Zhang et al. (2020) developed a deep reinforcement learning agent, able to select priority dispatch rules to solve the JSSP [19]. Liu et al. (2020) followed a similar deep reinforcement learning approach to solve both the static and dynamic JSSP [34]. Rather than only solving the JSSP, there have been also solutions for the optimization of the whole production system with the use of deep Q-learning, a very popular deep reinforcement learning technique in the last decade [35].

While all the technologically innovative techniques have helped to develop smarter and more efficient systems and tools, these solutions could also be integrated in an efficient way in the actual production system through a digital twins (DT) and can help in integrating such solution to increase productivity. Villalonga et al. (2021) proposed a framework for dynamic scheduling with the use of digital twins to represent actual production assets in order to enhance decision-making [36]. Zhang et al. (2021) use the digital twin concept to gather real-time data from the shop floor and realize an effective dynamic production scheduling [37]. To achieve real-time decision-making, the implementation of a digital twin appears a great potential, since uncertain and dynamic events are addressed effectively. Dynamic interactive scheduling method can be enhanced and strengthened by the use of DT [26, 38]. However, digital twin concept can also be implemented to support production scheduling in an offline mode, such as the offline simulation of a production system. This gives the ability to train scheduling agents in more dynamic environments and respond to uncertainties even when they have not yet been identified. Nevertheless, a main challenge in implementing production scheduling solutions and digital twins is the lack of a well-defined data model. A solution to this issue can be offered by the Asset Administration Shell (AAS) concept. AAS is basically a method to represent data in a defined architecture [13, 39]. While in other problems there is some effort made by the literature to implement AAS concept, in production scheduling it is not explored.

The need to explore and address well-defined standards for production optimization agents is clearly revealed when there is a need for cooperation between different production agents, in order to formulate a multi-agent system. Researchers from a variety of fields have given multi-agent systems (MASs) a great deal of attention as a way to break down complicated problems into smaller jobs. Individual tasks are assigned to agents, which are autonomous entities. Using a variety but well-defined inputs, each agent chooses the most appropriate plan of action to complete the task [40]. Agents make decisions based on the information provided in the environment they are integrated and choose their actions proactively or reactively [41]. In manufacturing, multi-agent systems have gathered attention of many researchers during recent years. MAS can limit the complexity of order scheduling in production systems through a cooperative multi-agent system for production control optimization [42]. A similar approach was followed for the implementation decentralized scheduling algorithms in a test-bed environment [43]. A scheduling strategy to assist a manufacturing system experiencing learning and forgetting was supported by a multi-agent system to carry out the scheduling tasks in conventional production systems in close to real-time, and a simulation was utilized for validation [44].

While the multi-agent systems implementation methods have been explored in recent years, further investigation to address challenges is required. For example, the use of standards in a scheduling multi-agent system is something crucial, in order to develop systems that could be easily transformed to a "plug & play" application. In addition, agents that control or implement different applications and software should follow a hierarchical implementation to achieve better multi-agent system utilization and agents' distribution. Lastly, if external applications are controlled through a multi-agent system functionality, Application Programming Interface (API) and standards are almost inevitable for the proper scheduling MAS integration for the actual production system. The implementation of the scheduling multi-agent system proposed in this work addresses the aforementioned issues and gives the opportunity for a more flexible implementation of scheduling algorithms, with different functionalities and heterogenous optimization techniques.

3 Multi-Agent System Framework

3.1 System Architecture

The architecture presented in Fig. 1 merges numerous Industry 4.0 technologies within a single framework with the goal of creating quality decision-making support for the production manager in his/her daily tasks. Specifically, there is used a (1) user interface for production manager interaction, (2) a multi-agent system for decentralized production scheduling, (3) a production digital twin for performance validation, and (4) Asset Administration Shell concept for the description of production information and agents as assets within the I4.0 environment.

The first aspect of the proposed framework is defining the information exchange mechanisms and the corresponding information model to pass data over the different components. This is one of the interoperability issues associated with enterprise software as it is usual to utilize different information format and structures for the same information context. In this architecture the AAS is used in order to represent production information, such as work orders, process plan, and production resources. However, AAS is a metamodel, and although it may specify some abstract modelling objects and interaction mechanisms, it does not specify the detailed model to be used for the description of the asset. In other words, there could be more than one AAS descriptions for the same asset, structuring differently the same asset components and behaviors. To this end, there is a whole other topic of choosing the "right" information model for describing the production data so that it is achieved standardization over the information exchange. However, this is not within the

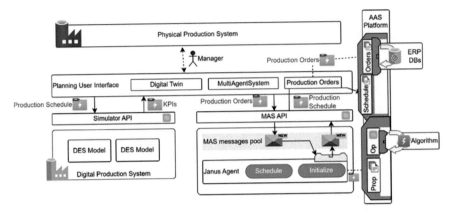

Fig. 1 Framework Architecture of all modules and their interactions

scope of this framework, and although the AAS is used for exchanging information between ERP software and the agents, the underlying model is not standardized.

As displayed in Fig. 1, information from the enterprise resource planning (ERP) are described within AASs for the corresponding work orders that the manager is called to satisfy within the following production period. This type of information is restored from the user interface (UI), allowing the user (in this case the production manager) review the workload of the upcoming days. The connection between the AAS and the ERP is performed via an ERP-to-AAS connector so that the proposed UI platform depends on the AAS model rather than the specific ERP information model structure. The UI rather than visualization of production information, it is also an enabler for interaction of the user with the MAS as well as the production digital twin. It is important to highlight that, unlike other systems, the integration of decision-making results to the actual system is not a trivial task. In practice human interferences is required to review and apply the production plan.

The exchange of information between the UI and the MAS is achieved via a MAS API, which is in practice a way of passing and receiving data regarding the production workload and status. The MAS is responsible for handling the data and provide scheduling decisions for the user given the current production scenario. There are multiple AI agents that were developed to address this problem each one giving its own benefits for the user. The reason for using more than one agent for a scheduling problem arises due to complexity of the problem, the user requirements, as well as the problem itself. Scheduling problems are widely diverse with respect to the environment, constraints, objectives, and equivalently the optimization methods are usually compatible with a small portion of the overall set of scheduling problems available. To this end, there cannot be a monolithic approach capable of addressing all production scheduling problems without lacking on satisfying the user requirements. In order to address this issue, there was proposed the concept of a meta-scheduling agent, which in practice was a compound of multiple AI scheduling agents each one providing different optimization attributes.

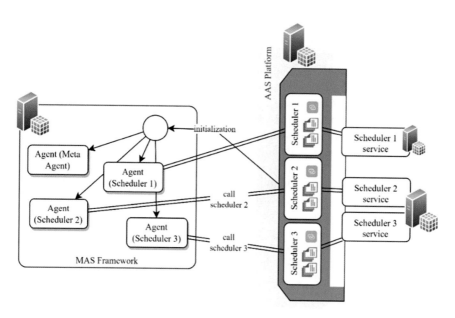

Fig. 2 MAS implementation based on the AAS description for the agent, showing interaction between agents and services/schedulers as well as the agent spawn procedure

The AAS was developed for the description of agent, which was retrieved by the MAS framework in order to deploy the corresponding entities and bring the algorithms to life. The AAS model for the meta-agent was consisted of a toolbox of optimization methods, with the description of connection dependencies as well as capabilities and skills provided by the specific method. During the initialization, there were spawned individual entities within the MAS, each one carrying a specific set of skills (operations) corresponding to the AAS operations. It is important to highlight though that the deployment of the agent within the MAS with the actual algorithm runtime may differ. Specifically, the MAS is operating within a single framework, which is usually a local installation of all the partial components, and in this case the deployment of the algorithms is better to be remote. Figure 2 illustrates this aspect for an example case of a scheduling agent. It can be displayed that a scheduling agent AAS may contain more than one scheduling methods, which are spawned as individual agents within the MAS framework. On top of that a meta-scheduling agent is spawned within this framework in order to support the scheduler selection and orchestration process within the MAS. The scheduling algorithms, however, may be deployed in different remote servers depending on the case. While a scheduling operation is requested from one of the schedulers, the AAS interfaces support the communication between the agent installation and the actual algorithm.

In the previous architecture, it is important to clarify the need for the meta-agent as well as the requirement for generating multiple agent entities within the MAS framework. In essence, the notion of an agent, as an independent entity, is useful when there is achieved some kind of communication within a network of agents. This type of communication is achieved usually via a MAS framework implementation, which allows all the messages exchange and events to be broadcasted between the inner entities seamlessly. As such, a MAS framework facilitates the interaction between the agents; however, the implementation of the agent logic does not have to be within the same software component as the MAS. This is because, usually the MAS is a unique software component with all of its agents and events operating within the same software container. It thus makes sense not to include complex computational process (such as scheduling) within the same resources.

To this end, the actual optimization processes are kept aside from the agent interfaces within the MAS. However, the reason that the scheduling agent AAS that contains multiple scheduling methods is not spawned within the single agent in the MAS is dues to easier management of the different scheduling operations. Although this is most of a design decision, it is easier to distribute a network of agents, each one responsible for a specific scheduling method, because on the contrary side, all scheduling requests independent of method selection would flow through the same agent, making it less efficient to work with two different scheduling requests in parallel. The meta-agent is thus present to support the selection of the algorithm based on the scheduling problem and allocate the optimization process to the different agents. In practice, this specific agent is aware of the different scheduling methods available within the system and is capable to analyze the request before selection.

In order to accurately assign the scheduling problems to the scheduling algorithms, there were used a problem classification method based on three notations: *environment*, *constraints*, and *objectives*. This notation method is widely used in the description of scheduling problems and has the ability to classify any type of problem. The *environment* expresses the production system equipment and process flow, the *constraints* express the job-related characteristics or specific equipment/buffer requirements, while the *objectives* have to do with the criteria that the scheduler need to optimize. The following are some examples for each case:

- *Environment*: job shop, flexible job shop, parallel machines, single machine, flow shop, flexible flow shop, conveyor line, batch machine (e.g., oven), and so on.
- *Constraints*: job release time, block (no buffer capacity before the machine), deadline, sequence-dependent setup time, recirculation, stochastic processing time, and so on.
- *Objectives*: makespan, flowtime, tardiness, energy consumption, and so on.

In order to classify the problem based on these notations in an automatic way, the meta-agent was enriched with different rules (per characteristic per notation) in order to check whether the specific type of problem complies with these conditions. For example, identifying a Job Shop schedule has to contain exactly one route per product and no alternatives. As a results, in cases that the agent was given with a schedule request that did not specify the type of scheduler to use, these rules were applied and the scheduler that complied with the rules was selected. In some cases,

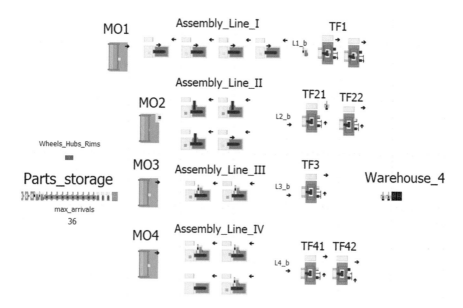

Fig. 3 Simulation model of the wheels assembly department utilized as a digital twin to apply the schedule outcome from the MAS and observe performance

more than one scheduler would comply with the rules and more than one responses may be produced.

It is also important to highlight that within a request there was used a specific information structure for providing the production data and similarly the scheduling outputs were contained within a specific scheduling response. The structure of the information may vary based on the implementation and thus it was not specified within this section. There are different alternative standards also to be used, while in some cases a specific ERP data model could be also utilized. In any case, this is another important aspect that is not specified within the chapter. However, the methodology remains the same, with the exception that the problem classification should be applied to a different model.

The digital twin was the final component of the architecture and ensured that information is validated in a close-to-reality scenario and the system performance is approved by the user. The production schedule was received by the MAS and then sent (on-demand) to the digital twin in order to calculate its performance (see Figs. 3 and 4). This step was executed before the schedule was displayed in detail to the user as there could be one of multiple competitive schedules available for a single case from different schedulers. The reason for using a digital representation of the production system was to give the ability to the user to evaluate the resulted schedule.

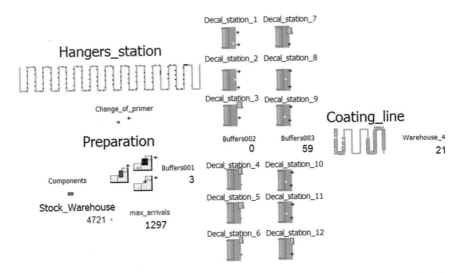

Fig. 4 Simulation model of the painting department utilized as a digital twin to apply the schedule outcome from the MAS and observe performance

3.2 Paint Shop Scheduling Agents

3.2.1 Mathematical Optimization

The paint shop scheduling agents were designed in order to be able to give solution to the Paint Shop Scheduling Problem (PSSP) as it can be found in the literature. This problem addresses the sequence of the items entering the painting line of the factory in order to optimize the performance indicators. This problem is different from other scheduling problems as it usually encounters higher detail in the combination of items and sorting before entering the line. The line itself is usually a moving conveyor of carriers with some specific spatial constraints, setup delays due to color, and a constant speed. The objective is to find the optimal combination of the items within the "bill of material" of the products and sequence them in order to comply with the desired performance.

Figure 5 illustrates the PSSP in a simplistic way. As it can be seen the goal is to create a schedule – sequence and combination of items – for entering the painting line so as to create the maximum utilization of the line, which will reflect in reducing the makespan for the system. There are some requirements, however, that the decision-making system needs to comply with in order to be in line with the physical characteristics of the system. The following aspects were taken into consideration:

- The conveyor speed is constant and the carriers are equally spaced along the line. This ensures that the input and output rate of the line is also constant.

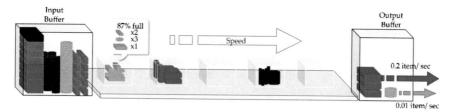

Fig. 5 PSSP graphical representation

- Each carrier has a unique capacity (100%) similar of all the carriers that cannot be exceeded.
- Items with different color cannot be placed within the same carrier. This is because in most cases the items are painting all together within the painting cabins.
- In cases where two consecutive colors are placed within the line there should be a setup delay, expressed in empty carriers so that the operators have the time to setup the new color.
- Each item type occupies a specific percentage of the carrier and can be mixed with others as long as the max capacity is not violated.
- In cases where an item cannot fit into one container/carrier, then it will be used the next consecutive carrier in order to hold the remaining capacity of the item. It is made the exception, however, that no item needs more capacity than two carriers.

Based on the proposed conversions, the following mathematical formulation can be created:

Sets:

P Set of production orders that need to be painted
I Set of different items (types) that need to be painted
C Set of color codes

Parameters:

$q_{p,i}$ Quantity of items $i \in I$ included in production order $p \in P$, $q_{p,i} \in \mathbb{Z}_{\geq 0}$
c_i Capacity a carrier required in order to carry item $i \in I$; $c_i \in \mathbb{R}_{>0}$
$d_{p,p'}$ Setup delay required between the items of two different production orders $(p,p') \in P$, $d_{p,p'} \in \mathbb{Z}_{\geq 0}$

Auxiliary variables:

s_i Defines whether item $i \in I$ has a size higher than one carrier, $s_i \in \{0, 1\}$
$a_{c,t}$ Defines whether this color $c \in C$ has entered the line on time $t \in \mathbb{Z}_{\geq 0}$, $a_{c,t} \in \{0, 1\}$
$e_{p,t}$ Defines whether this item $i \in I$ has entered the line on time $t \in \mathbb{Z}_{\geq 0}$, $e_{p,t} \in \{0, 1\}$

Decision variable:

$x_{p,i,t}$ The number of items $i \in I$ from product $p \in P$ that will enter the line on time t, $x_{p,i,t} \in \mathbb{Z}_{\geq 0}$,

Counters:

n_t Number of timesteps available in the schedule
n_i Number of item types in I-set
n_p Number of products in P-set
n_c Number of colors in C-set

Constrains:
First, in a feasible schedule, we need to ensure that all items enter the resource exactly once at some point during production time. This can be covered from the following linear equality:

$$\sum_{t=0}^{\infty} x_{p,i,t} = q_{p,i} \bullet (1 + s_i), \forall p \in P, \forall i \in I$$

Number of constrains : $n_p{}^* n_i$

Limitation for not allocating into the same carrier more than the items that can hold based on its capacity can be achieved via the following inequality:

$$\sum_{\forall p \in P} \sum_{\forall i \in I} c_i \bullet \left(1 - s_i \frac{1}{2}\right) \bullet x_{p,i,t} \leq 1, \forall t \in \mathbb{Z}_{\geq 0}$$

Number of constrains : n_t

In addition, for cases when items require more than one carrier, it needs to be placed in two consecutive carriers. This can be ensured by the following nonlinear expression:

$$\sum_{t=0}^{N} \sum_{\forall p \in P} \sum_{\forall i \in I} s_i \left(x_{p,i,t} \bullet x_{p,i,t+1}\right) \geq 1$$

Number of constrains : 1

In order to transition to the linear version of the above expression, we start from the logical expression:

$$\left(x_{p,i,t} > 0\right) \wedge \left(x_{p,i,t+1} > 0\right) \vee \left(x_{p,i,t-1} > 0\right)$$

Then, we define two auxiliary variables to carry the outputs of the above logical expressions:

$$z_{p,i,t} = \left(x_{p,i,t+1} > 0\right) \vee \left(x_{p,i,t-1} > 0\right)$$

$$y_{p,i,t} = \left(x_{p,i,t} > 0\right) \wedge z_{p,i,t}$$

Then, we use the linear expressions for $\wedge$ and $\vee$ operators:

$$\left.\begin{array}{c} z_{p,i,t} \geq x_{p,i,t+1} \\ z_{p,i,t} \geq x_{p,i,t-1} \\ z_{p,i,t} \leq x_{p,i,t+1} + x_{p,i,t-1} \\ y_{p,i,t} \leq x_{p,i,t} \\ y_{p,i,t} \leq z_{p,i,t} \\ y_{p,i,t} \geq x_{p,i,t} + z_{p,i,t} - 1 \end{array}\right\} \forall p \in P, \forall i \in I, \forall t \in \mathbb{Z}_{\geq 0} \mid s_i > 0$$

Number of constrains : $6 n_t{}^* \, n_p{}^* \, n_i$

In addition, when changing color between subsequence items the setup delay must be applied. This can be achieved by the following linear inequalities:

$$a_{c,t} \geq \frac{\sum_{\forall p} \sum_{\forall i} \left(x_{p,i,t} \bullet f_{p,c}\right)}{\sum_{\forall p} \sum_{\forall i} q_{p,i}}, \forall t \in \left[0, \mathbb{Z}_{\geq 0}\right], \forall c \in C$$

$$a_{c,t} + a_{c',t'} \leq 1 + \frac{|t - t'|}{d_{c,c'} + 1} \forall \left(c, c'\right) \in C, \forall t \in \mathbb{Z}_{\geq 0}, \forall t' \in \left[t, t + d_{c,c'}\right] \mid c \neq c'$$

Number of constrains : $n_c{}^* \, n_t + n_t{}^* \, (delay)^* \, (n_{\dot{c}} - 1)^2$

The same constraint can be achieved via the following nonlinear equation:

$$\sum_{t=0}^{N} \left(a_{c,t} \bullet \sum_{t'=t}^{t+d_{c,c'}} \left(1 - a_{c',t'}\right)\right) = 0, \forall \left(c, c'\right) \in C \mid c \neq c'$$

Number of constrains : $(n_c - 1)^2$

Objective function:
L_{total} The total flowtime of the production:

$$L_{\text{total}} = \sum_{t=0}^{\infty} \left(t \bullet a_{c,t}\right)$$

L_{weighted} The total weighted flowtime of the painting line:

$$L_{\text{weighted}} = \sum_{t=0}^{\infty} \sum_{\forall i \in I} \left(t \bullet w_p \bullet e_{p,t} \right)$$

where $x_{p,i,t} \leq q_{p,i} \bullet e_{p,t}, \forall p \in P, \forall i \in I, \forall t \in \mathbb{Z}_{\geq 0}$

$\lambda_{i,k}$ The output (production) rate for an item type in a specific interval can be defined as a moving average in the series of allocations for an item:

$$\lambda_{i,k} = \frac{\sum_{t=k\,L}^{(k+1)L} \sum_{\forall p \in P} x_{p,i,t}}{L}, k = 0, 1, 2, .., n_t$$

Figure 5 shows where the requirements for this objective come from, and what implications could come from missing to apply this objective. It is clear that in this example, missing to produce the items at the average rates that are departed from the buffer afterwards will cause an overflow, and in this case it is illustrated that the circle needs to be at a much slower output rate than the cube item.

The ways that this can be applied are more than one, specifically the user may require this to be a constrain to the scheduler, by means that at no times this rate is exceeded, which can be applied by the next inequality, or via the objective function trying to approach a specific value, yet this does not necessarily ensure that this value will not deviate in the final results.

$$\lambda_{i,k} \leq \lambda_i^{\text{desired}}, k = 0, 1, 2, .., N, \forall i \in I$$

$$\min \left\{ \sum_{\forall i \in I} \sum_{k=0}^{\infty} \left(\lambda_{i,k} - \lambda_i^{\text{desired}} \right)^2 \right\}$$

The above mathematical formulation is a very complex optimization problem to solve for a real-scale production problem. As shown in the results section, real-scale industrial problems may require scheduling up to 20,000 items from different orders, colors, and types, making the problem extremely difficult to solve in a considerably short time frame. Thus, following the previous mathematical formulation, three different versions were formed, each utilizing the expressions presented above differently.

- The first model is the nonlinear version (MINLP) of the problem, which applies the nonlinear inequality constraints presented above. This allows lower number of constrains (thus lower memory utilization), but a very complex solution space that most of the times requires more sophisticated optimization algorithms and more demanding computational delay.

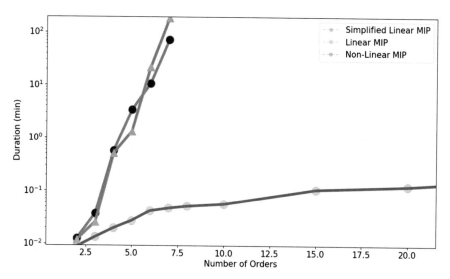

Fig. 6 Actual diagram from example case displaying the CPU delay differences of the modeling approaches as the number of orders increase

- The linear version (MILP) was also considered where only the linear constrains are utilized, improving the computational demand but increasing the requirements for memory utilization in computational resources.
- The last one was a simpler form of the linear version (two-stage MILP) in which constrain #4 is removed from the model, running the optimization only for mixing the items of order that acquire the same color. This allows a much faster response horizon, since there were no setup constrains to apply in the schedule. In a second stage, once the allocation of items is achieved, the optimization process is repeated, but this time it schedules the sequence of colors as a function of minimizing the setup delay. In this way, the model manages to reduce the solutions space and the constrains limitations. The problem with this model, however, is that it decreases the flexibility of the solution as a trade-off for lower CPU time because it is not capable of providing good solutions for the production rate issue.

In Fig. 6, the implications of the different models on the CPU duration of the computational resources is clearly shown. The graphs are in a logarithmic scale in the Y-axis and is clearly shown that both MILP and MINLP cannot outperform the simplified-MILP, which can cover up to a very high number of items (100 orders are usually 18,000 items) in a relatively short period of time (20 min).

Indices:

Discrete sets: n_p, n_i, n_c, n_t
Decision variables: $n_p^2 + n_i (n_p + 1)$
Input variables: $n_p^2 + n_i (n_p + 1)$

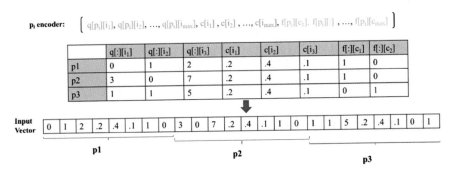

Fig. 7 Input layer (vector) encoding mechanism displaying an example for how the tables are reshaped into a single dimensional vector

$X[p_i]$ encoder: $\left[X[p_i][i_1][t_1], ..., X[p_i][i_{max}][t_1], X[p_i][i_1][t_2], ..., X[p_i][i_{max}][t_2], ..., X[p_i][i_1][t_{max}], ..., X[p_i][i_{max}][t_{max}] \right]$

	$X[:][i_1][t_1]$	$X[:][i_2][t_1]$	$X[:][i_3][t_1]$	$X[:][i_1][t_2]$	$X[:][i_2][t_2]$	$X[:][i_3][t_2]$	$X[:][i_1][t_3]$	$X[:][i_2][t_3]$	$X[:][i_3][t_3]$
p1	0	1	2	0	1	0	2	0	1
p2	3	0	7	2	0	4	1	0	1
p3	1	1	5	2	4	1	0	1	0

Output = | 0 | 1 | 2 | 0 | 1 | 0 | 2 | 0 | 1 | 3 | 0 | 7 | 2 | 0 | 4 | 1 | 0 | 1 | 1 | 1 | 5 | 2 | 4 | 1 | 0 | 1 | 0 |

i1 t1 p1 p2 p3

Fig. 8 Output layer (vector) encoding mechanism displaying the output can be encoded into the allocation per product per item type per timestep

Auxiliary variables: $n_i + n_t (n_i + n_c)$
Constrains: $n_p n_i + 6n_t n_i n_p + n_c n_t + n_t d (n_c + 1)$

3.2.2 Data-Driven Optimization

In order to avoid this long CPU delays and demanding RAM utilization, the utilization of data-driven (i.e., ML) approaches was investigated in order to rather predict the output of the scheduler. First, a feed-forward neural network (FFNN) was developed, which uses as an input information over the workload data (i.e., orders, items, colors, and sizes), as well as produces the sequence of orders/items allocated into the painting line. The input layer to the model was based on the same parameters that were used for the mathematical formulation of MIP models, resulting in the following encoded input vector $(\overline{x})$, while also the decision variable $x_{p,i,t}$ was described the output vector $(\overline{y})$ (Figs. 7 and 8):

$$\bar{x} = \left[\left(\left(\left(q\,[p]\,[i]\,\forall i \in I \right), \left(f\,[p]\,[c]\,\forall c \in C \right) \right) \forall p \in P \right), c\,[i], d\,[c]\,[c'] \right]$$

$$\bar{y} = [x\,[\forall p]\,[\forall i]\,[\forall t]]$$

$$L_{\bar{x}} = p_{\max} i_{\max} + p_{\max} c_{\max} + i_{\max} + (c_{\max})^2$$

$$L_{\bar{y}} = i_{\max}\, t_{\max}\, p_{\max}$$

$L_{\bar{x}}$ and $L_{\bar{y}}$ shown in the above equation represent the dimensions of the input and output layers, respectively. In contrast to the model-based methodologies, neural networks are consisted of a static number of I/O parameters, which contradicts to the arbitrary scale of the scheduling problems. In order to address this issue, the encoded input considers a prefixed maximum number of orders ($p_{\max}$), items ($i_{\max}$), color codes ($c_{\max}$), and production duration ($t_{\max}$). For cases where less than this maximum number is provided as an input, the encoder generates additional orders so as to fulfil the I/O layers of the neural network (NN) although sets the items quantity to zero, which will have no effect on the allocation process; for cases with more than the maximum numbers, the model is unable to encode the input. The number of neurons per layer and the total number of trainable parameters for the whole NN model are provided by the following formula:

$$\text{neurons\#} = 2^k L_{\bar{x}}$$

$$\text{params\#} = \sum_{\forall k: \text{layer}} \left[2^k L_{\bar{x}} \left(2^{k-1} L_{\bar{y}} + 1 \right) \right]$$

Another data-driven approach has been developed that treats the scheduling output as a time series from which the next allocations can be predicted based on known previous values of the sequence. As such, the whole production schedule can be generated in a recursive manner, reducing the model's prediction variables which improves the accuracy as well as avoiding any limitations regarding the problem's scalability. Similar to the above-mentioned approach, the ultimate objective is the prediction of $x_{p,i,t}$, for all the given orders, items, and timesteps; however, in this model a prediction is only applied for one timestep and is repeated for all the output sequence. The input features of the LSTM neural network consist of a dynamic and a static part. The dynamic part as presented below are the features that change as moving in the time axis.

The allocation of all items over a specific timestep (carrier) is given by the following vector:

$$\bar{x}\,[t_i] = [x\,[\forall p]\,[\forall i]\,[t_i]], t_i \in \mathbb{N}$$

$$x[t_i] = \left(\begin{array}{c} X[p1][i1][t1], X[p1][i2][t_i], ..., X[p1][i_{max}][t_i], X[p2][i1][t_i], X[p2][i2][t_i], ..., X[p2][i_{max}][t_i], ..., X[p_{max}][i1][t_i], \\ X[p_{max}][i2][t_i], ..., X[p_{max}][i_{max}][t_i] \end{array} \right)$$

$x[t_i].length = i_{max} * p_{max}$

$$Q[t_i] = \left(\begin{array}{c} q[p1][i1] - sum(X[p1][i1][t] \text{ for t in } [0, t_i]), ..., q[p1][i_{max}] - sum(X[p1][i_{max}][t] \text{ for t in } [0, t_i]), ..., q[p_{max}][i1] - \\ sum(X[p_{max}][i1][t] \text{ for t in } [0, t_i]), ..., q[p_{max}][i_{max}] - sum(X[p_{max}][i_{max}][t] \text{ for t in } [0, t_i]) \end{array} \right)$$

$Q[t_i].length = i_{max} * p_{max}$

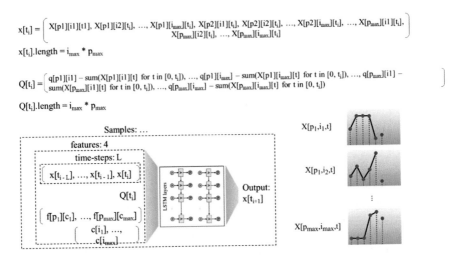

Fig. 9 Overview of LSTM RNN I/O layers design and how the specific input is derived as well as how the output is represented

The following defines a variable that provides the number of remaining items of an order at that given timestep, given the sequence of previous allocation selections:

$$Q[p][i][t_i] = q[p][i] - \sum_{t=0}^{t_i} x[p][i][t]$$

Given the above formula, the following vector is defined:

$$\overline{Q}[t_i] = [Q[\forall p][\forall i][t_i]], t_i \in \mathbb{N}$$

Moreover, similar to the feed-forward NN model, some static information of the workload revealing the colors, sizes, and setup delays must also be provided. The final configuration of the input layer is shown in Fig. 9.

Figure 9 presents the format of the I/O model, which is required for the LSTM model. Unlike MIP models, this method arises a problem in defining the first allocations ($\overline{x}[0:L]$) as it requires historical information of a window (L), which are not defined as the face a totally new schedule request. This problem is more apparent in the training procedure as multiple scheduling results from different workloads merged together into a single sequence to train the LSTM model. This issue was addressed by adding L number of timesteps in the beginning of each schedule, where L is the window of previous allocations that the model is using for the prediction. These timesteps contained zero allocations of items and were responsible only for fulfilling the input layer of the LSTM neural network model (Fig. 10).

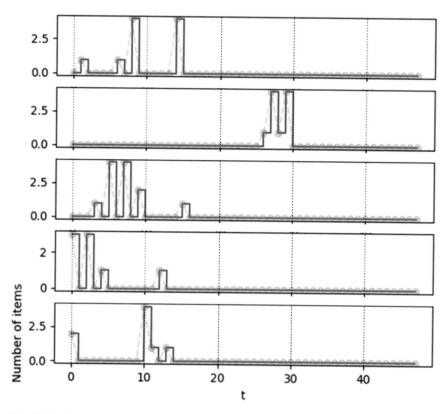

Fig. 10 Each graph (row) shows the total number of allocations from a product over time. Each graph contains two lines for display purposes

3.3 Deep Reinforcement Learning Scheduling Agent

The deep reinforcement learning (DRL) agent was selected to solve the dynamic scheduling problem (DSP). According to Chien and Lan [45], the DSP is susceptible to a number of uncertainties, including machine failures, the arrival of new, urgent jobs, and changes to job due dates. In the literature there are several articles on the DSP [46–49]. DRL agent is also combined with DNNs and deep Q-network to approximate of a state action value function [50, 51]. The proposed DRL agent is combined with a discrete event simulator in order for training and testing the DRL model. In details, the DES that was used is the Witness Horizon from Lanner [52]. The DRL and DES communicated via API, where the API is provided by Witness Horizon. In addition, except from API files, text files were used to exchange data among the DES and DRL (see Fig. 11). The concept that was used for the DRL agent is to purpose task allocation to resources via the use of dispatch rules. In the literature there are several research works that study the use of RL agent combined with dispatch rules [53].

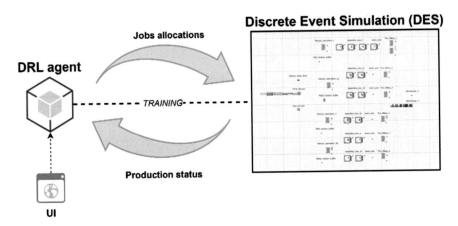

Fig. 11 DRL agent operation architecture

The Q-learning is an off-policy temporal difference algorithm and is based on the idea of Q-function [54]. In the following equation, the $Q^{\pi}(s_t, a_t)$ is the expected return of the discounted sum of rewards at state s_t by taking action a_t:

$$Q^{\pi}(s_t, a_t) = \max_{\pi} E\left[r_{t+1} + \gamma\, r_{t+2} + \gamma^2\, r_{t+3} + \ldots | s_t = s, a_t = a, \pi \right]$$

The main concept of the Q-learning is to use the Bellman equation as a value iteration update. The agent in a decision point t in a state $s_t \in S$ selects an action $a_t \in A$ according to a policy π. Taking the action a_t the agent gets to a state s_{t+1} with transition probability $p(s_{t+1}|s_t, a_t) \in P(S \times A \rightarrow S)$ and reward $r_t \in R$. Additionally, γ is a discount factor at each timestep t. Also, a is a learning rate, where $0 < a \leq 1$. The objective for the agent is to find the optimal policy π^* that maximizes the expected sum of rewards. The Q-leaning has some limitations when the environment is huge. For that reason, the deep Q-network (DQN) concept was used. Coupled RL with deep learning techniques Q-tables can be replaced with Q-function approximator with weights [55]. In order to solve the DSP problem, due to the fact that the environment is huge, DRL DQN concept was used. Let us denote as $Q(s, a; \theta_i)$ the approximate value using deep convolutional neural network. Additionally, the ψ_i are the weights at iteration i of the Q-network. The experiences are denoted as $e_t = (s_t, a_t, r_t, s_{t+1})$ where each time t are stored to a dataset $D_t = \{e_1, \ldots, e_t\}$. Chosen uniformly at random an instance from the pool of stored instances, a Q-learning update is applied of each experience $(s, a, r, s') \sim U(D)$.

$$L_i(\theta_i) = E_{(s,a,r,s') \sim U(D)}\left[\left(r + \gamma \max_{a'} Q\left(s', a'; \theta_i^-\right) - Q(s, a; \theta_i)\right)^2 \right]$$

θ_i are the weights of the Q-network at the iteration i and θ_i^- are the network weights used to compute the target in iteration i. Target network parameters (θ_i^-) are updated with the Q-network parameters every c step, where c is a constant number.

The state is a tuple of feature that characterizes a given input. This chapter contains the stats of the resources (down, busy, and available), the stats of the tasks (waiting, pending, on-going, and finished), and finally a list with the quantities or the product orders. Moreover, an action describes the dispatch rule that is selected by the DRL agent to propose the task allocation over resources.

3.4   Heuristic Optimization

The hierarchical scheduler is a decision-making module for extracting an efficient order of required tasks [56]. The problem that the scheduler solves is the resource allocation problem [57], where the problem seeks to find an optimal allocation of a discrete resource units to a set of tasks. The heuristic algorithm is based on the scientific research of [58]. It is based on the depth of search concept, except the number of layers for which the search method looks ahead. The main control parameters are the decision horizon (DH), the sampling rate (SR), and the maximum number of alternatives (MNA). In each decision point, a decision tree is created based on the DH, SR, and MNA. Figure 12 shows the nodes $A_1 \ldots A_N$ that represents decision point where a task is assigned to an operator. The proper selection of MNA, DH, and SR allows the identification of a good solution. For example, it is proven in [59] that the probability of identifying an alternative of good quality (i.e., utility value within a range Δ with respect to the highest utility value) is increasing with the MNA and Δ. The pseudocode of the algorithm is defined as follows [60]:

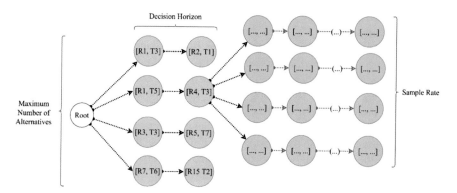

Fig. 12 Search methodology example in tree-diagram showing the generation of different branches and layers based on the MNA, DH, and SR

Algorithm:
Three adjustable parameters, MNA, SR, DH
  Initialize: MNA, SR, DH
    while full schedule is not generated
      Generate MNA-alternative-branches of allocations for DH-steps in the future.
      for each branch **in** alternatives:
        Generate SR sub-branches of allocations from DH-step and forward.
        Calculate average score of SR sub-branches on each MNA branch.
        Select alternative with the highest score.
        Store allocations of the alternative for up to DH-steps
      return: best alternative

For each decision tree, the algorithm returns a list with valid task-resource allocations [61–64]. MNA and DH control the breadth and DH the depth of the search, respectively. On the other hand, SR is used to direct the search toward alternatives that can provide better quality solutions. Thus, the quality of the solution depends on the selection of the MNA, DH, and SR.

4 Case Study

The proposed multi-agent scheduling framework was implemented, validated, and evaluated for a case study from the bicycle production industry. For this work and the deployment of AI scheduling agents in a production environment capable of producing optimized long- or short-term scheduling, two departments were chosen: painting and wheels assembly department. As already mentioned, there were different types of scheduling agents. The purpose of using a multi-agent system for the deployment of various scheduling applications is twofold. The first reason is that with the realization of a multi-agent system, one can use an integrated solution without affecting the other entities of the production system. An algorithm can be developed separately, as a stand-alone application in a multi-agent system. The second reason is that with a multi-agent system, there is the possibility for automated cooperation between different applications to coordinate multiple assets or functionalities. To combine both benefits of a multi-agent system capable of solving different scheduling problems and combine its assets to solve more than one scheduling problems at one, the implementation method proposed in this work follows the deployment of different scheduling algorithms integrated in a multi-agent system.

The multi-agent system for the scheduling agents was developed with the use of JANUS, an efficient and flexible agent-oriented programming framework that gives the opportunity for easy and fast deployment of virtual assets. JANUS multi-agent system framework is compatible with the programming language SARL and also with JAVA. In this multi-agent system, there are four main concepts that need to be defined before the deployment of any agent: agents, events, capacities, and skills. The agent instance stands for all the operating sequences required for a specific batch of functionalities and operations to happen when the agent needs to operate. Agents' communication and behavior is controlled by events, which are predefined patterns that allow all the agents in the framework to interact one with another. The term capacity refers to an abstract description for an implementation in skills, which is used to define reusable capabilities of agent patterns without defining implementation details. Lastly, the concept of skills is a manner of implementing a capability, which allows exchange and modification of implementations based on own or adapted skills without modifying the agent's behavior or the template agent's characteristics. To address the scheduling multi-agent system using the JANUS framework, the scheduling agents are modeled as agents in the JANUS framework, capable of spawning and operating under the control of a meta-agent, which is the orchestrator agent inside the multi-agent system. The scheduling agents have specific skills, related to the problem-solving algorithm and the meta-agent concept was integrated, in order to realize an automated and distributed cooperation of the different agents inside the multi-agent system, when there is a scheduling request. The user is able to interact with the multi-agent system in the backend of a UI, developed for the scheduling tasks visualization.

In practice, the meta-agent receives the scheduling request from the UI. This scheduling request is modeled in an AAS, as already described in previous sections, and the meta-agent is able to spawn the corresponding scheduling agent to solve a particular scheduling problem. A scheduling agent is the parent "class" in JANUS that implements events, skills, and methods, and can also consist of local variables. Each one of the scheduling agents accommodated three MAS events:

- "Initialization," where the scheduling agent has been spawned by the meta-agent during the initialization of the framework and waits for a scheduling request notification from the meta-agent. During the initialization of the agent, specific scheduling agent parameters are defined and initialized, able to serve a specific scheduling request type in the future.
- "Scheduling request," where the meta-agent is requested to notify the corresponding scheduling agent in order for the required scheduling computation to be performed. After this event call, specific skills and operations are performed by a scheduling agent in order for the scheduling algorithm to calculate the schedule.
- "Schedule response," where the output of the scheduling task is emitted to all other agents of the multi-agent system. When the scheduling agent finishes its operation, the event notices every other agent in the framework that can listen to this event.

When a scheduling request reaches the multi-agent system, the meta-agent is responsible for identifying the correct scheduler based on the request from AAS. This AAS also contains information that, in addition to the scheduling task information, will indicate the required scheduling agent, which is capable of performing the scheduling task based on some predefined characteristics. After the scheduling request, the meta-agent performs simple filtering in the provided information in the AAS to choose the corresponding scheduling agent. Each scheduling agent has its own input format. Since the JANUS meta-agent is responsible for the orchestration of the scheduling task, it will pass the information to the appropriate scheduling agent capable of performing the correct algorithm to compute the schedule.

As it was mentioned above, the reason for realizing a multi-agent system is that a scheduler can be developed as a stand-alone application. Hence, to give the ability to each scheduling agent to perform its scheduling skills without the development of the algorithmic part inside the multi-agent system framework, interfaces were utilized to perform the scheduling algorithms through the scheduling agents' skills. Moreover, a REST API was used for the agents to be able to reach out the scheduler's endpoint and pass information to the algorithm. On the other hand, since there is not a certain point in time that the resulted response is expected, RabbitMQ message exchange channel was utilized for the scheduler's responses. Of course, this is a design decision, and other protocols can be used to pass information around the different entities.

In the case study, three different scheduling agents were implemented in the multi-agent system, each one with its own characteristics and functionalities. To validate the aforementioned multi-agent system implementation, the schedulers developed and utilized were the following: (1) heuristic multi-objective scheduling framework, (2) mixed integer programming (MIP) model optimizer for production scheduling [65], and (3) a deep reinforcement learning (DRL) scheduler for dynamic production scheduling. The first two agents were utilized to solve the scheduling problem for the painting department of the bicycle industry, while the third one was utilized to solve the scheduling problem of the wheels assembly department. The first two agents were deployed with the goal of optimizing the scheduling sequence of a painting line. The DRL agent was deployed with the goal of solving the dynamic scheduling problem of a production system with uncertainties included.

The scheduling agents were deployed in the multi-agent system to support the scheduling task of the bicycle production system. Nevertheless, since the application should be used by a production manager in an industrial environment, a UI was required. The UI was developed with the scope to showcase the scheduling task with all the mandatory assets in an efficient and user-friendly manner. The UI consists of the scheduling task formulation tab, where the production manager selects the orders that need to be scheduled and chooses the corresponding scheduler. There is also a feature where the user can run all the supported schedulers and compare the results before one actually can apply the schedule in the real production system. Results are shown in another tab, and this is a common table for the scheduled production orders. In addition, there is the opportunity for the user to show some production

Fig. 13 UI multi-agent system tab

KPIs through the digital twin tab where a DES run of the resulted scheduling is performed.

To validate the whole framework performance, discrete event simulation (DES) was utilized. Two DES models were developed, representing the production environments of the two departments from a bicycle production system. These DES models were used to showcase the results of the scheduling request that the multi-agent framework handled, as well as for the actual operation of the agent for solving the dynamic scheduling problem. The heuristic and the MIP schedulers were deployed for the painting department whereas the DRL scheduler was deployed for the wheels assembly department. JANUS multi-agent system spawned all three scheduling agents when the necessary information for accessing them is provided within the AAS definition after the scheduling request formulated in the UI. As such, the user could choose any of the scheduling agents and, using the toolbox of schedulers provided in the UI, address similar or different kinds of problems. The user sent scheduling operations to the multi-agent system in an abstract manner without the need to specify the corresponding problem. After the scheduling request arrival, the meta-agent was responsible to spawn the required scheduling agent. Seamless integration between the SARL software and the individual schedulers was achieved.

The resulted framework implementation showed great potential in achieving multi-agent scheduling optimization. The UI (Figs. 13 and 14) allows the user to evaluate the resulted scheduling through the use of DES. Production KPIs are presented and through the evaluation of the system performance on each occasion, one can decide if the resulting schedule is efficient. Manual tests were made in collaboration with the production manager, and the results were validated for their accuracy and precision. Hence, the proposed scheduling multi-agent system implementation for the bicycle production industrial environment can effectively handle the workload distribution among its different scheduling agents in order to propose the most appropriate production sequence.

Table 1 summarizes the agent results from testing the framework over some real-life examples of the industrial use-case. The results do not directly translate on

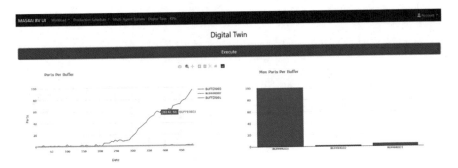

Fig. 14 UI digital twin tab

business KPIs and are the log from the schedulers. This is why the digital twin component is necessary to reflect how these solutions fit into the overall production scenario and inspect the performance.

5   Conclusion

In conclusion, the multi-agent system (MAS), digital twin (DT), Asset Administration Shell (AAS) concept, and artificial intelligence (AI) technology are part of the Industry 4.0, and more and more researchers and industrial experts aim to combine these technologies. Digital manufacturing is an important step for industries and researchers, where there are many gaps and challenges to overcome. Digitalization will enable automation, increase efficiency, real-time decision-making, flexibility, and adaptability in industries. This work proposes a MAS framework that was developed for the bicycle industry using the concept of AAS, DT, and MAS for the production scheduling problem. A mathematical optimization, deep reinforcement learning, heuristic algorithm, and deep learning algorithm have been developed to address the identified problems. The key contribution of this work is the use of the DT to accurately simulate the production environment and increase the efficiency of the developed AI agents. The AAS concept is also used to guarantee interoperable data transfer within MAS. Future research directions could be considered the continuous exploitation of the DT and AI integrations. Moreover, the AAS technology was used to fully parameterize the agents and the production environment on the simulator.

Acknowledgments This work has received funding from the European Union's Horizon 2020 research and innovation program under grant agreement No. 957204 MAS4AI. The dissemination of results herein reflects only the authors' view, and the Commission is not responsible for any use that may be made of the information it contains.

Table 1 Table with results from the different agents used to solve the use-case scheduling problems

Level	Optimization plugin	Assignments no. (average)		Reaction time (min) (average)	Options	Performance (average)
		Orders	Items			
Resource (painting line)	Conveyor scheduler (simple MILP, Gurobi)	98	18,500	22	2 [Cores] 12 [Threads]	5% [MIP Gap] 8000 [Hangers]
Resource (painting line)	Conveyor scheduler (MILP, Gurobi)	12	1500	71	2 [Cores] 32 [Threads]	7% [MIP Gap] 700 [Hangers]
Resource (painting line)	Conveyor scheduler (MINLP, Gurobi)	12	1500	180	2 [Cores] 32 [Threads]	7% [MIP Gap] 700 [Hangers]
Resource (painting line)	Conveyor scheduler (FFNN)	20	2000	0	20 [p_{max}], 16 [i_{max}], 1000 [t_{max}], 20 [c_{max}]	1000 [Hangers] 27 [Constrain Violations]
Resource (painting line)	Conveyor scheduler (LSTM RNN)	50	9000	0	50 [p_{max}], 16 [i_{max}], 50[c_{max}]	5000 [Hangers] 5 [Constrain Violations]
Resource (preparation)	Factory scheduler	8000 [Tasks]		70	0.4% [MNA], 0.15% [SR], 2 [DH]	70% [Utility]
Resources (assembly lines)	Department scheduler	50 [Orders], 1000 [Jobs], 2500 [Tasks]		2	1 shift [DH]	−7% [Makespan] 70–75% [Utilization]

References

1. Chryssolouris, G.: The design of manufacturing systems. In: Manufacturing Systems: Theory and Practice, pp. 329–463 (2006). https://doi.org/10.1007/0-387-28431-1_5
2. Rodammer, F.A., White, K.P.: A recent survey of production scheduling. IEEE Trans. Syst. Man Cybern. 18(6), 841–851 (1988). https://doi.org/10.1109/21.23085
3. Lawler, E.L., Lenstra, J.K., Rinnooy Kan, A.H.G., Shmoys, D.B.: Chapter 9 Sequencing and scheduling: algorithms and complexity. In: Handbooks in Operations Research and Management Science, vol. 4, no. C, pp. 445–522 (1993). https://doi.org/10.1016/S0927-0507(05)80189-6
4. Tao, F., Zhang, H., Liu, A., Nee, A.Y.C.: Digital twin in industry: state-of-the-art. IEEE Trans. Industr. Inform. 15(4), 2405–2415 (2019). https://doi.org/10.1109/TII.2018.2873186
5. Alexopoulos, K., Nikolakis, N., Chryssolouris, G.: Digital twin-driven supervised machine learning for the development of artificial intelligence applications in manufacturing. Int. J. Comput. Integr. Manuf. 33(5), 429–439 (2020). https://doi.org/10.1080/0951192X.2020.1747642
6. Jiang, Y., Yin, S., Li, K., Luo, H., Kaynak, O.: Industrial applications of digital twins. Phil. Trans. R. Soc. A. 379(2207) (2021). https://doi.org/10.1098/RSTA.2020.0360
7. Enders, M., Enders, M.R., Hoßbach, N.: Dimensions of digital twin applications-a literature review completed research. Accessed 16 June 2023. [Online]. Available: https://www.researchgate.net/publication/359715537
8. Rasheed, A., San, O., Kvamsdal, T.: Digital twin: values, challenges and enablers from a modeling perspective. IEEE Access. 8, 21980–22012 (2020). https://doi.org/10.1109/ACCESS.2020.2970143
9. Park, K.T., Son, Y.H., Ko, S.W., Do Noh, S.: Digital twin and reinforcement learning-based resilient production control for micro smart factory. Appl. Sci. 11(7), 2977 (2021). https://doi.org/10.3390/APP11072977
10. Plattform Industrie 4.0 - Reference Architectural Model Industrie 4.0 (RAMI4.0) - an introduction. https://www.plattform-i40.de/IP/Redaktion/EN/Downloads/Publikation/rami40-an-introduction.html. Accessed 16 June 2023.
11. Wei, K., Sun, J.Z., Liu, R.J.: A review of Asset Administration Shell. In: IEEE International Conference on Industrial Engineering and Engineering Management, pp. 1460–1465 (2019). https://doi.org/10.1109/IEEM44572.2019.8978536
12. Arm, J., et al.: Automated design and integration of Asset Administration Shells in components of Industry 4.0. Sensors. 21(6), 2004 (2021). https://doi.org/10.3390/S21062004
13. Wagner, C., et al.: The role of the Industry 4.0 asset administration shell and the digital twin during the life cycle of a plant. In: IEEE International Conference on Emerging Technologies and Factory Automation, ETFA, pp. 1–8 (2017). https://doi.org/10.1109/ETFA.2017.8247583
14. Cavalieri, S., Salafia, M.G.: A model for predictive maintenance based on Asset Administration Shell. Sensors. 20(21), 6028 (2020). https://doi.org/10.3390/S20216028
15. Ocker, F., Urban, C., Vogel-Heuser, B., Diedrich, C.: Leveraging the Asset Administration Shell for agent-based production systems. IFAC-PapersOnLine. 54(1), 837–844 (2021). https://doi.org/10.1016/J.IFACOL.2021.08.186
16. Chryssolouris, G., Alexopoulos, K., Arkouli, Z.: Artificial intelligence in manufacturing systems. Stud. Syst. Decis. Control. 436, 79–135 (2023). https://doi.org/10.1007/978-3-031-21828-6_4/COVER
17. De Simone, V., Di Pasquale, V., Miranda, S.: An overview on the use of AI/ML in manufacturing MSMEs: solved issues, limits, and challenges. Proc. Comput. Sci. 217, 1820–1829 (2023). https://doi.org/10.1016/J.PROCS.2022.12.382
18. Shahgholi Zadeh, M., Katebi, Y., Doniavi, A.: A heuristic model for dynamic flexible job shop scheduling problem considering variable processing times. Int. J. Prod. Res. 57(10), 3020–3035 (2018). https://doi.org/10.1080/00207543.2018.1524165

19. Chen, X., An, Y., Zhang, Z., Li, Y.: An approximate nondominated sorting genetic algorithm to integrate optimization of production scheduling and accurate maintenance based on reliability intervals. J. Manuf. Syst. **54**, 227–241 (2020). https://doi.org/10.1016/J.JMSY.2019.12.004
20. Essien, A., Giannetti, C.: A deep learning model for smart manufacturing using convolutional LSTM neural network autoencoders. IEEE Trans. Industr. Inform. **16**(9), 6069–6078 (2020). https://doi.org/10.1109/TII.2020.2967556
21. Wang, L., Pan, Z., Wang, J.: A review of reinforcement learning based intelligent optimization for manufacturing scheduling. Compl. Syst. Model. Simul. **1**(4), 257–270 (2022). https://doi.org/10.23919/CSMS.2021.0027
22. Negri, E., Fumagalli, L., Macchi, M.: A review of the roles of digital twin in CPS-based production systems. Proc. Manuf. **11**, 939–948 (2017). https://doi.org/10.1016/J.PROMFG.2017.07.198
23. Jones, D., Snider, C., Nassehi, A., Yon, J., Hicks, B.: Characterising the digital twin: a systematic literature review. CIRP J. Manuf. Sci. Technol. **29**, 36–52 (2020). https://doi.org/10.1016/J.CIRPJ.2020.02.002
24. Boschert, S., Rosen, R.: Digital twin-the simulation aspect. In: Mechatronic Futures: Challenges and Solutions for Mechatronic Systems and Their Designers, pp. 59–74 (2016). https://doi.org/10.1007/978-3-319-32156-1_5/COVER
25. Botín-Sanabria, D.M., Mihaita, S., Peimbert-García, R.E., Ramírez-Moreno, M.A., Ramírez-Mendoza, R.A., de Lozoya-Santos, J.: Digital twin technology challenges and applications: a comprehensive review. Remote Sens. **14**(6), 1335 (2022). https://doi.org/10.3390/RS14061335
26. Negri, E., Pandhare, V., Cattaneo, L., Singh, J., Macchi, M., Lee, J.: Field-synchronized digital twin framework for production scheduling with uncertainty. J. Intell. Manuf. **32**(4), 1207–1228 (2021). https://doi.org/10.1007/S10845-020-01685-9/FIGURES/16
27. Jhunjhunwala, P., Atmojo, U.D., Vyatkin, V.: Applying skill-based engineering using OPC-UA in production system with a digital twin. In: IEEE International Symposium on Industrial Electronics (2021, June). https://doi.org/10.1109/ISIE45552.2021.9576342
28. Fuchs, J., Schmidt, J., Franke, J., Rehman, K., Sauer, M., Karnouskos, S.: I4.0-compliant integration of assets utilizing the Asset Administration Shell. In: IEEE International Conference on Emerging Technologies and Factory Automation, ETFA, pp. 1243–1247 (2019, Sept). https://doi.org/10.1109/ETFA.2019.8869255
29. Pribiš, R., Beňo, L., Drahoš, P.: Asset Administration Shell design methodology using embedded OPC unified architecture server. Electronics. **10**(20), 2520 (2021). https://doi.org/10.3390/ELECTRONICS10202520
30. Chen, S., Pan, Q.K., Gao, L.: Production scheduling for blocking flowshop in distributed environment using effective heuristics and iterated greedy algorithm. Robot. Comput. Integr. Manuf. **71**, 102155 (2021). https://doi.org/10.1016/J.RCIM.2021.102155
31. Montiel, L., Dimitrakopoulos, R.: A heuristic approach for the stochastic optimization of mine production schedules. J. Heuristics. **23**(5), 397–415 (2017). https://doi.org/10.1007/S10732-017-9349-6/FIGURES/17
32. Aghelinejad, M.M., Ouazene, Y., Yalaoui, A.: Production scheduling optimisation with machine state and time-dependent energy costs. Int. J. Prod. Res. **56**(16), 5558–5575 (2017). https://doi.org/10.1080/00207543.2017.1414969
33. Jélvez, E., Morales, N., Nancel-Penard, P., Cornillier, F.: A new hybrid heuristic algorithm for the precedence constrained production scheduling problem: a mining application. Omega (Westport). **94**, 102046 (2020). https://doi.org/10.1016/J.OMEGA.2019.03.004
34. Liu, C.L., Chang, C.C., Tseng, C.J.: Actor-critic deep reinforcement learning for solving job shop scheduling problems. IEEE Access. **8**, 71752–71762 (2020). https://doi.org/10.1109/ACCESS.2020.2987820
35. Waschneck, B., et al.: Optimization of global production scheduling with deep reinforcement learning. Proc. CIRP. **72**, 1264–1269 (2018). https://doi.org/10.1016/J.PROCIR.2018.03.212
36. Villalonga, A., et al.: A decision-making framework for dynamic scheduling of cyber-physical production systems based on digital twins. Annu. Rev. Control. **51**, 357–373 (2021). https://doi.org/10.1016/J.ARCONTROL.2021.04.008

37. Zhang, M., Tao, F., Nee, A.Y.C.: Digital twin enhanced dynamic job-shop scheduling. J. Manuf. Syst. **58**, 146–156 (2021). https://doi.org/10.1016/J.JMSY.2020.04.008

38. Fang, Y., Peng, C., Lou, P., Zhou, Z., Hu, J., Yan, J.: Digital-twin-based job shop scheduling toward smart manufacturing. IEEE Trans. Industr. Inform. **15**(12), 6425–6435 (2019). https://doi.org/10.1109/TII.2019.2938572

39. Inigo, M.A., Porto, A., Kremer, B., Perez, A., Larrinaga, F., Cuenca, J.: Towards an Asset Administration Shell scenario: a use case for interoperability and standardization in industry 4.0. In: Proceedings of IEEE/IFIP Network Operations and Management Symposium 2020: management in the Age of Softwarization and Artificial Intelligence, NOMS 2020 (2020, April). https://doi.org/10.1109/NOMS47738.2020.9110410

40. Dorri, A., Kanhere, S.S., Jurdak, R.: Multi-agent systems: a survey. IEEE Access. **6**, 28573–28593 (2018). https://doi.org/10.1109/ACCESS.2018.2831228

41. Cardoso, R.C., Ferrando, A.: A review of agent-based programming for multi-agent systems. Computers. **10**(2), 16 (2021). https://doi.org/10.3390/COMPUTERS10020016

42. Dittrich, M.A., Fohlmeister, S.: Cooperative multi-agent system for production control using reinforcement learning. CIRP Ann. **69**(1), 389–392 (2020). https://doi.org/10.1016/J.CIRP.2020.04.005

43. Egger, G., Chaltsev, D., Giusti, A., Matt, D.T.: A deployment-friendly decentralized scheduling approach for cooperative multi-agent systems in production systems. Proc. Manuf. **52**, 127–132 (2020). https://doi.org/10.1016/J.PROMFG.2020.11.023

44. Renna, P.: Flexible job-shop scheduling with learning and forgetting effect by multi-agent system. Int. J. Ind. Eng. Comput. **10**(4), 521–534 (2019). https://doi.org/10.5267/J.IJIEC.2019.3.003

45. Chien, C.F., Bin Lan, Y.: Agent-based approach integrating deep reinforcement learning and hybrid genetic algorithm for dynamic scheduling for industry 3.5 smart production. Comput. Ind. Eng. **162**, 107782 (2021). https://doi.org/10.1016/J.CIE.2021.107782

46. Mohan, J., Lanka, K., Rao, A.N.: A review of dynamic job shop scheduling techniques. Proc. Manuf. **30**, 34–39 (2019). https://doi.org/10.1016/J.PROMFG.2019.02.006

47. Wen, X., Lian, X., Qian, Y., Zhang, Y., Wang, H., Li, H.: Dynamic scheduling method for integrated process planning and scheduling problem with machine fault. Robot. Comput. Integr. Manuf. **77**, 102334 (2022). https://doi.org/10.1016/J.RCIM.2022.102334

48. Yan, Y., Wang, Z.: A two-layer dynamic scheduling method for minimising the earliness and tardiness of a re-entrant production line. Int. J. Prod. Res. **50**(2), 499–515 (2011). https://doi.org/10.1080/00207543.2010.543171

49. Muhamadin, K., Bukkur, M.A., Shukri, M.I., Osama, Elmardi, M.: A review for dynamic scheduling in manufacturing. Type: Double Blind Peer Reviewed Int. Res. J. Publ. Glob. J. Online. **18**, 25 (2018)

50. Hu, L., Liu, Z., Hu, W., Wang, Y., Tan, J., Wu, F.: Petri-net-based dynamic scheduling of flexible manufacturing system via deep reinforcement learning with graph convolutional network. J. Manuf. Syst. **55**, 1–14 (2020). https://doi.org/10.1016/J.JMSY.2020.02.004

51. Chang, K., Park, S.H., Baek, J.G.: AGV dispatching algorithm based on deep Q-network in CNC machines environment. Int. J. Comput. Integr. Manuf. **35**(6), 662–677 (2021). https://doi.org/10.1080/0951192X.2021.1992669

52. WITNESS Simulation Modeling Software | Lanner. https://www.lanner.com/en-gb/technology/witness-simulation-software.html. Accessed 16 Jun 2023.

53. Zhang, C., Song, W., Cao, Z., Zhang, J., Tan, P.S., Chi, X.: Learning to dispatch for job shop scheduling via deep reinforcement learning. Adv. Neural. Inf. Proc. Syst. **33**, 1621–1632 (2020)

54. Mnih, V., et al.: Asynchronous methods for deep reinforcement learning. PMLR, 1928–1937 (2016) Accessed 16 June 2023. [Online]. Available: https://proceedings.mlr.press/v48/mniha16.html

55. Mnih, V., et al.: Human-level control through deep reinforcement learning. Nature. **518**(7540), 529–533 (2015). https://doi.org/10.1038/nature14236

56. Kousi, N., Koukas, S., Michalos, G., Makris, S.: Scheduling of smart intra – factory material supply operations using mobile robots. Int. J. Prod. Res. **57**(3), 801–814 (Feb. 2018). https://doi.org/10.1080/00207543.2018.1483587
57. Katoh, N., Ibaraki, T.: Resource allocation problems. In: Handbook of Combinatorial Optimization, pp. 905–1006 (1998). https://doi.org/10.1007/978-1-4613-0303-9_14
58. Chryssolouris, G., Dicke, K., Lee, M.: On the resources allocation problem. Int. J. Prod. Res. **30**(12), 2773–2795 (2007). https://doi.org/10.1080/00207549208948190
59. Chryssolouris, G., Papakostas, N., Mourtzis, D.: A decision-making approach for nesting scheduling: a textile case. Int. J. Prod. Res. **38**(17), 4555–4564 (2010). https://doi.org/10.1080/00207540050205299
60. Michalos, G., Makris, S., Mourtzis, D.: A web based tool for dynamic job rotation scheduling using multiple criteria. CIRP Ann. **60**(1), 453–456 (2011). https://doi.org/10.1016/J.CIRP.2011.03.037
61. Lalas, C., Mourtzis, D., Papakostas, N., Chryssolouris, G.: A simulation-based hybrid backwards scheduling framework for manufacturing systems. Int. J. Comput. Integr. Manuf. **19**(8), 762–774 (2007). https://doi.org/10.1080/09511920600678827
62. Kousi, N., Michalos, G., Makris, S., Chryssolouris, G.: Short – term planning for part supply in assembly lines using mobile robots. Proc. CIRP. **44**, 371–376 (2016). https://doi.org/10.1016/J.PROCIR.2016.02.131
63. Michalos, G., Fysikopoulos, A., Makris, S., Mourtzis, D., Chryssolouris, G.: Multi criteria assembly line design and configuration – An automotive case study. CIRP J. Manuf. Sci. Technol. **9**, 69–87 (2015). https://doi.org/10.1016/J.CIRPJ.2015.01.002
64. Alexopoulos, K., Koukas, S., Boli, N., Mourtzis, D.: Resource planning for the installation of industrial product service systems. IFIP Adv. Inf. Commun. Technol. **514**, 205–213 (2017). https://doi.org/10.1007/978-3-319-66926-7_24/FIGURES/5
65. Siatras, V., Nikolakis, N., Alexopoulos, K., Mourtzis, D.: A toolbox of agents for scheduling the paint shop in bicycle industry. Proc. CIRP. **107**, 1156–1161 (2022). https://doi.org/10.1016/j.procir.2022.05.124

A Manufacturing Digital Twin Framework

Victor Anaya, Enrico Alberti, and Gabriele Scivoletto

1 Introduction

1.1 Definition, Usages, and Types of Digital Twins

The manufacturing industry is continuously evolving, and digital twin (DT) technology has become a prominent driving force in this transformation. DTs play a crucial role in optimizing manufacturing processes, increasing productivity, and enhancing product quality.

A digital twin (DT) is a digital representation of a physical entity or process modeled with the purpose to improve the decision-making process in a safe and cost-efficient environment where different alternatives can be evaluated before implementing them. The digital twin framework (DTF) for manufacturing is a set of components for making and maintaining DTs, which describe the current state of a product, process, or resource.

DTs have pace momentum due to their seamless integration and collaboration with technologies such as IoT, machine learning (ML) algorithms, and analytics solutions. DTs and ML solutions benefit in a bidirectional way, as DTs simulate real environments, being a source of data for training the always data-eager ML algorithms. DTs are a source of data that would be costly to acquire in other conditions such as private data tied to legal and ethical implications, data labeling, complex data cleaning, abnormal data, or data gathering that require intrusive

V. Anaya (✉)
Information Catalyst SL, Xativa, Spain
e-mail: victor.anaya@informationcatalyst.com

E. Alberti · G. Scivoletto
Nextworks SRL, Pisa, Italy
e-mail: e.alberti@nextworks.it; g.scivoletto@nextworks.it

© The Author(s) 2024
J. Soldatos (ed.), *Artificial Intelligence in Manufacturing*,
https://doi.org/10.1007/978-3-031-46452-2_10

processes. In the other direction, ML models are a type of simulation technique that can be used to simulate processes and other entity behaviors for the DT. Some of the algorithms that can be used for simulation are deep learning neural networks, time-series-based algorithms, and reinforcement learning.

DTs are not specific software solutions, but they are a range of solutions that support the improvement of physical products, assets, and processes at different levels and different stages of the lifecycle of those physical assets [3]. Therefore, in the manufacturing domain, DTs can have different scopes such as [1] the following:

1. Process level—recreates the entire manufacturing process. Plant managers use process twins to understand how the elements in a plant work together. Process DT can detect interaction problems between processes at different departments of a company.
2. System level—monitors and improves an entire production line. System-level DTs cover different groups of assets in a specific unit and can be used for understanding and optimizing assets and processes involved in the production of a specific product [12].
3. Asset level—focuses on a single piece of equipment or product within the production line. Asset DTs can cover cases such as the optimization of energy consumption, the management of fleet performance, and the improvement of personnel assignment based on skills and performance.
4. Component level—focuses on a single component of the manufacturing process, such as an item of a product or a machine. Component-level twins help to understand the evolution and characteristics of the modeled component, such as the durability of a drill or the dynamics of a fan.

The creation of a digital copy of a physical object offers significant advantages throughout its entire life cycle [4]. This includes the design phase, such as product design and resource planning, as well as the manufacturing phase, such as production process planning and equipment maintenance. Additionally, during the service phase, benefits include performance monitoring and control, maintenance of fielded products, and path planning. Finally, during the disposal phase, the digital replica can facilitate end-of-life reuse, remanufacturing, and recycling efforts.

A DL framework is a toolkit that allows developers the creation of specific DT instances, and as such is a complex system composed of several tools such as data gathering and synchronization platforms, multi-view modelers, simulator engines, what-if analytic reporting, and vast integration capabilities.

Data collection is a crucial aspect of feeding a digital twin in the manufacturing industry. To ensure the accuracy and reliability of the DT's representation, a robust and efficient data collection platform is essential. Such platforms must possess certain characteristics to meet the requirements of flexibility, availability, and support manufacturing communication protocols, while also ensuring efficiency and security. The data collection platform must have the ability to adapt to different types of data inputs, including sensor readings, machine data, process parameters, and environmental variables. This adaptability enables comprehensive data gathering, capturing a holistic view of the manufacturing process and all its interconnected elements.

Ensuring availability is essential for a data collection platform to be effective. Manufacturing operations typically operate continuously, demanding a constant flow of real-time data. The platform should guarantee uninterrupted data acquisition, seamlessly handling substantial data volumes promptly. It should offer dependable connectivity and resilient infrastructure to prevent data gaps or delays, thereby maintaining synchronization between the DT and its physical counterpart.

To connect to a range of devices, machines, and systems, support for manufacturing communication protocols is crucial. Networked devices that adhere to specific protocols are often utilized in production environments. The data collection platform should therefore be able to interact via well-established protocols like OPC-UA[1], MQTT[2], or Modbus[3]. Rapid data transfer, synchronization, and seamless integration are all made possible by this interoperability throughout the production ecosystem.

Finally, security is of utmost importance in data collection for DTs. Manufacturing data often includes sensitive information, trade secrets, or intellectual property. The data collection platform must implement robust security measures, including encryption, access controls, and data anonymization techniques, to protect the confidentiality, integrity, and availability of the collected data. This ensures that valuable manufacturing knowledge and insights remain protected from unauthorized access or malicious activities.

The rest of this chapter is organized as follows: in the next subsection, we explain DT usages in the manufacturing sector. In Sect. 2, we present the digital twin framework. In Sect. 3, we present the case study and the methodology to experimentally evaluate the proposed method. In Sect. 4, we discuss the conclusion.

1.2 Digital Twin in Manufacturing

Digital twin technology has a wide range of applications in manufacturing, including predictive maintenance, quality management, supply chain management, and customer experience. This technology can help predictive maintenance breakthrough data fatigue and turn data into a competitive advantage [7]. By monitoring equipment data in real time, the DT can predict equipment failures before they occur, reducing downtime and increasing productivity. In a study, DTs of well-functioning machines were used for predictive maintenance, and the discrepancies between each physical unit and its DT were analyzed to identify potential issues before they become critical [8].

---

[1] OPC Unified Architecture. https://opcfoundation.org/about/opc-technologies/opc-ua/
[2] MQTT. https://mqtt.org/
[3] Modbus. https://modbus.org/

- *Predictive Maintenance and Process Optimization*: DTs enable manufacturers to monitor equipment performance and predict potential failures or malfunctions, leading to timely maintenance and reduced downtime. Additionally, DTs can optimize manufacturing processes by simulating different scenarios and identifying bottlenecks and inefficiencies [9].
- *Quality Control and Inspection*: DTs can play a critical role in quality control and inspection processes in manufacturing. By creating a virtual replica of the manufactured product, DTs can detect deviations from the desired specifications and suggest corrective actions to ensure optimal quality [10]. Additionally, DTs can help in automating inspection processes, reducing human error, and increasing efficiency [11].
- *Production Planning and Scheduling*: By simulating the production environment, DTs can assist in creating optimized production schedules and plans, considering various constraints such as resource availability, lead times, and capacity utilization [13]. DTs can also support real-time adjustments to the production plan, allowing manufacturers to adapt to unforeseen events or disruptions [14].
- *Workforce Training and Skill Development*: The integration of DT technology in manufacturing can facilitate workforce training and skill development. By simulating the production environment and processes, DTs enable workers to practice and enhance their skills in a virtual setting, reducing the learning curve and minimizing the risk of errors during real-world operations. Furthermore, DTs can provide personalized training and feedback based on individual performance, promoting continuous improvement [15].
- *Supply Chain Integration and Visibility*: DTs can enhance supply chain integration and visibility in manufacturing by providing real-time information and analytics about various aspects of the supply chain, such as inventory levels, lead times, and supplier performance [16]. This increased visibility enables better decision-making and collaboration among supply chain partners, ultimately improving the overall efficiency and responsiveness of the supply chain.

2 knowlEdge Manufacturing Digital Twin Framework

2.1 Digital Twin Standardization Initiatives

There are many articles referencing potential DT architectures, which provide different forms of naming for the main components and layers of the DT architecture [2, 5, 6].

Most of those DT architectures summarize a DT from a mathematical point of view as a five-dimensional model defined as follows [4]:

$$DT = F(PS, DS, P2V, V2P, OPT) \tag{1}$$

where DT refers to digital twin, that is expressed as a function (F) aggregating: the physical system (PS), the digital system (DS), an updating engine that synchronizes the two words (P2V), a prediction engine that runs prediction algorithms (V2P), and an optimization dimension containing optimizers (OPT).

One of the most relevant initiatives to standardize a DT's main building blocks is the one proposed by ISO 23247 [17] comprising a DT framework that partitions a digital twinning system into layers defined by standards. The framework is based on the Internet of Things (IoT) and consists of four main layers:

- *Observable Manufacturing Elements*: This layer describes the items on the manufacturing floor that need to be modeled. Officially, it is not part of the framework, as it already exists.
- *Device Communication Entity*: This layer collates all state changes of the observable manufacturing elements and sends control programs to those elements when adjustments become necessary.
- *Digital Twin Entity*: This layer models the DTs, reading the data collated by the device communication entity and using the information to update its models.
- *User Entities*: User entities are applications that use DTs to make manufacturing processes more efficient. They include legacy applications like ERP and PLM, as well as new applications that speed up processes.

On the other hand, the Digital Twin Capabilities Periodic Table (CPT) [16] is a framework developed by the Digital Twin Consortium to help organizations design, develop, deploy, and operate DTs based on use case capability requirements. The CPT is architecture and technology agnostic, meaning it can be used with any DT platform or technology solution. The framework clusters capabilities around common characteristics using a periodic-table approach:

The CPT framework clusters capabilities into the following main clusters:

- *Data Management*: This cluster includes capabilities related to data access, ingestion, and management across the DT platform from the edge to the cloud.
- *Modeling and Simulation*: This cluster includes capabilities related to the creation of virtual models and simulations of real-world entities and processes.
- *Analytics and Artificial Intelligence*: This cluster includes capabilities related to the use of analytics and artificial intelligence to analyze data and generate insights.
- *Visualization and User Interface*: This cluster includes capabilities related to the visualization of digital twin data and the user interface used to interact with the DT
- *Security and Privacy*: This cluster includes capabilities related to the security and privacy of DT data and systems
- *Interoperability and Integration*: This cluster includes capabilities related to the integration of DT systems with other systems and the interoperability of DT data with other data sources.

ISO 23247 and the Digital Twin Capabilities Periodic Table are generic frameworks that are worth taking into consideration when developing a digital twin framework because they provide a consistent and structured approach to digital twin implementation. Section 2.3 presents the alignment carried out between the knowlEdge Digital Twin Framework and the ISO 23247.

2.2 knowlEdge Digital Twin Framework

The knowlEdge DT framework is a toolkit solution composed of a set of modeling, scheduling, visualization, analysis, and data gathering and synchronization components that is capable to create instances of manufacturing digital twins at different scopes and phases of the product, process and asset lifecycle.

The components composing the solution (see Fig. 1) are described as follows:

- *Sensor Reader Interface*: This interface is composed of the set of field protocols needed for connecting the pilots' sensors to the knowlEdge Data Collection Platform [18]. The interface has to be aware of the details of the protocols in terms of networking, configuration, and specific data model.
- *Sensor Protocol Adapter*: Once one data has been read, the sensor protocol adapter can distinguish whether the data is meaningful for the Data Collection Platform or has to be collected and presented as raw data.
- *Unified Data Collector*: The module is responsible to add the semantic to the lower-level object and make them available to the upper level.
- *Data Model Abstractor*: The Data Model Abstractor unifies the different information models that depend on the specific field protocol to hide that information when the data is presented to the real-time broker.
- *Data Ingestion*: This interface is responsible for offering different mechanisms to communicate with the DTs' framework, such as MQTT or REST API services.
- *Platform Configurator*: The platform configurator exposes a REST API for the configuration of all the internal and external modules. Examples of configurations are the topic where the platform publishes the data, the configuration of the platform when a new sensor is been plugged into the system, its information model, etc.
- *DT Designer UI Interface*:
  - DT Domain Model UI: this is the UI interface that allows an IT person or a skilled operator to define the DT domain data model, that is, the digital assets containing the model, with their features and to assign to them their behavior and its graphical representation. This UI will provide subsections to specify simulation services.
  - DT Visual Editor: this component allows to edit 3D elements that will be used to animate 3D visualization when needed.
- *DT UI Interface*: It is the end-user UI set of interfaces used for running simulations and visualizing results through reports.

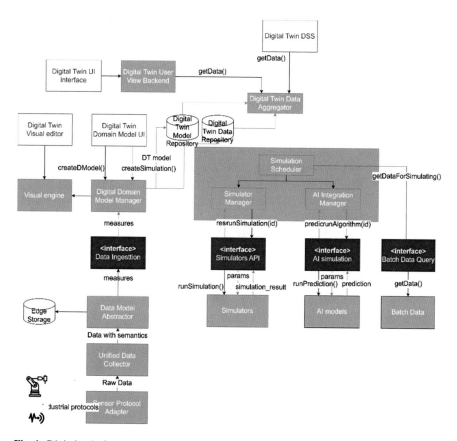

Fig. 1 Digital twin framework architecture

- *DT User View Backend*: It is the backend engine that according to the decision view of the digital twin can represent the different widgets (indicators, tables, 3D view) that were defined in design time.
- *Digital Domain Model Manager*: This is the main backend of the DT. It is in charge to create new DT instances based on data model definitions and connect them to existing simulators and other AI algorithms (such as reinforcement learning for production scheduling, neural networks for simulating the energy consumption of manufacturing machines). Domain Data Models contain the digital entities that will be part of the digital twin model, that is, the machines, resources, places, and people. The Digital Domain Model Manager will support the decomposition of digital elements in their parts through trees, and their connection with the physical objects through the knowlEdge real-time brokering component.
- *DT Data Aggregator*: It is the backend component in charge of maintaining the DT model synchronized with the physical counterparts and offering APIs with

the rest of the components of the architecture. One of its components is the context broker, which is based on the FIWARE Stellio Context Broker[4].

- *3D Visualization Engine*: This component can render 3D scenes of the simulations when a design is provided. Their results can be embedded into dashboards used by the operators when running simulations.
- *Behavior Manager*: This component is in charge of keeping a linkage with endpoints of the algorithms that define the behavior of digital things, for instance, a linkage to the knowlEdge HP AI component that provides a REST API to the repository of knowledge AI algorithms that are to be tested using the DT. This subcomponent is also in charge of keeping a repository of linkages to simulators and other algorithms through a REST API that can be third-party solutions provided by external providers. The behavior manager has a scheduler engine that runs simulations according to time events or data condition rules that are executed against the DT model that is being filled with data from the IoT devices.

2.3   knowlEdge Digital Twin Framework Alignment with Current Initiatives

The importance of aligning the knowlEdge DT framework to the ISO 23247 standard on DTs cannot be overstated. The ISO 23247 series defines a framework to support the creation of DTs of observable manufacturing elements, including personnel, equipment, materials, manufacturing processes, facilities, environment, products, and supporting documents. Aligning the knowlEdge DT framework to the ISO 23247 standard can help ensure that the framework follows recognized guidelines and principles for creating DTs in the manufacturing industry. The following are the specific blocks of the ISO 23247 standard that have been aligned to the knowlEdge DT framework:

- The knowlEdge DT has considered the terms and definitions provided by ISO 23247 standard to ensure that the framework is consistent with the standard.
- The knowlEdge DT provides many of the ISO 23247's functional entities (see Fig. 2, where the different colors are used to emphasize which functional entity from the ISO 23247 is covered by each functional block of the knowlEdge DT). It supports all the ISO 23247 functional entities based on the rest of the components provided by the knowlEdge project. This ensures that the framework meets the needs of the manufacturing industry.
- The knowlEdge DT provides a Graphical DT Domain Data Modeler Editor that has been customized with the ISO 2347 Observable Manufacturing Elements, so users define the digital things using the exact terminology of the standard,

---

[4] https://stellio.readthedocs.io/en/latest/index.html

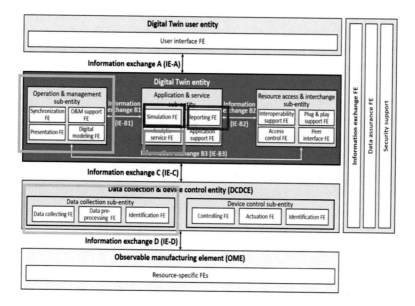

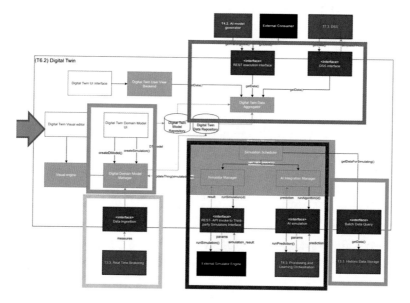

Fig. 2 ISO 2347 vs. digital twin framework architecture alignment

such as Personnel, Equipment, Material, Manufacturing Processes, Facilities, Environment, Products, and Supporting Documents.

- The knowlEdge DT provides integration mechanisms that make its usage in the application and services described in ISO 2347 possible.

3 knowlEdge Digital Twin for Process Improvement

DTs have become an essential tool for improving shop floor processes in the manufacturing industry. One specific application of a DT is for scheduling process improvement. By using a DT, manufacturers can optimize their production schedules to improve efficiency and reduce costs. The following is a description of how the DT framework was applied to a dairy company within the knowlEdge project to improve the management, control their processes, and automatize the scheduling of the weekly production of yoghurt.

The knowlEdge Data Collection Platform (DCP) is used to connect to the shop floor for gathering production and demand data. The platform was integrated with various sensors and devices to collect data in real time. The DCP was also used to collect data from various sources, such as the company's ERP[5]. By collecting data from various sources, manufacturers can get a complete picture of their production and demand data. The data was passed through the data collection platform for filtering, formatting and normalization. This assured the proper quality of data and ensured that the DT is accurate and reliable (see Fig. 3).

```
{
    "timestamp":"<Timestamp of msg reception>",
    "originalTimestamp":"<Original timestamp in payload (optional)>",
    "sourceType":"<same as in topic specs>",
    "sourceID":"<same as in topic specs>",
    "infoType":"<same as in topic specs>",
    "dataType":"<same as in topic specs>",
    "dataItemID":"<same as in topic specs>",
    "metricTypeID":"<identifies how to interpret the metricValue>",
    "metricValue":"<Value>",
    "measureUnit":"<Measure Unit related to the Value (optional)>"
}
```

Fig. 3 Standard format for exchanging information among knowledge components

---

[5] Enterprise Resource Planning.

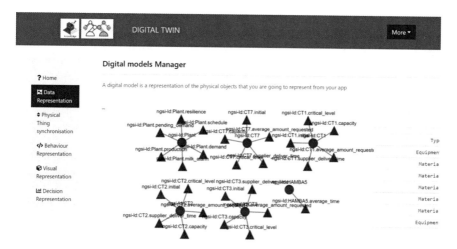

Fig. 4 Digital twin framework model editor

Fig. 5 Digital twin dashboard generated with the framework

The processed data is pushed through an MQTT broker to the DT framework. The DT uses the data to model the behavior and performance of the manufacturing process. The DT framework was used to model the plant using ISO 2347 Observable Manufacturing Elements concepts (see Fig. 4).

The DT framework was also used to simulate different mechanics of scenarios to identify opportunities for process improvement, specifically for computing a production schedule based on metaheuristic rules provided by the company and for simulating the execution of the schedules based on what-if scenarios, allowing the manufacturing operators to select the optimal production plan according to a range of aspects such as timestamp, resource occupation, uncertainty resilience, or customer satisfaction. Figure 5 shows a partial view of the dashboard generated using the DT Decision View and populated with the information and the results of the DT simulators.

4 Conclusions

Digital twin technology has the potential to revolutionize the manufacturing industry by optimizing processes, increasing productivity, and enhancing product quality. By leveraging advanced digital techniques, simulations, and hybrid learning-based modeling strategies, DT technology can help overcome the challenges faced by traditional manufacturing methods and pave the way for the next generation of smart manufacturing.

This chapter has presented the knowlEdge DT framework, an open-source toolkit of DT modules supporting the modeling of physical assets and processes, and the execution of functional and AI-based simulators for the execution of what-if scenarios for improving the decision-making process. The tool has been used to also for the generation of synthetic data for training AI algorithms. It is composed of a set of modules as a DT Data Modeler, 3D twin modeler, IoT Ingestion Connector, Simulator/AI Manager and Repository, Event Scheduler, DT Live Dashboard, and the Data Collection Platform.

The DT Framework proposed was successfully used for creating a manufacturing DT instance for generating weekly manufacturing schedules based on a rule-based simulator and a discrete event simulator. The company where it was applied has improved their reactiveness to incidents occurring on the shop floor, optimizing the rescheduling process accordingly.

As more case studies and practical implementations emerge, the true potential of DT technology in manufacturing will become increasingly apparent, driving further transformation and innovation in the industry.

Acknowledgment This work has received funding from the European Union's Horizon 2020 research and innovation programs under grant agreement No. 957331—KNOWLEDGE. This paper reflects only the authors' views, and the Commission is not responsible for any use that may be made of the information it contains.

References

1. Stavropoulos, P., Mourtzis, D.: Digital twins in industry 4.0. In: Design and Operation of Production Networks for Mass Personalization in the Era of Cloud Technology, pp. 277–316. Elsevier, Amsterdam (2022)
2. Ogunsakin, R., Mehandjiev, N., Marin, C.A.: Towards adaptive digital twins architecture. Comput. Ind. **149**, 103920 (2023)

3. He, B., Bai, K.J.: Digital twin-based sustainable intelligent manufacturing: a review. Adv. Manuf. **9**, 1–21 (2021)
4. Thelen, A., Zhang, X., Fink, O., Lu, Y., Ghosh, S., Youn, B.D., et al.: A comprehensive review of digital twin—part 1: modeling and twinning enabling technologies. Struct. Multidiscip. Optim. **65**(12), 354 (2022)
5. Kim, D.B., Shao, G., Jo, G.: A digital twin implementation architecture for wire+ arc additive manufacturing based on ISO 23247. Manuf. Lett. **34**, 1–5 (2022)
6. Shao, G., Helu, M.: Framework for a digital twin in manufacturing: Scope and requirements. Manuf. Lett. **24**, 105–107 (2020)
7. Farhadi, A., Lee, S.K., Hinchy, E.P., O'Dowd, N.P., McCarthy, C.T.: The development of a digital twin framework for an industrial robotic drilling process. Sensors. **22**(19), 7232 (2022)
8. Zhong, D., Xia, Z., Zhu, Y., Duan, J.: Overview of predictive maintenance based on digital twin technology. Heliyon (2023)
9. Hassan, M., Svadling, M., Björsell, N.: Experience from implementing digital twins for maintenance in industrial processes. J. Intell. Manuf., 1–10 (2023)
10. Lee, J., Lapira, E., Bagheri, B., Kao, H.A.: Recent advances and trends in predictive manufacturing systems in big data environment. Manuf. Lett. **1**(1), 38–41 (2013)
11. Rodionov, N., Tatarnikova, L.: Digital twin technology as a modern approach to quality management. In: E3S Web of Conferences, vol. 284, p. 04013. EDP Sciences (2021)
12. Fang, Y., Peng, C., Lou, P., Zhou, Z., Hu, J., Yan, J.: Digital-twin-based job shop scheduling toward smart manufacturing. IEEE Trans. Ind. Inf. **15**(12), 6425–6435 (2019)
13. Lu, Y., Liu, C., Kevin, I., Wang, K., Huang, H., Xu, X.: Digital twin-driven smart manufacturing: connotation, reference model, applications and research issues. Robot. Comput. Integr. Manuf. **61**, 101837 (2020)
14. Ogunseiju, O.R., Olayiwola, J., Akanmu, A.A., Nnaji, C.: Digital twin-driven framework for improving self-management of ergonomic risks. Smart Sustain. Built Environ. **10**(3), 403–419 (2021)
15. Ivanov, D., Dolgui, A., Sokolov, B.: The impact of digital technology and Industry 4.0 on the ripple effect and supply chain risk analytics. Int. J. Prod. Res. **57**(3), 829–846 (2019)
16. Capabilities Periodic Table – Digital Twin Consortium. Digital Twin Consortium. Published August 8, 2022. https://www.digitaltwinconsortium.org/initiatives/capabilities-periodic-table/. Accessed 2 June 2023
17. ISO 23247-2: ISO 23247-2: Automation Systems and Integration – Digital Twin Framework for Manufacturing – Part 2: Reference Architecture. International Organization for Standardization, Geneva (2021)
18. Wajid, U., Nizamis, A., Anaya, V.: Towards Industry 5.0–A Trustworthy AI Framework for Digital Manufacturing with Humans in Control. Proceedings http://ceur-ws.org. ISSN, 1613, 0073 (2022)

Reinforcement Learning-Based Approaches in Manufacturing Environments

Andrea Fernández Martínez, Carlos González-Val, Daniel Gordo Martín, Alberto Botana López, Jose Angel Segura Muros, Afra Maria Petrusa Llopis, Jawad Masood, and Santiago Muiños-Landin

1 Introduction

Over the past few decades, there has been a significant surge in the digitalization and automation of industrial settings, primarily driven by the adoption of Industry 4.0 principles. At its essence, Industry 4.0 aims to establish a world of interconnected, streamlined, and secure industries, built upon fundamental concepts such as the advancement of cyber-physical systems (CPS) [1–3], the Internet of Things (IoT) [4–6], and cognitive computing [7].

Computer numerical control machines (CNCs) play a pivotal role in aligning with the principles of Industry 4.0 [8–10], facilitating automated and efficient manufacturing of intricate and high-quality products. They have revolutionized various industries such as woodworking, automotive, and aerospace by enhancing automation and precision. By automating industrial processes, CNCs reduce the need for manual labor in repetitive and non-value-added activities, fostering collaboration between machine centers and human operators in factory settings [11]. Moreover, CNCs' modular design and operational flexibility empower them to perform a wide range of applications with minimal human intervention, ensuring the creation of secure workspaces through built-in security measures. These machines often incorporate advanced sensing and control technologies, optimizing their performance and minimizing downtime.

In parallel with the rapid adoption of CNCs in the market, simulation techniques have evolved to meet the industry's latest requirements. The emergence of the digital

A. Fernández Martínez · C. González-Val · D. G. Martín · A. B. López · J. A. S. Muros
A. M. P. Llopis · J. Masood · S. Muiños-Landin (✉)
AIMEN Technology Centre, Smart Systems and Smart Manufacturing Group, Pontevedra, Spain
e-mail: andrea.fernandez@aimen.es; carlos.gonzalez@aimen.es; daniel.gordo@aimen.es;
alberto.botana@aimen.es; jose.segura@aimen.es; afra.pertusa@aimen.es;
jawad.masood@aimen.es; santiago.muinos@aimen.es

© The Author(s) 2024
J. Soldatos (ed.), *Artificial Intelligence in Manufacturing*,
https://doi.org/10.1007/978-3-031-46452-2_11

twin (DT) [3] concept has particularly contributed to advancing cyber-physical systems (CPS) by establishing seamless coordination and control between the cyber and physical components of a system [12]. While there is no universally accepted definition of digital twin, it can be understood as a virtual representation of a physical machine or system. DTs offer numerous advantages for controlling and analyzing the performance of physical machines without the need for direct physical intervention. By conducting research and testing on virtual representations instead of physical machines, specialists can experiment and evaluate process performance from technological and operational perspectives, conserving physical resources and avoiding associated costs such as energy consumption, operational expenses, and potential safety risks during the research and development stages [13, 14].

The exponential growth of data generated by machines, coupled with the integration of information from their digital twins [3], has opened up new possibilities for data-driven advancements [12]. These developments leverage state-of-the-art analysis techniques to optimize processes in an adaptive manner. In the realm of robotics and automation, reinforcement learning has emerged as a foundational technology for studying optimal control. Reinforcement learning [15–17], a branch of Artificial Intelligence (AI), revolves around analyzing how intelligent agents should behave within a given environment based on a reward function. The underlying principle of RL draws inspiration from various fields of knowledge, including social psychology. In RL algorithms, intelligent agents interact with their environment, transitioning between states to discover an optimal policy that maximizes the expected value of their total reward function. These algorithms hold tremendous potential for overcoming existing limitations in the control of robotic systems at different length scales [18], offering new avenues for advancements in this field.

A significant challenge in the realm of mobile industrial machinery lies in designing path trajectories that effectively control robot movement [19–22]. Traditionally, computer-aided design (CAD) and computer-aided manufacturing (CAM) systems are utilized to generate these trajectories, ensuring adherence to security specifications such as maintaining a safe distance between the robot-head and the working piece to prevent product damage. However, these path trajectories often exhibit discontinuities and encounter issues in corners and curves due to the mechanical limitations of the physical machinery. Moreover, factors such as overall distance or the number of movements between two points, as well as the possibility of collisions among the robot's moving parts, are not efficiently optimized by these systems.

The optimization of path trajectories becomes increasingly complex as the number of movement dimensions and potential options for movement increases. In this context, reinforcement learning emerges as a promising solution for addressing high-dimensional spaces in an automated manner [23, 24], enabling the discovery of optimal policies for controlling robotic systems with a goal-oriented approach. Reinforcement learning algorithms offer the potential to tackle the challenges associated with path trajectory optimization, providing a framework for finding efficient and effective movement strategies for robotic systems [25]. By leveraging reinforcement learning techniques, mobile industrial machinery can navigate complex environments and optimize their trajectories, taking into account multiple dimensions of movement and achieving superior control and performance through an enhanced perception that represents the knowledge developed by reinforcement learning systems.

2 Reinforcement Learning

Reinforcement learning (RL) [15, 16] is widely recognized as the third paradigm of Artificial Intelligence (AI), alongside supervised learning [26, 27] and unsupervised learning [28]. RL focuses on the concept of learning through interactive experiences while aiming to maximize a cumulative reward function. The RL agent achieves this by mapping states to actions within a given environment, with the objective of finding an optimal policy that yields the highest cumulative reward as defined in the value function.

Two fundamental concepts underpin RL: trial-and-error search and the notion of delayed rewards. Trial-and-error search involves the agent's process of selecting and trying different actions within an initially uncertain environment. Through this iterative exploration, the agent gradually discovers which actions lead to the maximum reward at each state, learning from the outcomes of its interactions.

The concept of delayed rewards [15, 26] emphasizes the consideration of not only immediate rewards but also the expected total reward, taking into account subsequent rewards starting from the current state. RL agents recognize the importance of long-term consequences and make decisions that maximize the cumulative reward over time, even if it means sacrificing immediate gains for greater overall rewards.

By incorporating trial-and-error search and the notion of delayed rewards, RL enables agents to learn effective policies by actively interacting with their environment, continuously adapting their actions based on the feedback received, and ultimately maximizing cumulative rewards.

Reinforcement learning (RL) problems consist of several key elements that work together to enable the learning process. These elements include a learning agent, an environment, a policy, a reward signal, a value function, and, in some cases, a model of the environment. Let's explore each of these elements:

1. **Learning agent**: The learning agent is an active decision-making entity that interacts with the environment. It aims to find the optimal policy that maximizes the long-term value function through its interactions. The specific approach and logic employed by the learning agent depend on the RL algorithm being used.
2. **Environment**: The environment is where the learning agent operates and interacts. It can represent a physical or virtual world with its own dynamics and rules. The environment remains unchanged by the actions of the agent, and the agent must navigate and adapt to its dynamics to maximize rewards.

3. **Policy**: The policy determines the behavior of the learning agent by mapping states in the environment to the actions taken by the agent. It can be either stochastic (probabilistic) or deterministic (fixed), guiding the agent's decision-making process.

4. **Reward signal**: The reward signal is a numerical value that the agent receives as feedback after performing a specific action in the environment. It represents the immediate feedback obtained during state transitions. The goal of the agent is to maximize the cumulative rewards over time by selecting actions that yield higher rewards.

5. **Value function**: The value function represents the expected total reward obtained by the agent in the long run, starting from a specific state. It takes into account the sequence of expected rewards by considering the future states and their corresponding rewards. The value function guides the agent in estimating the desirability of different states and helps in decision-making.

6. **Model** (optional): In some cases, RL algorithms incorporate a model of the environment. The model mimics the behavior of the environment, enabling the agent to make inferences about how the environment will respond to its actions. However, in model-free RL algorithms, a model is not utilized.

In a typical reinforcement learning (RL) problem, the learning agent interacts with the environment based on its policy. The agent receives immediate rewards from the environment and updates its value function accordingly. This RL framework is rooted in the Markov decision process (MDP) [28], which is a specific approach used in process control.

RL has been proposed as a modeling tool for decision-making in both biological [29] and artificial systems [18]. It has found applications in various domains such as robotic manipulation, natural language processing, and energy management. RL enables agents to learn optimal strategies by exploring and exploiting the environment's feedback. Inverse RL, which is based on hidden Markov models, is another extensively studied topic in the field. Inverse RL aims to extract information about the underlying rules followed by a system that generate observable behavioral sequences. This approach has been applied in diverse fields including genomics, protein dynamics in biology, speech and gesture recognition, and music structure analysis. The broad applicability of RL and its ability to address different problem domains make it a powerful tool for understanding and optimizing complex systems in various disciplines.

Through iterative interactions, the agent adjusts its policy and value function to optimize its decision-making process and maximize cumulative rewards. Figure 1 illustrates the common interaction flow in an RL problem.

As previously mentioned, the learning agent is situated within and interacts with its environment. The environment's state reflects the current situation or condition, defined along a set of possible states denoted as S. The agent moves between states by taking actions from a set of available actions, denoted as A. Whenever the agent chooses and performs an action α from A, the environment E undergoes a transformation, causing the agent to transition from one state S to another S', where

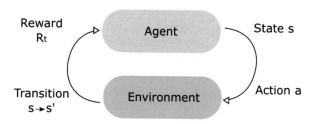

Fig. 1 Reinforcement learning classic feedback loop

$(S, S') \epsilon S$. Additionally, the agent receives a reward γ based on the chosen action α. The ultimate objective of the agent is to maximize the expected cumulative reward Rt over the long term, which can be estimated and reestimated through the learning process of the agent to include and adapt to the new knowledge acquired.

A significant challenge in RL is striking the right balance between exploration and exploitation. On one hand, it is advantageous for the agent to exploit its existing knowledge gained from past experiences. By selecting actions that have previously yielded high rewards, the agent aims to maximize the cumulative reward over time. On the other hand, exploration is crucial to enable the agent to discover new states and potentially identify better actions, thus avoiding suboptimal policies. Different RL algorithms employ various approaches to address this trade-off.

A fundamental characteristic of MDPs and RL is their ability to handle stochastic influences in the state–action relationship. This stochasticity is typically quantified by a transition function, which represents a family of probability distributions that describe the potential outcomes resulting from an action taken in a particular state.

By knowing the transition function, the agent can estimate the expected outcomes of applying an action in a state by considering all possible transitions and their corresponding probabilities. This analysis allows the agent to assess the desirability or undesirability of certain actions.

To formalize this process, a value function U is defined [16]. The value function assigns a numerical value to each state, representing the expected cumulative reward the agent can achieve starting from that state and following a specific policy. It serves as a measure of the desirability or utility of being in a particular state.

The value function helps guide the agent's decision-making process by allowing it to compare the potential outcomes and make informed choices based on maximizing the expected cumulative reward over time.

$$U^*(s) = \max_{\pi} E \left(\sum_{t=0}^{\infty} \gamma^t r_t \right)$$

Indeed, the parameter γ in the value function equation is widely referred to as the discount factor. It plays a pivotal role in regulating the importance of future events during the decision-making process, considering the delayed nature of rewards. By adjusting the discount factor, one can determine the relative significance of immediate rewards compared to future rewards. In the optimal value function equation, the discount factor appears to discount future rewards geometrically [15, 16]. This means that rewards obtained in the future are typically weighted less compared to immediate rewards. However, it's important to note that the specific value and impact of the discount factor depend on the chosen model of optimality. There are three main models of optimality that can be considered: finite horizon, infinite horizon, and average reward. In the infinite horizon model, which we are focusing on here, the discount factor is used to discount future rewards geometrically.

In the equation, the policy function π represents the mapping from states to actions and serves as the primary focus of optimization for the RL agent. It determines the action to be taken in each state based on the agent's acquired knowledge. The asterisk (*) symbol signifies the "optimal" property of the function being discussed, indicating that the equation represents the value function associated with the optimal policy.

One can extend the expression of the optimal value by writing the expected value of the reward using the transition function T as

$$U^*(s) = \max_a \left[r\,(s, a) + \gamma \sum_{s \in S} T\left(s, a, s'\right) U^*\left(s'\right) \right]$$

This is the Bellman equation, which is a fundamental concept in dynamic programming. It encompasses the maximization operation, highlighting the non-linearity inherent in the problem. The solution to the Bellman equation yields the policy function, which determines the optimal actions to be taken in different states.

$$\pi^*(s) = \operatorname*{argmax}_a \left[r\,(s, a) + \gamma \sum_{s \in S} T\left(s, a, s'\right) U^*\left(s'\right) \right]$$

As mentioned above, it returns the action to be applied on each state so that once converged it returns the best action to be applied on each state.

The value function within a MDP can be also expressed or summarized in a matrix that stores the value associated with an action a in a given state s. This matrix is typically called Q-matrix and is represented by

$$U^*(s) = \max_a Q^*\,(s, a)$$

so that the Bellman equation results

$$Q^*\,(s, a) = r\,(s, a) + \gamma \sum_{s \in S} T\left(s, a, s'\right) \max_a Q^*\left(s', a'\right)$$

The opening to what is commonly known as Q-learning is generally facilitated by this approach. It is important to note that the system's model, also known as the transition function, may either be known or unknown. In the case of model-based Q-learning, the model is known, while in model-free Q-learning, it is not. When dealing with an unknown model, the temporal differences approach has proven to be an effective tool for tackling strategy search problems in actual systems. In this approach, the agent is not required to possess prior knowledge of the system's model. Instead, information is propagated after each step, eliminating the need to wait until the conclusion of a learning episode. This characteristic renders the implementation of this method more feasible in real robotic systems.

2.1 Toward Reinforcement Learning in Manufacturing

Teaching agents to control themselves directly from high-dimensional sensory inputs, such as vision and speech, has long been a significant challenge in RL. In many successful RL applications in these domains, a combination of hand-crafted features and linear value functions or policy representations has been utilized. It is evident that the performance of such systems is heavily dependent on the quality of the feature representation employed.

In recent years, deep learning has witnessed significant progress, enabling the extraction of high-level features directly from raw sensory data. This breakthrough has had a transformative impact on fields such as computer vision and speech recognition. Deep learning (DL) techniques leverage various neural network architectures such as convolutional networks, multilayer perceptrons, restricted Boltzmann machines, and recurrent neural networks. These architectures have successfully employed both supervised and unsupervised learning approaches. Given these advancements, it is reasonable to inquire whether similar techniques can also benefit reinforcement learning (RL) when dealing with sensory data.

The advancements in deep learning (DL) [30] have paved the way for deep neural networks (DNNs) to automatically extract compact high-dimensional representations (features). This capability is particularly useful for overcoming the dimensional catastrophe, commonly encountered in domains such as images, text, and audio. DNNs possess powerful representation learning properties, enabling them to learn meaningful features from raw data. Deep reinforcement learning (DRL) [31] refers to a class of RL algorithms that leverage the representation learning capabilities of DNNs to enhance decision-making abilities.

The algorithm framework for DNN-based RL is illustrated in Fig. 2. In DRL, the DNN plays a crucial role in extracting relevant information from the environment and inferring the optimal policy $\pi*$ in an end-to-end manner. Depending on the specific algorithm employed, the DNN can be responsible for outputting the Q-value (value-based) for each state–action pair or the probability distribution of the output action (policy-based). The integration of DNNs into RL enables more efficient and effective decision-making by leveraging the power of representation learning.

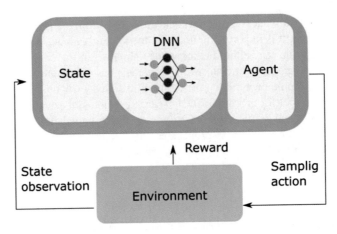

Fig. 2 Representation of the deep reinforcement learning (DRL) feedback loop

Such scenario made much more accessible or tractable classic problems in manufacturing frameworks to RL approaches. In the following section we present two cases where RL has migrated to DRL to address specific problems within manufacturing environments.

3 Deep Reinforcement Learning in Virtual Manufacturing Environments

In this section, we present two distinct examples that demonstrate the advancements made in RL, specifically in the context of deep reinforcement learning (DRL). These examples involve the application of RL within a virtual environment, which allows for the development of strategies that can later be translated to real systems. This approach opens up the possibility of deploying this technology in manufacturing environments.

One key advantage of utilizing virtual environments is that it mitigates the significant amount of learning episodes typically required by RL agents to develop an optimal policy. In a real manufacturing system, the time needed to explore numerous strategies would make the process highly inefficient for reaching an optimal solution. Moreover, certain strategies may introduce risks, such as safety concerns, that cannot be easily managed or assumed in a manufacturing environment. For instance, in the second example, robotic systems operating collaboratively may pose safety risks, while in the first example, machines with high power consumption may introduce operational risks.

By leveraging virtual environments [32, 33], RL techniques can be effectively applied to develop optimal strategies while minimizing risks and reducing the time and costs associated with experimentation in real systems. This approach enables the integration of RL technology into manufacturing environments, paving the way for enhanced efficiency, productivity, and safety [34–36]. Considering these issues, the development of digital environments (such as simulation platforms or digital twins) has been taken as the ideal scenario to train RL agents until the systems reach certain level of convergency or, in other words, trustworthiness. Once, certain strategies have reached a reasonable point, they can be tested in real scenario, and even a reduced optimization process can take place at that point to finally find the optimal strategy in the real context.

In this sense, the quality of the virtual context or the divergency with respect to the real process, becomes critical for the achievement of an optimal strategy later in the real world. However, the optimal digitalization of processes is out of the scope of this chapter.

The two scenarios presented in this section address the optimization of two different systems. The first one is the optimization of trajectories in a CNC cutting machine designed for different operations over large wood panels. The problem in this case is the optimization of the path between two different operations (cutting, drilling, milling, etc.). The second scenario faces the robotic manipulation of a complex material. In particular, the problem is the manipulation of fabric by two robotic arms in order to reduce wrinkles. These problems have been addressed within specifically developed digital environments to deliver optimal strategies that are later tested in the real system.

3.1 CNC Cutting Machine

The digital twin (DT) of the physical CNC presented here was developed and shared by one of the partners along the MAS4AI Project (GA 957204) within the ICT-38 AI for manufacturing cluster for the development of the RL framework. The DT is built on X-Sim, and it incorporates the dynamics of the machine, its parts, and the effects on the working piece, simulating the physical behavior of the CNC. The CNC of our study was a machining center for woodworking processes, more specifically for cutting, routing, and drilling.

The machine considered (not shown for confidentiality issues) consists of a working table in which wood pieces are located and held throughout the woodworking process and a robot-head of the CNC, which is responsible for performing the required actions to transform the raw wood piece into the wood product (Fig. 3).

The model-based machine learning (ML) agent developed in this context aims at optimizing the path trajectories of the five-axes head of a CNC within a digital twin environment. Currently, the DT CNC enables the 3D design of the wood pieces by users, creating all the necessary code to control the physical CNC.

Controlling a five-axes head automatically in an optimized way is yet an engineering challenge. The CNC must not only perform the operation required to transform the wood piece into the desired product, but it must also avoid potential collisions with other parts of the machine, control the tools required by the head

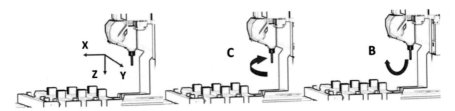

Fig. 3 Movements of the robot head in its five-axes

of the robot, and keep times short reducing unnecessary movements to enhance productivity and save energy, while ensuring the machine integrity and safety of operators throughout the process, as well as high-quality products.

In this context, a model-based ML agent based on a DRL framework was trained to optimize the path trajectories of the five-axes head of the CNC in order to avoid potential collisions while optimizing the overall time operation. The difficulty of working in a five-dimensional space (due to the five-axis robot head) is increased by the dimensions of the working table of the CNC, which goes up to 3110 × 1320 mm. In the DT environment, the measurement scale is micrometers, resulting in more than 4,105,200 million states to be explored by the agent in a discrete state–action space only in the plane XY of the board, without considering the extra three-axes of the robot head. This complex applicability of discrete approaches is the reason why only a continuous action space using Deep Deterministic Policy Gradient (DDPG) [37] is shown here.

The model-based AI CNC agent was trained to work considering different operations. The ultimate goal of the agent is to optimize the path trajectories between two jobs in a coordinated basis considering the five-axes of the CNC head. For this reason, the inputs of the model are the coordinates of the initial location, i.e., state of the five-axes head, and the destination location or a label representing the desired operation to be performed by the CNC. The agent returns the optimized path trajectory to reach the goal destination by means of a set of coordinates representing the required movements of the five-axes head. Currently, the agent has been trained separately to perform each operation independently. In a future stage, a multi-goal DRL framework was explored in order to enhance generalization.

Different operations and different learning algorithms were explored during the development of the deep RL framework, including 2-D, 3-D, and 5-D movements of the five-axes head of the CNC, different path trajectories to be optimized, and different learning algorithms including Q-learning [15, 38], deep Q-learning (DQL) [39], and DDPG.

As seen previously, Q-learning is a model-free, off-policy RL algorithm that seeks to find an optimal policy by maximizing a cost function that represents the expected value of the total reward over a sequence of steps. It is used in finite Markov decision processes (stochastic, discrete), and it learns an optimal action-selection policy by addressing the set of optimal actions that the agent should take in order to maximize the total reward (R_t). The algorithm is based on an agent, a set

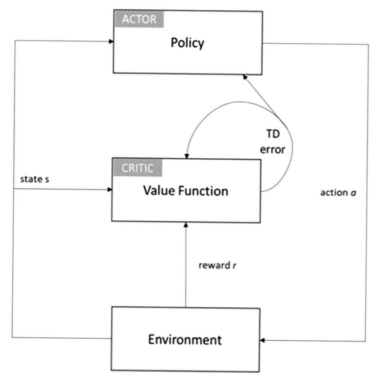

Fig. 4 Actor–critic architecture workflow [41]

of actions A, a set of states S, and an environment E. Every time the agent selects and executes an action $a \in A$, the environment E is transformed, and the agent transitions from one state, s, to another, s', with $(s, s') \in S$, receiving a reward r according to the action selected.

DDPG is an off-policy algorithm that simultaneously learns a Q-function and a policy based on the Bellman equation in continuous action spaces. DDPG makes use of four neural networks, namely an actor, a critic, a target actor, and a target critic. The algorithm is based on the standard "actor" and "critic" architecture [40], although the actor directly maps states to actions instead of a probability distribution across discrete action spaces.

In order to solve the problem of exhaustively evaluating all possible actions from a continuous action space, DDPG learns an approximator to $Q(s, a)$ by means of a neural network, the critic $Q_\theta(s, a)$, with θ corresponding to the parameters of the network (Fig. 4).

Q_θ learns from an experience replay buffer that serves as a memory for storing previous experiences. This replay buffer contains a set D of transitions, which includes the initial state (s), the action taken (a), the obtained reward (r), the new state reached (s'), and whether the state is terminal or not (d). In other words, each transition is represented as (s, a, r, s', d), where s, a, r, s', and d are elements of the set D.

In order to evaluate the performance of Q_θ in relation to the Bellman equation, the mean-squared Bellman error (MSBE) can be computed. The MSBE quantifies the discrepancy between the estimated Q-values produced by Q_θ and the values predicted by the Bellman equation. It is typically calculated by taking the mean squared difference between the Q-value estimate and the expected Q-value, using the current parameters θ. The MSBE provides a measure of how well the Q-function approximated by Q_θ aligns with the optimal Q-values as defined by the Bellman equation. Minimizing the MSBE during training helps the DRL algorithm converge toward an optimal Q-function approximation.

$$ L = \frac{1}{N} \sum \left(Q_\theta\,(s, a) - y(r, s', d) \right)^2 $$

$$ y\,(r, s', d) = r + \gamma\,(1 - d)\,Q_{\theta\text{target}}\left(s', \mu_{\phi\text{target}}\left(s'\right)\right) $$

where the $Q_{\theta\text{target}}$ and $\mu_{\phi\text{target}}$ networks are lagged versions of the Q_θ (critic) and μ_ϕ (actor) networks to solve the instability of the minimization of the MSBE due to interdependences among parameters. Hence, the critic network is updated by performing gradient descent considering loss L. Regarding the actor policy, it is updated using sampled policy gradient ascent with respect to the policy parameters by means of:

$$ \nabla_\phi \frac{1}{N} \sum_{s \in D} Q_\theta\left(s, \mu_\phi(s)\right) $$

Finally, the target networks are updated by Poliak averaging their parameters over the course of training:

$$ \theta_{Q\text{target}} \leftarrow \rho\theta_Q + (1 - \rho)\,\theta_{Q\text{target}} $$
$$ \phi_{\mu\text{target}} \leftarrow \rho\phi_\mu + (1 - \rho)\,\phi_{\mu\text{target}} $$

During training, uncorrelated, mean-zero Gaussian noise is added to actions to enhance exploration of the agent. The pseudocode for training a DDPG algorithm would be as in Table 1.

The CNC AI agent has a multi-goal nature within the environment. First, the agent shall learn not to collide with other machine parts. Collisions correspond to a penalty of -1000 and the reset of the environment. Second, the agent shall learn how to reach a goal destination by exploring the environment, which has an associated reward of $+500$ and causes the reset of the environment as well. Third, the agent

Table 1 Pseudocode for training a DDPG algorithm

1. Randomly initialize critic $Q_\theta(s, a)$ and actor $\mu_\phi(s)$ with weights θ_Q and ϕ_μ.
2. Initialize target networks $Q_{\theta \text{target}}$ and $\mu_{\phi \text{target}}$ with $\theta_{Q \text{target}} \leftarrow \theta_Q$, $\phi_{\mu \text{target}} \leftarrow \phi_\mu$
3. Initialize replay buffer D
4. For episode=1, M do
    5. Initialize random process N for action exploration
    6. Receive initial state s
    7. For t=1, max-episode-length do
        8. Observe state s and select action $a = \mu_\phi(s)+N$ according to current policy and exploration noise
        9. Execute action a in environment E and observe reward r, new state s', and done signal d
        10. Store (s, a, r, s', d) in replay buffer D and set $s \leftarrow s'$
        11. If s' is terminal (done=True), reset environment E
        12. If update network, then
            13. For i=1, G updates do
                14. Sample a random minibatch of B samples (s, a, r, s', d) from D
                15. Compute targets [2]
                16. Update critic by minimizing L [3]
                17. Update policy by gradient ascent [4]
                18. Update target networks [5, 6]
            19. End for
        20. End if
21. End for

shall optimize its policy considering the operational time and quality of the path. The operational time is calculated based on the distance that the robot head needs to travel to reach the goal destination following the proposed path. The quality of the path is calculated based on the number of actions needed to reach the destination. The former two aspects are favored by an extra reward of +100 to the agent (Fig. 5).

Figure 6 shows four exemplary paths found by the agent in a 2-D movement for visualization. In this problem, the agent needs to learn how to move from right to left without colliding with any machine parts. Since the action space is continuous, the goal destination does not correspond to a specific coordinate but to a subspace of the environment. The circles draw on the paths represent the subgoal coordinates proposed by the agent (each action correspond to a new coordinate to which the robot head is moved). From the figure, it can be seen that the yellow and pink trajectories comprise more actions and a longer path than the blue and green trajectories. Although these latter contain the same number of actions (three movements), the green trajectory requires a shorter path, and thus is preferred.

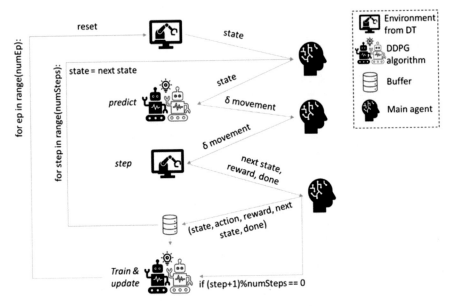

Fig. 5 Information flow in the DDPG framework during training phase

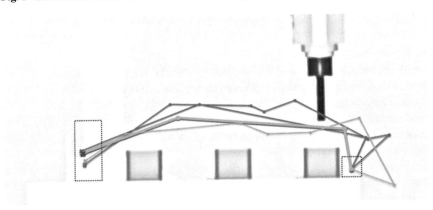

Fig. 6 Exemplary trajectories for simplicity in a 2-D movement. Circles in the path correspond to set of coordinates obtained after performing the action proposed by the framework in the environment. The total length of the trajectory and the number of actions are considered in the reward function. The red square corresponds with the starting point, while the red rectangle corresponds to the final region (target)

3.2 Robotic Manipulation of Complex Materials

The challenge of manipulating complex materials involves the identification of measurable quantities that offer insights into the system, which can then be utilized to make informed decisions and take appropriate actions [42, 43]. This essentially involves combining a perception system with a decision-making process. The Markov decision process (MDP) framework, as described earlier, is well-suited to address the task of defining optimal strategies for material manipulation. RL is particularly suitable for this purpose due to its probabilistic nature. RL accommodates the inherent uncertainty associated with characterizing the state of complex materials, which often presents challenges in traditional approaches.

The first step for the robotic manipulation of a fabric is the definition of the information required to perform the optimal actions for the manipulation of a material [44–46]. For that purpose, the state of the system needs to be characterized, a prediction needs to be done to infer what is the next state of the system under the application of a given action, and a criterium to decide what action to take for a given target needs to be chosen.

To address these points, on the one hand, a procedure for the generation of synthetic data has been deployed to generate automatically thousands of synthetic representations of a fabric and their transitions under the application of certain actions. On the other hand, in order to exploit such tool and build a solution based on data, a neural network has been developed and trained. Given that in a real system, in the real scenario for this work, a point cloud camera was used to detect the fabric, the entropy of the cloud has been calculated and taken as a reference magnitude to evaluate the goodness of a transition in terms of wrinkledness reduction.

In order to quantify the amount of knowledge in the system, we use entropy as a measurement of the information of the system. Using entropy maps, the wrinkledness of the fabric has been characterized. This entropy maps are calculated from the distributions of normal vectors within local regions using the classic form of information entropy for a distribution as follows:

$$H(X) = -\sum_{i=1}^{n} p(x_i) \log p(x_i)$$

Entropy is usually thought as a measurement of how much information a message or distribution holds. In other words, how predictable such a distribution is. In the context of the work presented here, entropy gives an idea about the orientation of the normal vectors of a given area of points taken from a reference one.

In order to address the massive complex manipulation of fabrics to reduce wrinkles over the surface, a specific digital environment has been developed as it is the *clothsim* environment made available as open access as part of the results of the MERGING Project. The detailed description of the simulation can be found on its own repository and is out of the scope of this chapter.

The *clothsim* environment has been used for the cloth initial random config-uration and later the training of the system. The learning routine has suggested different actions considering the Q-values (Q-matrix) by an *argmax* function. After the application of the actions, *clothsim* returned the transition of the fabric and the values of the Q-matrix have been updated following a Q-learning update rule where a reward function stablishes how good or bad the action was attending to

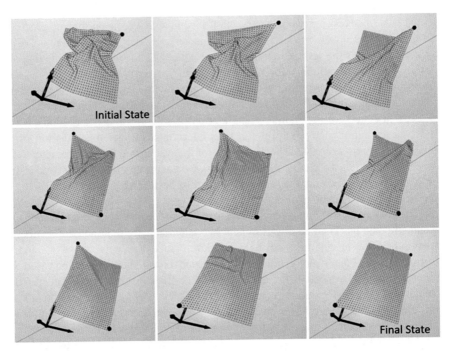

Fig. 7 Example of fabric manipulation using the developed environment for training under a complex manipulation task. The red and blue points indicate grab and pull points, respectively

the calculated entropy for each state. The DQL procedure has been trained built on a ResNet18 architecture as a backbone with an input shape (224, 224, 4) and the RL has been set with the following characteristic parameters: the gamma factor has been initially set as 0.75 and epsilon has variated from 1 to 0.3 in 8000 steps. During the whole learning process, a log file captures the global status of the knowledge acquired by the system.

Figure 7 shows a large example of the fabric manipulation in such a virtual environment (*clothsim*) in order to reduce the wrinkledness represented the entropy, which is estimated based on phi, theta, and z coordinates of the normal vectors of the points.

The images show the status of the fabric during the application of actions selected based on the values of the Q-matrix. These values are developed during the training procedure where the system tries to stablish, for given state, which action drives the maximum entropy reduction. This means that the entropy is the metric used during the whole process for the wrinkle minimization in the fabric (Fig. 8).

In order to select the actions, the knowledge of the system is encoded in the classic Q-matrix, which is inferred by the system for a given state. Such codification is done using a 6×6 matrix that considers corners to manipulate and directions that can be taken within the fabric manipulation. The final outcome of the procedure is a set of three points: one static point to fix the fabric, a second point that represents

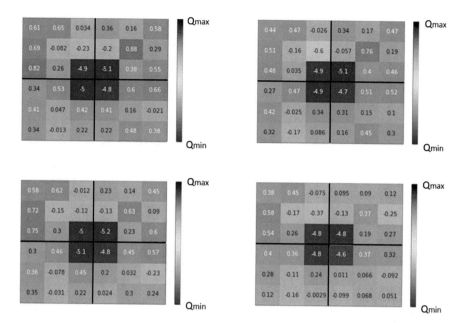

Fig. 8 Different examples of Q-matrix which is estimated for each state in order to drive an action attending at the position of its maximum value. The minimum values a very small displacement of the corners, meaning a very low reduction of the entropy so that they show low Q-values

the corner to be manipulated, and a third point that represents the point where this last corner has to be placed (grab point, pull point initial coordinates, pull point final coordinates). To decide what action to take, the system evaluates the Q-matrix and selects the action that corresponds with the maximum value on the matrix through an *argmax* function. The Q-values are updated to include a reward function that becomes positive as entropy decreases and remains negative otherwise. So, the Q-values hold the information of the reduction of the entropy that the system expects by the application of a given action. In a way that through the application of the action that corresponds with the maximum Q-value, the entropy reduction is also expected to be maximum in the long run (considering a whole episode).

To quantify the solution, we validate the use of the entropy as a metric for the wrinkledness of the fabric and its minimization as the target of the algorithm for the development of a strategy to reduce such wrinkledness. Fig. 9 shows how the entropy is reduced through the actions followed attending at the Q-matrix toward an acceptance distance from the target (plane).

The action-selection method, which serves as the system's plan to efficiently eliminate wrinkles from fabric, relies on a knowledge structure. This knowledge structure can be validated by evaluating the entropy variation resulting from the application of different actions, taking into account the information stored in the Q-matrix. In this process, various actions are applied to a given initial state, ordered based on their optimality as determined by the Q-matrix. By examining the entropy

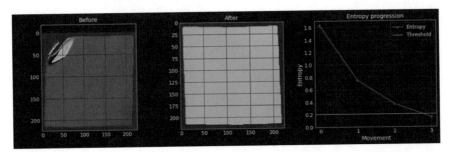

Fig. 9 Example of entropy evolution during synthetic fabric manipulation. The figure shows the initial state of the fabric and the final after different actions are applied. The red arrow (top-left corner) indicates the action suggested by the system. The right side of the figure shows the entropy evolution during the application of the actions. It can be seen how it decreases until a certain threshold (orange horizontal line) is crossed, meaning that the fabric is close to an ideal (plane) state

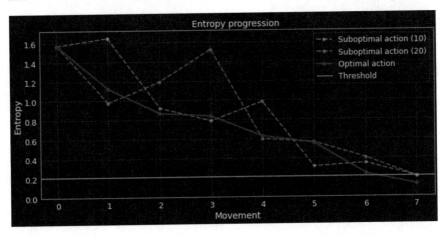

Fig. 10 Optimal path. Comparison of the entropy evolution when always the maximum of the Q-matrix is taken as optimal actions (blue curve) with entropy evolution under the selection of suboptimal actions (green and orange curves). The starting point for each transition is always considered as the state achieved by the application of the optimal (*argmax* from Q-values). Suboptimal actions drive more irregular behavior

variation, one can assess the effectiveness of the selected actions and validate the underlying knowledge structure.

Furthermore, the entropy's evolution along the optimal path can be compared with a scenario where suboptimal actions are taken, disregarding the Q-matrix. This comparison allows us to observe how the selection of the maximum Q-value drives a more robust curve toward entropy minimization. In Fig. 10, we can see that the entropy follows a decreasing curve toward the minimum entropy when the optimal actions are taken, considering the evolving state of the fabric. However, when suboptimal actions are consistently chosen for the same fabric state, the entropy

Fig. 11 Example of real fabric manipulation results. The figure shows three different steps during the manipulation following the actions suggested by the system. The sample manipulated is one from the real Use Case of the MERGING Project. In the figure, it can be clearly appreciated how the wrinkledness is reduced as the actions are applied

exhibits a more erratic behavior. This demonstration highlights how the developed solution offers a long-term optimal solution, ensuring a continuous reduction in entropy along the optimal path.

The strategies have been also tested in a real scenario, exploiting the information captured through a point cloud and following the outcomes suggested by the analysis of the Q-matrix associated with the given state. However, a complete description of the entire work behind this demonstration in terms of hardware is out of the scope of this chapter. Nevertheless, Fig. 11 shows the state of a fabric from an initial configuration to a final (ordered) state after the application of two actions to reduce the wrinkles.

4   Conclusions

We have provided an introduction to reinforcement learning as a robust AI technology for the development of complex strategies to be exploited by active agents.

For its deployment in manufacturing environments, RL applicability depends strongly on the digitalization of the process or its correct modeling. This is in order to provide a learning scenario to develop complex strategies to be demonstrated first under such digital strategy to be tested later in the real situation.

We have shown, as part of the results of two projects, examples of the application of RL in manufacturing. First, the application of Deep Deterministic Policy Gradient methods for the path optimization of a CNC machine in a digital twin. Second, deep Q-learning has been shown as a method for the development of optimal strategies related to the manipulation of fabric in a manufacturing environment. Showing results in a dedicated digital environment, as well as providing examples of the performance of the system, results in reducing the wrinkles on the fabric.

By utilizing reinforcement learning in a digital context, we have shown how to overcome the limitations posed by restricted training phases in manufacturing industries. Our research contributes to the development of effective strategies that can be tested and refined in digital environments before being applied to real-world systems. This approach allows for safer and more efficient exploration, enabling the optimization of manufacturing processes and performance.

The two manufacturing scenarios presented in this chapter highlight the potential and applicability of reinforcement learning in improving industrial processes. By bridging the gap between digital and real environments, we strive to advance the field of manufacturing and drive innovation in these sectors.

Overall, this research sheds light on the benefits of applying reinforcement learning in a digital context for manufacturing industries. It underscores the importance of leveraging digital environments to enhance training and strategy development, ultimately leading to improved performance and efficiency in real-world systems.

Acknowledgments This work has received funding from the European Union's Horizon 2020 research and innovation program under the MAS4AI Project (GA 957204) under the ICT-38 2020 – Artificial Intelligence for Manufacturing Topic. This work has also received funding from the MERGING Project (GA 869963).

References

1. Qi, Q., Tao, F.: Digital twin and big data towards smart manufacturing and industry 4.0: 360 degree comparison. IEEE Access. **6**, 3585–3593 (2018)
2. Fuller, A., Fan, Z., Day, C., Barlow, C.: Digital twin: enabling technologies, challenges and open research. IEEE Access. **8** (2020). https://doi.org/10.1109/ACCESS.2020.2998358
3. Wang, Z.: Digital twin technology. In: Bányai, T., Petrilloand, A., De Felice, F. (eds.) Industry 4.0 – Impact on Intelligent Logistics and Manufacturing. IntechOpen (2020)
4. Alexopoulos, K., Sipsas, K., Xanthakis, E., Makris, S., Mourtzis, D.: An industrial internet of things based platform for context-aware information Services in Manufacturing. Int. J. Comput. Integr. Manuf. **31**(11), 1111–1123 (2018). https://doi.org/10.1080/0951192X.2018.1500716
5. Kumar, S., Tiwari, P., Zymbler, M.: Internet of Things is a revolutionary approach for future technology enhancement: a review. J. Big Data. **6**, 111 (2019). https://doi.org/10.1186/s40537-019-0268-2
6. Saad, A., Faddel, S., Mohammed, O.: IoT-based digital twin for energy cyber-physical systems: design and implementation. Energies. **13**(18), 4762 (2020)
7. Wang, Y.: A cognitive informatics reference model of autonomous agent systems (AAS). Int. J. Cogn. Inform. Nat. Intell. **3**(1), 1–16 (2009c)
8. Usländer, T., Epple, U.: Reference model of Industrie 4.0 service architectures: basic concepts and approach. Automatisierungstechnik. **63**(10), 858–866 (2015). https://doi.org/10.1515/auto-2015-0017
9. Phuyal, S., Bista, D., Bista, R.: Challenges, opportunities and future directions of smart manufacturing: a state of art review. Sustainable Futures. **2**, 100023 (2020). https://doi.org/10.1016/j.sftr.2020.100023. ISSN 2666-1888
10. Ahuett-Garzaa, H., Kurfess, T.: A brief discussion on the trends of habilitating Technologies for Industry 4.0 and smart manufacturing. Manuf. Lett. **15**(Part B), 60–63 (2018). https://doi.org/10.1016/j.mfglet.2018.02.011
11. Martins, A., Lucas, J., Costelha, H., Neves, C.: CNC machines integration in smart factories using OPC UA. J. Ind. Inf. Integr. **34**, 100482 (2023)
12. Alexopoulos, K., Nikolakis, N., Chryssolouris, G.: Digital twin-driven supervised machine learning for the development of artificial intelligence applications in manufacturing. Int. J. Comput. Integr. Manuf. **33**(5), 429–439 (2020)

13. Grieves, M., Vickers, J.: Digital twin: mitigating unpredictable, undesirable emergent behavior in complex systems. In: Kahlen, F.-J., Flumerfelt, S., Alves, A. (eds.) Transdisciplinary Perspectives on Complex Systems, pp. 85–113. Springer, Cham (2017)
14. He, B., Bai, K.J.: Digital twin-based sustainable intelligent manufacturing: a review. Adv. Manuf. **9**(1), 1–21 (2021)
15. Sutton, R.S., Barto, A.G.: Reinforcement learning: an introduction, 2nd edn. The MIT Press (2014)
16. Kaelbling, L.P., Littman, M.L., Moore, A.W.: Reinforcement learning: a survey. J. Artif. Intell. Res. **4**, 237–285 (1996). https://doi.org/10.1613/JAIR.301
17. Jang, B., Kim, M., Harerimana, G., Kim, J.W.: Q-Learning algorithms: a comprehensive classification and applications. IEEE Access. **7**, 133653–133667 (2019). https://doi.org/10.1109/ACCESS.2019.2941229
18. Muiños-Landin, S., Fischer, A., Holubec, V., Cichos, F.: Reinforcement learning with artificial microswimmers. Sci. Rob. **6**(52), eabd9285 (2021)
19. Chen, G., Luo, N., Liu, D., Zhao, Z., Liang, C.: Path planning for manipulators based on an improved probabilistic roadmap method. Robot. Comput. Integr. Manuf. **72** (2021). https://doi.org/10.1016/j.rcim.2021.102196
20. Pohan, M.A.R., Trilaksono, B.R., Santosa, S.P., Rohman, A.S.: Path planning algorithm using the hybridization of the rapidly-exploring random tree and ant Colony systems. IEEE Access. **9** (2021). https://doi.org/10.1109/ACCESS.2021.3127635
21. Wei, K., Ren, B.: A method on dynamic path planning for robotic manipulator autonomous obstacle avoidance based on an improved RRT algorithm. Sensors (Switzerland). **18**(2) (2018). https://doi.org/10.3390/s18020571
22. Kang, J.G., Choi, Y.S., Jung, J.W.: A method of enhancing rapidly-exploring random tree robot path planning using midpoint interpolation. Appl. Sci. **11**(18) (2021). https://doi.org/10.3390/app11188483
23. Wawrzynski, P.: Control policy with autocorrelated noise in reinforcement learning for robotics. Int. J. Mach. Learn. Comput. **5**, 91–95 (2015)
24. Wawrzynski, P., Tanwani, A.K.: Autonomous reinforcement learning with experience replay. Neural Netw. **41**, 156–167 (2013).; Xie, J., Shao, Z., Li, Y., Guan, Y., Tan, J.: Deep reinforcement learning with optimized reward functions for robotic trajectory planning. IEEE Access. 7, 105669–105679 (2019). https://doi.org/10.1109/ACCESS.2019.2932257; Watkins, J.C.H.: Learning from Delayed Rewards. King's College, Cambridge (1989)
25. Damen, T., Trampert, P., Boughobel, F., Sprenger, J., Klusch, M., Fischer, K., Kübel, C., Slusallek, P. et al.: Digital reality: a model-based approach to supervised Learning from synthetic data. AI Perspect Adv. **1**(1), 2 (2019). https://doi.org/10.1186/s42467-019-0002-0
26. Liu, Q., Wu, Y.: Supervised learning. In: Encyclopedia of the Sciences of Learning, pp. 3243–3245 (2012). https://doi.org/10.1007/978-1-4419-1428-6_451
27. Hinton, G., Sejnowski, T.: Unsupervised Learning: Foundations of Neural Computation. MIT Press, Cambridge (1999). https://doi.org/10.7551/MITPRESS/7011.001.0001
28. White, D.J.: A survey of applications of Markov decision processes. J. Oper. Res. Soc. **44**(11), 1073–1096 (1993). https://doi.org/10.2307/2583870
29. Gustavsson, K., Biferale, L., Celani, A., Colabrese, S.: Finding efficient swimming strategies in a three-dimensional chaotic flow by reinforcement learning. Eur. Phys. J. E. **40**, 110 (2017)
30. Mnih, V., Kavukcuoglu, K., Silver, D., Graves, A., Antonoglou, I., Wierstra, D., Riedmiller, M.: Playing Atari with deep reinforcement learning. arXiv preprint arXiv, 1312.5602 (2013)
31. Sewak, M.: Deep Reinforcement Learning: Frontiers of Artificial Intelligence. Springer (2019). https://doi.org/10.1007/978-981-13-8285-7
32. Dröder, K., Bobka, P., Germann, T., Gabriela, F., Dietrich, F.: A machine learning-enhanced digital twin approach for human-robot-collaboration. In: 7th CIRP Conference on Assembly Technologies and Systems, vol. 76, pp. 187–192 (2018). https://doi.org/10.1016/j.procir.2018.02.010
33. Zayed, S.M., Attiya, G.M., El-Sayed, A., et al.: A review study on digital twins with artificial intelligence and internet of things: concepts, opportunities, challenges, tools and future scope. Multimed. Tools Appl. (2023). https://doi.org/10.1007/s11042-023-15611-7

34. Jazdi, N., Ashtari Talkhestani, B., Maschler, B., Weyrich, M.: Realization of AI-enhanced industrial automation systems using intelligent Digital Twins. Procedia CIRP. **97**, 396–400 (2020)
35. Hofmann, W., Branding, F.: Implementation of an IoT- And cloud-based digital twin for real-time decision support in port operations. IFAC-PapersOnLine. **52**(13), 2104–2109 (2019)
36. Bilberg, A., Malik, A.A.: Digital twin driven human–robot collaborative assembly. CIRP Ann. (2019). https://doi.org/10.1016/j.cirp.2019.04.011
37. Lillicrap, T.P., et al.: Continuous control with deep reinforcement learning. arXiv:1509.02971. (2016). https://doi.org/10.48550/arXiv.1509.02971
38. Maoudj, A., Hentout, A.: Optimal path planning approach based on Q-learning algorithm for mobile robots. Appl. Soft Comput. **97**, 106796 (2020). https://doi.org/10.1016/J.ASOC.2020.106796
39. Chen, X., 5-axis coverage path planning with deep reinforcement learning and fast parallel collision detection (2020). Available: https://smartech.gatech.edu/handle/1853/62825. Accessed 28 Aug 2022
40. Grondman, I., Busoniu, L., Lopes, G.A.D., Babuška, R.: A survey of actor-critic reinforcement learning: standard and natural policy gradients. IEEE Trans. Syst. Man Cybern. Part C Appl. Rev. **42**(6), 1291–1307 (2012). https://doi.org/10.1109/TSMCC.2012.2218595
41. Chen, L., Jiang, Z., Cheng, L., Knoll, A.C., Zhou, M.: Deep reinforcement learning based trajectory planning under uncertain constraints. Front. Neurorobot. **16**, 80 (2022). https://doi.org/10.3389/FNBOT.2022.883562
42. Jiménez, P., Torras, C.: Perception of cloth in assistive robotic manipulation tasks. Nat. Comput. **19**, 409–431 (2020). https://doi.org/10.1007/s11047-020-09784-5
43. Colomé, A., Torras, C.: Dimensionality reduction for dynamic movement primitives and application to bimanual manipulation of clothes. IEEE Trans. Robot. **34**(3), 602–615 (2018). https://doi.org/10.1109/TRO.2018.2808924
44. Cusumano-Towner, M., Singh, A., Miller, S., O'Brien, J.F., Abbeel, P.: Bringing clothing into desired configurations with limited perception. In: Proceedings of IEEE International Conference on Robotics and Automation (ICRA), vol. 2011, pp. 1–8 (2011) http://graphics.berkeley.edu/papers/CusumanoTowner-BCD-2011-05/
45. Hamajima, K., Kakikura, M.: Planning strategy for task of unfolding clothes. Robot. Auton. Syst. **32**(2–3), 145–152 (2000). https://doi.org/10.1016/S0921-8890(99)00115-3
46. Hou, Y.C., Sahari, K.S.M.: Self-generated dataset for category and pose estimation of deformable object. J. Rob. Netw. Artif. Life. **5**, 217–222 (2019). https://doi.org/10.2991/jrnal.k.190220.001

A Participatory Modelling Approach to Agents in Industry Using AAS

Nikoletta Nikolova, Cornelis Bouter, Michael van Bekkum, Sjoerd Rongen, and Robert Wilterdink

1 Introduction

With the increasing number of assets being digitized, all of which are expected to be interoperable with each other, the need for well-designed information models grows. Such models describe a given entity in a machine-readable way and enable interpreting, usage and reasoning with data from previously unknown devices. The usual approach to make these information models is through a heavy standardization process in which a large number of organizations all have to come to a common understanding of definitions and approaches, after which all parties implement this understanding in their systems [17]. These standards are usually well-scoped and documented, making them easier to use. For developers, this provides a steady base to build upon without fear of the implementation becoming outdated quickly due to a newly released version of a standard. Unfortunately, these standardization

Cornelis Bouter, Michael van Bekkum, and Sjoerd Rongen contributed equally to this work.

N. Nikolova · C. Bouter · M. van Bekkum (✉)
Data Science, Netherlands Organisation for Applied Scientific Research (TNO), Den Haag, The Netherlands
e-mail: nikoletta.nikolova@tno.nl; cornelis.bouter@tno.nl; michael.vanbekkum@tno.nl; sjoerd.rongen@tno.nl; robert.wilterdink@tno.nl

S. Rongen
Data Ecosystems, Netherlands Organisation for Applied Scientific Research (TNO), Den Haag, The Netherlands

R. Wilterdink
Advanced Computing Engineering, Netherlands Organisation for Applied Scientific Research (TNO), Den Haag, The Netherlands

© The Author(s) 2024
J. Soldatos (ed.), *Artificial Intelligence in Manufacturing*,
https://doi.org/10.1007/978-3-031-46452-2_12

efforts tend to be slow, often taking multiple years before agreement is reached, for example, ISO standards take 18 to 36 months to be developed [8].

Because of their slow development nature, formal standardization processes are not universally suitable for all applications. For example, when developing a new product, it is not feasible to delay the product launch by months, or even years, because an agreement needs to be reached on the information model. As such, currently, during development, there is little alignment done. Instead, the product developer creates a model with the information that they consider necessary. This is similar to what has been happening in the development of linked data ontologies, where there are few formal standards standardized by W3C (RDF, OWL, SHACL) defining the metamodel. A large number of common use case-specific models made by institutions unrelated to W3C have become *de facto* standards, such as Friend of a Friend,[1] Prov-o,[2] Schema.org.[3] These *de facto* standards are commonly used by individuals designing their models and enable easier alignment between a large number of use case-specific models. This results in a large number of datasets that mostly use domain-specific ontologies, while they can still be related to each other as done with the Linked Open Data Cloud.[4]

We believe these *de facto* standards are vital to quickly defining the semantic model. This belief appears to be shared with organizations in the manufacturing industry, as they have developed their own metamodel to ensure domain models can be aligned with each other. The standardized Asset Administration Shell (AAS) defines how to structure the information model of your asset without specifying what your asset is or what its information model is, cf. [2]. This provides a structure for various organizations to express interoperable models, which may become *de facto*o standards. However, in the Asset Administration Shell community, we detect a lack of tools and methodologies that allow these bottom-up developed AAS submodels to rise to the level of *de facto* standards. This leaves us with formal standards (such as those published by the Industrial Digital Twin Association (IDTA) [7]) and lots of use case-specific models which lack the reuse of domain-relevant models which aids their interoperability.

In this paper, we build upon earlier work for AAS modelling practices [3] and agent modelling approaches [10] to present a set of tools, aiming to help inter-operability and standardization. We build on lessons learned from the MAS4AI[5] European project in which software agents were described using an AAS utilizing a reusable model of describing software agents. The methodology referred to in this paper was developed within the European DIMOFAC[6] project. In this work, we aim

---

[1] http://xmlns.com/foaf/0.1/.

[2] https://www.w3.org/TR/prov-o/.

[3] https://schema.org/.

[4] https://lod-cloud.net/.

[5] https://www.mas4ai.eu.

[6] https://dimofac.eu/.

to aid in developing better bottom-up models of sufficient quality to become *de facto* standards and can then be adopted. For this, we present tooling and a methodology.

The rest of the paper starts with an overview of the relevant background and previous research. Afterward, we explain in three consecutive sections the model, methodology, and software tools we developed. Lastly, we provide an analysis of how our work can be used, what its limitations are, and what the next steps are.

2 Background

To achieve interoperability between collaborating parties, an agreement on semantics is needed. An information standard is required to effectively connect systems, organizations, and work. Commonly this is done via formal standardization processes such as those facilitated by ISO, governments, or large sector organizations. It can also be done on a meta-level with a higher level of abstraction. An example is the usage of RDF [15], OWL [12], and SHACL [14] as a standardized way to express domain language in a machine-readable format. In these cases, the metamodel has gone through a strict extensive standardization procedure, but the domain models based on it can be freely published by their authors.

Industry 4.0 introduced the meta-language AAS [13]. It is a semantic model, which provides a structure to describe assets in a machine-readable way including a standardized interaction interface. The standard is young, resulting in a lack of an established community, which actively implements it, or a uniform process for designing, defining, or implementing an AAS. To enable such a task, it is important to consider three main elements: (1) what the models represent; (2) how can they be created; (3) how can they be standardized.

When cogitating on what information models represent, we consider not only physical but also digital assets. Particularly, as agents are becoming more widely adopted in industrial applications, it is imperative to have an AAS modelling strategy. Recent works, such as the one by [11] propose a way to model two types of agents—product and resource agents. Our previous work [10] provides a more general solution to this by presenting a use case-agnostic general agent model structure, which we further extend in this paper.

When developing AAS models, the current approach is focused on using the AAS library provided by the IDTA [7], as it contains the set of the currently standardized AAS models available. As such, this is the first place many AAS users will check when looking for a model. However, the available library is not a complete overview of all AAS submodels that have been made and introducing a new submodel to it is time-consuming. Although the IDTA's contribution to standardization is indispensable for the development of a thriving AAS ecosystem, its approach is top-down. That is to say, when having developed a domain-specific AAS submodel it goes through multiple checks and reviews before being published [1]. While this brings the advantage of high-quality models and a steady base for developers to build their solutions, it is a slow process which is bottlenecked by the speed at which information models can be reviewed and consensus can be reached. To be able to achieve widely adopted semantic interoperability, it is needed to facilitate also a bottom-up approach.

Currently, there is a lack of common standardization, which is universally used, as shown in the work of [9]. To tackle this, there are multiple different approaches employed toward developing a common method—[5] proposes a solution using OPC-UA, [4] defines an interoperability framework for digital twins, and [6] offer a solution using ontologies. We tackle interoperability from the perspective of what is needed for a standardization approach to happen and how can it be implemented. Specifically, we look at the topic of agents and how their AAS modelling process can be standardized. We develop a set of tools, which can be used to simplify and streamline the process of modelling with AAS. To the best of our knowledge, there is no current common process. We place all those developments in the context of real Industry 4.0 applications.

3   AAS Model for an Agent

Standardization and interoperability require a common approach toward information representation and more particularly modelling. When modelling manufacturing environments, it is important to consider not only physical but also digital assets. More specifically, one may consider these digital assets to be agents responsible for a piece of logic, as such it is essential to have a way to model such an asset. To address this, we provide a general model structure, which can be used when creating a model of an agent using the asset administration shell.

3.1   General Model Structure

The general model structure follows the work we presented in [10]. We provide a set of general and standard submodels which aim to provide a structure, which can be followed when creating any agent model. The model is shown in Fig. 1, in which we add the *Parameterization* and *Configuration* submodels to replace the *Communication* submodel from earlier work. The purpose of this change is to address the difference between the information locally needed for an agent to be defined, and the data, defined by the framework where it is deployed. This creates a concrete distinction between what type of information is needed in the submodels and where it comes from.

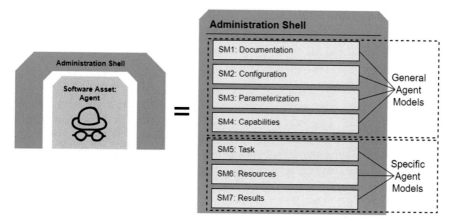

Fig. 1 General agent structure, which contains seven different distinct submodels, split into two categories

3.2 Generic Submodels

The set of generic submodels contains information, which is considered task agnostic and is needed for correct tracking and connection of the developed agent. The definitions follow from the work of [10].

- **Documentation:** The Documentation submodel contains information, which is relevant for describing any background and specifications for the agent. Example properties for this would be details about the developer, version, required software, language, algorithms, etc.
- **Configuration:** The Configuration submodel contains information, which is needed/provided by the framework or agent management system, where the agent will live. It provides information such as where the proper interface connections can be found, how to configure, etc. This is information which is determined by the framework and is usually shared between multiple agents.
- **Parameterization:** The Parameterization submodel contains information regarding the exact parameters that an agent needs to be able to initialize. Those parameters are determined by the exact algorithm and construction of the agent and are specific for a particular agent type.
- **Capabilities:** The Capabilities submodel contains information about what the agent can do. This can be done in various ways, including in combination with RDF [16].

3.3 Specific Submodels

Specific Submodels [10] are those, which depend on the use case. They contain information, which is determined by the exact situation and setting where the agent is to execute its work.

- **Task:** The task submodel provides a description of the exact task that the agent has to execute. Depending on the case there can be a single task such as *"Moving,"* but it can also be a set of multiple sub-tasks.
- **Resources:** The resources submodel aims to wrap in one place the connections to all resources and corresponding relevant information from them, which the agent would need. This would, for example, contain properties such as *"Machine Capability"* and *"Operator Availability."*
- **Results:** The results submodel presents the type of results and corresponding details that the agent provides after its task is executed. There can be multiple results such as *"Analysis," "Plan,"* etc.

3.4 Usage

The concept of the general agent structure is to serve as a base skeleton model, which provides a clear split and indication regarding what type of information needs to be contained in an agent model. The aim is to use the structure, when creating models of new agents, starting from concretely specifying the submodels to filling them in. To provide an approach for this, we have defined a methodology based on the work of [3], described in the next section.

4 Methodology for Developing an AAS

When creating a rich semantic information model there are several aspects to consider. To support proper modelling practices, we suggest the usage of a well-defined process which ensures no steps are missed.

We propose a methodology based on earlier work by [3], which we extend to make it more applicable when modelling not only physical assets but also software assets. The methodology is additionally extended by identifying four phases, described below and visually represented in Fig. 2.

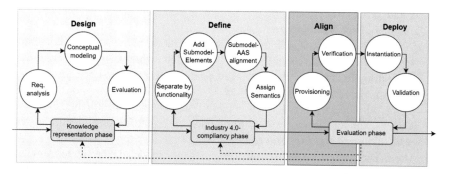

Fig. 2 Updated AAS Development Methodology, visualizing the four separate steps—design, define, align, deploy

4.1 Phases

The methodology is separated into four phases covering the implementation of a set of AAS models. The methodology starts with the Design phase which covers the knowledge representation necessary to describe the factory concepts underlying this problem. This phase requires no special knowledge about the AAS or other Industry 4.0 technologies. It is an application-independent phase of modelling the relevant factory assets and their properties. The domain expert is safeguarded from having to consider Industry 4.0 requirements. A further specification of the consideration of the four sub-steps is described in [3].

The methodology concludes with an evaluation phase separated into two parts: (1) the alignment of the existing data models being used in the factory with the developed AAS models and (2) the deployment of the tooling for which the AAS interoperability was established. The former is called verification since it is a process we can formally check: if the factory data elements have been aligned with the AAS data elements, both are aligned. The deployment is the validation step because it cannot be formally verified. We consider the AAS model validated when the tool the user had in mind at the start of the process can be built using the AAS models.

4.2 Agent Modelling

In the case of agent modelling, the standard process described above is directly applicable thanks to the two added states—provisioning and instantiation. Furthermore, the process can be split into four distinct stages, referring to the different implementation stages of agent development.

1. **Design:** The design stage is the use case specification moment. This is, for example, the point in time when we aim to identify the types of agents which

would be relevant/useful to the particular situation, such as *product agent, resource agent, planning agent,* etc.

2. ***Define:*** The define stage is the template generation phase. At this moment the developers can create/choose standard submodels, which can be used to provide a skeleton structure for the developed agents. More particularly, the aim is to fill the general agent structure, which was described in Fig. 1. For the General Agent Submodels, this would mean identifying the template structure to be used, whereas, for the Specific Agent Submodels, it would require identifying their exact definition and expected content.

3. ***Align:*** The align stage focuses on filling in the submodels in the created templates with any needed properties, mainly details such as exact parameters needed for the agent to operate properly. At the end of this stage, it is expected that the Agent templates contain all the relevant information.

4. ***Deploy:*** Lastly, the deployment stage takes care of the spawning of the agent. That includes filling all properties (such as Agent ID, Task ID, etc.) through the framework or agent management system, which is used. Since multiple entities of a single type of agent can be active, this is the phase where they are created and the corresponding models are filled.

5   AAS Model Repository

One of the key components of interoperability is enabling seamless sharing and distribution of developments. In the case of AAS models, there is currently no official software, which provides a simple and user-friendly interface for visualization and sharing. This is important since the correct handling of this process can enable collaboration between different parties and support the smooth distribution of work. Therefore, to close the loop, we developed an online public repository (https:// admin-shell-library.eu/). It focuses on providing a way to share models between parties and visualize them in a user-friendly manner.

5.1   Functionality

The main functionalities of the repository are (1) visualization and (2) distribution, both focusing on making the AAS models' development and cooperation easier. Currently, the majority of development happens behind closed doors in silos, with distribution and information sharing only happening at the last step of the process. It is very important to enable ways for cooperation, especially since the increased interest in the AASs also creates accidental duplication of work.

One of the main challenges for distribution is the lack of direct sharing possibilities when it comes to AAS models. Presently, the standard process requires downloading from a source (such as GitHub) and running an extra program (such as AAS package explorer) to open and view a model. This can hinder the development process, since more steps are required and hence more time and focus are needed to review a model. What the AAS repository enables is direct link sharing. Once a developer uploads their model online, it is possible to get a link, which leads to a web page visualization of the AAS. The link can be shared with other parties and removes the need for any software installation, which can significantly simplify the sharing and collaboration process, which are key for creating general and reusable models.

In general, a full AAS contains multiple Submodels, which each contain several (nested) SubmodelElements. For each of these Submodels and SubmodelElements, semanticIds, descriptions and Concept Descriptions should be maintained. Moreover, multi-language SubmodelElements exists, which can have several values attached for the various supported languages. Because of this size and complexity, a model can have, working collaboratively on an AAS template should be supported with proper visualization. Currently, the most commonly used tool is also the one used for model creation—the package explorer. While this is useful software, it can be unintuitive for non-accustomed users and increase complexity when looking at a model (especially for external parties, who do not work with this program). The repository provides a web interface, which removes the need for separate software tooling for viewers.

5.2 Working Principle

The working principle of the repository is visualized in Fig. 3. The figure visualizes the different interactions that a user can have with the repository and the high-level corresponding processing steps.

- *Upload:* A user can upload a model to the repository by filling in the metadata and providing the corresponding .aasx file. This model will be checked for errors by the system and if no errors are found, it would be uploaded to the repository.
- *Modify:* A user can modify the metadata of a model or upload a newer .aasx version at any point.
- *Visualize:* The model visualizes multiple details of the .aasx model. It provides collapsible submodels, each containing the corresponding elements. There is a possibility to show concept description and most importantly example data. This makes it easier for a viewer to understand the whole process of the models. If example data contains references to other models included in the repository, they can easily be followed via Web links. Especially for complex composite models, this can make a significant difference.
- *Share:* Any of the models can be shared via a link, which directly links to the relevant AAS file.
- *Access:* Model access is managed by a combination of groups/users and roles. Each model can have several groups or individual users assigned. Additionally,

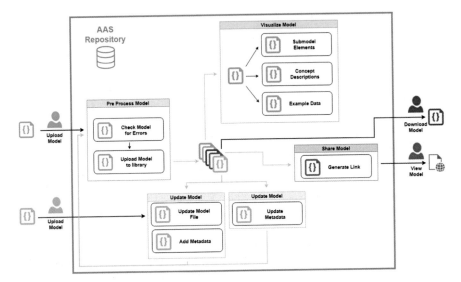

Fig. 3 Repository working principle, where black arrows represent interaction with the user and grey represent internal connections

for each user or group a role must be specified, i.e., guest, member, or editor. The combination of group/user and role then determines the effective access level. This ensures that during each step in the development process, the models can be shared with appropriate access levels and no sensitive information is publicly available.

6   Discussion

In this paper, we presented three ways of improving the creation of bottom-up standards which may rise to the level of *de facto* standards. Firstly, we presented the notion of a generic asset structure, such as for software agents, consisting of multiple submodels. Secondly, we presented a methodology on how to create your AAS models, and finally, we propose a repository for these templates which may in the findability and adoption of good AAS submodels.

6.1   Maturity of the AAS

Although we have presented a number of ways to improve the usage of the AAS and to ensure semantic interoperability we do not believe this will on its own lead to wide-spread adoption and an active ecosystem of AAS users. We believe it is important to nurture an environment in which proper AAS modelling can thrive. However, before this technology blossoms, we believe there are still several improvements that may be made. This is to be expected given the immaturity of the AAS technology but that does not make them less pressing to tackle.

Models

To construct reusable and interoperable models, it is essential to provide clear templates and corresponding specifications. Currently, the AAS implementations do not allow for specifying what components are needed for a minimal implementation. This is especially important for future uses, as it would provide a way for users to know what is the least information they need to fill in for their model to be viable.

Model Constraints

To make the usage of semantic models more beneficial to business users and give implementers more guidance on what parts of the AAS model are mandatory and which are optional, it would be beneficial to add the ability to constrain the instantiation of the AAS model. A basic implementation of this would support defining cardinality constraints to the AAS model, defining which properties are mandatory, and which are optional. This could then be extended with value contents and potentially even some simple reasoning, to implement basic business logic in the model. For example, the intended use of a creation date and maintenance date of an asset may make it impossible to maintain the asset before it was created. However, the current AAS standard does not allow for a way to express such constraints. In semantic modelling, we have been in a similar situation not that long ago with the adoption of the Web Ontology Language (OWL), a powerful modelling language which did not get adopted as quickly in business as it could have been due to a lack of easily implemented value constraints. The creation of the Shape Constraint Language (Shacl) largely solved this problem and has increased the adoption of RDF in business contexts.

Standardization

The methodology and the repository synergetically combine to improve the bottom-up standardization of the AAS. The methodology facilitates the identification of additional submodels that are lacking among what is currently available, as well as aiding the user in following the AAS paradigm. The repository facilitates the sharing of work-in-progress submodels with other users, such that the community can swiftly adopt new work. This approach complements the top-down approach of a comprehensive IDTA standardization procedure.

In top-down standardization, the authority of the standardization body ensures the quality of the approved models. When adopting a free-for-all approach of submitting submodels to the repository, the quality should be ensured differently. Firstly, the methodology functions as a way to increase the shared knowledge of the modelling paradigm. Secondly, statistics about the usage of the various models can indicate which models turn into *de facto* standards, thereby giving a measure of their quality. Additional research and implementation work may be needed to tailor the repository for this purpose.

The increased accessibility of the repository can aid the various groups who can be expected to share their models. The methodology is primarily intended for user integrators leveraging the AAS for improved interoperability to support a factory application. Another audience that may use the repository is the machine manufacturer who wants to share instantiated and specialized templates for their machines. These submodels similarly benefit from a bottom-up approach without a mandatory procedure.

Collaborative Modelling

Currently, the process of making an AAS model is primarily a solitary effort in which the modeller still needs to actively try to reach out for input. However, the whole point of standards is that they align between different parties. As such, adding more tools to facilitate collaborative modelling would be beneficial for the creation of a proper AAS. The presented repository and development methodology already aid in this, but we believe further steps could be taken to support a shared model discussion and increase iteration speed during the development process.

7 Conclusion

The methodology and the repository described in this paper provide us with means to develop AAS models as *de facto* standards for non-physical assets from bottom-up, community-based efforts. The MAS4AI project has shown that the methodology can easily accommodate particular use case requirements by allowing for straightforward bottom-up extensions to the models. An industry use case in MAS4AI on implementing planning agent software based on ISA-95 models has shown that bottom-up standardization is instrumental. The required models were created by applying the aforementioned methodology in a collaborative effort by modelling experts, domain experts, and software developers and have led to successful integration and deployment in the use case.

The use of the repository has similarly led to clear benefits in both the MAS4AI and the DIMOFAC project: sharing the AAS model templates with all stakeholders in an easy-to-use, intuitive way has made them more accessible to all parties and has promoted discussion and feedback on their contents, thus ensuring more widespread support for the models.

Acknowledgments The research leading to these results has been funded by the European Commission in the H2020 research and innovation program, under Grant agreement no. 957204, the "MAS4AI" project (Multi-Agent Systems for Pervasive Artificial Intelligence for Assisting Humans in Modular Production Environments) and under grant agreement No. 870092, the DIMOFAC project (Digital & Intelligent MOdular FACtories).

References

1. 4.0 PI: Structure of the asset administration shell. continuation of the development of the reference model for the Industrie 4.0 component (2016) . Tech. rep. https://industrialdigitaltwin. org/wp-content/uploads/2021/09/01_structure_of_the_administration_shell_en_2016.pdf
2. Bader, S., Barnstedt, E., Bedenbender, H., et al.: Details of the Asset Administration Shell. Part 1—The exchange of information between partners in the value chain of Industrie 4.0 (Version 3.0RC02) (2022)
3. Bouter, C., Pourjafarian, M., Simar, L., et al.: Towards a comprehensive methodology for modelling submodels in the industry 4.0 asset administration shell. In: 2021 IEEE 23rd Conference on Business Informatics (CBI), vol. 02, pp. 10–19 (2021)
4. Budiardjo, A., Migliori, D.: Digital twin system interoperability framework. Tech. rep. Digital Twin Consortium, East Lansing, Michigan (2021). https://www.digitaltwinconsortium.org/pdf/ Digital-Twin-System-Interoperability-Framework-12072021.pdf
5. Cavalieri, S.: A proposal to improve interoperability in the industry 4.0 based on the open platform communications unified architecture standard. Computers 10(6), 70 (2021). https:// doi.org/10.3390/computers10060070, https://www.mdpi.com/2073-431X/10/6/70, number: 6 Publisher: Multidisciplinary Digital Publishing Institute
6. Huang, Y., Dhouib, S., Medinacelli, L.P., et al.: Enabling semantic interoperability of asset administration shells through an ontology-based modeling method. In: Proceedings of the 25th International Conference on Model Driven Engineering Languages and Systems: Companion Proceedings. Association for Computing Machinery, New York, NY, USA, MODELS '22, pp. 497–502 (2022). https://doi.org/10.1145/3550356.3561606
7. IDTA: Industrial digital twin association (2023). https://industrialdigitaltwin.org/en/
8. ISO: Target date planner (2023). https://www.iso.org/files/live/sites/isoorg/files/developing_ standards/resources/docs/std%20dev%20target%20date%20planner.pdf
9. Melluso, N., Grangel-González, I., Fantoni, G.: Enhancing industry 4.0 standards interoperability via knowledge graphs with natural language processing. Comput. Ind. 140, 103676 (2022). https://doi.org/10.1016/j.compind.2022.103676, https://www.sciencedirect. com/science/article/pii/S0166361522000732
10. Nikolova, N., Rongen, S.: Modelling agents in industry 4.0 applications using asset administration shell. In: Proceedings of the 15th International Conference on Agents and Artificial Intelligence (2023). https://doi.org/DOI:10.5220/0011746100003393
11. Ocker, F., Urban, C., Vogel-Heuser, B., et al.: Leveraging the asset administration shell for agent-based production systems. IFAC-PapersOnLine 54(1), 837–844 (2021). https://doi.org/https://doi.org/10.1016/j.ifacol.2021.08.186, https://www.sciencedirect. com/science/article/pii/S2405896321009563. 17th IFAC Symposium on Information Control Problems in Manufacturing INCOM 2021
12. OWL Working Group: Owl 2 web ontology language document overview (2012). Tech. rep., W3C. https://www.w3.org/TR/owl2-overview/
13. Plattform Industrie 4.0: Details of the asset administration shell from idea to implementation (2019). https://www.plattform-i40.de/IP/Redaktion/EN/Downloads/Publikation/vws-in-detail-presentation.pdf
14. RDF Data Shapes Working Group: Shapes constraint language (SHACL) (2017). Tech. rep., W3C. https://www.w3.org/TR/shacl/
15. RDF Working Group: RDF 1.1 primer (2014). Tech. rep., W3C. https://www.w3.org/TR/rdf11-primer/
16. Rongen, S., Nikolova, N., van der Pas, M.: Modelling with AAS and RDF in industry 4.0. Comput. Ind. 148, 103910 (2023). https://doi.org/https://doi.org/10.1016/j.compind.2023. 103910, https://www.sciencedirect.com/science/article/pii/S016636152300060X

17. Toussaint, M., Krima, S., Feeney, A.B., et al.: Requirement elicitation for adaptive standards development. IFAC-PapersOnLine **54**(1), 863–868 (2021). https://doi.org/ https://doi.org/10.1016/j.ifacol.2021.08.101, https://www.sciencedirect.com/science/article/ pii/S2405896321008508. 17th IFAC Symposium on Information Control Problems in Manufacturing INCOM 2021

I4.0 Holonic Multi-agent Testbed Enabling Shared Production

Alexis T. Bernhard, Simon Jungbluth, Ali Karnoub, Aleksandr Sidorenko, William Motsch, Achim Wagner, and Martin Ruskowski

1 Introduction

Nowadays, globalization changes the manufacturing environment significantly. Global business signifies global competition and intends a need for shorter product life cycles, whereas consumer-oriented businesses foster customized products. However, rigidly connected supply chains have proven vulnerable to disruptions. Therefore, requirements are changing focusing on adaptability, agility, responsiveness, robustness, flexibility, reconfigurability, dynamic optimization and openness to innovations. Isolated and proprietary manufacturing systems need to move ahead with decentralized, distributed, and networked manufacturing system architectures [20]. Cloud Manufacturing could be a solution for highly diversified and reconfigurable supply chains. Liu et al. define it as "[...] a model for enabling the aggregation of distributed manufacturing resources [...] and ubiquitous, convenient, on-demand network access to a shared pool of configurable manufacturing services

A. T. Bernhard (✉) · A. Sidorenko · W. Motsch · A. Wagner
Deutsches Forschungszentrum für Künstliche Intelligenz GmbH (DFKI), Kaiserslautern, Germany
e-mail: alexis.bernhard@dfki.de; aleksandr.sidorenko@dfki.de; william.motsch@dfki.de; achim.wagner@dfki.de

S. Jungbluth · A. Karnoub
Technologie-Initiative SmartFactory KL e.V., Kaiserslautern, Germany
e-mail: simon.jungbluth@smartfactory.de; ali.karnoub@smartfactory.de

M. Ruskowski
Deutsches Forschungszentrum für Künstliche Intelligenz GmbH (DFKI), Kaiserslautern, Germany

Technologie-Initiative SmartFactory KL e.V., Kaiserslautern, Germany
e-mail: martin.ruskowski@smartfactory.de; martin.ruskowski@dfki.de

J. Soldatos (ed.), *Artificial Intelligence in Manufacturing*,
https://doi.org/10.1007/978-3-031-46452-2_13

that can be rapidly provisioned and released with minimal management effort or interaction between service operators and providers" [22].

SmartFactory[KL] shares their vision with the terminology *Production Level 4* and Shared Production (SP) [2] that represents reconfigurable supply chains in a federation of trusted partners that dynamically share manufacturing services over a product lifecycle through standardized Digital Twins. Hence, it represents an extension of Cloud Manufacturing.

This vision requires sharing data on products and resources. Production data contains sensitive information, leading to fear of data misuse. Thus, additionally to interoperability and standardization, data sovereignty is one major aspect when building a manufacturing ecosystem. In [15], a technology-dependent solution is presented. The authors propose using Asset Administration Shells (AAS) and the concepts of Gaia-X to exchange data in a self-sovereign interoperable way. In this sense, AAS aims to implement a vendor-independent Digital Twin [10], while Gaia-X creates a data infrastructure meeting the highest standards in terms of digital sovereignty to make data and services available. Key elements are compliance with European Values regarding European Data Protection, transparency and trustability [3]. Nevertheless, the authors of [15] focus more on the business layer and the data exchange process rather than intelligent decision-making processes. Due to their characteristics, Multi-Agent System (MAS) seems to be a promising natural choice for implementing the logic and interactions among different entities.

In our contribution, we present the structure of MAS for a modern production system to cope with the upper mentioned challenges. We focus on the intra-organizational resilience of the shop-floor layer to provide the necessary flexibility to enable an SP scenario. In Sect. 2, the considered concepts for the usage of MAS in industrial applications are described. Section 3 describes the architecture of the applied MAS and Sect. 4 gives an overview about the characteristics of a plant specific holonic MAS and implements a prototype in the demonstrator testbed at SmartFactory[KL].

2 State of the Art

To use MAS in modern production environments, MAS needs to manage the complexity of fractal control structures of the shop-floor layer. Therefore, we elaborate on the control structures of industrial manufacturing systems (see Sect. 2.1). In Sect. 2.2, terminology like skill-based approach and cyber-physical production systems (CPPS) are mentioned to deal with the encapsulation of production environment and separate the implementation from the functionality. Section 2.3 compares Agents and Holons and relates the upper mentioned concepts with the holonic paradigm.

2.1 Control Architectures in the Manufacturing Domain

Recently, computerization of industrial machines and tools led to hardware equipped with some kind of software-controlled intelligence. Digitalization trends enable and empower flexibility, shifting static to dynamic optimization. Contributions like [15, 23, 30] list the need for common information models, standardized interfaces, real-time and non-real-time cross-layer communication, separation of concern, flexibility, semantics, intelligence, scalability, inter-enterprise data exchange, collaborative style of works, privacy and self-sovereignty.

Leitão and Karnouskos [20] identify three principal types of control structures in industrial manufacturing systems: centralized, (modified) hierarchical and (semi-)heterarchical architectures. A *centralized* architecture has only one decision-making entity at the root of the system. It handles all planning and control issues, and the other entities have no decision power. The centralized architecture works best in small systems where short paths lead to effective optimization in a short amount of time. A *hierarchical* organization of control distributes the decision-making overall hierarchical levels. Higher levels make strategic-oriented decisions, and lower levels focus on simple tasks. Such architectures can be efficient in static production environments and are more robust to failures than centralized control structures. The hierarchical architectures are typical for the computer-integrated manufacturing paradigm. In contrast to the master-slave flow of control in the hierarchical structures, the *heterarchical* architectures rely on cooperation and collaboration for decision-making. In fully heterarchical architectures, there are no hierarchies at all and each entity is simultaneously master and slave. Such an organization is typical for default MASs. These control structures are highly flexible, but the global optimization goals are hard to achieve because knowledge and decisions are localized by each agent and require a number of interactions between the agents to make them global. This was one of the major criticism of classical MAS architectures. This led to the invention of semi-heterarchical control structures, which are also called loose or flexible hierarchies. Lower levels of such architectures should react quickly to disturbances and make fast decisions. These levels are characterized by hierarchical organization and mostly reactive agents. The higher levels appreciate the flexibility of heterarchical structures and intelligent decision-making by deliberative agents. Semi-heterarchical control structures are typical for so-called holonic architectures.

2.2 Cyber-Physical Production Systems

A flexible and modular production environment manifests in the concept of Cyber-Physical Production Modules (CPPMs), which provide standardized interfaces to offer different functionalities as a service [17]. Therefore, the skill-based approach aims to encapsulate the production module's functionalities and decouples them

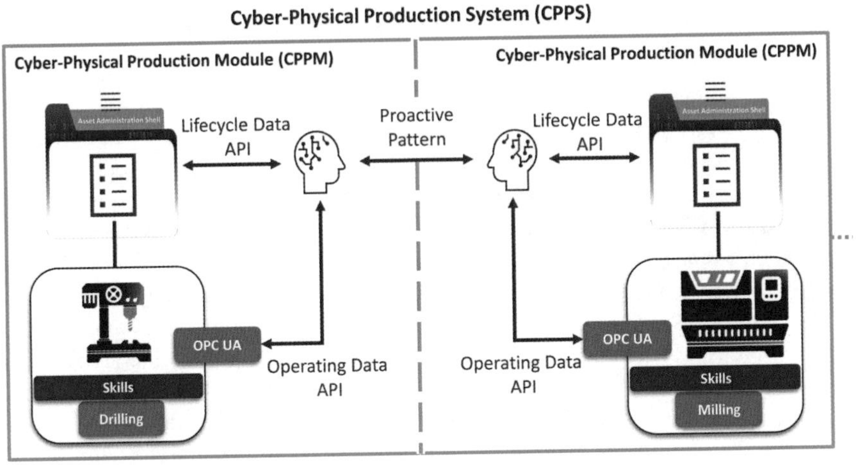

Cyber-Physical Production System (CPPS)

Fig. 1 Representation of a cyber-physical production system

from the specific implementation with the aim of increasing the flexibility. As visualized in Fig. 1, CPPMs can be combined into a CPPS to perform control tasks and react to information independently. CPPS connects the physical and digital worlds and react to information from the environment and environmental influences [8].

CPPMs require a need for a vendor-independent self-description to perform production planning and control. AAS represents an approach to achieve this standardized digital representation, where submodels (SM) are used to describe domain-specific knowledge [10]. The *Plattform Industrie 4.0* proposes an information model composing the concepts of capabilities, skills and services as machine-readable description of manufacturing functions to foster adaptive production of mass-customizable products, product variability, decreasing batch size, and planning efficiency. The Capability can be seen as an "implementation-independent specification of a function [. . .] to achieve an effect in the physical or virtual world" [25]. Capabilities are meant to be implemented as skills and offered as services in a broader supply chain. From the business layer, the service represents a "description of the commercial aspects and means of provision of offered capabilities" [25]. From the control layer, *Plattform Industrie 4.0* defines a Production Skill as "executable implementation of an encapsulated (automation) function specified by a capability" [25] that provides standardized interfaces and the means for parametrization to support their composition and reuse in a wide range of scenarios. The skill interface is mostly realized with OPC UA, since it has proven itself in automation technology. Topics such as OPC UA and AAS can lead to confusion regarding the separation of concerns. AAS is used as linkage to the connected world and lifecycle management in adherence to yellow pages, whereas OPC UA is applied for operative usage.

2.3 Agents and Holons

The study of MAS began within the field of Distributed Artificial Intelligence. It investigates the global behavior based on the agent's fixed behavior. The studies compromise coordination and distribution of knowledge. In this context, Leitão and Karnouskos define an agent as "[...] an autonomous, problem-solving, and goal-driven computational entity with social abilities that is capable of effective, maybe even proactive, behavior in an open and dynamic environment in the sense that it is observing and acting upon it in order to achieve its goals" [20]. MAS is a federation of (semi-)autonomous problem solvers that cooperate to achieve their individual, as well as global system's goals. To succeed, they rely on communication, collaboration, negotiation, and responsibility delegation [20]. MAS was motivated by subjects like autonomy and cooperation as a general software technology, while the emergence in the manufacturing domain has been growing recently.

The holonic concept was proposed by Koestler to describe natural beings that consist of semi-autonomous sub-wholes that are interconnected to form a whole [16]. Holonic Manufacturing System (HMS) is a manufacturing paradigm proposed at the beginning of the 1990s as an attempt to improve the ability of manufacturing systems to deal with the evolution of products and make them more adaptable to abnormal operating conditions [7]. Holonic production systems are fundamentally described in the reference architecture PROSA, with the aim of providing production systems with greater flexibility and reconfigurability [32]. A Holon is an autonomous, intelligent, and cooperative building block of a manufacturing system for transformation, transportation, storing and / or validating information and physical objects [33]. As shown in Fig. 2, a Manufacturing Holon always has an information processing part and often a physical processing part [4]. Holons join holarchies that define the rules for interaction between them. Each Holon can be simultaneously a part of several holarchies and as well as a holarchy itself. This enables very complex and flexible control structures, also called flexible hierarchies. It is important to note that the cooperation process also involves humans, who might enter or exit the Holon's context [4]. In summary, HMS can be seen as an analogy to CPPS, where skills provide the control interface to the physical processing parts.

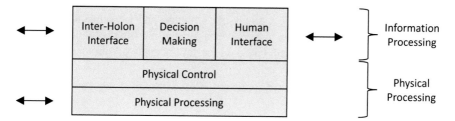

Fig. 2 General Architecture of a Holon in accordance to [7]

Previous research investigated the terminology between Agents and Holons. Giretti and Botti [7] perform a comparative study between the concepts of Agents and Holons. The authors explain that both approaches differ mainly in motivation but are very closely related, and a Holon can be treated as a special type of agent with the property of recursiveness and connection to the hardware. Subsequently, the authors define a recursive agent. One form of a MAS can be a holonic MAS, where there is a need for negotiation to optimize the utility of individual agents as well as the utility of their Holons [1]. The choice is more determined by the point of view. On the other hand, Valckenaers [31] explains that HMS and MAS frequently are seen as similar concepts, but it is important to be aware of contrasting views. In the author's opinion, it no longer holds to see MAS as a manner to implement HMS. HMS researches' key achievement is the absence of goal-seeking. The authors explicit distinguish between intelligent beings (Holons) to shadow the real-world counterpart and intelligent agents as decision makers. We want to sharpen the view and raise awareness of the different wording. Nevertheless, we treat a Holon as a special type of Agent.

The ADACOR architecture presents an example of a holonic MAS that provides a multi-layer approach for a distributed production and balancing between centralized and decentralized structures to combine global production optimization with flexible responses to disturbances [19].

Regarding modular production systems, the concept of Production Skills of CPPMs is important for usage in a multi-agent architecture to interact with a hardware layer [28]. To interact in a dynamic environment, agents furthermore need an environment model to describe the agent's knowledge and a standardized communication. The former serves to describe the respective domain of the agent, to configure the agent's behavior and aggregate information from external knowledge basis or other agents [20]. It requires a standardized information model to ensure autonomous access. In adherence to CPPS, AAS is suitable to describe the agents' properties, i.e., communication channels, physical connection, identification and topology. The latter might be direct or indirect. Direct communication means an exchange of messages. Known communication languages like Knowledge Query and Manipulation Language (KQML) or Agent Communication Language (ACL) rely on speech acts that are defined by a set of performative (agree, propose). Indirect communication relies on pheromone trails of ants, swarm intelligence, the concept of a blackboard, or auctions [20]. Interactions between AAS is mentioned as I4.0 Language (I4.0L) that is specified in VDI/VDE 2193. It defines a set of rules compromising the vocabulary, the message structure and the semantic interaction protocols to organize the message sequences to the meaningful dialogues. I4.0L implements the bidding protocol based on the contract network protocol [29].

2.4 Agent Framework

Agent Frameworks ease development and the operation of large-scale, distributed applications and services. For us, it is especially important that the framework is open, modular, extensible and fosters the holonic concept. [21] and [24] list and discuss a number of agent platforms, while the work of [5] evaluates five different agent languages and frameworks. The results imply that especially SARL language running on Janus platform is superior to other systems in aspects concerning the communication between agents, modularity and extensibility. The biggest advantage to other languages is that there are no restrictions regarding the interactions between agents. These positive effects are balanced by the fact that debugging is limited to local applications and, above all, the transfer between the design to the implementation is very complicated. SARL supports dynamic creation of agent hierarchies and implements holonic architecture patterns. Finally, SARL is chosen to implement our MAS.

In [27], the authors explain that SARL Agents represent Holons. In the following, we prefer using the notation of a Holon. SARL uses the concepts of Behaviors and Holon Skills to define the entities. As explained in [27], a "[. . .] Behavior maps a collection of perceptions [. . .] to a sequence of Actions", whereas a Holon Skill implements a specification of a collection of Actions to make some change in the environment. A Capacity provides an interface for the collection of all actions. Though, the focus lies on reusability and lowering of complexity. To avoid confusion with the different meanings of the term skill, the following explicitly refers to Holon Skill or Production Skill. Through the usage of Holon Capacities and Skills, SARL follows modularization in analogy to Capabilities and Production Skill of resources.

3 Towards an Architecture for Modern Industrial Manufacturing Systems

The challenge for SP is represented in a need for an architecture that enables flexibility, customizability and copes with dynamic optimization and decision-making. The information model of Capabilities-Skills-Services of the *Plattform Industrie 4.0* [25] promises standardization to foster shop-floor-level reconfiguration and dynamic planning. Hence, AAS proves strength regarding the interoperability of system elements. The work of [14] demonstrates a way to use AAS to (re)plan and execute the production and make production resources interchangeable. The authors use AAS as Digital Twin for assets like products and resources. MAS controls the production modules of a work center. This interconnection follows the idea of encapsulation and modularity to enable flexible and technology-independent access to production resources. Encapsulation intends to increase reconfigurability at the shop-floor level. Resources should not perform rigid operations but be assigned tasks that they can process themselves. In [15], an SP network is presented that enables data exchange in a self-sovereign interoperable manner. We implement that by combining concepts like Gaia-X, AAS and I4.0L. These technologies seem promising to enable a cross-company supply chain. In [15], we focus on the business layer communication to share data and explain the necessary components to build a data ecosystem. Nevertheless, there needs to be a component that enables logic to connect the shop floor to the connected world. Thus, [15] is our basis to realize inter-organizational communication and the following focuses on the intra-organizational resilience of the shop-floor.

3.1 Multi-agent System Manufacturing Architecture

Holonic MAS seems a promising pattern to wrap the factory's granularity and build complex and resilient systems. It is important to note that the technology to communicate with a customer and other factories might change over time or might differ for individual customers or factories. Consequently, to be technology-independent, our MAS does not include an explicit connection technology. MAS accumulates some of its resource capabilities into services and provides these services to the external world following the principles of service-oriented systems [9]. Besides, the holonic MAS supervises a production system to plan and execute the production and connects a factory to an SP network. Inspired by the upper mentioned concepts, a modern factory is represented in Fig. 3.

With reference to the described aspects, the system consists of three main Holons as displayed in Fig. 3. The basic structure follows the general idea of PROSA as described in Sect. 2. The difference between the presented MAS and their architecture is three-fold. First, the tasks of the management of the products are fulfilled by AAS instead of having a Product Holon in PROSA. A more detailed discussion about this change is discussed in 5. Second, the management of orders by

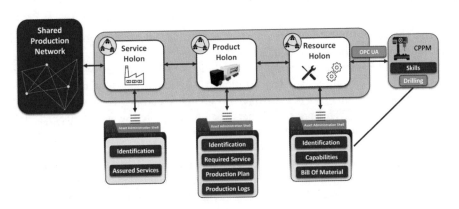

Fig. 3 The structure of the multi-agent system

the Order Holon is shifted from the Order Holon to the Product Holon. The reason for the new name is that our Product Holon takes care of orders and connects to Product AAS and thus encapsulates two tasks. Third, another Holon called Service Holon is added as an external representation of the factory layer as additional use case to PROSA which does not examine a SP scenario in detail and represents the factory as centralized HMS which spawns Order Holons based on the requests. However, we prefer to decouple the task of spawning Holons and the representation of the factory to achieve a higher resilience and flexibility.

Besides these changes, all three Holons are equipped with their own AAS to expose their self-description. This includes an SM Identification to identify the Holon in the MAS and an SM towards the interfaces of the Holon to be able to communicate with different communication technologies and to implement the communication technology independence. In addition, the SM Topology of each Holon describes the internal structure. Part of the structure are all aggregated Sub-Holons of the Holon. This SM eases the initialization of each Holon, especially in case of a restart.

In the following two subsections, we present details about the Service Holon in Sect. 3.2 and about the Product Holon in Sect. 3.3. For the Resource Holon, we will give a more detailed description in Sect. 4. The Resource Holon executes production steps by controlling and managing all resources. Each resource is via OPC UA connected to one Resource Holon, which controls its execution. For that purpose, the Resource Holon uses the Resource AAS to store information about the resource. This includes an SM Identification, the provided capabilities of the resource and the SM Bill of Material. In addition, the Resource Holon deals with tasks like lifecycle management of resources, human-machine interaction, resource monitoring and handling multiple communication technologies between the resources.

3.2 Service Holon

The Service Holon displayed on the left side of Fig. 3 manages and provides the services of the factory to the connected world. One example of a service is the 3D-printing service, which offers the capability of producing customized products using a fused deposition modeling process. Besides offering services, the Service Holon enables the MAS to process external tasks like ordering a specific product or executing a provided service. Besides that, the Service Holon takes care of the disposition of external services, products or substitutes. Therefore, the Service Holon has the ability to communicate with the SP network via an asynchronous event-based communication based on I4.0L. This implements VDI/VDE 2193 [34, 35] as the chosen communication standard. A more detailed use case for the connection to SP networks is applied in [15].

In the context of external SP Networks, the Service Holon represents the interface to the factory. For this reason, the Service Holon takes care of the Factory AAS. This AAS uses a unique identifier, a name and other general information to identify the factory. Furthermore, it contains a description of all assured services, the factory has to offer. Based on this service catalog, other network members can request an offered service.

In case of incoming production requests, the Service Holon processes these requests and requests the Product Holon to handle the request. The communication to the Product Holon is like the overall communication between all Holons in the present architecture, asynchronous and event-based.

3.3 Product Holon

The Product Holon takes care of the production process and subdivides production tasks into different subtasks. In this context, the holonic approach takes effect. Each Holon is responsible for one task and spawns for every subtask a Sub-Holon. Together, they display the production process in a tree-like manner. To handle these incoming tasks or derived subtasks, each Holon triggers an internal execution at the Resource Holon or requests an external execution from the Service Holon.

In the case of an internal execution, each Holon needs to check if the execution is feasible. Therefore, the Product Holon first matches the capabilities given by the resources to the desired capability to fulfill the (sub-)task. If both capabilities fit together, a feasibility check on the resources is triggered to simulate if a resource is able to perform the task under posed conditions (e.g., if the process can supply the desired product quality and evaluate estimated time, costs and consumption). After a successful feasibility check, the Product Holon spawns an AAS as Digital Twin for the product. The Product AAS contains information towards the product identification like a name and a unique identifier. If the AAS contains some subtasks, which require an external execution, the Product AAS contains a description of all required external services to execute the subtask. After starting the production process, the Product Holon further controls the process by triggering production steps or monitoring the current production state. To monitor the process, the Product Holon updates the corresponding Product AAS by adding logging data to the production log.

To illustrate the execution of the Product Holon, a model truck as sample product is ordered via the Service Holon. The truck is assembled out of two different semitrailers. The semitrailer_truck consists of a 3D-printed cabin pressed on a brick chassis called cab_chassis. Similar to the semitrailer_truck, the other semitrailer is built by mounting a 3D-printed or milled trailer onto a semitrailer_chassis. Figure 4 shows the corresponding product structure tree.

Each of the displayed components and component assemblies relates to a production step to produce the respective component. First, the components need to be manufactured, then the semitrailers are assembled from the components and at the end, the trailer is mounted on the cab_chassis to assemble the full truck. For each of the given components, an own Product AAS is spawned. For example, the truck Holon spawns on the highest level of the Bill of Material two semitrailer Holons.

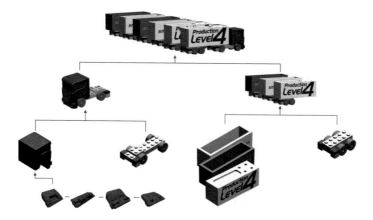

Fig. 4 The product tree of the model truck

Both Holons independently produce their related semitrailers and after completion, they report back to the truck Holon, which then controls the production of the full truck by controlling the assembly step of both semitrailers.

4 Execution System: Resource Holon

The Resource Holon takes care of the management of CPPMs. As visualized in Fig. 1, the Resource Holon serves as a proactive entity using the APIs of OPC UA as well as AAS and enables dynamic interactions. All AASs are deployed using the BaSyx middleware (v1.3) [18]. BaSyx is an open-source tool targeting to implement the specification of the AAS [10, 11] and provides additional services like storing AAS in databases, authorization and notifying the user with data change events. Holons collaboration fosters flexible execution and planning, while AAS and Production Skills provide interoperability. Next to these aspects, we want to emphasize resilience on the software layer. Resource Holons are deployed as Docker containers and managed by a GitLab repository that automatically creates container images and allows continuous deployment. Modern industrial environments need applications that are isolated and separated from the runtime environment to switch the underlying hardware and balance the load.

Our Resource Holon consists of an arbitrary number of Holons, where we distinguish between a Resource Holon of type Island and CPPM. The former is used to emphasize the existence of Sub-Holons, while the latter type is used to highlight the smallest possible entity that cannot be further partitioned. For interactions, it is not significant whether the Sub-Holon is an Island or a CPPM. We use this differentiation to separate the modules to build the Holons, i.e., to classify the Behaviors and Holon Skills for each type. Island Resource Holons are more responsible for lifecycle management and coordination, while CPPM Resource

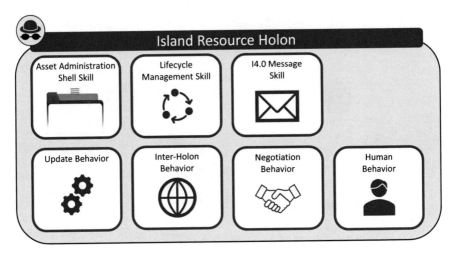

Fig. 5 The behaviors and skills of an Island resource Holon

Holons schedule and perform the concrete tasks. Furthermore, Island Resource Holons encapsulate the sum of all Sub-Holons, by providing proxy functionality.

4.1 Behaviors and Skills of a Holon

SARL Holons consist of a collection of Behaviors and Holon Skills. Island Resource Holons provide hierarchies and coordinate Sub-Holons, whereas CPPM Holons, as the smallest entity, are connected to physical assets. Figure 5 summarizes Island Resource Holon's Behaviors and Skills. Each Island Resource Holon has an AAS that describes the Holon's properties, configures and parametrize the system. The key aspects of AAS are to provide information on how to find a Holon, how a Holon is constructed, and how the Holon's interfaces are defined. This information is available in the SM Topology and Interface (see Sect. 3.1) and is also provided to Sub-Holons. The AAS Skill extracts the Holons' information as well as the environmental information. The Update Behavior is used to gather information about the contained Holons' states, update the topology when MAS changes, and synchronize this information with AAS. Since the Island Resource Holon manages Sub-Holons, the Lifecycle Skill allows to dynamically create, destroy or reboot Holons in its own holonic context. As SARL Janus provides message channels for all Holons in runtime, the communication with other Holons requires an external communication interface. The Inter-Holon Behavior allows the external communication to Holons in other runtimes, e.g., message exchange between the Resource Holon and the Product Holon via the open standard communication middleware Apache Kafka. For communication and understanding, an I4.0 Message Skill supports accordance with a standardized message model according to the

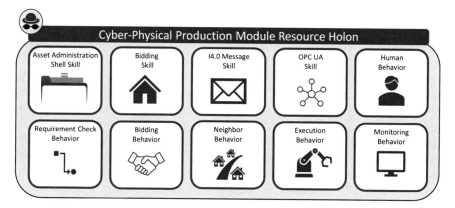

Fig. 6 The behaviors and skills of a CPPM resource Holon

VDI/VDE 2193-1 [35]. Communication, collaboration and negotiation are the key components for a successful process. Island Resource Holon responds to production requests of the Product Holon, initialize negotiations and sends production requests to Sub-Holons. In Negotiation Behavior, the Island Resource Holon verifies incoming messages. Depending on the request, Island Resource Holon forces Sub-Holons to follow a task or request the possible execution. The former is used when static optimization is applied, i.e., a global schedule shall be executed. An example of global scheduling is demonstrated in [13], in which we schedule value-adding as well as non-value-adding processes. The latter implements the bidding protocol to foster dynamic optimization. Negotiation behavior defines the duration of auctions and chooses incoming offers o based on max operator, i.e., for n incoming offers, the chosen offer calculates with $o_i = \max(o_1, \ldots, o_n)$. Next to software systems, Holons may interact with humans, intending a special treatment in terms of prioritization. Therefore, Human Behavior considers human knowledge and adjustments. We are not trying to accomplish fully automated plants and exclude humans from production. Instead, we want to support human decisions to benefit from experiences and intuition, as well as building factories for humans [36].

For CPPM Resource Holons some building blocks like the Asset Administration Shell Skill, the I4.0 Message Skill and the Human Behavior overlap (see Fig. 6). For interactions, the CPPM Resource Holon provides three Behaviors: Requirement Check, Bidding and Neighbor. Requirement Check acts in loose adherence to a method call to achieve control structures in a hierarchy. In case, Island Resource Holon fosters execution, the CPPM Resource Holon verifies if it can follow the call and starts or queues the task. In Bidding Behavior, the Bidding Skill is used to calculate a bid in the range between 0 and 1. The bid determines the desire to perform the job. This fosters dynamic optimization, while taking processing time, changeover times, deadlines, availability and resource's possible operations into account. The calculation process is achieved using a Reinforcement Learning algorithm. The basis of the Reinforcement Learning algorithm is described in [26], while a modified variant will be published in future work. The last interaction pattern is the neighbor behavior. CPPMs sense their environment, thus, CPPM Resource Holon can omit hierarchies and directly communicate with their physical neighbor to perform a complex task in a collaborative way. An example is a Pick&Place operation that usually requires the supply by transportation means. The CPPM Resource Holon has additional functions regarding the control and monitoring of Production Skills. The Execution Behavior builds an event-based sequence to reliably execute a Production Skill. In this context, it represents pipelines to set the Production Skill's parameters, verify compliance of all preconditions, tracks the execution state and manages postconditions. Therefore, the CPPM Resource Holon uses the OPC UA Skill, which allows access to a Production Skill interface directly deployed on the CPPM. Furthermore, CPPM Resource Holon has a Monitoring Behavior, which is used to check relevant sensor data, track the system status and update system-critical information. In the future, anomaly detection and supervision will also be implemented in this behavior.

4.2 Demonstrator Use Case

To demonstrate the application of the Resource Holon in a production environment, *Produktionsinsel_KUBA* of a real-world factory at *SmartFactory*[KL] is used (see Fig. 7).

Produktionsinsel_KUBA consists of three CPPMs named Connector Module, Quality Control Module and Conveyor Module as well as a *Produktionsinsel_SYLT*

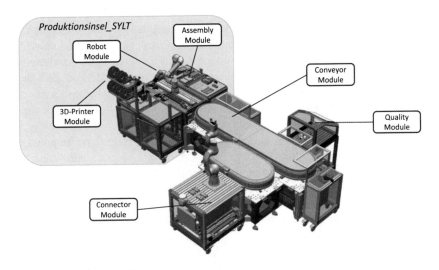

Fig. 7 *SmartFactory*[KL] real-world demonstration factory: *Produktionsinsel_KUBA*

enclosing a 3D Printer, a Robot and a Hand Assembly. The Connector Module serves as a supply and storage station for components parts. The Connector Module transfers the delivered components onto the Conveyor Module. The Conveyor Module transports the components to the *Produktionsinsel_SYLT*, where individual parts are mounted into a higher order assembly. Afterward, the quality of the product is checked, and the assembly is ejected at the Connector Module. Our exemplary production assembles a model truck shown in Fig. 4. In this context, *Produktionsinsel_KUBA* is incapable of producing all the components of the model truck on its own. Therefore, our scenario assumes that the required components of the model truck have already been manufactured in the sense of SP as explained in [15] and delivered to the Connector Module. The CPPMs can be positioned in an arbitrary layout. As a result, *Produktionsinsel_KUBA* offers different services depending on connected modules and increases flexibility. To provide a safe workspace, the CPPMs are mechanical locked. In this context, we refer to two or more mechanical connected CPPMs as neighbors. In [12], we present the development of our Conveyor Module to enable on-demand transportation. We mention five different coupling points on which CPPMs can be locked. The mechanism is realized with magnets and RFID sensors. Hence, the Connector Module, the Quality Control Module and the *Produktionsinsel_SYLT* are physically locked to the Conveyor Module, building a connected neighborhood. The RFID tag contains the CPPM's ID that allows to identify the locked module. Therefore, the self-description is accessible and CPPM Resource Holons can build peer-to-peer connections. Our production process starts when the Resource Holons gets the request to execute a production step from the Product Holon. The production step is described in a production plan following the metamodel of AAS. An example of a production plan is visualized in [14].

Based on the idea to encapsulate the manufacturing logic of production modules in Resource Holons, the interaction between the Island Resource Holon and the responding Sub-Holons is described. Our instantiated *Produktionsinsel_KUBA* Holon is visualized in Fig. 8. During the request to execute a production step, the *Produktionsinsel_KUBA* Resource Holon verifies the query in Negotiation Behavior and asks the Sub-Holons if they can perform the required effect. Since we omit global scheduling, dynamic optimization through collaboration takes place. The Sub-Holons compete to perform the effect while calculating a bid using the Bidding Skill. If the underlying CPPM is incapable of performing the action, the bid results in 0. Otherwise, the bid is calculated through Reinforcement Learning methods with a maximum value of 1. The *Produktionsinsel_KUBA* Resource Holon verifies the bids, by ignoring all offers $o < 0.2$ and distributes the steps. The CPPM Resource Holons' local decision leads to a global behavior, where the global allocation of resources is optimized by the bidding system. This procedure ensures local CPPM Resource Holon utilization and realizes on-demand scheduling. To perform a required effect, the CPPM Holons access Production Skills via the Execution Behavior to perform a change in the environment. During the execution of a task, Holons may cooperate to perform tasks they are unable to do on their own. As an example of the communication between different Holons, the Connector Holon

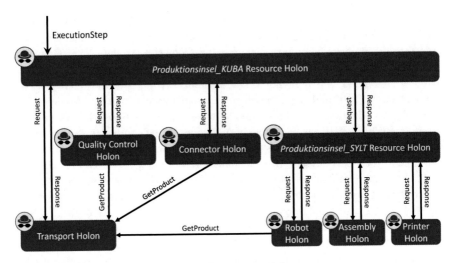

Fig. 8 Instantiation of *Produktionsinsel_KUBA* resource Holon

and the *Produktionsinsel_SYLT* Resource Holon communicate with the Conveyor Holon to order transportation means for a specified product. Conveyor Holon manages the orchestration of the transportation means by routing and queuing. This communication relies more on a call, since Conveyor Holon encapsulates on-demand transport.

This sequence describes the demonstration scenario at *Produktionsinsel_KUBA*. The combination of requesting and calling Holon's abilities leads to flexible control structures that allow manufacturing in a resilience way. As a result, we can handle static and dynamic requirements. Additionally, the modular structure in combination with usage of AAS and Production Skills foster interoperability and interchangeability at different layers of the factory, from the shop-floor to the connected world.

5   Conclusion

This chapter presents a MAS approach in the manufacturing domain, which enables a factory to control its resources, to define and manage products and to provide services to other SP participants. The MAS is based on a holonic approach and is subdivided into Holons, each taking care of one of these tasks. For us, a Holon is treated as a special type of Agent with additional characteristics regarding recursiveness and connection to the hardware. The MAS collaborates with and uses modern Industry 4.0 technologies such as Production Skills, AAS or OPC UA.

The presented MAS is enrolled on a demonstrator testbed at *SmartFactory*[KL], which is part of an SP scenario to produce model trucks. We divide our manufacturing system into three Holons. The Service Holon provides and retrieves services from the connected world. The Product Holon deals with modular encapsulated products to manage dependencies between individual parts and assemblies as well as controlling the production process. The Resource Holon encapsulates the layer of the production testbed and connects the virtual with the physical world. To guarantee autonomy, our Resource Holons use descriptions of AAS to gain knowledge about the environment and use CPPM's Production Skills to perform an effect in the physical world. We achieve a flexible and resilient system by providing communication patterns that allow hierarchical and heterarchical modularization.

However, the current state of our MAS is subject to different limitations. This means that it will be extended in the future to fulfill different other features and will solve different topics. One topic is to put more emphasize on product's lifecycle, while providing more complex planning systems to extract product's features, match capabilities and trace tender criteria. Another extension is planned on the monitoring system to embed a factory wide monitoring system to combine a supervision of the production process, factory level information like assured services and resource data. The last topic is to provide more generalized holonic patterns and give more insights about the Service Holon and the Product Holon.

As a result, we want to compare our architecture to other MAS systems, with a special focus on the applied technologies in our systems. One of these technologies is AAS. In comparison to full agent-based solutions, we typically replace one Holon (e.g., in PROSA the Product Holon). As a downside, this leads to more applied technologies in the system (due to the different technology stack) and thus to a more complex architecture. However, the AAS as manufacturing standard supports the interoperability between other factories and a simple data exchange format. Furthermore, we use one standardized data format to express all our knowledge to ease the internal usage of data via a system-wide interface.

Besides AAS as data format, we use SARL as Agent Framework. SARL itself is a domain-specific language, which leads to a couple of general advantages and disadvantages, as explained in [6]. We want to take up some of the listed problems and advantages and add a few more SARL language-specific arguments. First, SARL is especially designed to build MAS and includes an own metamodel to define the structure of a Holon. Besides that, SARL offers concepts to encapsulate certain functionality in Behaviors and Skills and leads to a modular system. One special feature of SARL is that Holons are able to control the lifecycle of other Holons, which is quite close to our applied concept of the MAS. Although SARL is functional suitable for us, SARL also has different disadvantages. For example, it is hard to find a documentation and help in the community if SARL specific problems occur. Unfortunately, developing SARL code is exhausting since general supported development environments do not always react in our desired response time.

Another difference between our MAS concept and other MAS concepts is the granularity of applied Holons. In many cases, each device (e.g., a robot or even in a smaller granularity like a sensor) has an own Holon. In our approach, a Holon connects to one CPPM, which encapsulates single resources like robot arms or 3D-printers. In this approach, a Holon accesses each CPPM by calling their provided skills. Having Holons on the device level leads to more holonic communication, and thus more resources and effort is required to handle the holonic communication. Moreover, MAS does not need to operate in real-time to perform actions without a delay. This is why, we decided to encapsulate internal communication inside a CPPM and keep time-critical and safety-critical tasks in the physical processing parts. Furthermore, Holons are independent of machine-specific control technologies, which increase the flexibility of the system towards resource-specific technologies. Finally, we want to mention that even for small holonic MAS, communication quickly becomes complex and lacks transparency. Using standardized technologies like OPC UA and AAS regains this transparency and supports application in the factory.

Acknowledgments This work has been supported by the European Union's Horizon 2020 research and innovation program under the grant agreement No 957204, the project MAS4AI (Multi-agent Systems for Pervasive Artificial Intelligence for Assisting Humans in Modular Production Environments).

References

1. Beheshti, R., Barmaki, R., Mozayani, N.: Negotiations in holonic multi-agent systems. Recent Advances in Agent-Based Complex Automated Negotiation pp. 107–118 (2016). https://doi.org/10.1007/978-3-319-30307-9_7
2. Bergweiler, S., Hamm, S., Hermann, J., Plociennik, C., Ruskowski, M., Wagner, A.: Production Level 4" Der Weg zur zukunftssicheren und verlässlichen Produktion (2022), https://smartfactory.de/wp-content/uploads/2022/05/SF_Whitepaper-Production-Level-4_WEB.pdf, (Visited on 18.05.2023)
3. Braud, A., Fromentoux, G., Radier, B., Le Grand, O.: The road to European digital sovereignty with Gaia-X and IDSA. IEEE network **35**(2), 4–5 (2021). https://doi.org/10.1109/MNET.2021.9387709
4. Christensen, J.: Holonic Manufacturing Systems: Initial Architecture and Standards Directions. In: First European Conference on Holonic Manufacturing Systems, Hannover, Germany, 1 December 1994 (1994)
5. Feraud, M., Galland, S.: First comparison of SARL to other agent-programming languages and frameworks. Procedia Computer Science **109**, 1080–1085 (2017). https://doi.org/10.1016/j.procs.2017.05.389
6. Fowler, M.: Domain-specific languages. Pearson Education (2010)
7. Giret, A., Botti, V.: Holons and agents. Journal of Intelligent Manufacturing **15**(5), 645–659 (2004). https://doi.org/10.1023/B:JIMS.0000037714.56201.a3
8. Hermann, J., Rübel, P., Birtel, M., Mohr, F., Wagner, A., Ruskowski, M.: Self-description of cyber-physical production modules for a product-driven manufacturing system. Procedia manufacturing **38**, 291–298 (2019). https://doi.org/10.1016/j.promfg.2020.01.038
9. Huhns, M.N., Singh, M.P.: Service-oriented computing: Key concepts and principles. IEEE Internet computing **9**(1), 75–81 (2005). https://doi.org/10.1109/MIC.2005.21
10. Industrial Digital Twin Association: Specification of the Asset Administration Shell Part 1: Metamodel, https://industrialdigitaltwin.org/wp-content/uploads/2023/06/IDTA-01001-3-0_SpecificationAssetAdministrationShell_Part1_Metamodel.pdf, (Visited on 28.06.2023)
11. Industrial Digital Twin Association: Specification of the Asset Administration Shell Part 2: Application Programming Interface, https://industrialdigitaltwin.org/wp-content/uploads/

2023/06/IDTA-01002-3-0_SpecificationAssetAdministrationShell_Part2_API_.pdf, (Visited on 28.06.2023)

12. Jungbluth, S., Barth, T., Nußbaum, J., Hermann, J., Ruskowski, M.: Developing a skill-based flexible transport system using OPC UA. at-Automatisierungstechnik **71**(2), 163–175 (2023). https://doi.org/10.1515/auto-2022-0115

13. Jungbluth, S., Gafur, N., Popper, J., Yfantis, V., Ruskowski, M.: Reinforcement Learning-based Scheduling of a Job-Shop Process with Distributedly Controlled Robotic Manipulators for Transport Operations. IFAC-PapersOnLine **55**(2), 156–162 (2022). https://doi.org/10.1016/j.ifacol.2022.04.186, https://www.sciencedirect.com/science/article/pii/S2405896322001872, 14th IFAC Workshop on Intelligent Manufacturing Systems IMS 2022

14. Jungbluth, S., Hermann, J., Motsch, W., Pourjafarian, M., Sidorenko, A., Volkmann, M., Zoltner, K., Plociennik, C., Ruskowski, M.: Dynamic Replanning Using Multi-agent Systems and Asset Administration Shells. In: 2022 IEEE 27th International Conference on Emerging Technologies and Factory Automation (ETFA). pp. 1–8 (2022). https://doi.org/10.1109/ETFA52439.2022.9921716

15. Jungbluth, S., Witton, A., Hermann, J., Ruskowski, M.: Architecture for Shared Production Leveraging Asset Administration Shell and Gaia-X (2023), unpublished

16. Koestler, A.: The ghost in the machine. Macmillan (1968)

17. Kolberg, D., Hermann, J., Mohr, F., Bertelsmeier, F., Engler, F., Franken, R., Kiradjiev, P., Pfeifer, M., Richter, D., Salleem, M., et al.: SmartFactoryKL System Architecture for Industrie 4.0 Production Plants. SmartFactoryKL, Whitepaper SF-1.2 **4** (2018)

18. Kuhn, T., Schnicke, F.: BaSyx, https://wiki.eclipse.org/BaSyx, (Visited on 07.07.2022)

19. Leitão, P., Colombo, A.W., Restivo, F.J.: ADACOR: A collaborative production automation and control architecture. IEEE Intelligent Systems **20**(1), 58–66 (2005). https://doi.org/10.1109/MIS.2005.2

20. Leitao, P., Karnouskos, S. (eds.): Industrial Agents: Emerging Applications of Software Agents in Industry. Kaufmann, Morgan, Boston (2015). https://doi.org/10.1016/C2013-0-15269-5

21. Leon, F., Paprzycki, M., Ganzha, M.: A review of agent platforms. Multi-paradigm Modelling for Cyber-Physical Systems (MPM4CPS), ICT COST Action IC1404 pp. 1–15 (2015)

22. Liu, Y., Wang, L., Wang, X.V., Xu, X., Jiang, P.: Cloud manufacturing: key issues and future perspectives. International Journal of Computer Integrated Manufacturing **32**(9), 858–874 (2019). https://doi.org/10.1080/0951192X.2019.1639217

23. Neubauer, M., Reiff, C., Walker, M., Oechsle, S., Lechler, A., Verl, A.: Cloud-based evaluation platform for software-defined manufacturing: Cloud-basierte Evaluierungsplattform für Software-defined Manufacturing. at-Automatisierungstechnik **71**(5), 351–363 (2023). https://doi.org/10.1515/auto-2022-0137

24. Pal, C.V., Leon, F., Paprzycki, M., Ganzha, M.: A review of platforms for the development of agent systems. arXiv preprint arXiv:2007.08961 (2020). https://doi.org/10.48550/arXiv.2007.08961

25. Plattform Industrie 4.0: Information Model for Capabilities, Skills & Services, https://www.plattform-i40.de/IP/Redaktion/EN/Downloads/Publikation/CapabilitiesSkillsServices.html, (Visited on 06.06.2023)

26. Popper, J., Ruskowski, M.: Using Multi-agent Deep Reinforcement Learning for Flexible Job Shop Scheduling Problems. Procedia CIRP **112**, 63–67 (2022). https://doi.org/10.1016/j.procir.2022.09.039

27. Rodriguez, S., Gaud, N., Galland, S.: SARL: a general-purpose agent-oriented programming language. In: 2014 IEEE/WIC/ACM International Joint Conferences on Web Intelligence (WI) and Intelligent Agent Technologies (IAT). vol. 3, pp. 103–110. IEEE (2014). https://doi.org/10.1109/WI-IAT.2014.156

28. Ruskowski, M., Herget, A., Hermann, J., Motsch, W., Pahlevannejad, P., Sidorenko, A., Bergweiler, S., David, A., Plociennik, C., Popper, J., et al.: Production bots für production level 4: Skill-basierte systeme für die produktion der zukunft. atp magazin **62**(9), 62–71 (2020). https://doi.org/10.17560/atp.v62i9.2505

29. Smith, R.G.: The contract net protocol: High-level communication and control in a distributed problem solver. IEEE Transactions on computers **29**(12), 1104–1113 (1980). https://doi.org/10.1109/TC.1980.1675516

30. Trunzer, E., Calà, A., Leitão, P., Gepp, M., Kinghorst, J., Lüder, A., Schauerte, H., Reifferscheid, M., Vogel-Heuser, B.: System architectures for Industrie 4.0 applications: Derivation of a generic architecture proposal. Production Engineering **13**, 247–257 (2019). https://doi.org/10.1007/s11740-019-00902-6

31. Valckenaers, P.: Perspective on holonic manufacturing systems: PROSA becomes ARTI. Computers in Industry **120**, 103226 (2020). https://doi.org/10.1016/j.compind.2020.103226

32. Van Brussel, H., Wyns, J., Valckenaers, P., Bongaerts, L., Peeters, P.: Reference architecture for holonic manufacturing systems: PROSA. Computers in Industry **37**(3), 255–274 (1998). https://doi.org/10.1016/S0166-3615(98)00102-X, https://www.sciencedirect.com/science/article/pii/S016636159800102X

33. Van Leeuwen, E., Norrie, D.: Holons and holarchies. Manufacturing Engineer **76**(2), 86–88 (1997). https://doi.org/10.1049/me:19970203

34. VDI/VDE-Gesellschaft Mess- und Automatisierungstechnik: Language for I4.0 components - Interaction protocol for bidding procedures, https://www.vdi.de/en/home/vdi-standards/details/vdivde-2193-blatt-2-language-for-i40-components-interaction-protocol-for-bidding-procedures, (Visited on 01.06.2023)

35. VDI/VDE-Gesellschaft Mess- und Automatisierungstechnik: Language for I4.0 Components - Structure of messages, https://www.vdi.de/en/home/vdi-standards/details/vdivde-2193-blatt-1-language-for-i40-components-structure-of-messages, (Visited on 01.06.2023)

36. Zuehlke, D.: SmartFactory–Towards a factory-of-things. Annual reviews in control **34**(1), 129–138 (2010). https://doi.org/10.1016/j.arcontrol.2010.02.008

A Multi-intelligent Agent Solution in the Automotive Component–Manufacturing Industry

Luis Usatorre, Sergio Clavijo, Pedro Lopez, Echeverría Imanol, Fernando Cebrian, David Guillén, and E. Bakopoulos (iD)

1 Introduction

The manufacturing industry is an ecosystem full of changes and variations where production conditions are never the same. As an example, the raw materials received from suppliers differ from one another, though within tolerances. Because of these differences, using the Asset Administration Shell[1] (AAS) [1] for each component/-part is not feasible. And similar variations appear in all areas of manufacturing, such as tool wearing, the statuses of production machines, and even operator decisions (Fig. 1).

On top of that, operator decisions are, most of the time, based on intuition and experience, not based on data analysis. One of the reasons is that manufacturing industry data are stored locally in departments and in silos, so the operator does not have access to them. Some of these data and their storage locations are as follows:

1. Scheduling data related to production orders, quantities, delivery times, etc. are usually stored in the enterprise resource-planning (ERP) system.

---

[1] Platform Industry 4.0—asset administration shell specifications (plattform-i40.de)

L. Usatorre (✉) · S. Clavijo · P. Lopez · E. Imanol
Fundacion TECNALIA R&I, Madrid, Spain
e-mail: Luis.usatorre@tecnalia.com; sergio.clavijo@tecnalia.com; pedro.lopez@tecnalia.com; imanol.echeverria@tecnalia.com

F. Cebrian · D. Guillén
Fersa Bearings, Zaragoza, Spain
e-mail: fernando.cebrian@fersa.com; david.guillen@fersa.com

E. Bakopoulos
LMS, Laboratory for Manufacturing Systems, Patras, Greece
e-mail: bakopoulos@lms.mech.upatras.gr

© The Author(s) 2024
J. Soldatos (ed.), *Artificial Intelligence in Manufacturing*,
https://doi.org/10.1007/978-3-031-46452-2_14

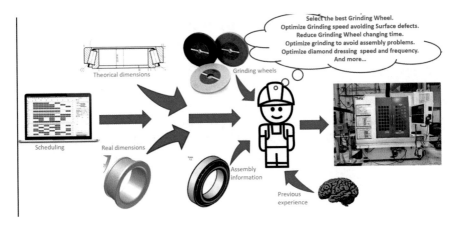

Fig. 1 Variables affecting operator decisions

2. Raw material suppliers, their characteristics, their arrival times, their quantities, etc. are usually stored in the suppliers' departmental repositories.
3. Nominal product characteristics can be found in the engineering department's drawings and repositories.
4. Quality-control results are usually stored either in the quality-control machine's memory or in the quality-control department's repositories.

This paper suggests that an improvement to production must balance various choices, not only technical but also economic. In existing production environments, several criteria are included and considered as part of determining the best solution to problems like storage size (e.g., economic costs, logistic costs), operator costs, production time, energy consumption, and more. When applying artificial intelligence (AI) in the manufacturing process, the criteria should be similar: Several agents with different goals should interact to determine the most holistic solution.

To this end, this paper illustrates that the multiagent method, based on a framework of distributed examples, leads to interactions between these agents in ways that improve the entire plant production system. Although the agents could work in isolation, the MAS4AI (the multiagent system for AI) is tasked with making them work together seamlessly and providing a better solution than they would produce by working separately.

The second goal is to address scalability in the implementation of AI solutions in the manufacturing industry: Agents need to be customized and specific to each manufacturing process (i.e., grinding) yet at the same time be generic enough for all machines and industries using this manufacturing process (i.e., all grinding machines in all shop floors). The MAS4AI proposal is for an agent-based Asset Administration Shell (AAS) approach to be deployed in every industry that wants to use a particular agent. In this way, an agent is composed of agent logic and the AAS. See the specifics for a case of using it at the Fersa plant, as depicted in Fig. 2.

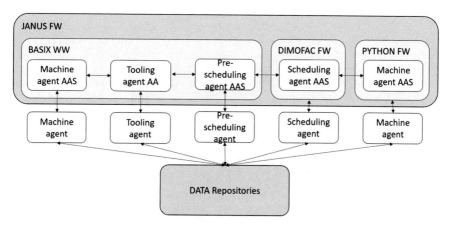

Fig. 2 Implementation of the agent-based AAS

1.1 Fersa's Pilot Plant

Fersa Bearings is a company that specializes in the design, manufacturing, and commercialization of bearings, mainly for the automotive market, but a minor percentage is dedicated to the industrial market. For the automotive market, Fersa Bearings manufactures bearings both for new vehicles (few references and production cases in long batches) and as spare parts, which implies many references and production cases in much-shorter batches.

The use case implemented at the Fersa plant focuses on the Z0 line, which is fully dedicated to manufacturing spare parts (short batches). In this line, one reference is composed of two sections manufactured in parallel that have to be assembled into one part. Therefore, quality is critical in these processes to ensure that the different parts perfectly match, meeting the final requirements.

The multiagent system improves Fersa's control over the processes in this manufacturing line, which improves end quality, communication and process coordination despite the variability among raw materials, tooling, and processes. Today, different actors interact in the physical world to optimize production holistically. The goal has been to apply AI in the manufacturing process—in a virtual world where several agents with different goals interact to determine the most holistic solution, as depicted in Fig 3.

At Fersa's pilot plant, we are considering having three agents interact: a tooling agent that optimizes the tooling selection, a machine agent that improves the machine parameters, and a scheduling agent that evaluates the manufacturing time for a production batch under certain circumstances, such tooling selection and machine parameters. This is presented in Fig. 4.

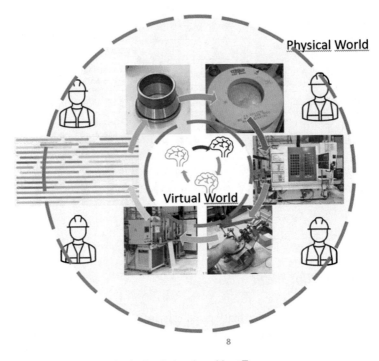

Fig. 3 Relationship between physical and virtual worlds at Fersa

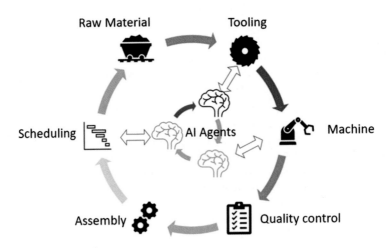

Fig. 4 Relation among agents at Fersa

2 Experimental Development

2.1 Data Aggregation

The challenge was that the Fersa use case generated a wide variety of heterogeneous data from different data sources. The decisions must take the following factors, among others, into consideration:

- Raw material characteristics
- BEAIN12 and RIFA6 machine parameters and historical data
- Quality-control stations' data
- Assembly stations' data
- Sensors' data
- Scheduling plans
  These data sources also appear in different types of formats:
- Portable data formats (PDFs)
- Comma-separated values (CSVs)
- Tabular data formats such as Extensible Markup Language (XML) and Excel (XLS)

On top of that, the data come from various locations:

- Shop floors (e.g., machine parameters)
- Suppliers (e.g., raw material characteristics)
- Management tools (e.g., ERPs and scheduling tools)

The first step in implementing Fersa's MAS4AI is to ensure that the data are properly connected, aggregated, and filtered.

Figure 5 presents the proposed solution to aggregating all the data sources: a—machine data, including ranges and grinding-wheel parameters; b—raw material data, including nominal and real data; c—other agents' data; and d—assembly and control stations' data. The central part shows the repositories where the aggregation takes place in MAS4AI for Fersa.

2.2 Tooling Agent

One of the main challenges is to find the most suitable griding wheel (GW) for a certain production process (reference and quantity). On one hand, a too-big GW will not be completely used; it will return to the warehouse as partially worn and be difficult to reuse. On the other hand, a too-small GW will stop functioning before the production cycle has finished, forcing a GW change, which stops production.

Our solution is to have the tooling agent check the Fersa GW warehouse in order to select the most suitable GW for the current production cycle (see selection criteria in 2.4). Most of the information needed by the agent is provided by the grinding

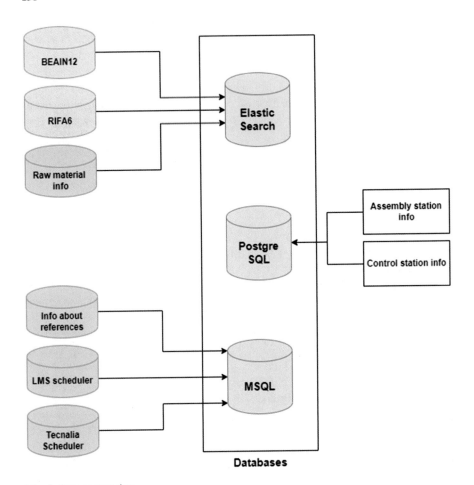

Fig. 5 Data aggregation

machine agent, which receives and optimizes the machine parameters of the current production cycle. Figure 6 shows the current workflow of this part of the Fersa system.

The reference for the production order and the suggested machine parameters for that production should be followed to determine the optimal grinding wheels. The length of change time for the GWs should always be minimized. The AAS type for the Fersa system was the resource one.

The main content of this agent can be found in the AAS of the GW agent, specifically in the submodel's production information. Figure 7 illustrates the reference for the production, the inputs used for the execution of the agent, and the list of GWs obtained for the tooling agent ASS.

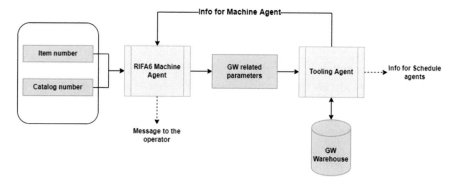

Fig. 6 Agent interaction logic

2.3 Machine Agent

Another challenge is manufacturing the highest number of good-quality parts (i.e., without burns; without derivatives; without conicity, ovality, or obliquity problems; and with Ha-T-1[2] diameters within tolerances) in the shortest amount of time. To accomplish this objective, the proposed solution takes into account the following actions that illustrate the agent's logic:

- It determines the GW diameter by taking into account the current parameters of the production. This determination is made in the short term, over periods of seven, fifteen, and thirty minutes.
- It provides critical values from the production cycle. These values are provided in the short term, over periods of seven, fifteen, and thirty minutes.
- It optimizes the initial machine parameters according to the batch reference.
- It detects deviations from the machine parameters during the execution of production.

All these objectives contribute in different ways to minimizing the time that a batch needs to be finished.

During the pilot operations at Fersa, assembly and quality stations fed the agent with information such as nominal and real values and the tolerances of the raw materials. The Fersa use case features two machine agents. While the real-time agent is related to the BEAIN12 machine, this machine agent relies on the RIFA6 machine. This agent calculates by using a series of machine-learning and deep-learning models. At the time when this chapter was written, the testing process involved three neural network structures with different numbers of hidden layers (two, three, and four), different numbers of neurons per layer (20, 15, and 10), and different optimization algorithms (Adam and Stochastic gradient descent (SGD)). The metrics used were mean absolute error (MAE) and mean squared error (MSE). For example, while deep-learning short-term prediction models for the Grinding

---

[2] Ha-T-1 is a process parameter (Axial tolerance).

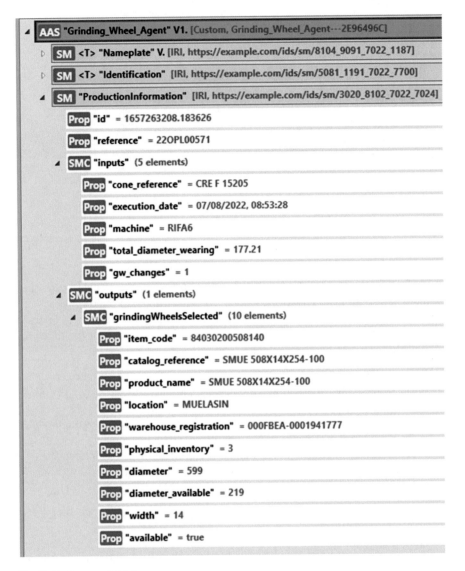

Fig. 7 Tooling agent AAS

Wheel (GW) diameter and positions have been fully developed and are in a testing phase, the optimization model for the initial parameters is still undergoing testing because it still needs historical data for past batch references that were not collected by Fersa before the project started. For the same reason, the model that predicts deviations from the machine parameters for a fixed batch is also still undergoing testing. Data are currently being collected to enable new models to be added and tested over the following months.

The main content of this agent can be found in the machine agent's AAS, specifically in the submodel's production information. Figure 8 presents the reference

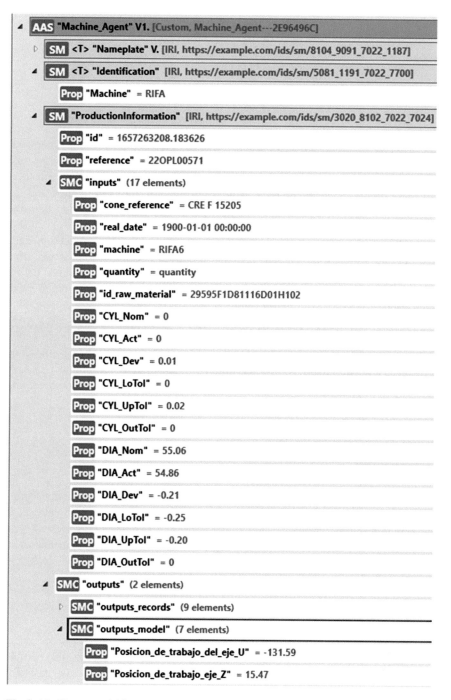

Fig. 8 Machine agent AAS

for the production, the inputs used for the execution of the agent, and the outputs obtained for the machine agent ASS.

2.4 Prescheduling Agent

In this case, the challenge is to determine the time required to produce a certain batch by factoring in the material, the tooling, and the machine parameters. Additional parameters to be considered include the number of wheel changes needed and the estimated time to complete these changes. The proposed solution considers all these inputs. Such inputs refer to item information: the item number and the catalog reference for the batch, the number of pieces to make (included in the production table from the Microsoft SQL Server (MSSQL) repository), Fersa's current inventory, the batch schedule, and information on the production once the item has entered the machine. The agent applies two logics: one in real time featuring the current status of the machine and another that was calculated before production. The output of the agent is based on the estimated time required to finish a production. The AAS for preschedule agent is the production-planning type.

The main content of this agent can be found in the prescheduling agent's AAS, in the submodel's production information. Figure 9 presents the reference of the production order, the inputs used for the execution of the agent (e.g., the cone reference and the quantity), and the outputs obtained (e.g., the number of GW changes or the estimated time required to produce the batch) for the prescheduling agent ASS.

2.5 Holon

The complete implementation of MAS4AI architecture is based on the concept of a holon as an abstraction in that it groups multiple agents and coordinates both control and information flow. A holon is also used in JANUS[3] [2]. The theoretical presentation of a holon is illustrated in Fig. 3, while its practical implementation at Fersa is presented in Fig 10.

3 Conclusion

This paper presents the ontology, semantics, and data architecture that permits multiagent interaction, and the RAMI 4.0 model was selected as the basis for designing and implementing the presented approach.

---

[3] Janus Agent and Holonic Platform (sarl.io)

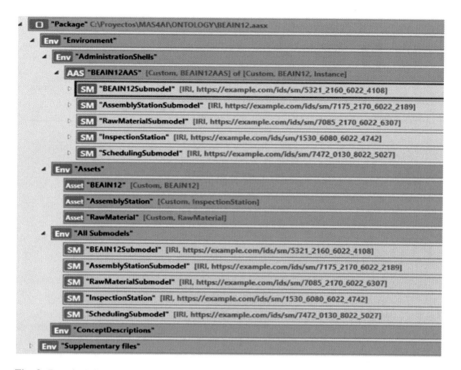

Fig. 9 Prescheduling agent AAS

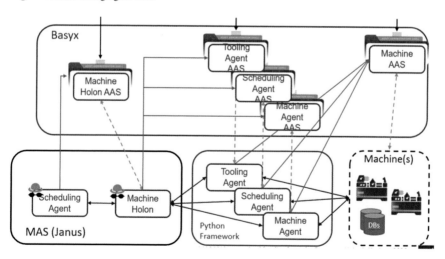

Fig. 10 Holon implementation at Fersa

The presented data architecture, depicted in Fig 11, permits the data analysis of raw materials, finished products, tooling characteristics and statuses, machine parameters, and external conditions, to minimize the influence of intuition and

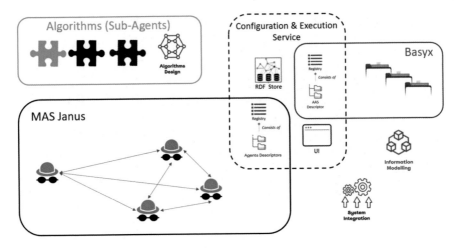

Fig. 11 MAS4AI FW and RA

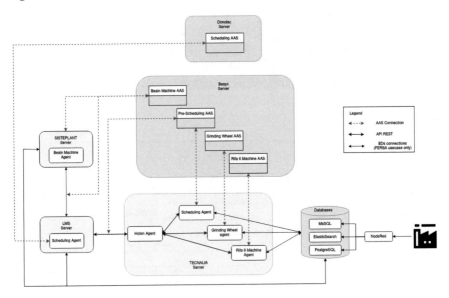

Fig. 12 MAS4AI FW and RA in Fersa

personal bias on decision-making in manufacturing and to improve production quality, throughput, and efficiency.

Figure 12 presents the generic MAS4AI Framework (FW) and Reference Architecture (RA) adapted for the Fersa pilot.

The pilot is scalable to many other sectors: From its use in the grinding process, MAS4AI RA and its framework can be easily moved and applied to manufacturing processes such as stamping, machining, compounding, and extruding, where the decisions are based on balancing factors such as tooling, machines, and scheduling requirements.

Acknowledgments This work received funding from the European Union's Horizon 2020 research and innovation program under grant agreement number 957204 (H2020-ICT-38-2020) (MAS4AI).

References

1. Sakurada, L., Leitao, P., De la Prieta, F.: Agent-based asset administration shell approach for digitizing industrial assets. IFAC-PapersOnLine. **55**(2), 193–198, ISSN 2405-8963, (2022). https://doi.org/10.1016/j.ifacol.2022.04.192
2. Galland S, Gaud N, Rodriguez S, et al.: Janus: another yet general-purpose multiagent platform. In: Proceedings of 7th Agent-Oriented Software Engineering Technical Forum (TFGASOSE-10); 2010 Dec; Paris, France

Integrating Knowledge into Conversational Agents for Worker Upskilling

Rubén Alonso, Danilo Dessí, Antonello Meloni, Marco Murgia, and Reforgiato Recupero Diego

1 Introduction

The supply and demand for labor, wherein employees give the supply and employers provide the demand, are referred to as the labor market or job market. They are a crucial part of every economy and are closely connected to the markets for capital, goods, and services. According to a recent International Labour Organization (ILO)

R. Alonso (✉)
R2M Solution s.r.l., Pavia, Italy

Programa de Doctorado, Centro de Automática y Robótica, Universidad Politécnica de Madrid-CSIC, Madrid, Spain
e-mail: ruben.alonso@r2msolution.com

D. Dessí
Knowledge Technologies for the Social Sciences Department, GESIS – Leibniz Institute for the Social Sciences, Cologne, Germany
e-mail: danilo.dessi@gesis.org

A. Meloni · M. Murgia
Mathematics and Computer Science Department, University of Cagliari, Cagliari, Italy
e-mail: antonello.meloni@unica.it; m.murgia98@studenti.unica.it

R. R. Diego
Mathematics and Computer Science Department, University of Cagliari, Cagliari, Italy

R2M Solution s.r.l., Pavia, Italy
e-mail: diego.reforgiato@unica.it; diego.reforgiato@r2msolution.com

© The Author(s) 2024
J. Soldatos (ed.), *Artificial Intelligence in Manufacturing*,
https://doi.org/10.1007/978-3-031-46452-2_15

265

assessment[1], the present global economic slump is likely to drive more employees to take poorer quality, poorly paid employment that lack job security and social safety, exasperating disparities that have already been made worse by the COVID-19 crisis. According to the same report, only 1.0% of the world's employment is expected to expand in 2023, which is less than half of the rate in 2022. The number of unemployed people worldwide is projected to increase by almost 3 million to 208 million in 2023, representing a 5.8% global unemployment rate. The limited labor supply in high-income nations is partly to blame for the moderate extent of this predicted increase. This would reverse the drop in unemployment seen globally between 2020 and 2022.

The COVID-19 epidemic has, therefore, recently had a significant effect on the world labor market. Additionally, new difficulties have arisen that are also negatively affecting the workplace, such as rapid increases in inflation, disruptions in the supply chain, and the conflict in Ukraine.[2] It is vital to take action to advance social justice by tackling issues like young employment, informality, productivity, and gender parity. To create a lasting and inclusive recovery that leaves no one behind, policymakers, companies, employees, and civil society must collaborate. This entails spending money on education and training, enhancing social safety nets, encouraging good work, and moving forward.

The industrial sector, and in particular the manufacturing sector, is not immune to this situation, since it is a very competitive labor market in which it is difficult to recruit top talents that are experienced with new technologies such as robotics or AI. The sector also suffers the fast pace of innovative technologies, making it difficult for both companies and employees, without adequate training and practice, to be up-to-date and competitive. The manufacturing sector needs to retain talent and adapt to new needs, through professional development activities and investment in employee growth, and at the same time be able to find quick solutions, such as cross-training, to be able to cover leaves of absence or production peaks. This calls for new training, retention, and recruitment strategies.

Since 2021, several industries and jobs have been predicted to grow by Career-Builder,[3] a 1995-founded employment website with operations in 23 countries and a presence in over 60 markets. CareerBuilder offers labor market information, talent management software, and other recruitment-related services. The listed employment spans a variety of industries, including leisure, arts, renewable energy, personal services, healthcare, and information technology. It is possible to upload and build a resume and look for one of many jobs posted by different organizations.

---

[1] https://www.ilo.org/wcmsp5/groups/public/---dgreports/---inst/documents/publication/wcms_865332.pdf

[2] https://ilostat.ilo.org/assessing-the-current-state-of-the-global-labour-market-implications-for-achieving-the-global-goals/.

[3] https://www.careerbuilder.com/.

Other systems providing similar services are Monster.com,[4] OfferZen,[5] LinkedIn,[6] Glassdoor,[7] JobStreet,[8] ZipRecruiter,[9] Dice,[10] G2 Deals,[11] Indeed Hiring Platform,[12] Hired,[13] Cord,[14] Circa,[15] Naukri,[16] to name a few.

Using these platforms to hunt for a job is simple, but getting chosen for the top chances is often a challenge. For instance, if the position requires a computer science specialist with fund-raising skills, a candidate with only computer science skills is unlikely to be chosen because they lack crucial qualifications like proposal writing experience or start-up development. For the aforementioned example, one should be aware that writing innovation proposals is a crucial skill for fulfilling fund-raising needs. It can be not easy to find answers to queries like these and information on how to develop the necessary skills. One possibility that has recently been exploited is the use of conversational agents [15]. Impressive outcomes have been seen in chats across a variety of areas since ChatGPT's release. Although ChatGPT claims to respond to follow-up queries, acknowledge mistakes, challenge false premises, and reject unsuitable requests, it frequently generates responses that are false and may deceive the user who interacts with it. Authors in [12] have further demonstrated this issue by pointing out related ChatGPT problems in the academic community. Even in this situation, ChatGPT partially made up some of the response's components, and the user was left in the dark as to which components are correct and which are incorrect. Therefore, even though ChatGPT occasionally offers accurate responses, it cannot be relied entirely upon and a different technique should be investigated.

The use of external domain knowledge can be the solution to the aforementioned issues. There are several lexical databases or online taxonomies that have been collected and designed by different organizations that can be relied on when asking for information in such a sense. As such, in this chapter, we will first list the most important existing conversational agents which can be leveraged within the labor domain for worker upskilling. Then, we illustrate and describe all the resources related to the labor domain. We will also provide a solution that integrates conversational agents with the information provided by one of such resources showing the benefits and impact that can be attained. The idea is to leverage such

---

[4] https://www.monster.com/.

[5] https://www.offerzen.com.

[6] https://www.linkedin.com/.

[7] https://www.glassdoor.com/.

[8] https://www.jobstreet.com/.

[9] https://www.ziprecruiter.co.uk.

[10] https://www.dice.com/.

[11] https://www.g2.com/deals.

[12] https://www.indeed.com/employers/hiring-platform.

[13] https://hired.com/.

[14] https://cord.co/.

[15] https://circaworks.com/.

[16] https://www.naukri.com/.

resources to boost existing conversational agents when asking questions in the underlying domain and overcoming their limitations.

The remainder of this chapter is organized as follows. Section 2 lists the scientific papers that have been published related to conversational agent technologies for the labor domain and worker upskilling. Section 3 will describe in depth the identified conversational agent systems and will illustrate the details of their functionalities and their main limitations, whereas lexical resources within the labor market domain that can be used to provide reliable domain knowledge are included in Sect. 4. A possible solution of a conversational agent integrated with one of such resources is presented in Sect. 5. Expected benefits and impact of the proposed solution are reported in Sect. 6. Section 7 ends the paper with conclusions and outlines future directions.

2 Related Work

Conversational agents can be found in a variety of domains such as mental health [9], lifestyle [5], and customer service [17] to support users of online platforms to cope with daily tasks and challenges. They make use of natural language as a personalization enabler, i.e., they allow a user to interact with complex systems using his/her own language rather than using a limited number of predefined options (e.g., command line, buttons in a web interface, etc.) [7]; this makes the interaction with the conversational agents similar to person-to-person interaction.

Recently, conversational agents have been released for the labor domain to support employers to find new employees as well as candidates who are looking for a job and are interested to improve their skills to be eligible. The reason is that the use of conversational agents can make the recruitment process and interviews more inclusive and efficient, and interviewees seem to be willing to trust the conversational agents in confiding information and listening to their words [18]. Conversational agents can support people during each stage of their career path starting from their early studies (e.g., in choosing a university or study path), through the search for a job (e.g., conversational agents that perform preliminary interviews), and during their employment (e.g., a conversational agent that guides employees in learning a new technology). For example, the conversational agent built on top of the IBM Watson suite[17] proposed in [10] presents a system to support new college students in the decision-making process to choose the right major on the basis of necessary skills and employment opportunities. Another recent conversational agent relevant to the labor domain is *GuApp*. It supports candidates to find a job in specific geographical areas using the content of the *Gazzetta Ufficiale*, the Italian public sector's official journal of records, and a knowledge graph built on top of the

---

[17] IBM Watson: https://www.itechbm.com/watson.

ISTAT website,[18] which provides a taxonomy of professions organized by sector, and DBpedia.[19]

In line with these advancements, the research community looked into the use of knowledge resources to inject knowledge into such systems and provide tools to support candidates in self-evaluating their CVs, and workers to improve their skills. For example, these novel tools can identify which skills a candidate should acquire to become eligible for a specific job. An example of such technologies is described in [16], where the authors looked into which tech skills are taught in colleges and which skills are truly required by the job market. They used Natural Language Processing (NLP) for detecting entities and keywords from computer science courses and job descriptions and provide recommendations for tech courses or topics for tech career development. In [2] a novel method to match the content of CVs with the O*NET database, a large database of work and worker characteristics and skill requirements, is proposed. The authors used advanced NLP technologies to automatically parse the content of CVs, state-of-the-art transformer models based on the *SentenceTransformers* framework [14] to encode the extracted information into latent representations, and a semantic layer to match the information from CVs to the O*NET database. This novel technology is provided to the public through a demo [3]. The same technology is also employed by the same authors to provide *STAR-BOT* [1], a conversational agent that can help explore the O*NET database using a novel grammar and transformer model to understand the user request and deliver information about jobs. In addition to this, *STAR-BOT* can also suggest educational online courses that can help the upskilling of workers if successfully passed.

3 Existing Conversational Agents

In this section, we will describe some state-of-the-art conversational agents that can be employed, among others, also within the labor domain.

ChatGPT Released in November 2022, ChatGPT[20] is an artificial intelligence (AI) chatbot created by OpenAI. It is a form of generative AI that allows users to enter prompts to receive humanlike images, text, and videos created by the AI. It is constructed on top of the foundational large language models (LLMs) GPT-3.5 and GPT-4 and has been fine-tuned utilizing both supervised and reinforcement learning strategies. OpenAI claims that "Reinforcement Learning from Human Feedback" (RLHF) was used to train ChatGPT. The model initially underwent supervised fine-tuning, in which OpenAI trainers acted as both a human user and an AI bot. To

---

[18] ISTAT: https://www.istat.it/en/.

[19] https://www.dbpedia.org/.

[20] https://openai.com/blog/chatgpt.

fine-tune the model for conversational usage, the trainers used this to develop a dialogue sequence that mimicked how people converse in real life. Later, ChatGPT was enhanced by developing a reward model to be used for the following phase of reinforcing learning. In order to produce responses, this entailed AI trainers engaged with the tool. Afterward, the responses were ranked according to their quality. With this information, there was a further fine-tuning phase called Proximal Policy Optimization. ChatGPT employs deep learning algorithms to assess incoming text prompts and produce responses based on patterns in the data it has been trained on. It can comprehend subtle language differences and deliver responses of the highest caliber because it has been trained on a huge corpus of literature, including books, papers, and websites. Users can give the chatbot feedback by clicking the "thumbs up" or "thumbs down" icons next to each response in order to help it learn. Users can also offer more textual comments to enhance and perfect upcoming dialogue. Users can ask a wide range of topics on ChatGPT, from straightforward inquiries to more difficult ones like, "What is the meaning of life?". ChatGPT is skilled in STEM fields and has the ability to troubleshoot or write code. There is no restriction on the kinds of queries that can be asked to ChatGPT. ChatGPT uses data only through 2021; therefore, it is unaware of events and data after that point. Additionally, because it is a conversational chatbot, users can request additional details or ask that it try again when producing content.

A list of limitations of ChatGPT is described in the following. It does not properly comprehend how intricate human language is. Words are generated using ChatGPT based on input. As a result, comments could come off as superficial and lacking in profundity. Moreover, ChatGPT could respond incorrectly if it does not fully comprehend the question. On top of that, responses may come off as artificial and robotic. The training data covers up to 2021 and, therefore, ChatGPT has no knowledge of what happened later. As ChatGPT is still being trained, providing comments when a response is erroneous is advised. ChatGPT may misuse terms like "the" or "and." Due to this, information must still be reviewed and edited by humans in order to make it read more naturally and sound more like human writing. ChatGPT does not cite sources and does not offer interpretation or analysis of any data or statistics. It is unable to comprehend irony and sarcasm. It can focus on the incorrect portion of a question and be unable to shift. For instance, if we ask, "Does a horse make a good pet based on its size?" and then we ask "What about a cat?". Instead of providing information regarding keeping the animal as a pet, ChatGPT might only concentrate on the animal's size. Because ChatGPT is not divergent, it cannot shift its response to address several questions in one.

Bing It is a chatbot created by Microsoft. Bing is integrated with Microsoft's Bing search engine and is made to respond to users' inquiries in a way that is specific to their search history and preferences.[21] The Bing AI chatbot may help with a variety of tasks, including question answering, providing recommendations,

---

[21] https://www.bing.com/?/ai.

presenting pertinent search results, and having casual conversations. Similar to ChatGPT, the new AI-powered application on Bing responds to user requests using a selection of words determined by an algorithm that has learned from scanning billions of text documents on the Internet. Bing AI indexes the entire web in order to produce a response. Because of this, the chatbot has access to the most recent news, information, and studies at the time the user submitted his/her query. As an AI model capable of error, OpenAI has already acknowledged that ChatGPT is prone to hallucinations and incorrect responses. Bing AI has made some efforts to address this problem. When the user asks a question in Bing AI, it will respond with footnotes that will take the user directly to the original source of the answer. Bing works on top of GPT4, the newest version of OpenAI's language model systems. It has also been integrated into Skype chat. Through a straightforward chat interface, this new experience is intended to give the user access to a rich and imaginative source of knowledge, inspiration, and solutions to user queries. Microsoft published in a blog post[22] that Bing was prone to getting off track, especially after "extended chat sessions" of 15 or more inquiries, but claimed that user feedback was helping it to make the chat tool better and safer.

Bard Google created the conversational generative AI chatbot and is still working on its release. It was created to compete with ChatGPT and other language models. Bard is based on the LaMDA (Language Model for Dialogue Applications) family of large language models. It was built on top of Google's Transformer neural network architecture, which was also the basis for ChatGPT's GPT-3.5 language model. Eventually, it will be employed to augment Google's own search tools as well as provide support and interaction for businesses. Differently from ChatGPT, Google Bard can use up-to-date information for its responses. Bard may occasionally provide inaccurate, misleading, or false information while presenting it confidently. For instance, Bard might provide developers with working code that is incomplete or fails to produce the correct output. Google Bard is already accessible via a waitlist, but it is difficult to predict when it will be accessible to everyone. However, Google's CEO stated that Google Bard would soon be used to improve Google Search, thus Bard might become more widely accessible shortly.

Chatsonic Chatsonic[23] is a GPT-4 based chatbot that tries to solve the main limitations of ChatGPT. Chatsonic uses Google to grasp information about the latest events. This makes this chatbot able to support users with timely answers and information about events that take place at any moment. It also provides a few sources harvested from the Internet used to generate the answer which can have a relevant impact on the trust that a user can give to the conversational agent. Chatsonic can generate images and can be integrated within the Chrome browser for efficiently working with everyday tools and web platforms such as Gmail, LinkedIn,

---

[22] https://blogs.bing.com/search/february-2023/The-new-Bing-Edge-%E2%80%93-Learning-from-our-first-week.

[23] https://writesonic.com/chat.

Twitter, etc. In the context of worker upskilling, it can be used to find online courses that can be attended to learn or improve specific skills. One of its main limitations is that each user has a budget of 10k words, and when finished, a premium account needs to be activated.

Copilot Copilot[24] is a language model that slightly differs from the others because it targets computer scientists and developers and their daily task of writing new source code. In fact, this model is trained on billions of lines of code. It allows one to write code faster, suggests more lines of code, proposes implementations of methods, and enables developers to focus on the overall approach instead of dispersing energies on trivial tasks. However, this tool is not perfect and it has several limitations; it often delivers code and method implementations that do not make sense and, additionally, it does not test the code that is written. Therefore, the developer has to verify that the automatically generated code is functional, as well as the quality of the delivered results. This is a crucial aspect for the accountability of the work, and for ensuring the quality of the developed software to customers. Nevertheless, Copilot is an interesting tool for the labor domain considering computer science-related areas and might be a plus in a large variety of companies in which software is mostly used for simple tasks.

ELSA-Speak English Language Speech Assistant[25] (ELSA) is an application based on artificial intelligence and speech recognition technologies designed to support users in learning English. It can be used to improve English pronunciation and provides lessons and exercises. The application provides real-time feedback to the users to enhance their speaking skills. ELSA-Speak can be employed within the labor market to prepare for job interviews (for example, candidates might be more confident to have an interview in English), to help non-native English speakers in their work, and to improve their English skills in order to increase opportunities for career advancement.

4   Skills, Competences, and Occupations

For some time now, different initiatives have been presented to help students, the workforce, and companies to meet the needs of the labor market. All of them have in common that they are initiatives supported by government agencies, are updated periodically, and offer the data in a public and open way. In this section, we will focus on three of the most relevant databases and information systems on occupations and competencies.

---

[24] https://github.com/features/copilot.

[25] https://elsaspeak.com/en/.

O*NET O*NET[26](Occupational Information Network) is a program sponsored by the U.S. Department of Labor/Employment and Training Administration, which brings together occupational information to understand how the changing labor environment impacts the workforce. One of the main outputs of the program is the O*NET database, a collection of hundreds of standardized descriptions related to more than 1000 occupations found in the USA. This collection is updated periodically through worker surveys supplemented by information from occupation experts. Each occupation in the O*NET database is associated with a set of skills, abilities, and knowledge and is linked to a variety of tasks and activities. The version 26.2 (February 2023) includes 35 `Skills`, 33 `Knowledge`, 52 `Abilities`, 41 `Work Activities`, 4127 `Tools Used`, 17,975 `Task Ratings`, and 8761 `Technology Skills`. Each of these entities allows for the categorization of each occupation, its compatibility, and the specific detection of the needs of the occupation. O*NET is based on a conceptual model that provides a framework to identify the most relevant information about occupations even across jobs and industries. This is useful to link titles referring to the same occupation or to search for intersectoral relationships.

ESCO ESCO[27] (European Skills, Competences, and Occupations) is the European reference terminology describing and classifying occupations and relates them to the most relevant skills and competencies. In addition to being freely and openly available, it is translated into 27 languages (including the 24 official EU languages) and connects to several international frameworks and classifications, such as the International Standard Classification of Occupations (ISCO)[28] or the European Qualifications Framework. ESCO v1.1.1 includes more than 3000 occupations, categorized into 10 groups, mapped with ISCO, and more than 13000 skills, subclassified into 4 types of concepts: knowledge, skills, attitudes and values, and language skills and knowledge. This hierarchy is partially based on the entities and elements of O*NET and NOC (another occupational resource described in the next section). This partial link with other databases, the mapping with ISCO, the connection with qualification levels, and above all, its multilingual support make it particularly attractive for applications that require adaptation to the languages that ESCO supports and especially for linking students and job seekers with employers and employment services.

NOC NOC[29](National Occupation Classification) is the national system for the description of occupations, in French and English, and aligned with the Canadian labor market situation. This open database is updated annually with major revisions every 10 years and is based on data from censuses, employment services, labor regulations, and public consultations among others, which are then analyzed and

---

[26] O*NET OnLine, National Center for O*NET Development, www.onetonline.org/.

[27] https://esco.ec.europa.eu/.

[28] https://www.ilo.org/public/english/bureau/stat/isco/isco08/.

[29] https://noc.esdc.gc.ca/.

processed, and discussed by working groups. Like the aforementioned resources, NOC also includes a hierarchy and structure of occupations and is linked to the Occupational and Skills Information System (OaSIS),[30] the Canadian database with detailed information on more than 900 occupations. It includes, among other information, the main duties of each occupation, together with the skills, abilities, and even the most common interests among the people who hold those occupations. Among the most interesting functionalities of NOC and OaSIS are the comparison between occupations, the link with Training, Education, Experience, and Responsibilities (TEER)[31] to understand the training and education required to perform the main duties of an occupation, and the search occupations by interests (e.g., investigative, artistic, or social).

In addition to the above, it is important to mention other international and national classifications that are either of reference or are applicable in other countries and are inspired by or refer to the above. The International Standard Classification of Occupations [6] (ISCO) is an international classification developed by the International Labor Organization of the United Nations. This classification and, in particular, its publications ISCO-88 and ISCO-08 define groups and titles of occupations associated with tasks. In general, these database and associated documents are less comprehensive than the three previous ones, but in certain occupational databases, connections with ISCO allow occupations to be linked between databases in different countries.

One of the most common criticisms of these classifications is that they are created and adapted in high-income countries. For this reason, initiatives to map competencies in other low/middle-income countries are also being undertaken. Two examples of these initiatives are STEP (Skills towards Employment and Productivity) and PIAAC (Programme for the International Assessment of Adult Competencies). STEP is a World Bank program to improve the understanding of skills (cognitive, non-cognitive, and technical) in the labor market and relates skills to education, through household surveys and employer-based surveys. It is a program of the Organisation for Economic Cooperation and Development (OECD) that analyzes socio-labor and educational characteristics and links them to competencies at different cognitive levels such as literacy, numeracy, and problem-solving. The first cycle of PIAAC started in 2011 analyzing 42 countries and its second cycle will run until at least 2029 covering 30 countries.

All these initiatives and databases provide a stable and updated background, based on reliable sources and in many cases that can be linked together, which allows us to propose ideas and proofs of concept such as the one mentioned in the following section.

---

[30] https://noc.esdc.gc.ca/Oasis/OasisWelcome.

[31] https://www.statcan.gc.ca/en/subjects/standard/noc/2021/introductionV1.

5 Proposed Solution

The goal of the solution we have proposed in recent papers [1–3] is to make available to users, who are not always technologically prepared, the information contained within any of the resources described in Sect. 4 in the form of natural language and to help them analyze which skills and personal knowledge are most useful for their application in any job.

One of the worker upskilling resources covered in Sect. 4, the O*NET database, contains many types of information. Our system uses the titles of the principal job categories that cover all the jobs in the North American job market, their descriptions, and the alternative titles for each one. Abilities, Skills, Knowledge, Work Activities, Tasks, Technology Skills, and Tools are the information that characterizes each job category. The number of elements for each job is constant for the first four categories and variable for the last three. The items in the first five present a score relative to their importance to the job they refer to, while the last two are lists of tools or technologies commonly used in each job category.

More in detail, the system we propose performs the following tasks:

1. Determine how much the user's resume is appropriate for the chosen job category.

   The system uses NLP techniques to analyze the user's resume and to extract its most important information. The database entities and the extracted content from the CV are compared using semantic similarity. The resume score for a job is the sum of the scores of the discovered elements divided by the maximum score obtainable for the job itself and a corrective factor required to avoid penalizing jobs with few items in the entities with a variable number of elements. The system draws attention to the lacking aspects of the CV and suggests ways for users to increase their knowledge of each one.

2. Determine the user's suitability for the selected job.

   The system prompts the user to enter his/her self-assessed level (none, little, medium, good, excellent) in terms of Abilities, Knowledge, Skills, Work Activities, and Tasks entities taken from the database and associated with the selected job. The system returns a percentage score, where 60% is the suitability threshold and a list of lacking knowledge or abilities that the user might improve to raise his/her score.

3. Determine what occupation the user is best suited for.

   The system prompts the user to enter his/her self-assessed level (none, little, medium, good, excellent) in terms of Abilities, Knowledge, Skills, and Work Activities entities taken from the database and returns a list with five suggested job categories.

4. Perform conversation with the user.

   To integrate the knowledge provided by O-NET into the conversational agent, we first chose a reduced set of questions that it must be able to answer. The base question templates are:

- Alternate names of jobs
- [*Abilities* | *Knowledge* | *Skills* | *Tasks* | *Tech-Skills* | *Tools used* | *Work activities*] that are important or necessary in any job
- Description of a job
- The similarities between a couple of jobs
- Which jobs require a specific [*Ability* | *Knowledge* | *Skill* | *Task* | *Tech-Skill* | *Tool* | *Work activity*]
- Recommendations on how to improve one [*Ability* | *Knowledge* | *Skill* | *Task* | *Tech-Skill* | *Tool* | *Work activity*]

Then we implemented the query models necessary to retrieve the data to generate the answers.

When the conversational agent recognizes a question in the user's input that is part of the set of known questions, it requests the database for the data necessary to prepare the answer in natural language.

The system uses semantic similarity to match user input with one of the predefined question templates. If the similarity value is not high enough, the system will answer without the database knowledge, returning a default answer, or using a generative pre-trained model. More details of our system can be found in [1–3].

6 Expected Challenges, Benefits, and Impact

In terms of challenges, employee engagement in training activities and work commitments is one of the key elements. Knowing the information about their job and potential growth paths, while reducing doubts about the stability of their job, allows their adaptation and evolution, and thus engagement. Interactive chatbots and conversational tools can simplify the task of obtaining this information, for example, through informal interactions and by simply relying on the chatbot as a companion that offers relevant and appropriate information, focused on the characteristics of each occupation.

As far as technical challenges are concerned, it is still necessary to reach an adequate balance between the use of large language models (e.g., ChatGPT) and models trained with concrete content based on validated or accepted reference information. One of the main problems of LLMs is the lack of determinism in the results [11] and the potential failures in the recommendations [13]. Both can be compensated by relying on databases such as the ones mentioned above and prioritizing the results based on the information in those databases.

It is important to also mention and consider the challenges of conversational agents related to privacy. While many of the ideas mentioned in this article do not require storing worker information, and in our use cases we have demonstrated this, there are several cases, such as group data discovery or candidate analysis, where the handling of worker data could end up in the processing of private data, aggregation of personal data and in some cases in profiling. The topic related to the concerns about user data's inappropriate use in user-chatbot interactions is being researched and surveyed [4, 8], and still there are several debates about it at the international level (e.g., in April 2023 Italy was the first Western country to block ChatGPT[32]). This calls for a critical analysis of the different uses of chatbots for upskilling and, above all, an alignment with the initiatives for the maintenance of the privacy of workers and users, from the General Data Protection Regulation to the Artificial Intelligence Act.

Regarding benefits and impact, at the scientific and technical level, and as can be seen in the state of the art, advances in conversational agents and language models are clear. All these advances have an impact on the development of solutions such as the one we present, for example, by providing better interaction capabilities or even the possibility of answering open-domain questions. At the economic level, as also discussed in the introduction, there is a demand for workers and a demand for knowledge for specific activities, and upskilling agents can facilitate this task. At the social level, it impacts workers and the unemployed, in terms of analyzing their strengths, detecting training or internship needs, and adapting their CVs to fit specific jobs.

From the worker's perspective, conversational AI agents based on standardized and reliable databases of occupational information have direct impacts and benefits.

On the one hand, these conversational solutions have a direct impact on workers' knowledge of their work activity. The workers themselves can at any time obtain reliable information about their job and learn about skills, activities, and knowledge relevant to the task they are carrying out. Moreover, the worker can ask the conversational agent which are the usual technical skills for a certain occupation or activity to imagine where they might be oriented.

On the other hand, it is beneficial for their training and career. The employees can detect points in which their profile would benefit from training and at the same time detect training needs and create training plans to adapt to the needs of their position, or to another position they plan to move to. This is one of the other benefits of these conversational agents since the workers can compare themselves with professions and positions in which they would be interested in changing or being promoted.

From a more organizational and business point of view, there are also a couple of areas where AI-based conversational agents such as the ones mentioned in this article can be beneficial. The part of assistance in hiring processes is a point in which these agents will have a direct impact: from providing candidates with information about the position, to pre-screening or determining the suitability of a candidate by using conversational agents to compare the candidate's CV with the baseline of the occupational reference database. Conversational agents are also beneficial in the detection of training needs for specific positions or the creation of group training plans, based on the needs of multiple workers.

---

[32] https://www.bbc.com/news/technology-65139406.

Therefore, the impact on the HR area is clear: talent management, discovery of training paths, HR resources, growth programs, etc. All of them benefit in the short term from conversational agents like the ones mentioned above. Proofs of concept and use cases conducted with conversational agents, based in our proposed solution, acting as virtual interviewers, skills gap detectors or as supporters for the selection of better training, demonstrate the potential of these technologies. Not only at the level of large technology companies but the aforementioned solutions can also make a difference at the level of SMEs in industrial and manufacturing environments. Where they may not have large HR departments but need reliable, objective, and consistent feedback for resource upskilling and solutions based on standardized information, starting from the same baseline and with clear and repeatable criteria.

7 Conclusions

This chapter explores the potential uses, benefits, and limitations of conversational agents within the labor market. The chapter guides the reader through the state-of-the-art research on this topic and presents which technologies are currently employed. Furthermore, the chapter analyzes conversational agents that are disrupting the labor market, where AI tools are expected to be employed for a large variety of tasks such as learning new skills, supporting document writing, code development, language skills improvement, and so on. Along with these conversational agents, we also introduce the reader to valuable sources of information which can be leveraged to build AI systems on top. Finally, we present how potential solutions can be used for specific use cases and discuss how all these technologies can impact the labor market.

As main lessons, we would like to mention the need for both the workforce and companies to adapt to the rapidly changing world of new technologies and how this can be beneficial, for example, by taking advantage of it to better understand occupations, retain talent, hire, gain job knowledge, or adapt to new activities. It is also important to consider the need to efficiently integrate large language models and validated and reliable databases. Finally, it is essential to consider and analyze the privacy implications of these solutions. It is possible to develop privacy-aware conversational systems for the upskilling of workers, but there is a possibility that certain solutions may abuse data. For this reason, it is critical to provide these systems with consideration of Human Centricity aspects and all the factors that compose it, including privacy.

In the medium term, it is expected that these technologies will be integrated into the regular workflow of employees and will have a positive impact on employee engagement and professional development. For this reason, we expect that this chapter will foster several visions to the reader about the role of AI-based conversational agents in the labor domain and worker upskilling which might bring unprecedented development in the field.

Acknowledgments This work has been partly supported by EU H2020 research and innovation programme project STAR—Safe and Trusted Human Centric Artificial Intelligence in Future Manufacturing Lines (Grant n. 956573).

We acknowledge financial support under the National Recovery and Resilience Plan (NRRP), Mission 4 Component 2 Investment 1.5—Call for tender No.3277 published on December 30, 2021 by the Italian Ministry of University and Research (MUR) funded by the European Union—NextGenerationEU. Project Code ECS0000038—Project Title eINS Ecosystem of Innovation for Next Generation Sardinia—CUP F53C22000430001—Grant Assignment Decree No. 1056 adopted on June 23, 2022 by the Italian Ministry of University and Research (MUR)

References

1. Alonso, R., Dessí, D., Meloni, A., Recupero, D.R.: Incorporating knowledge about employability into conversational agent. Appl. Intelligence submitted (2023)
2. Alonso, R., Dessí, D., Meloni, A., Recupero, D.R.: A novel approach for job matching and skill recommendation using transformers and the o*net database. Big Data Res. submitted (2023)
3. Alonso, R., Dessí, D., Meloni, A., Reforgiato Recupero, D.: A general and NLP-based architecture to perform recommendation: A use case for online job search and skills acquisition. In: Proceedings of the 38th ACM/SIGAPP Symposium On Applied Computing. ACM Special Interest Group on Applied Computing (SIGAPP) (2023)
4. Belen Saglam, R., Nurse, J.R.C., Hodges, D.: Privacy concerns in chatbot interactions: When to trust and when to worry. In: Stephanidis, C., Antona, M., Ntoa, S. (eds.) HCI International 2021 - Posters, pp. 391–399. Springer International Publishing, Cham (2021)
5. Fadhil, A., Gabrielli, S.: Addressing challenges in promoting healthy lifestyles: the al-chatbot approach. In: Proceedings of the 11th EAI International Conference on Pervasive Computing Technologies for Healthcare, pp. 261–265 (2017)
6. Hoffmann, E.: International Statistical Comparisons of Occupational and Social Structures, pp. 137–158. Springer US, Boston, MA (2003). https://doi.org/10.1007/978-1-4419-9186-7_8
7. Hussain, S., Ameri Sianaki, O., Ababneh, N.: A survey on conversational agents/chatbots classification and design techniques. In: Web, Artificial Intelligence and Network Applications: Proceedings of the Workshops of the 33rd International Conference on Advanced Information Networking and Applications (WAINA-2019) 33, pp. 946–956. Springer (2019)
8. Ischen, C., Araujo, T., Voorveld, H., van Noort, G., Smit, E.: Privacy concerns in chatbot interactions. In: Følstad, A., Araujo, T., Papadopoulos, S., Law, E.L.C., Granmo, O.C., Luger, E., Brandtzaeg, P.B. (eds.) Chatbot Research and Design, pp. 34–48. Springer International Publishing, Cham (2020)
9. Lee, M., Ackermans, S., Van As, N., Chang, H., Lucas, E., IJsselsteijn, W.: Caring for Vincent: a chatbot for self-compassion. In: Proceedings of the 2019 CHI Conference on Human Factors in Computing Systems, pp. 1–13 (2019)
10. Lee, T., Zhu, T., Liu, S., Trac, L., Huang, Z., Chen, Y.: CASExplorer: A conversational academic and career advisor for college students. In: The Ninth International Symposium of Chinese CHI, pp. 112–116 (2021)
11. Maddigan, P., Susnjak, T.: Chat2vis: Generating data visualisations via natural language using ChatGPT, codex and gpt-3 large language models. IEEE Access (2023)
12. Meloni, A., Angioni, S., Salatino, A., Osborne, F., Reforgiato Recupero, D., Motta, E.: Integrating conversational agents and knowledge graphs within the scholarly domain. IEEE Access **11**, 22468–22489 (2023). https://doi.org/10.1109/ACCESS.2023.3253388
13. Oviedo-Trespalacios, O., Peden, A.E., Cole-Hunter, T., Costantini, A., Haghani, M., Kelly, S., Torkamaan, H., Tariq, A., Newton, J.D.A., Gallagher, T., et al.: The risks of using ChatGPT to obtain common safety-related information and advice. Available at SSRN 4346827 (2023)

14. Reimers, N., Gurevych, I.: Sentence-Bert: Sentence embeddings using Siamese Bert-networks. In: Proceedings of the 2019 Conference on Empirical Methods in Natural Language Processing. Association for Computational Linguistics (2019). https://arxiv.org/abs/1908.10084
15. Singh, S., Beniwal, H.: A survey on near-human conversational agents. Journal of King Saud University - Computer and Information Sciences **34**(10, Part A), 8852–8866 (2022). https://doi.org/10.1016/j.jksuci.2021.10.013. https://www.sciencedirect.com/science/article/pii/S1319157821003001
16. Vo, N.N., Vu, Q.T., Vu, N.H., Vu, T.A., Mach, B.D., Xu, G.: Domain-specific NLP system to support learning path and curriculum design at tech universities. Comput. Educ. Artif. Intell. **3**, 100042 (2022)
17. Xu, A., Liu, Z., Guo, Y., Sinha, V., Akkiraju, R.: A new chatbot for customer service on social media. In: Proceedings of the 2017 CHI Conference on Human Factors in Computing Systems, pp. 3506–3510 (2017)
18. Zhou, M.X., Mark, G., Li, J., Yang, H.: Trusting virtual agents: The effect of personality. ACM Trans. Interact. Intell. Syst. (TiiS) **9**(2-3), 1–36 (2019)

Advancing Networked Production Through Decentralised Technical Intelligence

Stefan Walter and Markku Mikkola

1 Introduction

Networked production has been increasingly important for companies looking to stay competitive in today's global marketplace [1, 2]. Initially undertaken due to the requirements of profitability, many manufacturers have been focusing on their core competencies and outsourcing non-core functions to specialised suppliers [1, 3, 4]. The resulting fragmentation made increased digitalisation necessary.

By leveraging technology and digital platforms, companies can connect their manufacturing activities with other organisations, suppliers, customers, and resources, both domestically and globally, to achieve a competitive advantage [5–8]. Companies do so by accessing a broader range of suppliers and customers, reducing costs, increasing efficiency, and improving the quality of their products when networking. This also allows for greater flexibility in responding to changes in demand, supply chain disruptions, or market conditions and provides the required agility that is so important to counter economic volatility or political instability [9–11].

In addition, networked production can create opportunities for innovation and value creation, as companies can work with partners to develop new products or improve existing ones. By sharing expertise, knowledge, and resources, companies can capitalise on each other's strengths and capabilities to create new business opportunities and improve their competitive position [12, 13]. Companies can thus speed up the product development process and improve time-to-market. This position is also reflected in the concepts of networking in engineering, networked product/service engineering, and collaborative design, respectively [14].

S. Walter (✉) · M. Mikkola
VTT Technical Research Centre of Finland Ltd., Espoo, Finland
e-mail: stefan.walter@vtt.fi; markku.mikkola@vtt.fi

© The Author(s) 2024
J. Soldatos (ed.), *Artificial Intelligence in Manufacturing*,
https://doi.org/10.1007/978-3-031-46452-2_16

Thus, a significant benefit of networked production is the possibility for companies to move away from traditional sequential or linear process chains and adopt a more dynamic approach to manufacturing and support processes [15, 16]. By connecting their operations with other organisations and resources through digital platforms, companies can access a broader range of manufacturing capabilities, materials, and services. For example, companies can dynamically arrange and release manufacturing and support processes in case production requires a quick scale-up to meet a sudden increase in demand. Additionally, networked production allows and virtually demands that companies share information and collaborate with partners, suppliers, and customers in real-time. This enhances transparency and visibility across the supply chain, which can help to identify and mitigate potential bottlenecks or delays [17].

The functioning of networked production is bound to the management capabilities of the companies involved. Typically, effective management in networked production has required strong leadership and management expertise to ensure that the partners involved work together seamlessly and effectively. Companies needed to establish clear communication channels and protocols, define roles and responsibilities, and establish processes for monitoring and managing performance.

Moreover, the success of networked production depends on the compatibility of integrated, cross-organisational information chains [4]. Network partners need to ensure that they can exchange information seamlessly and securely with their partners, suppliers, and customers. This necessitated the adoption of common standards for data exchange [18]. In addition, data exchange must take place on a trusted basis and therefore requires corresponding communication security measures to protect sensitive information [19].

All this will become even more difficult with a wider expanding network. The increasing complexity of networked production presents a significant challenge to management, which requires new and intelligent approaches to effectively manage and optimise the manufacturing process. While managing networked production would require a holistic understanding of the entire value network involved, including suppliers, customers, internal operations, and other stakeholders, it is hardly possible with traditional corporate management methods based on hierarchical structures [20, 21]. This is also true for the increasingly demanded agile management involving the flexibility and adaptability to respond quickly to changes in customer demand, market conditions, or supply chain disruptions.

To effectively manage networked production, companies can leverage advanced technologies, such as artificial intelligence and data analytics, to develop new management approaches and enhance decision-making. Decentralised Technical Intelligence (DTI) is a new management approach that involves the integration of human and machine intelligence in decision-making processes. DTI allows for the creation of a network of interconnected systems, devices, and agents that operate in a decentralised and autonomous manner, reducing the need for leadership through hierarchy and potentially leading to increased efficiency and productivity [22]. The role of the workforce in this integrated system is crucial, and the human dimension adds a unique quality to the process [23].

To achieve advanced networked production with the help of DTI, a roadmap is necessary, which involves several elements linked to a vision, value promise, and development pathway. The roadmap should include major building blocks, major capabilities, characteristics, and processes typical for the effective implementation of DTI in networked production. By implementing this roadmap, companies can maintain a leadership position in future networked production, which will be crucial for maintaining a competitive edge in the global market.

The technological advancements in networked and self-controlling production can offer new opportunities for creating value, but it is important to ensure that these opportunities align with business goals and are economically viable. The business perspective provides a real-world corrective, meaning that any new technological solutions must also make sense from a business standpoint, considering factors such as profitability and feasibility. Thus, it is important to balance technological possibilities with business realities to ensure the long-term success of any innovation.

This article explores the concept of Decentralised Technical Intelligence and its meaning for advancing networked production as a competitive factor. It will also discuss possible building blocks for implementing an effective roadmap for advanced networked production through DTI. Finally, this article will emphasise the need for a business perspective to ensure that any technological advancements align with the economic viability and create real value.

2 Decentralised Technical Intelligence

Decentralised Technical Intelligence (DTI) is a concept that aims to support the deployment of decentralised, autonomous systems with embedded intelligence in manufacturing. It is a response to the management requirements of networked manufacturing and aims to enable decentralised decision-making and autonomous action while reducing centralised planning or making it redundant. It arose from a demand by the ManuFuture European Technology Platform to increase productivity and efficiency in future manufacturing [22, 24].

DTI focuses on several areas to leapfrog performance gains in terms of high performance, high quality, high resource efficiency, high speed, high flexibility, self-optimisation and self-control, to name just a few (Fig. 1). These performance gains stretch over different areas, including ICT technologies, architectures, platforms and standards, high-performance engineering, high-performance manufacturing systems, and high value-added networked production [22, 24].

According to the Plattform Industrie 4.0 [25], decentralised intelligence is grounded on the recognition that the European industry's strength is derived from a system of innovation and commerce fuelled by diversity, heterogeneity, and specialisation. These elements are integral to the European industrial society. A decentralised regime of open and adaptable systems creates optimal conditions for shaping the digital economy while adhering to the principles of a free and socially-oriented market economy [25].

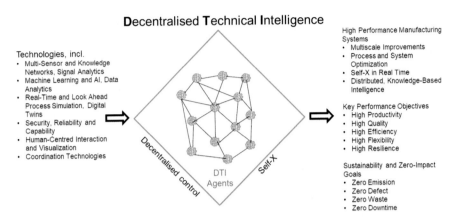

Fig. 1 Utilising technologies to enhance universal impact (adapted from Sautter [22])

Table 1 DTI as the next evolutionary step in industry performance (adapted from Sautter [22])

Control theory, systems engineering	Artificial intelligence	Cyber-physical systems	Agent-based/holonic manufacturing	Decentralised technical intelligence
Closed-loop control system incorporating sensors, actuators and controllers	Computer-enabled simulation of human-like cognitive abilities	Convergence of software systems and interconnected mechanical/electronic components enabled by a data infrastructure such as the Internet-of-Things	Independent and coordinated agents deliver flexibility, adaptability, agility and dynamic reconfigurability to manufacturing systems	Next transformative phase to revolutionise industry performance—exceeding current limitations through an interdisciplinary framework, including self-X capabilities

DTI is adopting this position and is developed as a systemic and interdisciplinary approach that represents a key factor for the macro- and multidimensional transformation of manufacturing systems. It builds on previous developments in systematic and automated manufacturing and can be seen as an evolutionary step towards more advanced manufacturing systems (Table 1). Earlier steps included for example control theory and systems engineering, involving feedback loops, artificial intelligence, cyber-physical systems as a connection of informational and electromechanical components, and holonic manufacturing, which incorporates autonomous and collaborative agents creating an adaptive and flexible manufacturing system [22].

By integrating human knowledge and experience with artificial intelligence and other advanced technologies, DTI aims to create a more intelligent and adaptable manufacturing system that can respond quickly to changing conditions and optimise performance. The concept recognises that humans and machines each have unique strengths and limitations, and that by combining their abilities and expertise, a more effective and efficient system can be created. This requires a multi-agent architecture, where different agents (including humans and machines) work together in a coordinated manner to achieve common goals. The goal is to create a system where each agent contributes to the overall intelligence of the system and helps to create a more efficient and effective manufacturing process.

Consequently, DTI consciously aims for holism and recognises the importance of humans in industrial manufacturing, which puts the concept in contrast to the earlier conceptualisations that were strictly technological in their orientation. DTI recognises that humans play a crucial role in manufacturing and seeks to empower them with the necessary tools and technologies to make informed and effective decisions. By integrating humans into the manufacturing process, DTI enables them to work alongside machines and systems more collaboratively and efficiently.

Sautter [22] compares DTI as being similar to our nervous system. The body's nerves make up part of the main control structure, which provides information on the environment and inner body functioning. This way, the autonomic nervous system controls many body functions and processes in a seemingly automatic way with little or no conscious involvement. Similarly, DTI ensures the smooth functioning and good performance of complex technical systems such as production systems. It allows the production system to execute tasks autonomously in interaction with the environment and keeps the system in optimal condition. DTI components can be used to make the system perform tasks autonomously and optimise its performance without central or conscious control. Therefore, DTI is concerned with self-organisation and self-steering aspects in production systems. Mere artificial intelligence, in turn, is a simulation of human intelligence processes by machines based on advanced data analytics and algorithms that emulate and simulate processes and systems. It is often used as a tool to support human decision-making.

The DTI concept can also be considered a forerunner or prerequisite for the industrial metaverse. The industrial metaverse, as a comprehensive digital twin of every relevant process or artefact in a factory or production network and including the human workforce, will make use of existing and developing technologies such as artificial intelligence and machine learning, extended reality, advanced data management structures and cloud and edge computing. These technologies create an interface between the real and digital worlds. This interface aims at offering fully immersive, real-time, interactive, persistent, and synchronous representations and simulations of complex systems such as machines, factories, cities, and logistics networks in a decentralised way [26]. While specific applications in the networked production realm are still to be explored, possibilities include interactive assembling, involving, for example, network partners such as suppliers and customers, collaborative development of products and services, their design, engineering and testing, product presentations and evaluations and product-related services, sales activities and tracking of material flows, and training of personnel [27].

3 Implications for Networked Production Management

Traditionally, the objectives of networked production included the cutting of logistics costs and high inventories, in particular, by improving the coordination and integration of production and logistics activities across the network, the shortening of lead times, especially by improving the availability of information and the flow of materials across the network to generate quick responses, the enhancement of service levels, for example by improving the coordination, collaboration and understanding between companies and of companies' needs in the network, and enhancing competitiveness, in particular, by increasing the capacity of industrial companies to operate globally in an agile manner through the network [1, 2, 4, 28, 29]. All of these remain true, and these aspects are joined by more recent perspectives on collaborative design and engineering and the management of services.

Achieving these goals requires collaboration among the companies involved. Therefore, all participants need to see benefits from their effort in collaborating. Although companies will try to maintain their autonomy as actors in the network, emerging technologies enabling DTI are likely to dilute the perception of organisational boundaries in the future. Hence, production will largely take place in non-hierarchical networks where decision-making is decentralised [30, 31]. This requires the development of methods and tools for collaborative planning, management, and optimisation of production resources, as well as distributed planning and scheduling [32–34]. Other important tools for management include monitoring production and equipment and their maintenance. Increasingly, tools also need to support the requirements of the circular economy.

It is essential to ensure interoperability between the different technologies and software solutions, business and production processes, and organisational structures [35]. All new methods and tools should follow the principle behind DTI, which allows operating in a decentralised manner so that companies can operate in several production networks simultaneously. Conventionally, these functions have been relying on regular updates to a central database with limited learning capacity built-in and neither with consideration of external data. For advanced management of networked production and its monitoring, decentralised ICT systems are needed

[36, 37]. These systems use distributed knowledge and integrate heterogeneous data sources to evaluate the state of monitored systems [14].

Competitiveness is based on various performance objectives such as quality, price, delivery service, responsiveness, and flexibility. With shorter product life cycles and a greater variety of models, manufacturers need to respond quickly to customers' needs and handle capacities flexibly. The technologies driving DTI allow companies to move towards decentralised operations, take advantage of available resources and, for example, be closer to their markets.

Taking advantage of available resources via decentralised operations and intense network interactions offers, in particular, opportunities to run make-to-order and engineer-to-order strategies [38, 39]. Both strategies begin their specific operations only after a customer has placed an order. It represents in some way an orientation towards lean management because it consequently aims at saving resources while avoiding producing to stock. On the other hand, the close customer relationship targets customer wishes, flexibility and responsiveness to individual needs and, therefore, reflects the need for customisation [4, 40]. The mentioned strategies will help companies maintain profitability whilst operating with smaller batch sizes and shorter lead times for customer benefit [14].

The technologies associated with decentralisation enable coordination and synchronisation of production and decision-making processes, which are required in non-hierarchical networks [41–44]. The production strategies mentioned above are most effective when typical production functions such as engineering, capacity planning, order management, manufacturing, and logistics are integrated comprehensively. DTI technologies allow the inclusion of network partners in all information flows and enable sharing of appropriate resources. When leveraging advanced technologies such as AI, machine learning and edge computing to enable decentralised decision-making and coordination of production processes, humans contribute their domain expertise, intuition, and creativity. They possess contextual knowledge and problem-solving abilities and can provide oversight, make strategic decisions and handle complex and non-routine tasks that require creativity, adaptability and critical thinking. AI, on the other hand, can assist humans by automating routine tasks, analysing large datasets, identifying patterns and anomalies and providing recommendations for optimised decision-making. DTI, thus, facilitates the integration of humans and AI through technologies like edge computing, where AI algorithms can be deployed on edge devices, and cloud computing, where AI models can be trained and deployed centrally. This allows for real-time decision-making at the edge and enables seamless coordination and synchronisation across the production network.

Case Study: Production Scheduling Pilot

In the context of production and supply chain management, coordination of processes and their consequences is crucial for maintaining robustness and resilience. The EU research project *knowlEdge - Towards Artificial Intelligence powered manufacturing services, processes, and products in an edge-to-cloud-knowledge continuum for humans [in-the-loop]* [45] aims to unlock the full potential of AI in manufacturing by developing AI methods, systems and data management infrastructure, with a focus on edge-cloud computing and collaboration between humans and AI. In a production scheduling use case, the project optimises supply chain planning, demand forecasting and production batch optimisation in an industrial environment. DTI technologies are integral to the solution.

Real-time data from production facilities is collected using sensors and data interfaces at the edge, enabling monitoring, analysis and informed decision-making. Machine learning algorithms analyse historical and real-time data to predict demand, optimise production schedules and identify bottlenecks. Process simulation and digital twins are used to simulate and predict future scenarios, allowing proactive adjustments to production schedules.

The integration of AI and data analytics empowers employees to make informed decisions, reducing errors and saving time. The AI system continuously learns and improves, leveraging insights from documented deviations, errors and solutions. User-friendly interfaces facilitate human-centred interaction and visualisation, enabling monitoring, process adjustments and coordination at the edge. This holistic approach enhances production, facilitates proactive decision-making and improves coordination and collaboration between humans and AI agents.

Essentially, this points to smart and collaborative manufacturing, which involves collaborative planning, management and operation of networked production. The approach deliberately utilises the network as a resource for distributing tasks and acquiring means of production and could also be seen as in line with what Jovane et al. call a plug-and-produce network of production services [14]. This way, the network can respond to immediate challenges in a much better way than under a regime of hierarchical planning and management. Thus, the resulting inter-organisational integration of companies reflects the regime of networked production of the future. This becomes particularly obvious when considering the self-organisational and autonomous operations of DTI agents, which, based on predefined process targets, can easily transect organisational boundaries to carry out tasks. Consequently, in a resource constraint world, the focus on networked production and the potential to expand the network when opportunities arise, companies can leverage the strengths of different network partners and achieve a more robust and resilient production, whilst unlocking new opportunities for growth and innovation [46–48].

Following the DTI logic, future production networks will be following a self-organising and self-forming paradigm [22]. Rather than being pure manufacturers of goods, companies will be more and more providers of service, which solve a problem for a user of the manufactured product [49–51]. The shift towards a self-forming and decentralised supply chain requires the development of a conceptual framework for a manufacturing-as-a-service (MaaS) approach. This approach involves the provision of value-added services throughout the entire life cycle of a product, including planning, installation, operation, and alterations as well as the recirculation of used product and associated raw materials [52, 53].

In line with the principles of decentralisation and networked production, manufacturing capabilities and resources are provided as on-demand services to companies or individuals. Essentially, MaaS can be facilitated through the integration of technologies such as edge computing, cloud computing and IoT. These technologies enable the connection and coordination of manufacturing resources across a network of distributed nodes.

The decentralisation of activities allows companies to define and execute self-contained service packages independently, supporting the flexibility and adaptability of the network [14]. This means that manufacturing tasks can be outsourced to specialised service providers who offer specific capabilities or expertise. These providers can be in different geographical locations, allowing for flexibility in production and leveraging the strengths of each partner. MaaS may include the integration and coordination of typical manufacturing processes, such as design and engineering, production planning, procurement, assembly, quality control and logistics. Additionally, DTI enables continuous monitoring and control of manufacturing operations, ensuring general visibility, heightened transparency and adaptability throughout the production process.

Additionally, there is a growing demand for "all-inclusive" solutions that incorporate both products and related services, which presents an opportunity for collaboration between larger and smaller companies to design and offer specialised services for new markets [54]. Maas in this context also enables smaller companies or individuals to access manufacturing capabilities that were traditionally available only to larger organisations.

4 Building Blocks and Implementation Roadmap

A technology roadmap is a strategic planning tool that helps organisations to align their technology development efforts with their business goals and objectives [55, 56]. It can also serve as a research agenda for continuous research. This way, the elements of the roadmap, building blocks, required intermediate steps and envisioned outcomes, can be further refined and adapted to new insights. Thus, through the roadmap the major capabilities, characteristics, and processes that are needed to create a successful DTI-enabled and networked production can be understood.

The roadmap for realising advanced networked production with the help of DTI is comprehensive and involves several building blocks, each with its vision, value promise and development pathway. To achieve the desired goals, it is important to have a clear pathway for each building block, divided into near-term (0–5 years), mid-term (5–10 years) and long-term (over 10 years) goals and performances (Fig. 2).

The first building block, *Universal Transparency*, aims to enable end-to-end transparency and accountability across value networks, generating added value through comprehensive information available for all involved. The goal is to enable end-to-end transparency and visibility in networked production. An increase in coordination can help to ensure that the different activities within the network are better aligned, which can lead to greater efficiency and better results [57, 58]. This is achieved using various technologies and tools that enable comprehensive availability of information and real-time monitoring of the network's performance. Universal transparency also promotes customer value by providing more visibility and transparency in the production process, which can help build trust [59]. By building trust, the different partners within the network can better collaborate and achieve their goals together. It also enables social sustainability by promoting accountability and compliance with legal requirements and ensuring that all participants in the network act ethically and responsibly [4]. The near-term goals include progress against sustainability and compliance goals, data management protocols, and information security protocols. Mid-term goals include data analysis for transparency goals involving x-tier suppliers, while the long-term goal is to achieve end-to-end transparency in value chains that extends beyond legal requirements.

Cooperate to Compete is another building block that involves knowledge-based continuous development and network innovation. It involves co-design, co-engineering, and co-production of customised products and services. This building block generates added value by capturing the benefits of collaboration, such as sharing knowledge and resources [60]. The integration of product development into the network is especially important during the early phases of development when collaboration between different actors in the production network can have a decisive influence on competitive factors, such as reducing costs, shorten time-to-market, improve product quality, increase flexibility in production, promote innovation, and develop sustainable solutions [4]. Near-term goals include hybrid systems for knowledge-based engineering and process development, while mid-term goals include inter-factory organisation for enhanced collaborative design and engineering. Long-term goals involve advanced interfaces and methodologies for domain knowledge capturing and situation/context-dependent systems for engineering and process development across factory boundaries.

Sustainable and Circular Operations is a building block that focuses on closed-loop cycles, greater efficiencies, and new ways of sourcing materials. The added value is particularly produced through resource efficiency, extended machine operation, and securing new resources. In terms of resource efficiency, the focus is on optimising the utilisation rate of machines and equipment, avoiding downtime and production disruptions and reducing waste and energy consumption [4, 14].

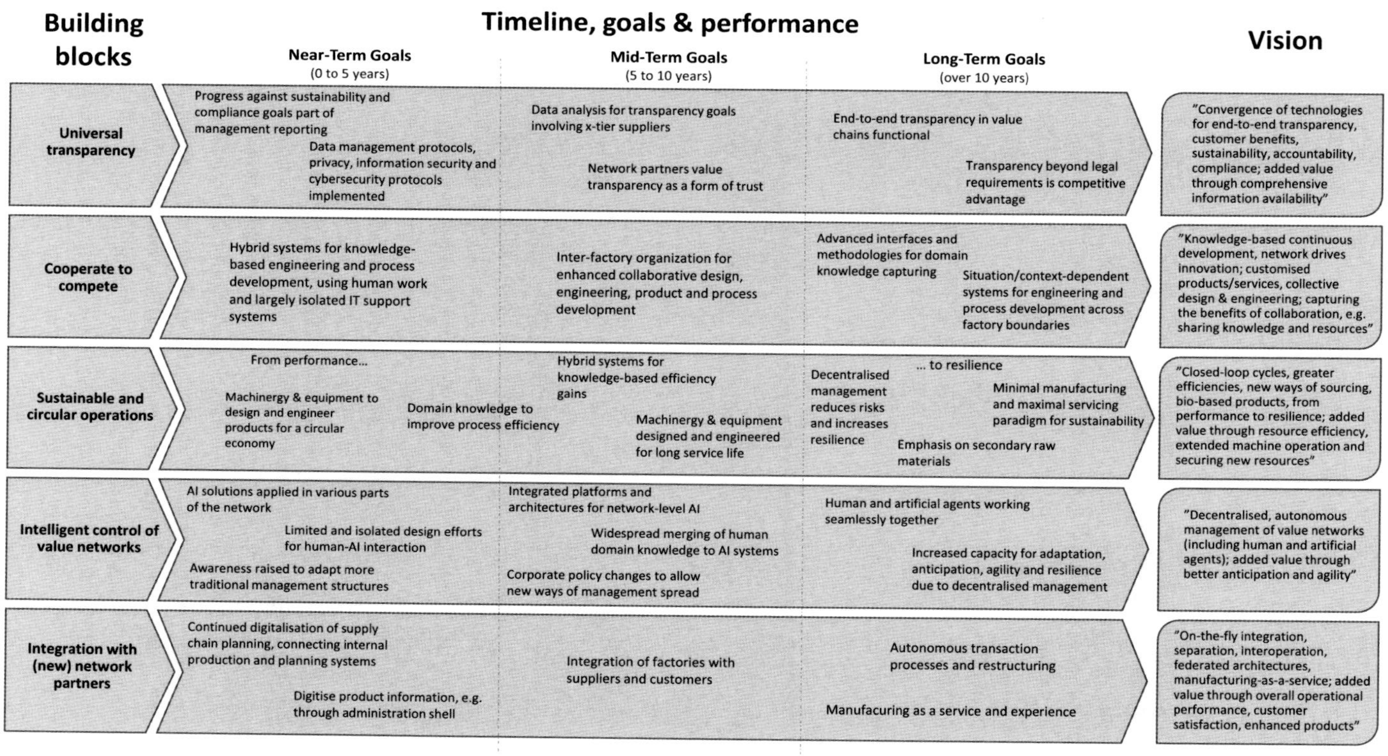

Timeline, goals & performance

Building blocks	Near-Term Goals (0 to 5 years)	Mid-Term Goals (5 to 10 years)	Long-Term Goals (over 10 years)	Vision
Universal transparency	Progress against sustainability and compliance goals part of management reporting; Data management protocols, privacy, information security and cybersecurity protocols implemented	Data analysis for transparency goals involving x-tier suppliers; Network partners value transparency as a form of trust	End-to-end transparency in value chains functional; Transparency beyond legal requirements is competitive advantage	"Convergence of technologies for end-to-end transparency, customer benefits, sustainability, accountability, compliance; added value through comprehensive information availability"
Cooperate to compete	Hybrid systems for knowledge-based engineering and process development, using human work and largely isolated IT support systems	Inter-factory organization for enhanced collaborative design, engineering, product and process development	Advanced interfaces and methodologies for domain knowledge capturing; Situation/context-dependent systems for engineering and process development across factory boundaries	"Knowledge-based continuous development; network drives innovation; customised products/services, collective design & engineering; capturing the benefits of collaboration, e.g. sharing knowledge and resources"
Sustainable and circular operations	From performance...; Machinery & equipment to design and engineer products for a circular economy; Domain knowledge to improve process efficiency	Hybrid systems for knowledge-based efficiency gains; Machinery & equipment designed and engineered for long service life	... to resilience; Decentralised management reduces risks and increases resilience; Minimal manufacturing and maximal servicing paradigm for sustainability; Emphasis on secondary raw materials	"Closed-loop cycles, greater efficiencies, new ways of sourcing, bio-based products, from performance to resilience; added value through resource efficiency, extended machine operation and securing new resources"
Intelligent control of value networks	AI solutions applied in various parts of the network; Limited and isolated design efforts for human-AI interaction; Awareness raised to adapt more traditional management structures	Integrated platforms and architectures for network-level AI; Widespread merging of human domain knowledge to AI systems; Corporate policy changes to allow new ways of management spread	Human and artificial agents working seamlessly together; Increased capacity for adaptation, anticipation, agility and resilience due to decentralised management	"Decentralised, autonomous management of value networks (including human and artificial agents); added value through better anticipation and agility"
Integration with (new) network partners	Continued digitalisation of supply chain planning, connecting internal production and planning systems; Digitise product information, e.g. through administration shell	Integration of factories with suppliers and customers	Autonomous transaction processes and restructuring; Manufacturing as a service and experience	"On-the-fly integration, separation, interoperation, federated architectures, manufacturing-as-a-service; added value through overall operational performance, customer satisfaction, enhanced products"

Fig. 2 DTI roadmap for advanced networked production

Sustainability-wise, the emphasis is on minimising the environmental impact of production processes by reducing waste and emissions, using alternative sources for raw materials, such as secondary sources, and considering circular economy principles for material and resource flows. Near-term goals include machinery and equipment to design and engineer products for a circular economy. Mid-term goals include hybrid systems for knowledge-based efficiency gains, and machinery and equipment designed and engineered for long service life. Long-term goals involve decentralised management for reduced risks and increased resilience, minimal manufacturing and maximal servicing paradigm for sustainability, and emphasis on secondary raw materials.

Intelligent Control of Value Networks involves decentralised, autonomous management of value networks, comprising various agents including human and artificial agents. This building block generates added value through better anticipation and agility. In a learning cognitive production network, assets in the network are interconnected and can self-optimise. Depending on the context, control is carried out automatically or autonomously within the collaboration between humans and artificial intelligence [61]. Predictions can be made based on data sources and amounts that would be hidden from humans alone [4]. Near-term goals include AI solutions applied in various parts of the network, while mid-term goals involve integrated platforms and architectures for network-level AI. Long-term goals involve human and artificial agents working seamlessly together, and increased capacity for adaptation, anticipation, agility, and resilience due to decentralised management.

Finally, *Integration with (New) Network Partners* (or their dissolution of redundant relationships) involves on-the-fly integration, separation, and interoperation, federated architectures, and manufacturing-as-a-service [62]. This building block creates added value through overall network performance, operational efficiency, customer satisfaction, and enhanced products. Companies can expand their capabilities and access new markets, assets, technologies and resources. This leads to enhanced network performance and will also involve strong collaboration between large companies and smaller, niche manufacturers [14]. Near-term goals include continued digitalisation of supply chain planning, connecting internal production and planning systems, and digitalisation of product information. Mid-term goals involve integration of factories with suppliers and customers, while long-term goals involve autonomous transaction processes and restructuring, and manufacturing-as-a-service and experience.

The roadmap for realising advanced networked production with the help of DTI is complex considering, for example, the variety of technologies involved and requires a long-term vision. In addition, it is important to recognise and exploit new business models and opportunities for creating value, while keeping in mind the viability of the proposed solutions from a business perspective.

5 DTI Deployment from Business and Organisational Perspective

The industrial evolution towards networked and self-controlling manufacturing concepts enables and generates groundwork for new business models, as well as sets new requirements for the skills and expertise of the personnel. In addition to the need to maximise profits, new challenges such as acting properly in the face of environmental damage cannot be ignored. It is therefore a matter of recognising and exploiting new opportunities for creating value throughout a product's life cycle. Due to its cross-company nature, networked manufacturing offers the appropriate conditions for this, for example, partnerships for knowledge-based and sustainable extraction of necessary resources, manufacturing and product development, and innovation development. At the same time, new business models must also recognise new forms of work and the development of skills, learning and education.

Digitalization of companies' and organisations' processes enables the separation of data and information from their physical/real operational environment. This in turn enables new ways to organise work within and between organisations, which also leads to new business model opportunities. In practice, the so-called cyber-physical systems can be implemented where physical processes are monitored by creating a virtual counterpart of the physical world in which decentralised decisions can be made by communicating and cooperating with each actor in real-time using the Internet and IoT components [63].

From an organisation point of view, the DTI vision brings about changes to the distribution of work between human employees and computer systems. More and more human processing can be automated to be done by smart ICT systems. This provides, on the one hand, an opportunity to increase the efficiency of the organisation, and on the other hand, develop new kinds of data-based services and new ways of connecting customers, partners, and suppliers. Furthermore, the skills and competencies of the personnel must be developed and adapted to meet the new requirements the DTI deployment entails. In general DTI increases the connectivity of the different stakeholders, which means that the actors need to broaden their perspective on their business environment to include partners that are further away from them in the value chain, e.g., they need to consider the operations of their customer's customer. More systemic thinking is required and skills and competences to realise it. An example of the business opportunities this enables is a condition based maintenance service concept, where a component or sub-system provider is connected to the equipment in use either directly or via OEM's systems to follow the product performance during its life cycle. All these connections, data availability and security issues need to be understood by all partners in the connected system. This requires enhancement of two kinds of competences: (1) competences to understand the technological system and its operation, and (2) competences to understand the operational environment of the end-product at end-user location.

In the digital world, the DTI envisions new services, such as platforms, that are required to enable the myriad of potential connections of digital data flows between different stakeholders. This digital world forms its market environment where new platform providers as well as traditional business ecosystem actors with different resources, competitive advantages and strategic aims are positioning themselves. The digital marketplace is much more flexible and dynamic than the traditional as digital relationships between actors are easier to establish and terminate. Digital platforms have a specific characteristic of aiming to orchestrate a large number of business actors, i.e., their business is based on connecting other actors. This challenges the traditional strategic planning of organisations rooted in the resource-based view of the firm, which focuses on achieving competitive advantage by owning and controlling resources [64]. These resources should have the so-called VRIN characteristics, i.e., be valuable, be rare, be inimitable, or be non-substitutable. Digital platform business models also challenge the positioning framework of strategic management [3], which focus on competitive advantage by raising structural barriers to entry and increasing bargaining power. The digital marketplace enables more dynamic competition, where, e.g., disruptive innovation from start-ups and entry by firms from different industry sectors can bring completely new competitive pressure to existing business models [65]. Instead of competition between firms, the competition is between business ecosystems consisting typically of one or two central (platform) companies and other firms providing complementary goods and services. In this kind of context, the value capture component of a business model should balance the focal firm's profits and the profitability of the ecosystem partners [65].

While creating new business opportunities the dynamic digital marketplace creates new challenges as well. Many new digital solutions and services are based on combining data from different sources, i.e., from different stakeholders. The knowledge and capabilities of different stakeholders are combined to deliver a specific offering or solution. This can create a challenging business landscape where the perspective on specific assets and complementary assets may be viewed differently by different stakeholders, especially between platform providers and linked service providers [64]. This can bring about risk, especially to the linked parties of the digital ecosystem around the platform. On the one hand, the platform provider may be enticed to integrate successful solutions provided by the linked parties into its offering. On the other hand, the platform operator may stop providing required data sources to the linked parties, thus destroying their digital business model. Naturally, the platform provider has to consider the risk these kinds of actions can cause to its business reputation [64].

Another challenge posed by the digital integration of businesses is the inclusion of the smaller actors in the value chains and business ecosystems, especially considering their digitalization skills and competencies [66]. SMEs are often an important part of a value chain in the manufacturing sector. Thus, to take full advantage of the DTI their integration is essential. Lack of resources and limited competencies are generic barriers for SMEs to implement new technologies. On the one hand, these barriers can be overcome by introducing scalable technology solutions that are easier to be deployed by SMEs [66]. On the other hand, new services and service providers can be established to support the SMEs in digitalization, from understanding the systemic change it brings to their business to running specific technology services for them [67].

6 Discussion

This article discusses the use of Decentralised Technical Intelligence (DTI) to improve networked production. While the primary focus is on the network, the transition is built on a multilevel perspective that spans from technical processes on the factory floor to networking and business operations at the upper level. This requires the integration of various micro and macro levels in real-time through both company internal and external networks of information and communication technologies, necessitating international standards.

Consequently, it is essential to develop and integrate a multilevel architecture for control and management, which includes implementing decentralised intelligence through edge-cloud solutions. This integration allows for the effective combination of physical levels with ICT control structures.

This development implies that organisations involved in the networked production will integrate more and their roles may transform into functional specialisation or strategic competencies within the network. This integration leads to a synergistic effect among collaborating companies, resulting in significant efficiency gains and cost savings across different categories of costs. Achieving advanced networked production through DTI requires following the described roadmap and implementing the building blocks of DTI in a way that is transparent, trustworthy, cooperative, sustainable, and intelligent.

The figure (Fig. 3) presented in this article illustrates the approach's goal to surpass the current fragmented approaches that mainly involve incremental

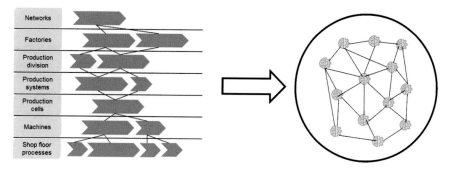

Fig. 3 Progress towards advanced networked production based on DTI (adapted from ManuFuture [24] and Sautter [22])

improvements. Despite witnessing progress within the different levels of the production system and the interconnections between these levels, disciplinary boundaries often limit us. With DTI, we can surpass these limitations by enabling autonomous activities of DTI agents, which include both humans and machines. This can lead to various benefits such as self-organisation, self-optimisation, and self-maintenance. All of these actions occur within a predefined performance space, with a focus on situational and contextual requirements.

Moreover, leveraging DTI technologies enables a shift away from strict hierarchies towards more flexible and adaptive operational management. With DTI, employees are empowered to make decisions based on their expertise and the real-time information provided by AI systems. This empowerment allows for quicker response times, as employees can take immediate action and adapt to unforeseen events without waiting for approval from higher levels of management.

Thus, DTI involves converging towards heterarchy, which refers to a non-hierarchical organisational structure where decision-making authority and control are distributed among various nodes or agents rather than being concentrated at the top of a hierarchical chain. Heterarchies emphasize collaboration, decentralised decision-making and empowered employees [68].

As a result, DTI facilitates a systemic and transcending approach that dissolves the boundaries of organisations and disciplines, allowing for a fundamental transition in production. DTI catalyses leapfrogging performance, as Sautter [22] describes it. DTI also provides a novel understanding of process quality, flexibility, and resilience, enabling the production system to adapt to changing situations.

From a business implementation perspective, it is essential to develop incentive systems that support the collaboration of the different stakeholders in the network. Each actor should have a clear business value proposition to engage in the activities DTI entails, i.e., collaboration, commitment to common goals, and sharing of knowledge and information. This may require, e.g., further development of pricing methods and mechanisms for knowledge and data.

7 Conclusion

DTI has wide-ranging implications for long-term development in industrial production. The concept has the potential to alter operative management and enhance flexibility, adaptability, robustness, and resilience in production networks.

Understanding the benefits will coincide with a surge in strategic management efforts, research, development and deployment of technologies, which provide an underpinning to the decentralisation of intelligence and agents' actions and responses.

DTI is an overarching visionary concept for the future of manufacturing. Implementing it is a long-term process that requires a mindset change from all the actors in manufacturing ecosystems. Changing all at once is not viable, instead, incremental development steps in collaboration with relevant actors in the manufacturing networks should be taken.

The roadmap serves as a guiding framework that reflects the need for incremental development in the implementation of DTI and emphasises the importance of taking progressive steps forward while allowing for continuous refinement and adjustment. Stakeholders can align their efforts towards realising the long-term vision of enhanced flexibility, adaptability, robustness and resilience in production networks. This will position them at the forefront of the future of manufacturing.

Acknowledgments The research leading to these results has received funding from Horizon 2020, the European Union's Framework Programme for Research and Innovation (H2020/2014-2020) under grant agreement no. 957331 (www.knowlEdge-project.eu).

References

1. Bowersox, D.J., Closs, D.J., Cooper, M.B., Bowersox, J.C.: Supply Chain Logistics Management, 4th edn. McGraw-Hill, New York, NY, USA (2016)
2. Christopher, M.: Logistics and Supply Chain Management: Strategies for Reducing Cost and Improving Service, 2nd edn. Prentice-Hall, Harlow, UK (1998)
3. Porter, M.E.: Competitive Advantage. Free Press, New York (1985)
4. Werner, H.: Supply Chain Management: Grundlagen, Strategien, Instrumente und Controlling, 7th edn. SpringerGabler, Wiesbaden (2020)
5. Ketchen Jr. D.J., Hult, G.T.M.: Bridging organization theory and supply chain management: The case of best value supply chains. J. Oper. Manag. **25**, 573–580 (2007)
6. Lii, P., Kuo, F.I.: Innovation-oriented supply chain integration for combined competitiveness and firm performance. Int. J. Prod. Econ. **60**, 142–155 (2016)
7. McDougall, N., Wagner, B., MacBryde, J.: Leveraging competitiveness from sustainable operations: frameworks to understand the dynamic capabilities needed to realise NRBV supply chain strategies. Supply Chain Manag. **27**, 12–29 (2022)
8. Sakuramoto, C., Di Serio, L.C., de Vicente Bittar, A.: Impact of supply chain on the competitiveness of the automotive industry. RAUSP Manag. J. **54**, 205–225 (2019)
9. Maozhu, J., Wang, H., Zhang, Q., Zeng, Y.: Supply chain optimization based on chain management and mass customization. Inf. Syst. e-Bus. Manag. **18**, 647–664 (2020)
10. Tukamuhabwa, B.R., Stevenson, M., Busby, J., Zorzini, M.: Supply chain resilience: definition, review and theoretical foundations for further study. Int. J. Prod. Res. **53**(18), 5592–5623 (2015)
11. Christopher, M., Holweg, M.: Supply chain 2.0: managing supply chains in the era of turbulence. Int. J. Phys. Distrib. **41**(1), 63–82 (2011)
12. Al-Omoush, K.S., de Lucas, A., del Val, M.T.: The role of e-supply chain collaboration in collaborative innovation and value-co creation. J. Bus. Res. **158**, 113647 (2023)
13. Kähkänen, A.-K., Lintukangas, K.: The underlying potential of supply management in value creation. J. Purchas. Supply Manag. **18**(2), 68–75 (2012). Vision 20/20: Preparing Today for Tomorrow's Challenges
14. Jovane, F., Westkäper, E., Williams, D.: The ManuFuture Road. Towards Competitive and Sustainable High-Adding-Value Manufacturing. Springer, Berlin (2009)
15. Bak, O.: Understanding the stimuli, scope, and impact of organizational transformation: The context of eBusiness technologies in supply chains. Strateg. Change **30**, 443–452 (2021)
16. Zhang, J., Xu, J., Liu, Y.: Complex adaptive supply chain network: The state of the art. In: 2009 Chinese Control and Decision Conference, pp. 5643–5647 (2009)

17. Walter, S.: AI impacts on the performance in supply chains. In: The 34th Annual NOFOMA Conference June 8–10, 2022 - Reykjavík, Iceland, pp. 1–14. The Nordic Logistics Research Network (NOFOMA) (2022)
18. Sendhil Kumar, R., Pugazhendhi, S.: Information sharing in supply chains: An overview. Procedia Eng. **38**, 2147–2154 (2012)
19. Ahmad, I., Rodriguez, F., Kumar, T., Suomalainen, J., Kumar, S., Walter, S., Asghar, M.Z., Li, G., Papakonstantinou, N., Ylianttila, M., Huusko, J., Sauter, T., Harjula, E.: Communications security in Industry X: A survey. TechRxiv. Preprint (2023)
20. Baesens, B., Bapna, R., Marsden, J.R., Vanthienen, J., Leon Zhao, J.: Transformational issues of big data and analytics in networked business. MIS Q. **40**(4), 807–818 (2016)
21. Surana, A., Kumara, S., Greaves, M., Raghavan, U.N.: Supply-chain networks: a complex adaptive systems perspective. Int. J. Prod. Res. **43**(20), 4235–4265 (2005)
22. Sautter, B. (ed.): Decentralised Technical Intelligence (DTI) for increased manufacturing performance. Going beyond the limits of today in an interdisciplinary approach. White paper. Version 1.0. ManuFUTURE, Brussels (2022)
23. Jarrahi, M.H.: Artificial intelligence and the future of work: Human-AI symbiosis in organizational decision making. Bus. Horizons **61**(4), 577–586 (2018)
24. ManuFUTURE High-level Group: ManuFUTURE Strategic Research Agenda SRIA 2030. For a competitive, sustainable and resilient European manufacturing. ManuFUTURE, Brussels (2019)
25. Plattform Industrie 4.0: 2030 Vision for Industrie 4.0. Shaping Digital Ecosystems Globally. Federal Ministry for Economic Affairs and Energy (BMWi), Berlin (2019)
26. Lin, Z., Xiangli, P., Li, Z., Liang, F., Li, A.: Towards metaverse manufacturing: A blockchain-based trusted collaborative governance system. In: The 2022 4th International Conference on Blockchain Technology, ICBCT'22, pp. 171–177, New York, NY, USA, 2022. Association for Computing Machinery
27. Siahaan, B.P., Simatupang, T.M., Okdinawati, L.: Logistics landscape for metaverse. In: 2022 IEEE International Conference of Computer Science and Information Technology (ICOSNIKOM), pp. 1–6 (2022)
28. Thomas, D.J., Griffin, P.M.: Coordinated supply chain management. Eur. J. Oper. Res. **94**(1), 1–15 (1996)
29. Chandra, C., Kumar, S.: Supply chain management in theory and practice: a passing fad or a fundamental change? Ind. Manag. Data Syst. **100**(3), 100–114 (2000)
30. Leitão, P.: Agent-based distributed manufacturing control: A state-of-the-art survey. Eng. Appl. Artif. Intell. **22**(7), 979–991 (2009). Distributed Control of Production Systems
31. Almeida, R., Toscano, C., Lopes Azevedo, A., Maia Carneiro, L.: A Collaborative Planning Approach for Non-hierarchical Production Networks, chapter 9, pp. 185–204. Wiley (2012)
32. Shen, W., Wang, L., Hao, Q.: Agent-based distributed manufacturing process planning and scheduling: a state-of-the-art survey. IEEE Trans. Syst. Man Cybern. C (Appl. Rev.) **36**(4), 563–577 (2006)
33. Straube, F., Beyer, I.: Decentralized planning in supply networks. In: 2006 IEEE International Technology Management Conference (ICE), pp. 1–8 (2006)
34. Andrés, B., Poler, R., Hernández, J.E.: An operational planning solution for SMEs in collaborative and non-hierarchical networks. In: Hernández, J.E., Liu, S., Delibašić, B., Zaraté, P., Dargam, F., Ribeiro, R. (eds.), Decision Support Systems II - Recent Developments Applied to DSS Network Environments, pp. 46–56, Berlin, Heidelberg, 2013. Springer, Berlin, Heidelberg (2013)
35. Bousdekis, A., Mentzas, G.: Enterprise integration and interoperability for big data-driven processes in the frame of industry 4.0. Front. Big Data 4 (2021)
36. Marques, M., Agostinho, C., Zacharewicz, G., Jardim-Gonçalves, R.: Decentralized decision support for intelligent manufacturing in industry 4.0. J. Ambient Intell. Smart Environ. **9**(3), 299–313 (2017)

37. Fortino, G., Savaglio, C., Palau, C.E. de Puga, J.S., Ganzha, M., Paprzycki, M., Montesinos, M., Liotta, A., Llop, M.: Towards Multi-layer Interoperability of Heterogeneous IoT Platforms: The INTER-IoT Approach, pp. 199–232. Springer International Publishing, Cham (2018)
38. Stadtler, H.: Supply chain management and advanced planning—basics, overview and challenges. Eur. J. Oper. Res. **163**(3), 575–588 (2005). Supply Chain Management and Advanced Planning
39. Stavrulaki, E., Davis, M.: Aligning products with supply chain processes and strategy. Int. J. Logist. Manag. **21**(1), 127–151 (2010)
40. Amrani, A., Zouggar, S., Zolghadri, M., Girard, P.: Towards a collaborative approach to sustain engineer-to-order manufacturing. In: 2010 IEEE International Technology Management Conference (ICE), pp. 1–8 (2010)
41. Ayel, J.: Decision coordination in production management. In: Castelfranchi, C., Werner, E. (eds.), Artificial Social Systems, pp. 295–310, Berlin, Heidelberg, 1994. Springer, Berlin, Heidelberg (1994)
42. Chandra, C., Kumar, S.: Enterprise architectural framework for supply-chain integration. Ind. Manag. Data Syst. **101**(6), 290–304 (2001)
43. Zou, X., Pokharel*, S., Piplani, R.: Channel coordination in an assembly system facing uncertain demand with synchronized processing time and delivery quantity. Int. J. Prod. Res. **42**(22), 4673–4689 (2004)
44. Chankov, S.M., Becker, T., Windt, K.: Towards definition of synchronization in logistics systems. Procedia CIRP **17**, 594–599 (2014). Variety Management in Manufacturing
45. Alvarez-Napagao, S., Ashmore, B., Barroso, M., Barrué, C., Beecks, C., Berns, F., Bosi, I., Chala, S.A., Ciulli, N., Garcia-Gasulla, M., Grass, A., Ioannidis, D., Jakubiak, N., Köpke, K., Lämsä, V., Megias, P., Nizamis, A., Pastrone, C., Rossini, R., Sànchez-Marrè, M., Ziliotti, L.: knowledge project–concept, methodology and innovations for artificial intelligence in industry 4.0. In: 2021 IEEE 19th International Conference on Industrial Informatics (INDIN), pp. 1–7 (2021)
46. Camarinha-Matos, L.M., Afsarmanesh, H.: Collaborative networks. In: Wang, K., Kovacs, G.L., Wozny, M., Fang, M. (eds.), Knowledge Enterprise: Intelligent Strategies in Product Design, Manufacturing, and Management, pp. 26–40, Boston, MA, 2006. Springer US (2006)
47. Camarinha-Matos, L.M., Afsarmanesh, H., Galeano, N., Molina, A.: Collaborative networked organizations – concepts and practice in manufacturing enterprises. Comput. Ind. Eng. **57**(1), 46–60 (2009). Collaborative e-Work Networks in Industrial Engineering
48. Romero, D., Molina, A.: Collaborative networked organisations and customer communities: value co-creation and co-innovation in the networking era. Prod. Plann. Control **22**(5–6), 447–472 (2011)
49. Brax, S.: A manufacturer becoming service provider – challenges and a paradox. Manag. Serv. Q. **15**(2), 142–155 (2005)
50. Cohen, M.A., Agrawal, N., Agrawal, V.: Winning in the aftermarket. Harv. Bus. Rev. (2006)
51. Isaksson, O., Larsson, T.C., Rönnbäck, A.Ö.: Development of product-service systems: challenges and opportunities for the manufacturing firm. J. Eng. Des. **20**(4), 329–348 (2009)
52. Gao, J., Yao, Y., Zhu, V.C.Y., Sun, L., Lin, L.: Service-oriented manufacturing: a new product pattern and manufacturing paradigm. J. Intell. Manuf. **22**, 435–446 (2011)
53. Fisher, O., Watson, N., Porcu, L., Bacon, D., Rigley, M., Gomes, R.L.: Cloud manufacturing as a sustainable process manufacturing route. J. Manuf. Syst. **47**, 53–68 (2018)
54. Wahlström, M., Walter, S., Salonen, T-T., Lammi, H., Heikkilä, E., Helaakoski, H.: Sustainable Industry X – a Cognitive Manufacturing Vision. VTT Technical Research Centre of Finland, Espoo (2020)
55. Bray, O.H., Garcia, M.L.: Technology roadmapping: the integration of strategic and technology planning for competitiveness. In: Innovation in Technology Management. The Key to Global Leadership. PICMET '97, pp. 25–28 (1997)
56. Phaal, R., Farrukh, C.J.P., Probert, D.R.: Technology roadmapping—a planning framework for evolution and revolution. Technol. Forecast. Soc. Change **71**(1), 5–26 (2004). Roadmapping: From Sustainable to Disruptive Technologies

57. Shih, S.C., Hsu, S.H.Y., Zhu, Z., Balasubramanian, S.K.: Knowledge sharing–a key role in the downstream supply chain. Inf. Manag. **49**(2), 70–80 (2012)
58. Barreto, L., Amaral, A., Pereira, T.: Industry 4.0 implications in logistics: an overview. Procedia Manuf. **13**, 1245–1252 (2017). Manufacturing Engineering Society International Conference 2017, MESIC 2017, 28–30 June 2017, Vigo (Pontevedra), Spain
59. Min, S., Zacharia, Z.G., Smith, C.D.: Defining supply chain management: In the past, present, and future. J. Bus. Logist. **40**(1), 44–55 (2019)
60. Varela, L., Putnik, G., Romero, F.: The concept of collaborative engineering: a systematic literature review. Prod. Manuf. Res. **10**(1), 784–839 (2022)
61. Jones, A.T., Romero, D., Wuest, T.: Modeling agents as joint cognitive systems in smart manufacturing systems. Manuf. Lett. **17**, 6–8 (2018)
62. Deshmukh, R.A., Jayakody, D., Schneider, A., Damjanovic-Behrendt, V.: Data spine: A federated interoperability enabler for heterogeneous iot platform ecosystems. Sensors **21**(12), (2021)
63. Monostori, L., Kádár, B., Bauernhansl, T., Kondoh, S., Kumara, S.R.T., Reinhart, G., Sauer, O., Schuh, G., Sihn, W., Ueda, K.: Cyber-physical systems in manufacturing. Cirp Ann. Manuf. Technol. **65**, 621–641 (2016)
64. Paredes-Frigolett, H., Pyka, A.: The global stakeholder capitalism model of digital platforms and its implications for strategy and innovation from a Schumpeterian perspective. J. Evolut. Econ. **32**, 463–500 (2022)
65. Teece, D.J., Linden, G.: Business models, value capture, and the digital enterprise. J. Organiz. Des. **6**, 1–14, 2017
66. Ghobakhloo, M., Iranmanesh, M., Vilkas, M., Grybauskas, A., Amran, A.: Drivers and barriers of industry 4.0 technology adoption among manufacturing SMEs: a systematic review and transformation roadmap. J. Manuf. Technol. Manag. (2022)
67. Mikkola, M., Salonen, J.: Manufacturing SME's are not worried about novel technology, but people. In: ISPIM Connects Athens Conference: The Role of innovation: Past, Present, Future. Lappeenranta University of Technology (2022)
68. Filip, F.G., Zamfirescu, C.-B., Ciurea, C.: Computer-Supported Collaborative Decision-Making. Springer International Publishing, Cham (2017)

Part III
Trusted, Explainable and Human-Centered AI Systems

Wearable Sensor-Based Human Activity Recognition for Worker Safety in Manufacturing Line

Sungho Suh, Vitor Fortes Rey, and Paul Lukowicz

1 Introduction

The advent of Industry 4.0, introduced in 2011, aimed to revolutionize manufacturing by incorporating advanced technologies to achieve operational efficiency and productivity gains [24]. However, as the focus shifted from technology-driven advancements to a more human-centric approach, the concept of Operator 4.0 emerged. Operator 4.0 envisions a workforce assisted by systems that alleviate physical and mental stress while maintaining production objectives [14]. This shift in perspective laid the foundation for the development of Industry 5.0, a value-driven manufacturing paradigm that places worker well-being at the forefront of the production process [30].

Industry 5.0 encompasses two main visions: one involving human–robot collaboration and the other centered around a bioeconomy utilizing renewable biological resources. As the researchers explore the potential of Industry 5.0, it becomes crucial to investigate how technology can support the industry while prioritizing worker safety and productivity. This necessitates meeting the human needs outlined in the Industrial Human Needs Pyramid, ranging from workplace safety to the fulfillment of human potential through a trustworthy relationship between humans and machines.

In this context, wearable sensor-based human activity recognition (HAR) emerges as a vital component of Industry 5.0. By continuously and unobtrusively monitoring workers' activities, wearable sensors enable a synergistic collaboration between humans and machines [23, 29]. This collaboration enhances productivity

S. Suh (✉) · V. F. Rey · P. Lukowicz
German Research Center for Artificial Intelligence (DFKI), Kaiserslautern, Germany

Department of Computer Science, RPTU Kaiserslautern-Landau, Kaiserslautern, Germany
e-mail: sungho.suh@dfki.de; vitor.fortes_rey@dfki.de; paul.lukowicz@dfki.de

© The Author(s) 2024
J. Soldatos (ed.), *Artificial Intelligence in Manufacturing*,
https://doi.org/10.1007/978-3-031-46452-2_17

while empowering workers to unleash critical thinking, creativity, and domain knowledge. Simultaneously, machines autonomously assist with repetitive tasks, reducing waste and costs.

To foster the development of trustworthy coevolutionary relationships between humans and machines, interfaces must consider the unique characteristics of employees and organizational goals. An example of such collaboration is evident in the use of cobots, where machines share physical space, perceive human presence, and perform tasks independently, simultaneously, sequentially, or in a supportive manner [8].

Within this framework, this book chapter focuses on the crucial intersection of wearable sensor-based HAR and Industry 5.0. Especially, we explore the transformative role of sensor-based HAR in promoting worker safety and optimizing productivity in manufacturing environments. We aim to study the intricacies of wearable sensor technologies, sensor data fusion techniques, and advanced machine learning algorithms to effectively capture and interpret workers' activities in real time. By utilizing wearable sensors, manufacturers can gain real-time insights into the physical movements and behaviors of workers, enabling them to identify potential safety hazards, proactively intervene in hazardous situations, and implement preventive measures.

Moreover, the applications of sensor-based HAR extend beyond worker safety. The data collected from wearable sensors can facilitate the optimization of manufacturing processes by identifying bottlenecks, streamlining workflows, and minimizing errors [29]. By analyzing workers' activities, manufacturers can uncover insights into the ergonomics of workstations, leading to improvements in the job design and a reduction in musculoskeletal disorders. Furthermore, the knowledge derived from sensor-based HAR can inform training programs, enabling targeted interventions and skill development to enhance worker efficiency and job satisfaction. Likewise, the integration of HAR using wearable sensors in manufacturing environments aligns with the growing interest in smart manufacturing and Industry 4.0 [23, 29]. In the manufacturing line, HAR can be utilized to quantify and evaluate worker performance [28], understand workers' operational behavior [1], and support workers' operations with industrial robots [22]. Activity recognition for worker safety in the manufacturing line is becoming increasingly important, as it provides the ability to quickly identify workers' needs for assistance or prevent industrial accidents.

In this book chapter, we study the intricacies of wearable sensor technologies and their integration into manufacturing environments. We explore various sensor modalities, including inertial sensors, motion sensors, and body capacitance sensors, and discuss their relevance in capturing a comprehensive view of workers' activities. We also examine sensor data fusion techniques to effectively integrate and interpret data from multiple wearable devices, enabling a holistic understanding of workers' actions. The outcomes of this research shed light on the transformative capabilities of wearable sensor technologies and open new avenues for future research in this field, aligning with the principles of the Industry 5.0 paradigm.

To demonstrate the practical application of sensor-based worker activity recognition, we present a use case in a smart factory testbed. In this use case, we deploy and test HAR using wearable sensors to predict workers' movement intentions and plan optimal routes for mobile robots in a collaborative environment. By anticipating workers' actions, collision risks between workers and robots can be minimized, ensuring both high production levels and worker safety.

To further improve the performance of HAR, we explore deep learning techniques such as adversarial learning [26, 27] and contrastive learning [9, 12]. These approaches have shown promise in enhancing activity recognition by leveraging additional information and improving the generalization capabilities of the models.

In addition to traditional wearable sensor modalities, we introduce the use of body capacitance sensing as an alternative modality for HAR. Body capacitance sensing captures the electric field feature between the body and the environment, providing unique insights into body movement and environmental variations [6]. By combining body capacitance sensing with inertial measurement unit (IMU)-based activity recognition, we aim to extend the sensing capabilities and improve the accuracy of activity recognition systems.

Throughout this chapter, we showcase real-world applications of sensor-based worker activity recognition in manufacturing environments. From real-time monitoring of workers' movements to detecting unsafe actions and alerting workers and supervisors, the potential benefits are substantial. We also discuss the limitations and potential ethical considerations associated with wearable sensor systems, emphasizing the importance of privacy, data security, and worker consent.

In the following sections, we provide a comprehensive review of related works in Sect. 2. Section 3 presents the wearable sensor-based worker activity recognition in a manufacturing line with IMU sensors and body capacitance sensing module, including data fusion approaches and a use case in a smart factory testbed. Section 4 introduces deep learning techniques for improving the performance of human activity recognition, such as adversarial learning and contrastive learning. Finally, Sect. 5 concludes this book chapter.

2 Background

The Internet of Things (IoT) has revolutionized various industries, including manufacturing, by enabling the integration of physical devices and digital systems. In the context of smart factories, IoT technologies play a crucial role in creating intelligent and automated production environments. With the advancement of sensing technologies, it has become increasingly feasible to recognize human activities using various sensors such as IMU sensors [26], microphones [5], cameras [20], and magnetic sensors [19]. These sensors capture valuable data about human movements, interactions, and environmental factors, which can be utilized to enhance productivity, safety, and efficiency in smart factory environments.

Supervised machine learning techniques have been widely used for recognizing activities using labeled training data. However, collecting labeled data can be time-consuming and labor-intensive. To overcome this challenge, researchers have explored methods that reduce the effort required for data collection. Among the popular approaches are transfer learning, activity modeling, and clustering techniques. These methods leverage existing knowledge or unsupervised learning to recognize activities with minimal labeled data.

The concept of smart manufacturing, often associated with Industry 4.0, has gained significant attention in recent years [21]. Researchers have focused on recognizing and supporting factory work using sensor technologies [2, 3]. In the manufacturing domain, the use of sensors for activity recognition has been extensively explored. For example, Koskimaki et al. [15] utilized a wrist-worn IMU sensor and a K-Nearest Neighbors model to classify activities in industrial assembly lines. Maekawa et al. [17] proposed an unsupervised method for lead time estimation of factory work using signals from a smartwatch with an IMU sensor. These studies demonstrate the potential of sensors in recognizing worker activities in manufacturing environments, particularly with the application of machine learning techniques.

Sensor-based HAR is a crucial aspect of wearable technology. The IMU has been the dominant sensor in wearable devices, providing motion-sensing capabilities. However, the IMU's ability is limited to capturing the wearer's movement patterns and does not account for body–environment and body–machine interactions, which play critical roles in security and safety in manufacturing environments. To extend the motion-sensing ability of wearables, the researchers have explored alternative sensing sources, such as the body capacitance [6]. The body capacitance describes the electric field between the body and the environment, and variations in this field can provide valuable information for pattern recognition. For example, capacitive sensing has been used to detect touch patterns between human fingers and different objects. This alternative sensing approach benefits from low cost and low power consumption and extends the sensing ability beyond IMU-based activity recognition.

Machine learning models play a crucial role in wearable activity recognition, enabling the extraction of meaningful patterns and features from sensor data [13]. Classical machine learning methods extract handcrafted features from sensor data, such as time-domain and frequency-domain features [16]. These methods rely on expert knowledge and domain-specific feature engineering. In recent years, deep learning-based methods have gained significant attention in activity recognition [18]. Convolutional neural networks (CNNs), recurrent neural networks (RNNs), and hybrid models have been developed to capture temporal correlations and learn sensor representations for improved activity recognition accuracy. Additionally, multitask learning and generative adversarial learning have been introduced to address different data distribution problems and enhance recognition performance [4, 7]. These advancements in machine learning techniques have paved the way for more accurate and robust activity recognition in wearable devices.

To bridge the gap between wearable sensor-based HAR and the principles of Industry 5.0, this book chapter offers unique contributions and addresses existing research gaps. It presents a comprehensive case study on wearable sensor-based worker activity recognition in a manufacturing line with a mobile robot. The investigation includes the exploration and comparison of sensor data fusion approaches using neural network models to effectively handle the multimodal sensor data obtained from wearable devices. Additionally, several deep learning-based techniques are introduced to enhance the performance of human activity recognition. By harnessing the potential of wearable sensors for human activity recognition, this chapter provides valuable insights into improving worker safety on the manufacturing line, aligning with the principles of the Industry 5.0 paradigm. The outcomes of this research shed light on the transformative capabilities of wearable sensor technologies and open new avenues for future research in this field.

3 Wearable Sensor-Based Worker Activity Recognition in Manufacturing Line

3.1 Use Case at the SmartFactory Testbed

The SmartFactory Testbed, developed by the Technology Initiative SmartFactory KL, is a collaborative effort between the Department of Machine Tools and Controls (WSKL) at the TU Kaiserslautern and the Innovative Factory Systems (IFS) research unit at the German Research Center for Artificial Intelligence (DFKI). This non-profit organization focuses on advancing manufacturing technologies with industry specialists known as factory innovators.

The SmartFactory Testbed offers unique manufacturer-independent demonstrations that enable the development, testing, and deployment of innovative ICT technologies in a realistic industrial production environment. It serves as a vital test environment, particularly for the European GAIA-X subproject SmartMA-X, where flexible production systems can be arranged and integrated in highly customized configurations, enhancing the dynamism of the industrial environment.

In this chapter, we utilize the SmartFactory Testbed as a real-world deployment setting to evaluate and validate the Human Action Recognition module within a specific use case referred to as "Human Action Recognition and Prediction in the Respective Environment." The use case focuses on a worker's activity pipeline, including their presence in the pilot area and collaboration with different modules and robots across 20 diverse activities. The accurate prediction of human actions within the production lines is crucial, especially when workers interact with moving robots or when robots generate paths to avoid potential collisions in the layout. By leveraging information about the worker's location and their next anticipated action, the robot's movement path can be manipulated to minimize the risk of collisions. This proactive approach ensures the safety and reliability of the industrial environment, particularly when humans are present.

3.2   Data Acquisition

In this chapter, we conducted an experiment involving 12 volunteers, from diverse cultural backgrounds and genders, who wore a wearable sensing prototype that we designed, as well as an Apple Watch, currently the top-selling smartwatch [10], while performing various tasks that simulated typical worker scenarios during their daily work. These tasks included opening and closing doors, walking, checking parts inside a module, and interacting with a touch screen. To ensure robust results, the experiment consisted of five sessions, each lasting between 2 and 3 minutes. Some sessions were conducted and recorded in a different direction from the flow chart. Prior to participation, all volunteers signed an agreement in compliance with the university's committee for the protection of human subjects policies. The experiment was video recorded to enable further confidential analysis, including ground truth activity annotation. Both the observer and the participants adhered to an ethical and hygienic protocol in accordance with public health guidelines.

Figure 1 illustrates the wearable sensors attached to the participants. The prototype sensors were placed on both wrists while the Apple Watches were attached to both wrists, and an iPhone mini was attached to the left arm. The

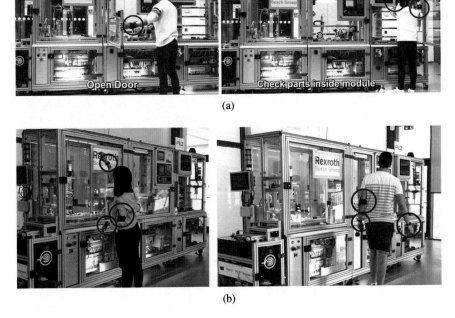

(a)

(b)

Fig. 1 The wearable sensors attached to the participants. (**a**) The wearable sensing prototype we designed on both wrists and (**b**) Apple Watches on both wrists and the iPhone mini on the left arm

Table 1 Comparison of the data distribution of activities in the prototype sensor and Apple Watch datasets

		Prototype	Apple
# of Subjects		12	12
# of Session		5	5
# of Channels per device		10	9
# of Devices		2	3
Sampling frequency		25 Hz	100 Hz
Activity	Null	14 m 9 s (9.9%)	29 m 43 s (17.5%)
	Pressing button	2 m 24 s (1.7%)	1 m 54 s (1.1%)
	Sliding doorlock	1 m 58 s (1.4%)	1 m 21 s (0.8%)
	Opening door	8 m 53 s (6.2%)	10 m 53 s (6.4%)
	Closing door	12 m 3 s (8.5%)	14 m 52 s (8.7%)
	Checking machines	47 m 6 s (33.1%)	48 m 53 s (28.7%)
	Walking	19 m 13 s (13.5%)	32 m 23 s (19.0%)
	Taking key	4 m 20 s (3.0%)	2 m 21 s (1.4%)
	Rotating key	10 m 14 s (7.2%)	6 m 47 s (4.0%)
	Placing key back	6 m 26 s (4.5%)	6 m 16 s (3.7%)
	Checking door lock	2 m 54 s (2.0%)	0 m 29 s (0.3%)
	Touching screen	12 m 44 s (8.9%)	14 m 11 s (8.3%)
	Total	142 m 24 s	170 m 3 s

collected data from the wearable sensing prototype consisted of 10 channels per sensor, resulting in a total of 20 channels, including three channels of acceleration, three channels of gyroscope, three channels of magnetometer data, and one channel of body capacitance data. The data collected using the Apple Watch and iPhone mini comprised 9 channels per sensor, totaling 27 channels, including 3 channels of acceleration, 3 channels of gyroscope, and 3 channels of magnetometer data. To synchronize the recorded video data, left wrist sensor data, and right wrist sensor data, we performed five claps at the start and end of each session. Based on the video data, we manually annotated the users' activities, resulting in 12 different activities, including a Null class. The prototype sensors provided IMU sensor data and body capacitance data with a sampling rate of 25 Hz, while the Apple Watch only provided IMU sensor data with a sampling rate of 100 Hz. This experiment allowed us to assess the performance of the sensor hardware and collect a dataset suitable for developing and testing algorithms for human intention recognition.

Table 1 presents the data distribution of activities in the prototype sensor and Apple Watch datasets. The prototype sensor dataset involved data collection for a total duration of 142 minutes and 24 seconds, while the Apple Watch dataset had a total duration of 170 minutes and 3 seconds. To focus on preventing collisions between workers and mobile robots, we decided to collect additional sensor data from the Apple Watches for the walking class after collecting data from the prototype sensor. As a result, the Apple Watch dataset had a higher proportion of walking class data compared to the prototype sensor dataset.

Table 2 Comparison results between Apple Watches and the prototype sensing module combining IMU and body capacitance sensors in terms of the testing accuracy and macro F1 score. The numbers are expressed in percent and represented as *mean ± std*

Method	Accuracy	Macro F1	Walking accuracy
Apple Watch	67.94 ± 2.55	43.02 ± 2.49	69.96 ± 4.39
The prototype sensing module	69.76 ± 3.09	52.40 ± 2.32	75.88 ± 2.55

To evaluate the prototype sensor hardware and neural network models for worker activity recognition, we annotated the user's activities based on the workflow of the use case at the SmartFactory Testbed. The workflow classified the sensor data into 12 activities, including Null, opening/closing doors, checking machines, walking, pressing buttons, and placing back keys.

In our application, each instance is a sliding window of sensor data. For the Apple Watch data, we used a window of length 100 (1 second) and a step size of 4 (0.04 seconds). For the prototype sensor data, we employed a window of length 25 (1 second) and a step size of 1 (0.04 seconds). That is, in both cases, we use data from 1 second, with a step size of 0.04 seconds, which maximizes the number of windows while keeping the step size suitable in both sensor scenarios.

3.3 Worker Activity Recognition Results

To process the collected sensor data, we used neural networks based on convolutional neural networks (CNNs) [25]. For training and validation of the neural network models, we employed a leave-one-session-out scheme. In each fold, one session was allocated for testing, another for validation, and the remaining three sessions for training.

Table 2 presents the comparison results between Apple Watches and the prototype sensing module, which combined IMU and body capacitance sensors, for the worker's activity recognition at the smart factory testbed. The evaluation metrics used include accuracy, macro F1 score [26], and the accuracy of the walking class, which is critical for preventing potential collisions between workers and mobile robots. The results demonstrate that the prototype-sensing module outperformed the Apple Watch in terms of accuracy, macro F1 score, and walking class accuracy.

4 Deep Learning Techniques for Human Activity Recognition Improvement

In this section, we focus on exploring advanced deep learning techniques to improve the performance of worker activity recognition in the manufacturing line for worker safety, building upon the findings presented in the previous section. Two specific

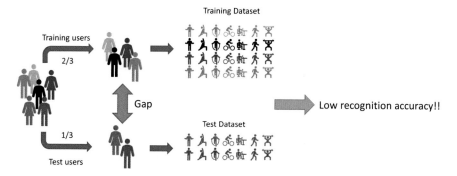

Fig. 2 Challenges in activity recognition: accounting for diverse behavior patterns across individuals

techniques, namely adversarial learning and contrastive learning, are discussed in detail. These techniques offer innovative approaches to improve the accuracy and robustness of the recognition system, addressing the challenges associated with complex and dynamic industrial environments [9, 27].

4.1 Adversarial Learning

Despite the successful digitalization of worker activities through wearable sensors and their recognition by simple CNN models, achieving generalization to unseen workers remains a significant challenge. Numerous studies have demonstrated that individuals perform the same activities in different ways, posing a challenge for activity recognition, despite the feasibility of user recognition [26], as illustrated in Fig. 2. This discrepancy becomes evident when evaluating performance by leaving out subjects rather than leaving out sessions.

To deal with this problem, we present an adversarial learning-based method for user-invariant HAR in this subsection, as illustrated in Fig. 3. Inspired by generative adversarial networks (GANs) [11], adversarial learning has been introduced as a technique to enhance the model's ability to discriminate between different worker activities and improve generalization. This technique, described in [26] and [27], employs four independent networks: a feature extractor, a reconstructor, an activity classifier, and a subject discriminator. The feature extractor maps sensor data to a common embedding feature space, while the reconstructor reconstructs the original signal from the embedding features. The activity classifier predicts activity labels based on the embedding features, and the subject discriminator differentiates between subjects based on the embedding features. The feature extractor and subject discriminator are trained by adversarial learning, with the subject discriminator aiming to distinguish subjects and the feature extractor aiming to deceive the subject discriminator. The method also incorporates a reconstruction loss to minimize the

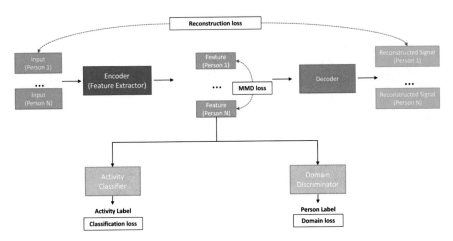

Fig. 3 The overall framework of the user-invariant HAR method using adversarial learning

difference between the original and reconstructed signals, along with a classification loss to train the feature extractor and activity classifier using activity labels. Additionally, the proposed method employs maximum mean discrepancy (MMD) regularization to align the distributions among source and target subjects, thereby enhancing the generalization of the embedding feature representation.

The user-invariant HAR method has demonstrated improvements of up to 7% in accuracy and 28% in macro F1 score compared to the baseline using the CNN model [26]. By using this adversarial learning technique, we expect to improve worker activity recognition in the manufacturing line for worker safety. For further details on the methods and experiments, please refer to [26] and [27].

4.2 Contrastive Learning

We tackle a common challenge in wearable HAR, where sensor locations that are commonly used for everyday wear provide inadequate information for accurate activity recognition. For instance, in our manufacturing work scenario, we found that sensors placed on the wrist and arm are not optimal for recognizing activities like walking, which would be better detected with an IMU sensor placed on the leg. This is a well-known limitation in HAR, where sensors deployed for long-term everyday use often result in poor or noisy data for the intended application.

To address this issue, we propose a method outlined in our work [9] that aims to improve the representation of the deployed sensors during training. The idea is to leverage additional sensors that are available only during the training phase to build a better representation of the target (deployed) sensor. This approach allows us to capture more relevant information about the activities being recognized, even if the

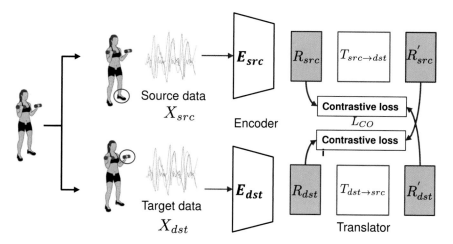

Fig. 4 Step 1: Training representations with paired sensor data by contrastive learning

target sensor alone may not provide sufficient information, we can improve it by guiding its representation with through contrastive learning with the source sensor.

In our proposed method, we collect temporally paired data from both the source and target sensors. The source sensor refers to a sensor that is available during training but not during deployment, while the target sensor is the sensor that will be deployed for activity recognition in real-world scenarios. Through contrastive learning, we learn a mapping between the representations of the source and target sensors. This process enables us to capture the relationship and similarities between the activities observed by the source and target sensors, enhancing the representation of the target sensor.

The training of representations and the mapping between them are depicted in Fig. 4. Each sensor's data are processed separately through deep neural network encoders to obtain their respective representations. The translation between representations is facilitated by translation networks, which learn to transform the source sensor representation to align with the target sensor representation. This contrastive learning step can be performed using unlabeled data, where the network learns to align the representations based on the temporal correspondence between the sensors.

Once the representations are learned, we proceed to the next step using labeled data from both sensors. In this step, we train our activity classifier using data either from the target sensor in its learned representation or from the source sensor by translating it to the target representation using the translation network. This joint training process allows the classifier to learn to recognize activities based on the enhanced representations from both sensors. The overall training process is illustrated in Fig. 5. For evaluation, we utilize only the target sensor data, as it represents the real-world deployment scenario. By applying the learned representations and the trained classifier to the target sensor data, we can more

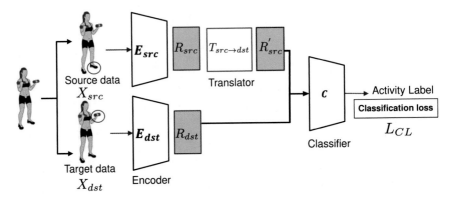

Fig. 5 Step 2: Training representations and classifier by minimizing classification loss

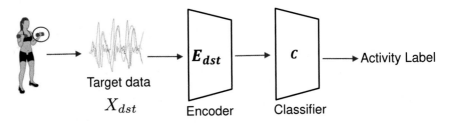

Fig. 6 Step 3: Testing with data of the target sensor only in order to evaluate the method

accurately recognize and classify activities in practical settings. The evaluation process is illustrated in Fig. 6.

We evaluated our method on two benchmark datasets for HAR: PAMAP2 and Opportunity. The results demonstrated significant improvements in activity recognition performance. We achieved an average macro F1 score increase ranging from 5 to 13 percentage points across all activities compared to traditional approaches that only rely on the target sensor. Notably, in specific scenarios where the source sensor provides highly informative data compared to the target sensor (e.g., recognizing walking with an ankle sensor as the source and a wrist sensor as the target), we observed even greater improvements, reaching up to 20 to 40 percentage points for certain activity classes.

This method using contrastive learning has important implications for real-world applications beyond IMU sensors. By using contrastive learning and knowledge transfer between sensors, our approach enables the development of more robust and accurate HAR systems. This has potential applications in various domains, including human–robot collaboration in manufacturing lines, where HAR plays a crucial role in improving productivity, quality assurance, and worker safety.

5 Conclusion

In this book chapter, we explored the role of wearable sensor-based HAR in promoting worker safety and optimizing productivity in manufacturing environments. We discussed the importance of worker safety and the potential benefits of using wearable sensors to monitor and recognize workers' activities on the manufacturing line. We presented a case study on wearable sensor-based worker activity recognition in a manufacturing line with a mobile robot, using sensor data fusion approaches and neural network models. By combining data from different sensor modalities, such as inertial sensors and body capacitance sensors, we were able to capture a comprehensive view of workers' activities and improve the accuracy of activity recognition. Furthermore, we introduced several deep learning-based techniques to enhance the performance of HAR, including adversarial learning and contrastive learning. These approaches have shown promise in improving activity recognition accuracy and generalization capabilities.

The use case in the SmartFactory Testbed demonstrated the practical application of sensor-based worker activity recognition in a real-world manufacturing environment. By accurately recognizing workers' activities and predicting their movement intentions, collision risks between workers and robots can be minimized, ensuring both worker safety and high production levels. Throughout this chapter, we discussed the potential benefits of sensor-based HAR beyond worker safety, including process optimization, ergonomics improvement, and worker training. We also highlighted the challenges and limitations associated with wearable sensor systems, such as privacy concerns and data security.

While our studies demonstrate the potential of wearable sensor-based HAR in manufacturing environments, we acknowledge that there are still challenges to overcome. These challenges include ensuring the robustness and reliability of sensor data, addressing issues related to real-time processing and inference, and managing privacy concerns in the collection and storage of personal data.

The results obtained from the adopted research can be applied in the practice of manufacturing companies by providing real-time insights into workers' activities. By accurately recognizing workers' activities and predicting their movement intentions, manufacturers can identify safety hazards, optimize processes, and enhance overall operational efficiency. The implementation of wearable sensor-based HAR systems can lead to improved worker safety, reduced workplace accidents, enhanced productivity, and more efficient resource allocation.

In conclusion, wearable sensor-based HAR holds significant potential for improving worker safety and productivity in manufacturing environments. By leveraging the wealth of data collected from wearable sensors and utilizing advanced machine learning techniques, manufacturers can gain real-time insights into workers' activities, identify safety hazards, optimize processes, and enhance overall operational efficiency. Future research in this field should focus on addressing the challenges and limitations of wearable sensor systems and exploring novel sensing modalities and machine learning approaches to further improve the performance of HAR in manufacturing environments.

Acknowledgments This work was supported by the European Union's Horizon 2020 program projects STAR under grant agreement number H2020-956573. Carl-Zeiss Stiftung also funded it under the Sustainable Embedded AI project (P2021-02-009).

References

1. Aehnelt, M., Gutzeit, E., Urban, B., et al.: Using activity recognition for the tracking of assembly processes: Challenges and requirements. WOAR **2014**, 12–21 (2014)
2. Al-Amin, M., Tao, W., Doell, D., Lingard, R., Yin, Z., Leu, M.C., Qin, R.: Action recognition in manufacturing assembly using multimodal sensor fusion. Procedia Manuf. **39**, 158–167 (2019)
3. Al-Amin, M., Qin, R., Tao, W., Doell, D., Lingard, R., Yin, Z., Leu, M.C.: Fusing and refining convolutional neural network models for assembly action recognition in smart manufacturing. Proc. IME C J. Mech. Eng. Sci. **236**(4), 2046–2059 (2022)
4. Bai, L., Yao, L., Wang, X., Kanhere, S.S., Guo, B., Yu, Z.: Adversarial multi-view networks for activity recognition. Proc. ACM Interact. Mobile Wearable Ubiquitous Technol. **4**(2), 1–22 (2020)
5. Bello, H., Zhou, B., Lukowicz, P.: Facial muscle activity recognition with reconfigurable differential stethoscope-microphones. Sensors **20**(17), 4904 (2020)
6. Bian, S., Lukowicz, P.: A systematic study of the influence of various user specific and environmental factors on wearable human body capacitance sensing. In: Body Area Networks. Smart IoT and Big Data for Intelligent Health Management: 16th EAI International Conference, BODYNETS 2021, Virtual Event, October 25–26, 2021, Proceedings, pp. 247–274. Springer, Berlin (2022)
7. Chen, L., Zhang, Y., Peng, L.: Metier: a deep multi-task learning based activity and user recognition model using wearable sensors. Proc. ACM Interact. Mobile Wearable Ubiquitous Technol. **4**(1), 1–18 (2020)
8. El Zaatari, S., Marei, M., Li, W., Usman, Z.: Cobot programming for collaborative industrial tasks: an overview. Robot. Auton. Syst. **116**, 162–180 (2019)
9. Fortes Rey, V., Suh, S., Lukowicz, P.: Learning from the best: contrastive representations learning across sensor locations for wearable activity recognition. In: Proceedings of the 2022 ACM International Symposium on Wearable Computers, pp. 28–32 (2022)
10. Global smartwatch shipments grow 9% yoy in 2022; price polarization seen in demand. https://www.counterpointresearch.com/global-smartwatch-shipments-grow-yoy-2022/, accessed: 2023-06-22
11. Goodfellow, I., Pouget-Abadie, J., Mirza, M., Xu, B., Warde-Farley, D., Ozair, S., Courville, A., Bengio, Y.: Generative adversarial nets. In: Advances in Neural Information Processing Systems, pp. 2672–2680 (2014)
12. Haresamudram, H., Essa, I., Plötz, T.: Contrastive predictive coding for human activity recognition. Proc. ACM Interact. Mobile Wearable Ubiquitous Technol. **5**(2), 1–26 (2021)
13. Janidarmian, M., Roshan Fekr, A., Radecka, K., Zilic, Z.: A comprehensive analysis on wearable acceleration sensors in human activity recognition. Sensors **17**(3), 529 (2017)
14. Kaasinen, E., Schmalfuß, F., Özturk, C., Aromaa, S., Boubekeur, M., Heilala, J., Heikkilä, P., Kuula, T., Liinasuo, M., Mach, S., et al.: Empowering and engaging industrial workers with operator 4.0 solutions. Comput. Ind. Eng. **139**, 105678 (2020)
15. Koskimaki, H., Huikari, V., Siirtola, P., Laurinen, P., Roning, J.: Activity recognition using a wrist-worn inertial measurement unit: a case study for industrial assembly lines. In: 2009 17th Mediterranean Conference on Control and Automation, pp. 401–405. IEEE, New York (2009)
16. Kwon, H., Abowd, G.D., Plötz, T.: Adding structural characteristics to distribution-based accelerometer representations for activity recognition using wearables. In: Proceedings of the 2018 ACM International Symposium on Wearable Computers, pp. 72–75 (2018)
17. Maekawa, T., Nakai, D., Ohara, K., Namioka, Y.: Toward practical factory activity recognition: unsupervised understanding of repetitive assembly work in a factory. In: Proceedings of the

2016 ACM International Joint Conference on Pervasive and Ubiquitous Computing, pp. 1088–1099 (2016)

18. Nakano, K., Chakraborty, B.: Effect of dynamic feature for human activity recognition using smartphone sensors. In: 2017 IEEE 8th International Conference on Awareness Science and Technology (iCAST), pp. 539–543. IEEE, New York (2017)

19. Pirkl, G., Hevesi, P., Cheng, J., Lukowicz, P.: mBeacon: accurate, robust proximity detection with smart phones and smart watches using low frequency modulated magnetic fields. In: Proceedings of the 10th EAI International Conference on Body Area Networks, pp. 186–191 (2015)

20. Pirsiavash, H., Ramanan, D.: Detecting activities of daily living in first-person camera views. In: 2012 IEEE Conference on Computer Vision and Pattern Recognition, pp. 2847–2854. IEEE, New York (2012)

21. Radziwon, A., Bilberg, A., Bogers, M., Madsen, E.S.: The smart factory: exploring adaptive and flexible manufacturing solutions. Procedia Eng. **69**, 1184–1190 (2014)

22. Roitberg, A., Somani, N., Perzylo, A., Rickert, M., Knoll, A.: Multimodal human activity recognition for industrial manufacturing processes in robotic workcells. In: Proceedings of the 2015 ACM on International Conference on Multimodal Interaction, pp. 259–266 (2015)

23. Rožanec, J.M., Novalija, I., Zajec, P., Kenda, K., Tavakoli Ghinani, H., Suh, S., Veliou, E., Papamartzivanos, D., Giannetsos, T., Menesidou, S.A., et al.: Human-centric artificial intelligence architecture for industry 5.0 applications. Int. J. Protein Res., **61**(20), 6847–6872 (2022). https://www.tandfonline.com/doi/full/10.1080/00207543.2022.2138611

24. Sanchez, M., Exposito, E., Aguilar, J.: Industry 4.0: survey from a system integration perspective. Int. J. Comput. Integr. Manuf. **33**(10–11), 1017–1041 (2020)

25. Sornam, M., Muthusubash, K., Vanitha, V.: A survey on image classification and activity recognition using deep convolutional neural network architecture. In: 2017 Ninth International Conference on Advanced Computing (ICoAC), pp. 121–126. IEEE, New York (2017)

26. Suh, S., Rey, V.F., Lukowicz, P.: Adversarial deep feature extraction network for user independent human activity recognition. In: 2022 IEEE International Conference on Pervasive Computing and Communications (PerCom), pp. 217–226. IEEE, New York (2022)

27. Suh, S., Rey, V.F., Lukowicz, P.: Tasked: transformer-based adversarial learning for human activity recognition using wearable sensors via self-knowledge distillation. Knowl.-Based Syst. **260**, 110143 (2023)

28. Tao, W., Lai, Z.H., Leu, M.C., Yin, Z.: Worker activity recognition in smart manufacturing using imu and semg signals with convolutional neural networks. Procedia Manuf. **26**, 1159–1166 (2018)

29. Wang, S., Wan, J., Li, D., Zhang, C.: Implementing smart factory of industrie 4.0: an outlook. Int. J. Distrib. Sens. Netw. **12**(1), 3159805 (2016)

30. Xu, X., Lu, Y., Vogel-Heuser, B., Wang, L.: Industry 4.0 and industry 5.0—inception, conception and perception. J. Manuf. Syst. **61**, 530–535 (2021)

Object Detection for Human–Robot Interaction and Worker Assistance Systems

Hooman Tavakoli, Sungho Suh, Snehal Walunj, Parsha Pahlevannejad, Christiane Plociennik, and Martin Ruskowski

1 Introduction: Why Object Detection in the Industrial Environment is Helpful?

The integration of Object Detection (OD) technology into industrial environments offers significant changes and improvements by addressing critical challenges and optimizing operations. By accurately detecting and recognizing objects in real time, OD systems play an important role in ensuring safety, streamlining workflows, and efficient assistance for humans. In today's complex industrial landscape, accurate OD is pivotal as it serves as the foundation for various safety mechanisms, such as identifying obstacles and hazardous materials, thereby reducing accidents and downtime.

Efficiency, productivity, and worker safety are paramount in industrial settings, making accurate object detection an essential component of worker assistance systems. Leveraging advanced algorithms, sensor fusion techniques, and machine learning methodologies, industries can achieve improved automation, enhanced human–robot interaction, and optimized processes. OD technology enables indus-

H. Tavakoli (✉) · S. Walunj · P. Pahlevannejad · C. Plociennik · M. Ruskowski
German Research Center for Artificial Intelligence (DFKI), Kaiserslautern, Germany

Technologie-Initiative SmartFactory, Kaiserslautern, Germany
e-mail: hooman.tavakoli_ghinani@dfki.de; snehal.walunj@dfki.de;
parsha.pahlevannejad_chaleshtori@dfki.de; christiane.plociennik@dfki.de;
martin.ruskowski@dfki.de

S. Suh
German Research Center for Artificial Intelligence (DFKI), Kaiserslautern, Germany

Department of Computer Science, RPTU Kaiserslautern-Landau, Kaiserslautern, Germany
e-mail: sungho.suh@dfki.de

© The Author(s) 2024
J. Soldatos (ed.), *Artificial Intelligence in Manufacturing*,
https://doi.org/10.1007/978-3-031-46452-2_18

trial modules and agents to perceive and analyze their surroundings, leading to optimized operations and safeguarding the well-being of workers.

By integrating OD technology with human assistance and collaboration mechanisms, safer and more productive interactions are fostered in industrial environments. Collaborative robots, or Cobots, are increasingly deployed to augment the capabilities of human operators and improve overall productivity. Leveraging object detection technology, Cobots accurately perceive and respond to the presence of humans, ensuring safe and seamless cooperation within shared workspaces. This prioritizes safety and productivity in human–robot interactions, driving innovation and efficiency in the industrial domain.

Assistance systems play an important role where manual work is prevalent. In industrial processes such as assembly, training, or maintenance processes, these systems support minimizing the workload of humans. There are different types of assistance systems, ranging from manual workstations equipped with cameras and displays to immersive assistance systems over head-mounted devices or smart devices. Computer vision techniques such as object detection help understand the worker's environment from visual data. This understanding can be used to enrich existing software systems with object information to achieve multiple goals. With applications ranging from healthcare to the automotive industry, object detection models such as Yolo [22] and Faster RCNN [7] have gained popularity on account of their real-time detection performance. For example, in [15], an Advanced Driver Assistance System (ADAS) is equipped with real-time object detection to provide safety and a better driving experience. In use cases such as ADAS systems encountering moving objects, real-time object detection becomes even more important. Object detection also finds applications for solving industrial problems such as quality inspection.

An Augmented Reality (AR)-based assistance system uses the context of the existing reality or real environment and intends to augment it with useful information. In order to capture the real-world context, camera sensors play an integral part in the AR system. They provide real-world data, and the display serves as the counterpart helping the user visualize the system together with augmented information. The input visual data are understood using deep learning methods such as object detection and pose estimation, the output of which can be used for solving a large spectrum of problems in the form of an assistance system.

Object detection outputs can be utilized to create effective assistance systems that provide real-time recommendations and guidance. By integrating OD with head-worn devices, workers can receive valuable information and instructions related to their tasks. For instance, an OD system can detect and recognize objects in the assembly place, or potential hazards in the worker's surroundings. The system can then analyze these data and provide recommendations, such as the next step in the assembly pipeline. It can also alert the worker to the presence of a hazardous object. These recommendations are displayed on the worker's head-worn device, providing immediate and personalized assistance. By leveraging OD in human assistance systems, workers can benefit from increased safety, improved efficiency, and enhanced situational awareness, ultimately leading to a more productive and secure work environment.

Supervised learning approaches such as object detection are highly dependent on data. A large amount of image data are required for training. Synthetically generated and labeled dataset offers several advantages in this regard. It simplifies and economizes the generation of large datasets, eliminating manual labeling and reducing human errors. Its flexibility allows for easy manipulation according to specific requirements, enabling researchers to control factors such as lighting, camera angles, and object placements. This enhances the model's ability to learn from diverse scenarios and improves its generalization capabilities.

Furthermore, synthetic data are well-suited for various computer vision tasks, including the detection of small or rare objects that may be challenging to capture in real-world datasets. Its ability to simulate objects of any scale or size makes it invaluable in training models. Synthetic data can be generated in large quantities, providing ample training samples for deep learning models that require extensive labeled data. Moreover, it allows for easy augmentation, increasing dataset diversity and variability. These advantages contribute to the development of more accurate and robust deep learning models in computer vision.

In the subsequent sections, we will delve into the following aspects:

1. Background: We will explore the utilization of object detection in industrial environments and its diverse range of applications.
2. Scenarios: We will discuss two pivotal scenarios within the industrial environment setting where object detection is employed for human–robot interaction and worker assistance systems. Moreover, the respective methodology for the worker assistance systems scenario is discussed, and the results are presented.
3. Ongoing and Future Work: We will highlight currently ongoing research that involves the exploration of continual learning techniques and dataset optimization through the combination of real and synthetic datasets.
4. Conclusion: The chapter will conclude by summarizing the key findings and contributions to the topic.

2 Background

The industrial environment is undergoing a profound transformation toward greater intelligence and autonomy, thanks to the emergence of machine learning (ML) approaches. With access to abundant data and powerful hardware resources, deep learning techniques and artificial intelligence methods have become increasingly valuable. In this context, leveraging AI-based methods within industrial settings offers significant potential for reducing human errors and enhancing safety [4], particularly in scenarios involving human–robot collaboration or shared workspaces with close proximity between these two crucial agents. The exploration of human–robot interaction in complex and unpredictable environments has become a prominent research area, where AI methods can be effectively harnessed [13]. Within this landscape, deep models serve as powerful tools that excel in addressing intricate challenges encountered in the industrial domain. Their ability to learn hierarchical representations directly from raw data positions them as ideal solutions for a wide range of applications, encompassing object detection, anomaly detection, predictive maintenance, quality control, and optimization.

Deep models, capable of addressing diverse challenges encountered in industrial settings, have emerged as highly effective tools. These encompass vision-based approaches [25], natural language processing (NLP) techniques [1], as well as human wearable sensor-based approaches for HRI [16], among others. Notably, both conventional and deep-learning-based ML models have recently been harnessed for video streaming analysis with the aim of object detection [6]. Deep models, characterized by their end-to-end nature, aim to address the challenge of laborious and time-consuming feature extraction from data [11].

Vision-based AI approaches have become increasingly valuable in addressing industrial problems due to their ability to interpret and analyze visual data in real time. These approaches leverage deep learning algorithms and computer vision techniques to process images or video streams captured by cameras or sensors installed in industrial settings. By employing vision-based AI, various industrial challenges can be effectively tackled. Hence, vision-based approaches are widely recognized for their significant value across diverse scenarios. One prominent vision-based approach is object detection, which involves the classification and localization of multiple objects in the target frames or images. It has proven to be highly applicable in various scenarios, contributing to tasks such as improving safety, facilitating human–robot interaction [25], aiding in error identification for workers, and optimizing task completion time.

Given the wide range of challenges and numerous common use cases encountered by vision-based AI approaches in industrial settings, it is imperative to thoroughly investigate various aspects of AI-based approaches, especially in industrial environments. These aspects encompass examining the architecture of deep learning models, conducting comprehensive analysis and assessment of the data sources, and addressing other significant factors to ensure optimal performance and effectiveness. In the following, we dig into some of these challenges and solutions.

2.1 Dataset

In the case of industrial scenarios, there is difficulty in terms of real-world data collection. Especially in vision problems, collecting images with diverse conditions and viewpoints, and also labeling them, is an effort-intensive as well as a time-consuming task. In some cases, the frames used as the source of input can originate from various cameras within the environment. These cameras can include ceiling-mounted cameras, robot cameras, as well as head-worn cameras of mixed reality devices like HoloLens that capture workers' point of view [24]. Also in some cases, the data collection cameras are different from the edge devices on which an object detection model needs to be deployed. In such a situation, synthetic data generation becomes inevitable. Also, CAD data play an eminent role in the complete product life cycle of a manufacturing product, and for the product as well as the machine, it is readily available. These CAD models can be exploited to generate synthetic datasets. However, using CAD data directly cannot serve as a reliable solution. The CAD data only resemble the real objects in geometry; however, they lack materials and texture. Thus, there exists a large amount of distinction in the appearance of the real and CAD data. If we use CAD-based synthetic images for training, which is to be tested on real-world objects, there is a problem of domain difference, i.e., the real and synthetic data domain. In an attempt to solve this issue, the technique of domain randomization is demonstrated in [19]. Domain randomization can be achieved by randomizing various aspects of the simulation scene such as the backgrounds, illumination, orientation of the objects, etc. [3].

2.2 Architectures

Numerous prominent object detection architectures, including the R-CNN pipeline [7], Fast-RCNN [8], and Faster-RCNN [9], SSD [12], You Only Look Once (YOLO) approaches [22], and its following versions such as YOLOv7 [28], provide well-defined pipelines for detecting objects across different scenarios. These architectures offer robust methodologies for accurately identifying and localizing objects within the visual data. State-of-the-art techniques, such as YOLO object detection, have particularly excelled in enabling real-time or near-real-time object detection. This capability is of utmost importance in industrial environments, where the ability to detect objects promptly is crucial for effective use cases. By leveraging these advanced object detection methods, industries can enhance their operational efficiency, safety, and decision-making processes.

2.3 Application in Industrial Environment

Object detection is highly practical and beneficial in various broad use cases. For instance, in [21], object detection pipelines were employed to automate logistic processes within industrial environments. Saeed et al. [23] addressed the challenge of detecting faults in industrial product images, particularly focusing on small-object detection. Another case study is conducted by Usamentiaga et al. [27] for evaluating the state-of-the-art deep-based object detection models as well as semantic segmentation in the scenario of automated surface inspection in metals. In the field of robotics, [2] researched the integration of an object detection CNN-based model to leverage a robot in a sorting task.

2.4 Challenges

It is obvious that object detection in the industrial environment can facilitate many tasks especially in which a robot and a human need to collaborate and work in a shared workplace, and promotes safety by facilitating interaction between humans and robots. However, object detection in the industrial environment does face some challenges. Object detection in a complex and unpredictable industrial environment, in which interference objects similar to the goal objects can be found easily due to the similarity in shape, size, and color, and random positioning and orientation of different objects make this detection much more difficult [2]. Also, detecting small objects poses a significant challenge due to their limited representation of features.

An overview of concepts and terminology related to artificial intelligence (AI) is given in ISO/IEC 22989:2022 Information Technology—Artificial intelligence—Artificial intelligence concepts and terminology" [10].

3 Scenarios

In this section, we explore the scenarios in both projects, STAR and InCoRAP for object detection in the factory environment with specific emphasis on safety in human–robot collaboration and interaction, and human assistance systems. We study the use cases in which object detection can be utilized for safety in the industrial environment. Additionally, we investigate the application of object detection in human assistance systems. Furthermore, we will explain our recent work on context-based object detection methods, particularly for small objects in the assembly use case within the industrial environment. We dive into the details of how context-based approaches can effectively detect and identify small objects in assembly scenarios.

3.1 Object Detection for Human–Robot Interaction in the STAR Project

Object detection for human detection can play a crucial role in enhancing safety in industrial environments. Although human detection can be considered as a sub-task of object detection, it is facing some more complexity due to the wide range of possible appearances on account of the articulated pose, and clothing, to name a few [20]. Hence, studying human detection is vital from a safety perspective, especially in a complex, unpredictable, and dangerous industrial environment. Here are a few ways human detection will be utilized:

- Enhancing Worker Safety: Implementing object detection systems that are capable of accurately identifying and tracking human presence can greatly contribute to worker safety. These systems enable the implementation of proactive safety measures to prevent accidents or potentially dangerous situations. For instance, if a worker enters a restricted area or approaches a hazardous machine, the object detection system can promptly detect their presence and trigger warnings that can be raised in the HoloLens head-worn device or automatically shut down the equipment to mitigate potential accidents.

– Collision Avoidance: Object detection can be utilized to detect the presence of humans in the vicinity of moving machinery or vehicles. This information can be used to alert operators or autonomous systems to slow down, change direction, or stop to avoid collisions and ensure worker safety.

Our pilot study aims to explore the practical application of ceiling cameras or robot cameras in hazardous accident scenarios occurring within an industrial environment. Our focus is specifically directed toward examining the collaboration between a moving robot and human workers. The moving robot's primary role is to assist in the transportation of objects from the warehouse to the workstation, as well as facilitating the transfer of objects between various production lines, which often occurs in an unpredictable timescale.

To ensure the safety of workers and prevent any potential accidents, we employ localization and classification from object detection techniques for both the robot and human agents. These techniques enable us to effectively localize the position of the robots and humans within the workspace through the ceiling cameras in the environment. By real-time monitoring and analyzing of the robot and human locations, we can promptly identify and mitigate potential collisions or unsafe situations, which leads to a secure and accident-free working environment.

Overall, object detection for human detection in industrial environments brings safety by enabling proactive measures and collision avoidance. It helps create a safer working environment, reduces the risk of accidents, and enhances the well-being of workers.

3.2 Object Detection for Manual Assembly Assistance System in InCoRAP

In the InCoRAP use case, the AR-based Assistance system observes the egocentric point of view of the worker. The goal is to observe worker activity in the assembly process in order to support them through a mobile robot collaboration. The object detection model is a part of the assistance used to observe the assembly state based on the detected objects. For AR applications, the head-mounted device: HoloLens2 is used. The detected objects are the observations that correspond to the assembly steps, and these observations are later used to support the worker.

Research focusing on the evaluation of AR systems has proven AR-based assistance systems advantageous over conventional instruction manuals [5]. Thus,

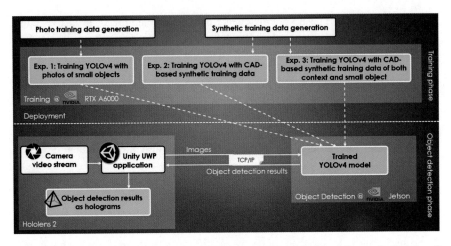

Fig. 1 The pipeline of the 3 experiments within the training phase with testing on the HoloLens2 frames transferred to the edge server for detecting target objects [25]

AR-based assistance places less workload on the worker or user compared to document-based guidance.

The field of small-object detection has garnered considerable research interest and has become increasingly popular. In vision-based object detection methods, the texture and arrangement of objects play a pivotal role in enabling the object detection pipeline to extract relevant features. It is worth noting that smaller objects may pose a challenge, as their size can be reduced significantly during the feature extraction process. For example, an object with dimensions of 32×32 pixels, after passing through five pooling layers in the VGG16 model, would be represented as a mere 1-pixel [17]. The concept of small objects, as defined in [26], encompasses objects sized at 32×32 pixels within the context of image analysis.

We specifically focus on small-object detection within an assembly scenario, where workers assemble various electrical components (e.g., buttons, resistors, LED, buzzers) on a breadboard to create a final product (Fig. 3b) [25]. In this scenario, a robot assists the worker by delivering parts from the warehouse, and an assistance system detects the current assembly steps and suggests the next probable part to be installed on the breadboard. The frames captured from the worker's point of view (POV) are seized using the HoloLens2, which the worker wears during the assembly process. The pipeline for this approach is illustrated in Fig. 1. In the testing phase, the frames are transmitted to the server where the object detection model performs inference. The detected objects provide information such as class identification and bounding box coordinates. This information is then communicated from the server to the HoloLens2 device using Unity communication. Subsequently, the HoloLens2 device generates holographic representations, displaying the class identification and bounding box information as augmented data. Workers in the environment can observe these holograms when they focus their gaze on the respective objects.

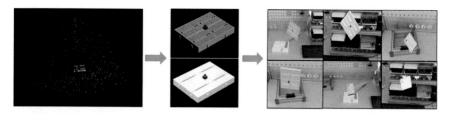

Fig. 2 Synthetic dataset generation pipeline [25]

3.3 Methodology: Context-Based Two-Step Object Detection

For computer vision problems such as image classification or object detection, the foremost task is the collection of suitable image data and then labeling it. However, this process can be automated using a game engine such as Unity [3]. In a Unity renderer space, there is a game camera that consists of the attributes of a physical camera. There are two types of views obtained in such a game development environment, one being the developer view and the other being the game view. These features allow for creating a simulation scene for synthetic image data generation. As introduced in [3], the Unity Perception allows for customization and user-defined feature development with the goal of synthetic data generation in Unity. We developed a scene similar to that of the Unity Perception Package that allows for multi-object detection dataset generation. The 3D models of the desired object are CAD models converted into Unity-compatible format and imported into the scene.

The synthetic dataset generation pipeline shown in Fig. 2 shows the CAD model of the object imported into a Unity scene. On importing into the scene, the CAD models are processed to achieve a photo-realistic appearance using the renderer features. The scene is developed in such a way that it simulates a systematic image-capturing process [3] to deliver the desired dataset. A script has been written for the game camera to capture the object images from various viewpoints and distances to the object during the simulation. From a set of background images, the backgrounds of the objects are randomized in the simulation. By rendering multiple views of the scenes, we generate a diverse set of synthetic images that mimic real-world conditions. Additionally, data augmentation techniques such as object rotation, scene illumination, and object occlusion are applied. Furthermore, the labeling tool within this Unity scene automatically annotates the generated images with bounding boxes, providing ground truth information for training and validation.

In Table 1, the benefits of using synthetic data were demonstrated in the context of installing small objects on a breadboard as an assembly process. Additionally, utilizing synthetic data presents advancements in the two-step detection approach. Table 1 illustrates the Mean Average Precision (mAP) for the detection of the small button on the breadboard for our three experiments. The mAP is a widely used evaluation metric in object detection tasks. It measures the accuracy and precision

Table 1 The mAP results for all 3 experiments for different Intersection of Union (IoU) are illustrated and confirmed that the two-step detection improved the mAP for the buttons, the target small object in the frames [25]

IoU	mAP (full size)						mAP (cropped)					
	0.01	0.10	0.20	0.30	0.40	0.50	0.01	0.10	0.20	0.30	0.40	0.50
Exp. 1	0.3%	0%	0%	0%	0%	0%	2.6%	0%	0%	0%	0%	0%
Exp. 2		44%	26%	4%	0.6%	0.03%						
Exp. 3		44%	26%	4%	0.6%	0.03%		70%	69%	58%	27%	8.5%

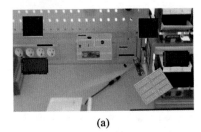 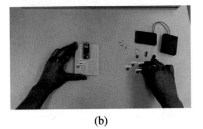

(a)　　　　　　　　　　　　　　　　　　　　　　(b)

Fig. 3 (**a**) Synthetic image for assembly objects, (**b**) object detection on a video captured from HoloLens2 for the corresponding real objects

of object detection algorithms by calculating the average precision for each class and then taking the mean across all classes. The mAP is calculated according to [18]. The first experiment is trained on the 295 conventional images from buttons and tested on the 90 images containing 221 tiny buttons. The same setup is applied in the 2nd experiment, but in the training phase, we utilized the 1300 synthetic images of the breadboard and 2500 images of tiny buttons installed on the breadboard. It is clear that the mAP for the second experiment is considerably higher than the first experiment, which is around 0% for almost all different IoUs. In the third experiment, we utilized the same training and testing data as in the second experiment. However, during the inference phase, we adopted a different approach. Instead of resizing the entire frame, we first detected the breadboard and then cropped it based on the context (specifically, the Breadboard). The resulting cropped frame was then forwarded to the YOLOv4 object detection pipeline to detect the buttons as small objects. Through the third experiment, we demonstrated how employing a context-based cropping approach on the frames leads to a significant improvement in mAP (Table 1). More details can be found in [25] (Fig. 3).

4　Ongoing Research

In this section, we present the current research endeavors conducted at our research center in the field of object detection. We will highlight the ongoing projects and studies that are directly aligned with our focus on advancing object detection techniques.

4.1 Hybrid Dataset

Considering the advantages that synthetic data offer with respect to reduced human effort and time consumption, it has gained research interest. However, it can encounter challenges related to optimizing various aspects of the scene, including CAD models, lighting conditions, backgrounds, and object textures. In our recent research, we are specifically investigating the utilization of real data in combination with synthetic data to enhance the precision of object detection. By leveraging real data in this manner, we aim to optimize the object detection performance and facilitate the creation of datasets that meet our specific requirements.

4.2 Continual Learning (CL)

An additional challenge in this chapter is the need to update previously trained object detection models to accommodate new tasks, rather than retraining the model from scratch with both old and new data. This process of continual learning aims to address the issue of catastrophic forgetting, where the model's performance on previous tasks significantly declines as it is trained on new tasks [29]. This presents an interesting avenue for further exploration within the field of study.

Both domain incremental learning and task incremental learning [14] offer potential research approaches that can be applied to our specific environmental scenarios. These methods enable the model to adapt to new tasks while retaining knowledge from previous tasks. Investigating and leveraging different continual learning techniques can contribute to the development of a more flexible and efficient object detection approach.

Figure 4 depicts the utilization of continual learning to optimize the object detection process using the synthetic data in our scenarios. The procedure begins by importing CAD models of new objects and subsequently generating synthetic image datasets. These datasets are then passed to the next phase, where the previously trained YOLOv7 model is retrained using the replay approach of continual object detection. Finally, in the evaluation phase, frames captured from HoloLens are processed by the new object detection model, and the performance of this model is assessed. Thus it is an iterative process, starting from importing synthetic data to dataset generation, model retraining, and evaluation phase, aiming to expand the range of OD classes by updating the previously trained model.

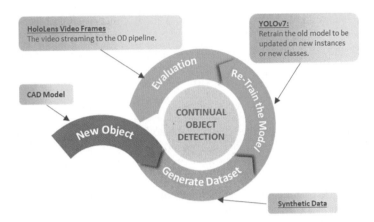

Fig. 4 The pipeline demonstrates the sequential process of generating synthetic data, conducting training and testing using the HoloLens2 World Camera, with the aim of establishing a continual learning pipeline for model updates

5 Conclusion

This chapter provides an exploration of object detection in industrial environments, specifically focusing on various scenarios involving human assistance systems, and safety requirements for human–robot collaboration. Additionally, we examine the application of object detection in conjunction with augmented reality devices, which offer intuitive communication interfaces for workers in these environments.

Moreover, we emphasize the substantial advantages gained from the incorporation of synthetic data in expediting the laborious data generation process and enhancing object detection outcomes. Through the utilization of synthetic data, we achieve improved efficiency and precision in object detection results. Additionally, we present our recent research focusing on the detection of small objects within industrial environments, which poses a significant challenge. We demonstrate that our approach significantly enhances the detection of small objects in manual assembly scenarios, resulting in notable improvements in performance.

Ultimately, this chapter not only sheds light on some solutions to object detection barriers in industrial settings but also paves the way for further exploration in related fields. This includes the development of sustainable practices within our scenarios and the establishment of a more generalized process that encompasses data generation and the preparation of object detection models.

Acknowledgments This work was supported by the European Union's Horizon 2020 program project STAR, under grant agreement numbers H2020-956573, and by the BMBF (German Federal Ministry of Education and Research) project InCoRAP (0IW19002).

References

1. Agnello, P., Ansaldi, S.M., Lenzi, E., Mongelluzzo, A., Roveri, M.: RECKONition: a NLP-based system for industrial accidents at work prevention. arXiv preprint arXiv:2104.14150 (2021)
2. BinYan, L., YanBo, W., ZhiHong, C., JiaYu, L., JunQin, L.: Object detection and robotic sorting system in complex industrial environment. In: 2017 Chinese Automation Congress (CAC), pp. 7277–7281 (2017). https://doi.org/10.1109/CAC.2017.8244092
3. Borkman, S., Crespi, A., Dhakad, S., Ganguly, S., Hogins, J., Jhang, Y.C., Kamalzadeh, M., Li, B., Leal, S., Parisi, P., et al.: Unity perception: generate synthetic data for computer vision. arXiv preprint arXiv:2107.04259 (2021)
4. Chen, J.H., Song, K.T.: Collision-free motion planning for human-robot collaborative safety under cartesian constraint. In: 2018 IEEE International Conference on Robotics and Automation (ICRA), pp. 4348–4354. IEEE, New York (2018)
5. Eversberg, L., Lambrecht, J.: Evaluating digital work instructions with augmented reality versus paper-based documents for manual, object-specific repair tasks in a case study with experienced workers. arXiv preprint arXiv:2301.07570 (2023)
6. Gallo, G., Di Rienzo, F., Ducange, P., Ferrari, V., Tognetti, A., Vallati, C.: A smart system for personal protective equipment detection in industrial environments based on deep learning. In: 2021 IEEE International Conference on Smart Computing (SMARTCOMP), pp. 222–227 (2021). https://doi.org/10.1109/SMARTCOMP52413.2021.00051
7. Girshick, R.: Fast R-CNN. In: Proceedings of the IEEE International Conference on Computer Vision, pp. 1440–1448 (2015)
8. Girshick, R.: Fast R-CNN. In: Proceedings of the IEEE International Conference on Computer Vision (ICCV) (2015)
9. Girshick, R., Donahue, J., Darrell, T., Malik, J.: Rich feature hierarchies for accurate object detection and semantic segmentation. In: Proceedings of the IEEE Conference on Computer Vision and Pattern Recognition, pp. 580–587 (2014)
10. ISO: Information technology—artificial intelligence—artificial intelligence concepts and terminology. Standard ISO/IEC 22989:2022, International Organization for Standardization (2022)
11. Krizhevsky, A., Sutskever, I., Hinton, G.E.: ImageNet classification with deep convolutional neural networks. Commun. ACM **60**(6), 84–90 (2017). https://doi.org/10.1145/3065386
12. Liu, W., Anguelov, D., Erhan, D., Szegedy, C., Reed, S., Fu, C.Y., Berg, A.C.: SSD: Single shot MultiBox detector. In: Computer Vision–ECCV 2016: 14th European Conference, Amsterdam, The Netherlands, October 11–14, 2016, Proceedings, Part I 14, pp. 21–37. Springer, Berlin (2016)
13. Liu, Z., Liu, Q., Xu, W., Liu, Z., Zhou, Z., Chen, J.: Deep learning-based human motion prediction considering context awareness for human-robot collaboration in manufacturing. Procedia CIRP **83**, 272–278 (2019). https://doi.org/https://doi.org/10.1016/j.procir.2019.04.080. https://www.sciencedirect.com/science/article/pii/S2212827119306948, 11th CIRP Conference on Industrial Product-Service Systems
14. Menezes, A.G., de Moura, G., Alves, C., de Carvalho, A.C.: Continual object detection: a review of definitions, strategies, and challenges. Neural Netw. (2023)
15. Murthy, J.S., Siddesh, G.M., Lai, W.C., Parameshachari, B.D., Patil, S.N., Hemalatha, K.L.: ObjectDetect: a real-time object detection framework for advanced driver assistant systems using YOLOv5. Wirel. Commun. Mob. Comput. **2022**, 10 (2022). https://doi.org/10.1155/2022/9444360
16. Neto, P., Simão, M., Mendes, N., Safeea, M.: Gesture-based human-robot interaction for human assistance in manufacturing. Int. J. Adv. Manuf. Technol. **101**, 119–135 (2019)
17. Nguyen, N.D., Do, T., Ngo, T.D., Le, D.D.: An evaluation of deep learning methods for small object detection. J. Electr. Comput. Eng. **2020**, 1–18 (2020)

18. Padilla, R., Passos, W.L., Dias, T.L., Netto, S.L., Da Silva, E.A.: A comparative analysis of object detection metrics with a companion open-source toolkit. Electronics **10**(3), 279 (2021)
19. Pasanisi, D., Rota, E., Ermidoro, M., Fasanotti, L.: On domain randomization for object detection in real industrial scenarios using synthetic images. Procedia Comput. Sci. **217**, 816–825 (2023). https://doi.org/https://doi.org/10.1016/j.procs.2022.12.278. https://www.sciencedirect.com/science/article/pii/S1877050922023560, 4th International Conference on Industry 4.0 and Smart Manufacturing
20. Paul, M., Haque, S.M., Chakraborty, S.: Human detection in surveillance videos and its applications-a review. EURASIP J. Adv. Signal Process. **2013**(1), 1–16 (2013)
21. Poss, C., Ibragimov, O., Indreswaran, A., Gutsche, N., Irrenhauser, T., Prueglmeier, M., Goehring, D.: Application of open source deep neural networks for object detection in industrial environments. In: 2018 17th IEEE International Conference on Machine Learning and Applications (ICMLA), pp. 231–236 (2018). https://doi.org/10.1109/ICMLA.2018.00041
22. Redmon, J., Divvala, S., Girshick, R., Farhadi, A.: You only look once: unified, real-time object detection. In: Proceedings of the IEEE Conference on Computer Vision and Pattern Recognition, pp. 779–788 (2016)
23. Saeed, F., Ahmed, M.J., Gul, M.J., Hong, K.J., Paul, A., Kavitha, M.S.: A robust approach for industrial small-object detection using an improved faster regional convolutional neural network. Sci. Rep. **11**(1), 23390 (2021)
24. Su, Y., Rambach, J., Minaskan, N., Lesur, P., Pagani, A., Stricker, D.: Deep multi-state object pose estimation for augmented reality assembly. In: 2019 IEEE International Symposium on Mixed and Augmented Reality Adjunct (ISMAR-Adjunct), pp. 222–227. IEEE, New York (2019)
25. Tavakoli, H., Walunj, S., Pahlevannejad, P., Plociennik, C., Ruskowski, M.: Small object detection for near real-time egocentric perception in a manual assembly scenario. arXiv preprint arXiv:2106.06403 (2021)
26. Torralba, A., Fergus, R., Freeman, W.T.: 80 million tiny images: a large data set for nonparametric object and scene recognition. IEEE Trans. Pattern Anal. Mach. Intell. **30**(11), 1958–1970 (2008)
27. Usamentiaga, R., Lema, D.G., Pedrayes, O.D., Garcia, D.F.: Automated surface defect detection in metals: a comparative review of object detection and semantic segmentation using deep learning. IEEE Trans. Ind. Appl. **58**(3), 4203–4213 (2022)
28. Wang, C.Y., Bochkovskiy, A., Liao, H.Y.M.: YOLOv7: trainable bag-of-freebies sets new state-of-the-art for real-time object detectors. arXiv preprint arXiv:2207.02696 (2022)
29. Wang, L., Zhang, X., Su, H., Zhu, J.: A comprehensive survey of continual learning: theory, method and application. arXiv preprint arXiv:2302.00487 (2023)

Boosting AutoML and XAI in Manufacturing: AI Model Generation Framework

Marta Barroso, Daniel Hinjos, Pablo A. Martin, Marta Gonzalez-Mallo, Victor Gimenez-Abalos, and Sergio Alvarez-Napagao

1 Introduction

The field of manufacturing is increasingly interested in adopting Artificial Intelligence (AI) due to its ability to revolutionize operations, improve efficiency, and drive innovation. AI algorithms can effectively analyze and extract valuable insights from data, enabling manufacturers to optimize processes, detect anomalies, and make data-driven decisions. In addition, they bring automation and predictive capabilities, by enabling the automation of tasks, such as quality control and predictive maintenance, leading to faster inspections, reduced downtime, and improved operational efficiency.

However, despite all these benefits, the adoption of AI from manufacturing side is not happening as quickly as expected [1], and it comes with its own set of challenges and difficulties. While it holds great promise, several challenges need to be addressed for successful integration into the manufacturing processes. These challenges include issues related to data availability and quality, integration with existing infrastructure, shortage of skilled personnel, ethical and regulatory considerations, and change management [2]. Manufacturing environments often entail complex data ecosystems, requiring proper data collection and preparation processes to ensure the availability of high-quality and relevant data usable to produce intelligent models. Furthermore, integrating these models with existing infrastructure may require system upgrades and compatibility assurance. The shortage of skilled professionals with simultaneous expertise in both AI and manufacturing processes hinders adoption, while ethical and regulatory concerns

M. Barroso (✉) · D. Hinjos · P. A. Martin · M. Gonzalez · V. Gimenez · S. Alvarez-Napagao
High Performance and Artificial Intelligence, Barcelona Supercomputing Center, Barcelona, Spain
e-mail: marta.barroso@bsc.es; daniel.hinjos@bsc.es; pablo.martin@bsc.es; marta.gonzalez@bsc.es; victor.gimenez@bsc.es; sergio.alvarez@bsc.es

© The Author(s) 2024
J. Soldatos (ed.), *Artificial Intelligence in Manufacturing*,
https://doi.org/10.1007/978-3-031-46452-2_19

require robust governance policies. Change management is also crucial, requiring cultural shifts and addressing employee concerns to foster a positive AI adoption environment.

To overcome these challenges, AI providers should also facilitate adoption through the development of user-centered tools, fostering partnerships between experts and manufacturing professionals, and showcasing the tangible benefits of implementing machine and deep learning algorithms. By making AI accessible, tailored, and demonstrably valuable, the manufacturing industry can overcome barriers and embrace these technologies to unlock their full potential for improving operational efficiency, productivity, and competitiveness.

In this matter, there are two research areas that can play significant roles in nurturing this adoption: Automated Machine Learning (AutoML) [3] and Explainable Artificial Intelligence (XAI) [4].[1] AutoML simplifies the process of developing models by automating tasks such as feature engineering, model selection, and hyperparameter tuning. This automation reduces the need for extensive data science expertise, making it easier for manufacturing professionals to leverage AI capabilities. By providing user-friendly interfaces, pre-built algorithms, and automated workflows, AutoML tools enable manufacturers to quickly and efficiently build accurate and robust intelligent models tailored to their specific requirements.

On the other hand, XAI addresses the critical need for transparency, interpretability,[2] and trust in AI systems. Manufacturing environments require clear explanations of model decisions, especially when it comes down to critical factors such as quality control, predictive maintenance, and process optimization. XAI techniques allow manufacturers to understand the inner workings of models, identify the factors influencing predictions, and detect potential biases or risks. By providing interpretability and explanations, XAI helps build trust in AI systems and facilitates their adoption by manufacturing professionals.

Currently, there are some large-scale proprietary solutions that aim to tackle the aforementioned issues by providing users with end-to-end solutions that operationalize the AI cycle with very limited prior technical knowledge. The most popular comes from big tech companies, e.g., Google AutoML-Zero [5], Microsoft Azure AI [6], Amazon SageMaker [7], and the open-source H2O [8].

Although using this type of platforms has been shown to be beneficial [9], they bring about some drawbacks that organizations should seriously consider. One major setback is the potential lack of transparency and control over the underlying algorithms and models. These platforms often abstract away the details of the machine learning process, making it difficult to understand and interpret how the models arrive at their predictions. This lack of transparency can be a concern,

---

[1] Explainability (XAI) focuses on providing meaningful and transparent explanations for the decisions or outputs of an AI system.

[2] Interpretability is concerned with understanding and making sense of how a model or algorithm operates. It focuses on the internal workings of the model, including the relationships between inputs and outputs, feature importance, or the overall behavior of the model.

especially in regulated industries or situations where interpretability is crucial for decision-making.

Another inconvenience is that in many cases using these tools requires contracting their providers' services and infrastructure. The user is then subject to the limitations and potential downtime of third-party software and hardware. Unexpected disruptions or system issues are out of the control of the user and can impact the availability and performance of AutoML platforms. Additionally, organizations may be locked into specific pricing models and contracts with the cloud provider, limiting flexibility and potentially leading to increased costs in the long run.

Data privacy and security are also important considerations when using AutoML platforms in Cloud. Organizations need to ensure that sensitive or proprietary data used for training models are protected and handled in compliance with relevant regulations. The transfer of data to the cloud, data storage, and data access controls should all be carefully evaluated to mitigate any risks associated with data privacy and security breaches.

Finally, we should consider *vendor lock-in* issues. Once an organization builds and deploys models on a specific platform, migrating those models to another platform or bringing the infrastructure in-house can be challenging and time-consuming. This can limit flexibility and hinder the ability to switch providers or adapt to changing business needs in the future.

To tackle previous obstacles and motivate the use of AI in manufacturing, we introduce our framework called the AI Model Generation.

This framework is part of an European project called knowlEdge—Toward AI-powered manufacturing services, processes, and products in an edge-to-cloud-knowlEdge continuum for humans in-the-loop and is intended to be used by people with no background in AI, including manufacturing engineers, plant managers, quality control specialists, and supply chain managers. Throughout the chapter, we will refer to it as AMG. AMG is responsible for the automatic creation of supervised AI models and is able to solve tasks based on various scenarios and input variables. Each of these stages constitutes a submodule by itself. After describing the main functionalities of the system (Sect. 2), we then describe the architecture (Sect. 3), the use cases (Sect. 4), and we describe each of the submodules that make it up (Sect. 5).

2 AI Model Generation Framework

This component is able to automatically generate AI models capable of solving user-defined tasks based on an initial configuration in which the data source, the type of problem, and the algorithm are specified. In addition, it enables the computation of the algorithms' training costs using a set of heuristics. Models can be efficiently deployed in different layers of the computer continuum (cloud, fog, and edge) and be saved using standards such as ONNX (Open Neural Network Exchange) and PMML (Predictive Model Markup Language).[3]

The initial configuration is defined in terms of data source, task, task setup, and strategy. Any dataset is made up of a number of variables and can come from three different data sources (local, static database, broker). See Sect. 5.1 for more information on data sources. The problem to be solved is formalized in a task object. A task has a name, a type (classification, regression, or optimization), an associated performance metric, an optional risk function, a set of input and output variables, and one or more associated execution settings. The configuration of a task is formalized in a task setup object, which contains information about the type of validation, training, and evaluation datasets, random seeds, and implemented strategies. A task configured with a task setup can train or run inference using one or more algorithms. The algorithm and the hyperparameters to use are encapsulated in an object called strategy. A strategy consists of the method name, the hyperparameters, the initial state, and the loss function of a given algorithm. As a result, models are generated, and a model is associated with a strategy, a set of metrics describing its performance, and the Docker image tag that deployed it.

On the top of that, the user can infer existing models or train new models by defining an initial configuration in JSON format. The next code 2 is an example of configuration file that uses a local dataset to perform training and inference.

```
{
    "task": "classification",
    "task_name": "mushrroom_classification",
    "method": {
        "strategy_list": ["randomForestClassifier"],
        "arguments": {
            "validation_type": "SPLIT",
            "validation_percentage": 0.2,
            "random_seed": 24,
            "risk_function": "risk_function",
            "performance_metric": "accuracy"
        }
    },
    "processing": {
        "arguments": {
            "dataset_name": ""
        },
        "orders": [
            {
                "order": 1,
                "action": "train",
```

---

[3] PMML and ONNX are standardized file formats that facilitate the interchangeability and interoperability of machine learning models between different software tools and platforms.

```
                        "read": {
                            "url": "~/home/datasets/mushrooms
                                .csv",
                            "type": "static",
                            "source_type": "tabular",
                            "connector": {
                                "name": "local",
                                "arguments": {}
                            },
                            "input_attributes": [],
                            "target_attributes": ["class"],
                            "from_i": 0,
                            "to_i": 0
                        }
                },
                {
                        "order": 2,
                        "action": "predict",
                        "read": {
                            "url": "~/home/datasets/mushrooms
                                .csv",
                            "type": "static",
                            "source_type": "tabular",
                            "connector": {
                                "name": "local",
                                "arguments": {}
                            },
                            "input_attributes": [],
                            "target_attributes": ["class"],
                            "from_i": 0,
                            "to_i": 0
                        }
                }
                ]
        },
        "modelrepo": {
            "url": "~/home/model_descriptors/"
        }
}
```

To specify what kind of operations we want to perform, we create different orders. An order can perform training or inference. We can also specify the columns to use, or leave the field blank if we want to use all of them. At the end of the process, the model output is provided in a different JSON and converted into a standard for future reuse. This file is stored in the directory specified in modelrepo.

3 System Architecture

At the structural level, the component is divided into the following elements depicted in Fig. 1:

- Python RESTful API: It is implemented through a Flask web service. Allows the user to make requests about existing models, tasks, task configurations, and strategies.
- Core functionalities module: It is responsible for implementing the functionalities of the component. In turn, each step of the AI cycle is implemented as a submodule.
- PostgreSQL Database: Relational database that stores information about datasets, tasks, task configurations, strategies, and models in order to reproduce results and keep a history of the models' evolution.
- Redis and Celery: Celery is a popular asynchronous task queue library in Python that allows for the distribution and execution of tasks across multiple workers. Redis is used as the result backend in Celery, which means it stores the results of completed tasks. After a task is executed by a Celery worker, the result is stored in Redis for retrieval by the Celery client.

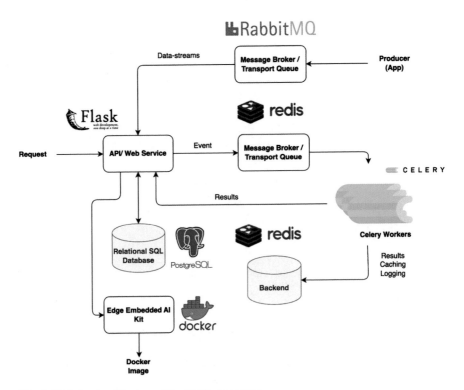

Fig. 1 High-level architecture of the AMG component

- RabbitMQ broker: It is a robust and feature-rich message broker that implements the Advanced Message Queuing Protocol (AMQP). After configuring the broker, several consumers subscribed to a queue can receive asynchronous messages. In order to establish connection to a specific queue, the configuration of the broker must be set.
- Edge Embedded AI Kit: This component is responsible for the deployment of models. All the models are containerized to enhance portability and scalability allowing their deployment across different platforms, such as local machines, cloud infrastructure, or edge devices.

4 Use Cases

In order to validate the component, several use cases have been defined for three pilot partners participating in the project:

- Dairy products company: A constraint optimization model has been implemented for production scheduling. They are interested in predicting the orders to be produced in a week window taking into account working machines, current orders, and product–machine compatibility.
- Manufacturing plastic fuel company: Several anomaly detection models are currently under testing. The goal is to predict anomalies in the production chain using manufacturing and quality data.
- Power transmission and drives company: Using an image dataset, they want to improve the quality of their assembly procedure automatic quality controls and thus reducing the rate of possible failures. Several defect detection models based on CNNs have been implemented.

Before testing these models in customer environments, we have the test machine in the LINKS[4] infrastructure in order to test these components and to correctly estimate the host specifications.

In addition, several configurations have been defined for the deployment of the component along the continuum depending on the pilot needs. We find two kinds of deployments:

- Training cloud-based/inference fog-based: In this scenario, AI models are trained in the cloud infrastructure of the client. This means that the training process, which often involves computationally intensive tasks such as processing large datasets and training complex models, is handled in their remote cloud environment. Once the model is trained, it is deployed and executed on fog devices or edge nodes for real-time inference in the manufacturing environment. The

---

[4] LINKS Foundation operates at the heart of the Turin research and innovation ecosystem, in a solid international network. Its target is to contribute to technological and socio-economic progress through advanced processes of applied research.

fog nodes, being closer to the edge devices, are responsible for performing the inference tasks on local data. This deployment scenario offers several advantages. First, cloud-based training allows for the efficient utilization of computational resources and can handle large-scale datasets and complex model architectures. It provides flexibility in terms of accessing various tools, libraries, and computing power required for model development and training. Additionally, centralized management in the cloud simplifies the process of model training, deployment, and updates. This deployment scenario offers several advantages. First, cloud-based training allows for the efficient utilization of computational resources and can handle large-scale datasets and complex model architectures. It provides flexibility in terms of accessing various tools, libraries, and computing power required for model development and training. Additionally, centralized management in the cloud simplifies the process of model training, deployment, and updates. However, there are some drawbacks to this deployment scenario. The latency introduced by transmitting data from fog devices to the cloud for training and then back to the edge for inference may not be suitable for real-time or time-sensitive applications. It also relies on reliable and high-bandwidth network connectivity between the edge/fog devices and the cloud, which may not always be available or practical. Furthermore, fog devices may have limited offline capabilities, as they may not be able to access the latest models or perform updates if they are disconnected from the cloud.

- Training and inference in fog: Both training and inference occur at the fog or edge devices themselves. Fog nodes have sufficient computational resources to handle training tasks, and the trained models are directly executed on the same devices. This scenario offers low latency as data processing and decision-making happen locally, without the need to communicate with the cloud. It also provides offline capability, making it suitable for environments with intermittent connectivity. Furthermore, fog-based deployment enhances privacy and security as sensitive data remain on the local devices, reducing the need to transmit it to external servers. However, the limited computational resources of fog devices can pose challenges for complex and large-scale training tasks. Managing and updating models across a distributed network of edge devices can also be more complex compared to cloud-based deployment. For this reason, techniques such as split learning along with the application of privacy-preserving encryption methods are advisable. By leveraging split learning, the fog devices can benefit from more efficient utilization of their limited resources. The data remain on the edge device, minimizing the need for data transmission and reducing latency. Only the model updates or gradients are transferred between the edge device and the server, significantly reducing the bandwidth requirements. This approach enables fog devices to participate in the training process without being overwhelmed by the computational demands.

5 Core Components

The overall logic of model generation can be broken down into the various steps that make up the AI life cycle: data retrieval model (data acquisition), automatic pre-processing module, cost computation module (estimation of training cost for a specific algorithm), automatic hyperparameter tuning module, automatic training, inference and standardization, explainability module (generation of local and global explanations), pipeline execution module (constitutes the main program that calls the rest of modules), and edge embedded AI kit (builds Docker images for model deployment). The implementation of the individual submodules is described in detail in the following sections.

5.1 *Data Retrieval Module*

Data retrieval stage involves the acquisition and collection of relevant data required for training and testing AI models. It is an essential step as the quality and comprehensiveness of the data directly impact the performance and effectiveness of the AI system. Typically, this process in turn includes: data identification, data collection, and data storage. At this point, we assume that the above processes have been performed and the data are ready to be consumed. Accepted data types include tabular data, images, or time series. Different types of connectors are offered to allow the user to upload their data:

- Local data connector: This allows files to be uploaded from the user's local file system. Files in csv format or training and evaluation directories for image datasets are taken into account. However, it is not mandatory that image directories be pre-partitioned for training, evaluation, and optional validation. If only one directory is uploaded, the component takes care of the partitioning (note that the data are split 70% training, 10% validation, and 20% evaluation).
- Broker connector: This connector is used to get real-time data from other databases, APIs, or sensors through a RabbitMQ broker. The user must format the data into an appropriate MQTT message format, such as JSON or plain text. Then publish the formatted data to the appropriate topic on the RabbitMQ broker. Later, the MQTT clients subscribed to this topic can receive the published data. Only certain topics are currently considered, but the component is easily scalable to add new ones.
- Connector for Edge/Cloud Apache Database: This connector has support for retrieving data from the Apache IoTDB (Database for Internet of Things), an IoT native database with high performance for data management and analysis, deployable on the edge and the cloud. Because of its lightweight architecture, high performance, and rich feature set, as well as deep integration with Apache Hadoop, Spark, and Flink, Apache IoTDB can meet the needs of massive data storage, high-speed data ingestion, and complex data analysis in IoT industrial

fields. Using the connection details, i.e., IP address, port and login information, data can be easily queried in the form of SQL statements.

Once the data are collected, it often requires pre-processing to ensure its quality and usability. This process involves tasks such as removing noise, normalizing, or standardizing variables. Other tasks such as formatting issues, fixing inconsistencies and outliers, and missing value analysis are not performed in this component because the management of missing data and outliers is highly user-dependent and has been handled by other components.

5.2 *Automatic Pre-processing Module*

Overall, the pre-processing stage aims to transform raw data into a clean, structured, and optimized form that is suitable for analysis by AI algorithms. As mentioned above, while data cleaning is out of our scope, our focus is put on data transformation, feature selection, feature engineering, and class imbalance handling.

It should be noted that pre-processing is highly dependent on the type of data you are working with. For tabular data, you can choose between normalization and standardization. In addition, label encoding is also applied. For feature selection and engineering, we use the AutoGluon [10] framework, specifically the AutoMLPipelineFeatureGenerator. This pipeline is able to handle most tabular data including text and dates adequately.

Pre-processing for image datasets includes, but is not limited to, data augmentation, label encoding, normalization, and rescaling. Since these processes depend on the neural network model to be used, they are only executed after the model has been defined, i.e., before training. This process was implemented using Keras framework. Specifically, it is worth noting that all Keras models with predefined architectures have an in-built method called `preprocess_input`, which is responsible for pre-processing a tensor or NumPy array representing a stack of images into the right format for the corresponding model. On top of that, data augmentation parameters are defined by the HyperImageAugment class implemented by the KerasTuner.[5]

Regarding time series data, we have generalized the steps to scaling and data transformation for stationarity conversion. This last step is really important since working with stationary data in time series analysis simplifies the modeling process, enhances interpretability, and ensures reliable and accurate analysis. Time stationarity is tested by applying the adfuller test for each feature. If data are non-stationary, the following transformations are considered: differencing, detrending, moving average, and power transformation. In order to ensure that the previous transformations can be applied, autocorrelation, non-zero mean checking, seasonality, and heteroscedasticity (Breusch–Pagan Lagrange Multiplier test) are tested

---

[5] KerasTuner is an open-source Python library that provides a user-friendly and efficient API for hyperparameter tuning of machine learning models built with Keras and TensorFlow.

for each feature. Statistical tests are imported from the stattools module from statsmodels.[6]

To overcome class imbalance issues when dealing with tabular datasets, we use a combination of over- and under-sampling methods implemented by the library imbalanced-learn [11]: SMOTETomek and SMOTEENN [12].

5.3 Cost Computation Module

This module is responsible for computing the cost of training a model and ensures that executed AI models are running with the desired behavior and performance, providing enough data to expose inefficiencies or poorly managed resources. The cost is computed using the performance and runtime history of previous models trained on the same algorithm. For this purpose, the execution of a number of representative models of the different algorithms was analyzed manually using tools such as Extrae, Paraver, and Dimemas, commonly used in the field of parallel computing and performance analysis. The scalability of each model has been analyzed on four machines with different architectures (MN4,[7] CTE-ARM,[8] CTE-AMD,[9] CTE-Power[10]), and Barcelona Supercomputing Center infrastructure.[11] The overall process is depicted in Fig. 2.

5.4 Automatic Hyperparameter Tuning Module

The Automatic Hyperparameter Tuning module provides automated methods for finding the best combination of hyperparameters that optimize the performance of machine learning and deep learning models. Hyperparameters are configuration settings that cannot be learned directly from the training data, such as the learning rate, regularization strength, or the number of hidden layers in a neural network, that are involved in the training process. They are not learned from the data but are set

---

[6] Stattools is a Python package that provides a collection of statistical tools and functions specifically designed for time series analysis and econometric modeling.

[7] MN4 is composed by 2 sockets Intel Xeon Platinum 8160 CPU with 24 cores each @ 2.10 GHz for a total of 48 cores per node 96GB of main memory, 2GB/core, 12x 8GB 2667MHz DIMM.

[8] CTE-ARM is composed by A64FX CPU @ 2.20GHz, 32GB of main memory HBM2, 0.7GB/core.

[9] CTE-AMD is composed by AMD EPYC 7742 @ 2.25GHz, 1024GB of main memory distributed in 16 dimms x 64GB @ 3200MHz, 8GB/core.

[10] CTE-Power is composed by 2x IBM Power 8335-GTH @ 2.4GHz (3.0GHz on turbo), 512GB of main memory distributed in 16 dimms x 32GB @ 2666MHz, 3.2GB/core.

[11] The Barcelona Supercomputing Center (BSC) is a research center located in Barcelona, Spain, dedicated to high-performance computing (HPC) and advanced scientific research.

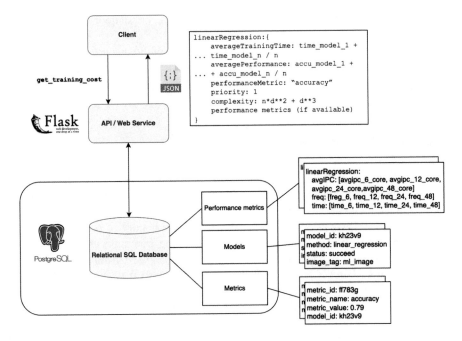

Fig. 2 Overview of the cost computation process

by the practitioner or determined through a process called hyperparameter tuning or optimization. However, determining their optimal values can be a time-consuming and computationally expensive process and requires extensive experimentation and testing. AutoML algorithms for hyperparameter tuning automate this process by automatically searching through a predefined range of hyperparameters and selecting the best combination based on performance metrics, such as accuracy or loss.

In case of machine learning algorithms, hyperparameter tuning is performed by RayTune library [13]. It allows to tune several machine learning frameworks (PyTorch, XGBoost, Scikit-Learn, TensorFlow, and Keras, etc.) by running state-of-the-art search algorithms such as population-based training (PBT) [14] and HyperBand [15]/ASHA [16]. In addition, RayTune further integrates with a wide range of additional hyperparameter optimization tools. However, it is worth noting that not all optimizers are available for all frameworks and image datasets must be treated as a numpy array without allowing the use of TF.data.Dataset[12] or other dynamic objects.

---

[12] TF.data.Dataset is a high-level interface for efficient and flexible data loading, pre-processing, and manipulation, making it easier to handle large datasets and build scalable machine learning pipelines with TensorFlow.

The machine learning models included in the component are implemented using the scikit-learn [17] library or similar libraries such as lightgbm [18] and XGBoost [19], reason for which we chose the Tune.Sklearn. According to the type of search, it implements: TuneGridSearchCV (grid search) and TuneSearchCV (random search). The latter with Bayesian search method was preferable to grid search because of its advantages [20]. Bayesian optimization is a method for hyperparameter optimization that uses Bayesian inference to efficiently search for the optimal set of hyperparameters. It is particularly useful when the evaluation of the objective function (e.g., model performance) is time-consuming or computationally expensive. After the search algorithm has been defined, the model and the parameter grid with the value range of the hyperparameters to be examined must be passed to the tuner along with other optional parameters, i.e., n_trials (the number of parameter settings tried by the tuner). All machine learning models are encapsulated in a class called `SklearnModelBuilder`. See Sect. 5.5 for more details. The best set of hyperparameters can be retrieved after fitting the tuner with the call `tuner.best_params_`.

In the case of deep learning models, the difficulty of passing complex data to the Ray tuner led us to use tuners specifically designed for hyperparameter tuning of neural networks, that is, the KerasTuner mentioned above. The tuner manages the hyperparameter search process, including model creation, training, and evaluation. Keras Tuner provides different kind of tunners based on the optimization strategy that is used to select hyperparameters. In addition, some tuners can be combined or aggregated to leverage the strengths of each. In our specific case, we define an instance of `BayesianOptimization` tuner with the metric to be optimized during (called objective) and the seed. At this point, we will also define the hyperparameters and their parameter space to carry out the search, by means of `KerasTunerHyperparameters` class. Additionally, we modify the tuner training flow so that we can also fine-tune training-associated hyperparameters such as batch size or epochs. In this flow, the hyperparameters for data augmentation, as implemented by the KerasTuner `HyperImageAugment` class mentioned before, can also be tuned. After calling the search method, the best selection of hyperparameters is retrieved with the call `get_best_hyperparameters`. Furthermore, it is also possible to obtain the trained model that maximizes the objective using `get_best_models` call.

In the end, both tuners are encapsulated in two different methods of the `HyperparameterTuner` class: `tune_ml` and `tune_dl`.

5.5 Automatic Training, Inference, and Standardization

This module builds a model that can make accurate predictions or classifications based on pre-processed data and the optimal choice of hyperparameters. Once a model has been trained, it can be deployed to a production environment where it can process new data and generate predictions or classifications. Models are

M. Barroso et al.

interoperable, and all of them have been wrapped in the convention of Sklearn objects to enable interoperability. Once a model has been trained, it can be directly converted into ONNX or PMML. PMML and ONNX are both model interchange formats that enable the portability and interoperability of machine learning models across different platforms and frameworks. They serve as standard representations for sharing, deploying, and executing machine learning models. PMML is a more general-purpose model interchange format that supports a broader range of predictive models beyond deep learning. It is used for various machine learning algorithms and techniques. ONNX, on the other hand, focuses specifically on deep learning models and their interchange between frameworks.

All machine learning models are instantiated within the method `_build_model` inside the `SklearnModelBuilder` class. This method passes the best selection of hyperparameters to the model. In order to perform any operation to a machine learning model, an instance of `SklearnModelBuilder` is created. A `SklearnModelBuilder` object can execute the following public methods:

- `fit`: Trains the algorithm using the training dataset. The builder already has defined the best selection of hyperparameters and the name of the algorithm to be trained.
- `predict`: Executes inferences process on evaluation dataset.
- `export`: Saves the model into the corresponding standard.
- `load`: Loads the model from the filename passed as argument. The filename must be a PMML or ONNX file.
- `explain_model`: Provides insights and explanations behind the model predictions or actions, enabling users to trust, validate, and effectively use the model properly.
- `explain_instance`: Focuses on explaining individual predictions or decisions made by the model, providing insights into which features or factors contributed most to a particular output. More information can be found in this matter in Sect. 5.6.

Deep learning models are implemented following two approaches. In case the user wants to perform hyperparameter tuning, the training is done according to the process described above. On the contrary, when the set of hyperparameters is initially clear, the system creates an object of the `KerasModelBuilder` class. All objects of this class have the same format as the objects of the `SklearnModelBuilder` class. Both classes inherit from the abstract `ModelBuilder` class. This allows us to keep a more homogeneous format while making the code more intuitive for the user.

5.6   Explainability Generation Module

Toward the end of the pipeline, we find the application of Explainable Artificial Intelligence (XAI). Its application is of utmost importance as it addresses key challenges in AI adoption. XAI enhances transparency, fostering trust between users and AI systems by providing insights into the decision-making process. It promotes ethical practices and accountability, ensuring that AI models operate with fairness and avoid biases or discriminatory outcomes. Furthermore, XAI supports compliance with regulatory standards that demand interpretability and explainability. Despite its significance, XAI is often skipped in the AI pipeline due to several factors. These include the complexity of implementing XAI techniques, the focus on achieving high performance without considering interpretability, and the trade-off between model complexity and explainability. Additionally, limited awareness and understanding of XAI among practitioners, along with the perception that explainability compromises predictive accuracy, may contribute to its omission.

To overcome these obstacles and motivate its application, explainability is embedded in the model. So it can be run once the model has been trained by calling the methods previously introduced: `explain_model` and `explain_instance`. The XAI methods are implemented by the library Dalex [21].[13] All the explanations are model-agnostic that allow us to provide information of these models without relying on their internal structure. As an example, the Dalex explainer generates, in a JSON or png format, feature importance analysis and visualizations such as partial dependence plots [22] and accumulated local effects (ALE) [22]. These visualizations offer intuitive representations of the model's behavior and enable users to explore the relationships between input features and the model's predictions. In order to explain concrete predictions, the explainer generates visualizations of the SHAP values [23], the ceteris paribus profile, and interactive breakdown plots. Although the number of XAI methods applied may seem sufficient because of the possibilities that Dalex offers, we are considering expanding the explainer further.

5.7   Pipeline Execution Module

The above steps are part of the so-called AI life cycle or AI pipeline. To deploy the model later, all these steps are encapsulated in the `execute_pipeline` method. This method is responsible for receiving the input configuration and creating the appropriate objects to perform data loading, pre-processing, hyperparameter optimization (optional), training and/or evaluation, and explainability (optional).

---

[13] Dalex is a Python library designed for model-agnostic explainability in machine learning. It provides a comprehensive set of tools and visualizations to understand and interpret the behavior of machine learning models.

As a result of executing the pipeline, a Docker image responsible for deploying the model is generated if it did not previously exist. Afterward, using the image, the container can be created to execute the pipeline by passing it the input configuration. In turn, the pipeline returns a JSON file with the execution result.

5.8 Edge Embedded AI Kit

This module is in charge of the model deployment. In this stage, it is assumed that the AI model is available and operational for use in a production environment. After developing and training a model, model deployment involves implementing the model into a system where it can generate predictions or perform specific tasks in real time.

The complete AI pipeline can be integrated seamlessly with the existing software or system architecture of the client using Docker images for the deployment. Docker helps to simplify the deployment process, improves flexibility, and ensures consistent and reliable execution of AI models. More precisely, it provides portability, allowing the model to run consistently across different environments without compatibility issues. Overall, there are two types of images, those that are in charge of deploying machine learning models and those for deep learning, both use the Python image as a base but use different dependencies. As endpoint, we use the `execute_pipeline` method described above.

Once the image has been created, it is stored in a private Docker registry. The creation and configuration of the Docker registry must be done by the user. The username, password, and ip address along with the port are part of the configuration of this component. In order to upload and download images, the Edge Embedded AI Kit has an API with push and pull methods, respectively. Those methods have been implemented using the Docker SDK for Python.[14]

6 Conclusions and Future Work

In this chapter, we presented the AI Model Generation (AMG) framework, which enables the creation of AI models for non-experienced users. AMG component has been designed in order to solve general-purpose problems supporting different types of data and allowing the deployment of machine learning and deep learning models. Ensuring the reproducibility of any type of analysis, model results and configuration are saved in a PostgresSQL Database. In addition, model descriptors are generated that allow easy loading and exporting of models by means of standardization (ONNX and PMML). All the models are interoperable, which will allow to add

---

[14] Docker SDK is a Python library for the Docker Engine API. It lets you do anything the Docker command does, but from within Python apps—run containers, manage containers, manage Swarms, etc.

new models or integrate other AutoML systems in the future. Furthermore, it enhances the scalability, promotes the interpretability, and simplifies long-term maintainability of the code.

On the other hand, it should be noted that the models are also containerized to ensure that they can be run in any environment. This gives the customer the freedom to transfer the model and use it wherever they want. Explainability is easy to apply, and all models have a set of explanatory techniques that can provide insight into the model's decision-making process. In that direction, as future work we believe that the platform is easily scalable and a possible research direction would be to analyze the performance of new models and applying new methods of explainability. It would also be worth comparing different search algorithms outside of Bayesian optimization. Additionally, it would make sense to consider new approaches such as Federated Learning (FE)[15] and Split Learning (SL)[16] in cases where privacy, data distribution, or network restrictions play an important role. However, building a general federated system that can effectively support most of machine learning and deep learning models is challenging due to the inherent heterogeneity of models, varying data distributions, and the need to address communication and privacy concerns. Models used in different domains or tasks have distinct architectures, training algorithms, and requirements, making it difficult to create a single framework that caters to all models. Furthermore, ensuring efficient communication and privacy preservation across participants adds complexity to the design and implementation of a general federated system.

Acknowledgments This work has been supported by the H2020 Framework Programme of the European Commission under Grant Agreement no. 957331.

References

1. Kinkel, S., Baumgartner, M., Cherubini, E.: Prerequisites for the adoption of AI technologies in manufacturing – Evidence from a worldwide sample of manufacturing companies. Technovation **110**, 102375 (2022)
2. Peres, R.S., Jia, X., Lee, J., Sun, K., Colombo, A.W., Barata, J.: Industrial artificial intelligence in industry 4.0 - systematic review, challenges and outlook. IEEE Access **8**, 220121–220139 (2020)
3. He, X., Zhao, K., Chu, X.: AutoML: a survey of the state-of-the-art. Knowl. Based Syst. **212**, 106622 (2021). ArXiv:1908.00709 [cs, stat]
4. Gohel, P., Singh, P., Mohanty, M.: Explainable AI: current status and future directions (2021). ArXiv:2107.07045 [cs]

---

[15] Federated learning is a machine learning approach that enables training models on decentralized data sources while preserving data privacy by keeping the data local and performing model updates collaboratively.

[16] Split learning is a distributed machine learning technique where the model is divided into segments, with the initial layers residing on the client device, and the remaining layers processed on a central server, allowing for privacy-preserving and resource-efficient training.

5. Real, E., Liang, C., So, D.R., Le, Q.V.: AutoML-Zero: evolving machine learning algorithms from scratch (2020). ArXiv:2003.03384 [cs, stat]
6. Salvaris, M., Dean, D., Tok, W.H., Salvaris, M., Dean, D., Tok, W.H.: Microsoft AI platform. Deep Learning with Azure: Building and Deploying Artificial Intelligence Solutions on the Microsoft AI Platform, pp. 79–98 (2018)
7. Das, P., Perrone, V., Ivkin, N., Bansal, T., Karnin, Z., Shen, H., et al.: Amazon SageMaker Autopilot: a white box AutoML solution at scale (2020). ArXiv:2012.08483 [cs]
8. LeDell, E.: H2O AutoML: Scalable Automatic Machine Learning (2020)
9. Singh, V.K., Joshi, K.: Automated machine learning (AutoML): an overview of opportunities for application and research. J. Inf. Technol. Case Appl. Res. 24(2), 75–85 (2022)
10. Erickson, N., Mueller, J., Shirkov, A., Zhang, H., Larroy, P., Li, M., et al.: AutoGluon-Tabular: Robust and Accurate AutoML for Structured Data (2020). ArXiv:2003.06505 [cs, stat]
11. Lemaitre, G., Nogueira, F., Aridas, C.K.: Imbalanced-learn: A Python Toolbox to Tackle the Curse of Imbalanced Datasets in Machine Learning (2016). ArXiv:1609.06570 [cs]
12. Batista, G.E.A.P.A., Prati, R.C., Monard, M.C.: A study of the behavior of several methods for balancing machine learning training data. ACM SIGKDD Explorat. Newsl. 6(1), 20–29 (2004)
13. Liaw, R., Liang, E., Nishihara, R., Moritz, P., Gonzalez, J.E., Stoica, I.: Tune: A Research Platform for Distributed Model Selection and Training (2018). ArXiv:1807.05118 [cs, stat]
14. Jaderberg, M., Dalibard, V., Osindero, S., Czarnecki, W.M., Donahue, J., Razavi, A., et al.: Population Based Training of Neural Networks (2017). ArXiv:1711.09846 [cs]
15. Li, L., Jamieson, K., DeSalvo, G., Rostamizadeh, A., Talwalkar, A.: Hyperband: A Novel Bandit-Based Approach to Hyperparameter Optimization (2018). ArXiv:1603.06560 [cs, stat]
16. Li, L., Jamieson, K., Rostamizadeh, A., Gonina, E., Hardt, M., Recht, B., et al.: A System for Massively Parallel Hyperparameter Tuning (2020). ArXiv:1810.05934 [cs, stat]
17. Buitinck, L., Louppe, G., Blondel, M., Pedregosa, F., Mueller, A., Grisel, O., et al.: API design for machine learning software: experiences from the scikit-learn project. In: ECML PKDD Workshop: Languages for Data Mining and Machine Learning, pp. 108–122 (2013)
18. Ke, G., Meng, Q., Finley, T., Wang, T., Chen, W., Ma, W., et al.: LightGBM: A Highly Efficient Gradient Boosting Decision Tree. In: NIPS (2017)
19. Chen, T., Guestrin, C.: XGBoost: A Scalable Tree Boosting System, pp. 785–794 (2016)
20. Turner, R., Eriksson, D., McCourt, M.J., Kiili, J., Laaksonen, E., Xu, Z., et al.: Bayesian Optimization is Superior to Random Search for Machine Learning Hyperparameter Tuning: Analysis of the Black-Box Optimization Challenge 2020 (2021). ArXiv:abs/2104.10201
21. Baniecki, H., Kretowicz, W., Piatyszek, P., Wisniewski, J., Biecek, P.: dalex: Responsible Machine Learning with Interactive Explainability and Fairness in Python (2021). ArXiv:2012.14406 [cs, stat]
22. Apley, D., Zhu, J.: Visualizing the effects of predictor variables in black box supervised learning models. J. Roy. Stat. Soc. B (Stat. Methodol.) 06, 82 (2020)
23. Lundberg, S., Lee, S.I.: A Unified Approach to Interpreting Model Predictions (2017)

Anomaly Detection in Manufacturing

Jona Scholz, Maike Holtkemper, Alexander Graß, and Christian Beecks

1 Introduction

Consider a machine that performs a simple operation in the production cycle of a product. On a typical day, this machine can perform the same motions over and over, with very little deviation. Then one day, one of its motors develops a problem and does not function properly anymore. As a result, the machine performs its operations incorrectly and damages the product. The longer it continues, the more damages accrue, potentially causing significant losses in damages and production delays. Detecting anomalous behaviors like this is important for quality control and safety. Fortunately with the rise of digitization and advanced analytics, industry has developed automated methods to accomplish just that.

We begin this chapter by introducing the concept of anomaly detection in more detail. For this purpose, we summarize different perspectives on that topic. We then discuss statistical methods to detect anomalies. The widely used techniques discussed in this section will offer a solid foundation for detecting anomalies and serve as a starting point for further research.

Next, we explore deep learning, a more advanced approach that employs artificial intelligence to detect anomalies. Deep learning has found success in many areas of machine learning including anomaly detection. Here we present a case study from the EU project knowlEdge [3], where an autoencoder was used to detect anomalies

J. Scholz (✉) · M. Holtkemper · C. Beecks
FernUniversity of Hagen, Hagen, Germany
e-mail: jona.scholz@fernuni-hagen.de; maike.holtkemper@fernuni-hagen.de;
christian.beecks@fernuni-hagen.de

A. Graß
Fraunhofer-Institut für Angewandte Informationstechnik FIT, Schloss Birlinghoven, Sankt,
Augustin
e-mail: alexander.grass@fit.fraunhofer.de

© The Author(s) 2024
J. Soldatos (ed.), *Artificial Intelligence in Manufacturing*,
https://doi.org/10.1007/978-3-031-46452-2_20

in a manufacturing process of fuel tanks. The autoencoder architecture is explained in depth and further illustrated by the case study.

Finally, we stress the importance of human involvement in the anomaly detection process. While AI has many capabilities, humans are essential in interpreting the results, refining the models and making informed decisions.

2 Anomaly Detection in Industry

As digitization progresses, the desire of manufacturing companies for more transparency about their machine and plant landscape is increasing. Once the data are available, suitable processes are needed to gain a deeper insight into the production processes. With the help of data and process analysis, irregularities and occurring disturbances in the production process flow can be viewed and analyzed in detail. These irregularities are referred to as anomalies, although there is no standard definition of the term in the literature [8]. For example, Zheng et al. [24] define an anomaly as "a mismatch between a node and its surrounding contexts," Lu et al. [13] as a "data object that deviates significantly from the majority of data objects", and Su et al. [18] as an "unexpected incidence significantly deviating from the normal patterns formed by the majority of the dataset."

Anomalies can be further specified depending on the context, such as described in Hasan et al. [10], where anomalies that occur in IoT datasets are divided into eight classes: Denial of Service (DoS), Data Type Probing (D.P), Malicious Control (M.C), Malicious Operation (M.O), Scan (SC), Spying (SP), Wrong Setup (W.S), and Normal(NL). Wu et al. [23], on the other hand, classify the anomalies that occur into three types that are more descriptive in nature, namely punctual, contextual, and collective anomalies. Here, a punctual anomaly represents a specific reference point with anomalous information, a collective anomaly characterizes a set of data records that have an anomalous character compared to the other data collected, and a continuous anomaly, which is a collective anomaly whose considered time period extends from a specific starting point to infinity [8].

The occurrence of anomalies can have many reasons. One common reason represents a change in the environment, such as a sudden increase in temperature, which is considered an abnormal condition by the sensor [8]; another reason may simply be due to a sensor error [8]; or it may be a malicious attack intended to weaken the computing power of an IoT network and thus intentionally cause the sensor to malfunction [15].

With the help of machine learning (ML) tools for anomaly detection, a variety of algorithms and methods are available to enterprises to identify anomalies [14]. Agrawal and Agrawal [2] divided the process of anomaly detection into three major phases, as shown in Fig. 1. Here, parameterization describes the preprocessing of data into a previously defined acceptable format, which in turn serves for the further training phase. In this, a model is created based on the normal or abnormal behavior of the system. Depending on the type of anomaly detection considered,

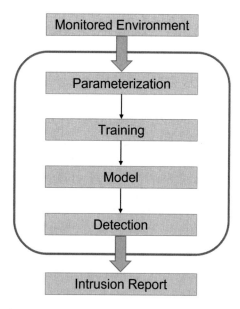

Fig. 1 Methodology of anomaly detection

different methods can be chosen, which can be either manual or automatic. The last phase is the detection phase, where finally the model is compared with the parameterized datasets. For example, if a predefined threshold is exceeded, an alarm can be triggered to draw attention to the anomalies.

Toshniwal et al. [21] further elaborate on the basic idea of ML algorithms. According to them, ML algorithms require input data for training, and in turn output labels are generated for test instances, where each feature represents a dimension. The input data represent the batch or real-time data of the data instances, where each data instance can be considered as a data point. Depending on the type of data, the input data are tagged or untagged. The output of the algorithm in turn has classes with which the instances are associated.

Clustering is used as one method for anomaly detection. For this, data are divided into groups of similar objects, where each group (cluster) consists of objects that are similar to each other and can be distinguished from objects in other groups [4]. Various clustering methods [1] can be used for anomaly detection, including partitional methods such as k-means or probabilistic methods such as EM.

In addition, outlier detection algorithms find patterns in data that do not follow a specific behavior. There are a variety of outlier detection schemes, such as the distance-based approach [4]. This is based on the nearest-neighbor algorithm and uses a distance-based metric to identify outliers [19].

Another possibility for the identification of anomalies is based on classification, which describes the problem of identifying the category of new instances based on a classification model learned on a training dataset that contains observations of known memberships of categories. The category represents the class label, and different observations may belong to multiple class labels. In machine learning, classification is considered an instance of supervised learning. An algorithm that resorts to the method of classification is called a classifier. It can predict class labels and distinguishes between normal and abnormal data in the case of anomaly detection [2]. Well-known classification methods form the Classification Tree, Fuzzy Logic, Naïve Bayes network, Genetic Algorithm, and Support Vector Machines.

The Classification Tree is also called decision tree because it resembles the structure of a flow chart. Here, the internal nodes form a test property, each branch represents the test result, and the leaves represent the classes to which an object belongs [22]. Fuzzy Logic, derived from fuzzy set theory, deals with approximate inference. In this method, data are classified using various statistical metrics and classified as normal or abnormal data according to the Fuzzy Logic rules [12].

The Naïve Bayes network is based on a probabilistic graph model where each node represents a system variable and each edge represents the influence of one node on another [2]. The Genetic Algorithm belongs to the class of Evolutionary Algorithms and generates solutions to optimization problems based on techniques inspired by natural evolution such as selection and mutation. The Genetic Algorithm is particularly robust to noise and is characterized by a high anomaly detection rate [12].

A Support Vector Machine (SVM) is a supervised learning method used for classification and regression. It is more widely used especially in the field of pattern recognition. A one-class SVM is based on examples that belong to a specific class and do not include negative examples [20].

3 Feature Selection and Engineering

As with all machine learning methods, feature selection plays an important role in anomaly detection. One important category of features in the context of manufacturing is sensory data from monitoring systems. For example, the operation of a robotic arm might create measurable vibrations during motor operations [16]. These vibration data could be used in the process of detecting anomalous behavior, which might act as a trigger to perform maintenance on the arm.

The importance of feature selection comes from the fact that irrelevant features can lead to decreased performance of the model. In addition, the computational complexity and data storage costs increase with the number of features. When developing models for feature detection, a large chunk of development time will typically be allocated to the selection and transformation of features.

Not all data can readily be used by a model. Continuing the example of our robotic arm, there may be noise in the data from other vibrations or inaccurate measurements. In these cases, we may have to preprocess or engineer features. This could involve running a noise filter on our vibration data to produce a clearer signal. However, it may also be the case that there is hidden information unavailable to the model due to a lack of complexity in its architecture. In the example of our robotic arm, performing frequency analysis leads to engineered features that may be more informative for a given model.

4 Autoencoder Case Study

In this case study, we will take a look at a real manufacturing process from an industry partner in the knowlEdge project. KnowlEdge is an EU project funded by the Horizon2020 initiative that advances AI powered manufacturing processes and services [3]. The scenario described here represents the production of fuel tanks for combustion engines in cars. More precisely, it defines the blow molding procedure during tank manufacturing, which is subject to a variety of individual steps controlled and observed by high-precision sensors for different metrics such as temperature, position, and energy consumption.

In Fig. 2, we see an overview of the blow molding process as exemplified by a water bottle. Essentially, it involves melting plastic and forming it into a preform, which resembles a test tube with a threaded neck. This preform is then placed into a mold cavity and air is blown into it, forcing the plastic to expand and take the shape of the mold. Once the plastic has cooled and solidified, the mold is opened and the finished bottle is ejected. This process can be highly efficient, allowing for the mass production of objects with consistent shapes and properties.

Unfortunately, as blow molding of complex shapes is a very sensitive process, not every production cycle is successful, and produced tanks can show defects indicated

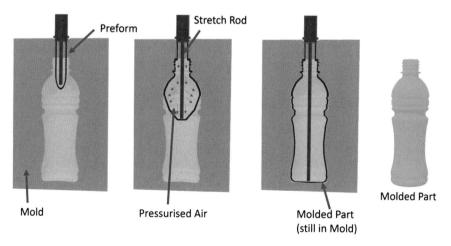

Fig. 2 An illustration of the blow molding process. The plastic is melted and shaped into a tube-like preform. The preform is then placed in a mold, inflated with air to take the mold's shape, and cooled. Once solidified, the finished bottle is ejected from the mold

by quality measures out of tolerance. To reduce these erroneous cycles already in early stages and thus to decrease additionally emerging costs, we will present a solution for anomaly detection that was implemented during the project. Although in this context different kinds of detection methods were considered including supervised approaches for already identified anomalous behaviors, in the following, we will focus on autoencoders [9] as a representative method for unsupervised anomaly detection that is applicable to a large quantity of potential use cases.

4.1 Autoencoders

An autoencoder is a neural network architecture that encodes and decodes data to learn a compact and efficient representation. For simplicity, imagine a neural network with several layers that simply tries to output whatever was put in (identity function). This is of course a non-challenging task for small datasets, if the layers in-between are big enough to maintain all the information. But what if you need a representation of many individual data instances? Since learning every instance is too complex for large datasets, the idea of autoencoders is to find a reduced description that generalizes well for big data, while still maintaining the essential information to reproduce given inputs. Considering the aforementioned architecture of a multi-layered neural network (as illustrated in Fig. 3), autoencoders split up the propagation of information into an encoder and decoder part. In the encoder, information generally flows through to a chain of shrinking layers with a final layer that equals a bottleneck to encode and therefore to learn a compressed version of the data. Subsequent layers, representing the decoder, are responsible to restore information from the generated encoding. The accuracy of the model is given by the reconstruction error, which is defined as a measure on how much the original data differ from its compressed approximation constructed via a propagation through the learned encoding.

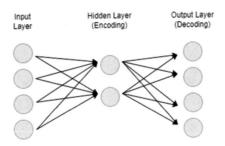

Fig. 3 A simplified example of an autoencoder with 1 hidden layer. There are fewer neurons in the middle than at the start and the end. This forces the model to learn a compressed representation of the data

4.2 Anomaly Detection for Blow Molding

The original task of fuel tank manufacturing comprises many steps including continuous and cyclic sub-processes such as extrusion or blow molding. Focusing on the cyclical blow molding process, more than 100 associated data attributes were analyzed. Attributes mainly corresponded to predefined and observed values for machine positions, temperatures, energy consumption, and pressures. As this information is recorded over time, the problem becomes a multivariate anomaly detection task for timeseries data. To better capture the inherent structure of timeseries data, a specific variant of autoencoders was utilized, called Long-Short-Term-Memory Autoencoders (or LSTM-Autoencoders) [11]. In the first step, all relevant process information was assigned to its associated machine cycle—indicated by a combination of two binary machine events—to produce input data for model training and evaluation. The resulting dataset is then used to learn a compressed representation of individual machine cycles. Since autoencoders are used to find a suitable representation of the provided information, the model tries to find an encoding, which reflects essential cycle information. With anomalies being defined as rarely occurring events, it makes use of the fact that a deviation from common behaviors is not regarded as characteristic information. As explained in the previous section, this deviation leads to higher reconstruction errors for cycles not following the usual patterns. A problem with unsupervised models is the lack of a ground truth. While theoretically one can try to make the model more and more complex to capture different behaviors of a system, the goal is not to consider anomalies as rare, but valid data entries. However, the question remains: how do we draw a line to distinguish which reconstruction errors indicate an anomaly and which do not? One potential solution, which was also used in the presented approach, is to make use of probability distributions—usually the normal distribution—to estimate the distribution of the reconstruction error for each data attribute. As a result, the reconstruction error can be compared against a predefined threshold derived from the overall dispersion, e.g., errors that are higher than three times the standard deviation. This simple method also regularizes the accuracy, as a poorly fitted model automatically leads to an increase in the overall reconstruction error and the associated spread. A representative illustration is given in Fig. 4. It shows individual machine cycles during a blow molding process, including modeled approximations and identified anomalies based on exceeded thresholds with respect to the reconstruction error.

4.3 Human-Enhanced Interaction

The lack of a ground truth, as is the case for unsupervised anomaly detection, leaves room for further enhancements to improve the accuracy of the system. One solution is to integrate expert feedback, also known as "humans-in-the-loop." A domain-

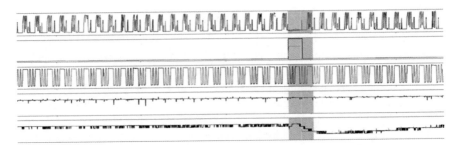

Fig. 4 A snippet from the original process data showing 5 attributes including identified anomalies. Each row presents a different attribute over time, with the original signal shown in black. Differently colored lines indicate distinct cycle approximations generated from the autoencoder (reproduced signals). Highlighted by the red-shaded rectangles, two cycles are regarded as anomalies, since for one or more dimensions, the reconstruction error is too high

specific verification of generated results enables us to label former predictions and to subsequently reuse this information as input for supervised approaches. In addition, human feedback can help to fine-tune predefined thresholds for the reconstruction error. If a domain expert notices that the approach detects too many false-positives (i.e., the number of identified anomalies, which are not anomalies according to the expert), thresholds for associated attributes can be reduced. The same holds for the opposite case, where the sensitivity of the system might be increased if no anomalies can be identified, although a manual inspection would reveal existing anomalies. Further information on how human feedback enhances approaches for unsupervised anomaly detection can be found in the literature [5–7, 17].

5   Conclusions

In the fifth industrial revolution, human-centered AI (HCAI) is becoming a key part of manufacturing processes. Anomaly detection can be used to improve quality control and elevate the human operator. There are several machine learning approaches to detect anomalies. Some of the methods we discussed are clustering, which groups similar data objects, and classification, which categorizes new instances based on known categories. In the scope of the EU project KnowlEdge, we have used autoencoders to detect manufacturing flaws in the production of fuel tanks. We have seen in detail how autoencoders work and how they efficiently solve the problem of anomaly detection for large datasets. The involvement of human expertise, or "humans-in-the-loop," is crucial for improving model performance and managing false-positive detections.

Acknowledgments This work has received funding from the European Union's Horizon 2020 research and innovation programme under grant agreement No. 957331—KNOWLEDGE.

References

1. Aggarwal, C.C., Reddy, C.K. (eds.): Data Clustering: Algorithms and Applications. CRC Press (2014)
2. Agrawal, S., Agrawal, J.: Survey on anomaly detection using data mining techniques. Procedia Comput. Sci. **60**, 708–713 (2015). Knowledge-Based and Intelligent Information & Engineering Systems 19th Annual Conference, KES-2015, Singapore, September 2015 Proceedings
3. Alvarez-Napagao, S., Ashmore, B., Barroso, M., Barrué, C., Beecks, C., Berns, F., Bosi, I., Chala, S.A., Ciulli, N., Garcia-Gasulla, M., et al.: Knowledge project–concept, methodology and innovations for artificial intelligence in Industry 4.0. In: 2021 IEEE 19th International Conference on Industrial Informatics (INDIN), pp. 1–7. IEEE (2021)
4. Berkhin, P.: A survey of clustering data mining techniques. Grouping Multidimensional Data: Recent Advances in Clustering, pp. 25–71 (2006)
5. Chai, C., Cao, L., Li, G., Li, J., Luo, Y., Madden, S.: Human-in-the-loop outlier detection. In: Proceedings of the 2020 ACM SIGMOD International Conference on Management of Data, pp. 19–33 (2020)
6. Chawla, A., Jacob, P., Farrell, P., Aumayr, E., Fallon, S.: Towards interpretable anomaly detection: Unsupervised deep neural network approach using feedback loop. In: NOMS 2022–2022 IEEE/IFIP Network Operations and Management Symposium, pp. 1–9. IEEE (2022)
7. Das, S., Wong, W.-K., Dietterich, T., Fern, A., Emmott, A.: Incorporating expert feedback into active anomaly discovery. In: 2016 IEEE 16th International Conference on Data Mining (ICDM), pp. 853–858. IEEE (2016)
8. DeMedeiros, K., Hendawi, A., Alvarez, M.: A survey of AI-based anomaly detection in IoT and sensor networks. Sensors **23**(3), 1352 (2023)
9. Goodfellow, I., Bengio, Y., Courville, A.: Deep Learning. MIT Press (2016)
10. Hasan, M., Islam, M.M., Zarif, M.I.I., Hashem, M.: Attack and anomaly detection in IoT sensors in IoT sites using machine learning approaches. Internet Things **7**, 100059 (2019)
11. Hochreiter, S., Schmidhuber, J.: Long short-term memory. Neural Comput. **9**(8), 1735–1780 (1997)
12. Kaur, H., Gill, N.: Host based anomaly detection using fuzzy genetic approach (FGA). Int. J. Comput. Appl. **74**(20), 5–9 (2013)
13. Lu, J., Wang, J., Wei, X., Wu, K., Liu, G.: Deep anomaly detection based on variational deviation network. Fut. Internet **14**(3), 80 (2022)
14. Ma, X., Wu, J., Xue, S., Yang, J., Zhou, C., Sheng, Q.Z., Xiong, H., Akoglu, L.: A comprehensive survey on graph anomaly detection with deep learning. IEEE Trans. Knowl. Data Eng. (2021)
15. Ngo, Q.-D., Nguyen, H.-T., Tran, H.-A., Pham, N.-A., Dang, X.-H.: Toward an approach using graph-theoretic for IoT botnet detection. In: 2021 2nd International Conference on Computing, Networks and Internet of Things, pp. 1–6 (2021)
16. Sharp, M.: Observations on developing anomaly detection programs with case study: Robotic arm manipulators 2019-02-13 2019
17. Smits, G., Lesot, M.-J., Yepmo Tchaghe, V., Pivert, O.: PANDA: Human-in-the-loop anomaly detection and explanation. In: Information Processing and Management of Uncertainty in Knowledge-Based Systems: 19th International Conference, IPMU 2022, Milan, Italy, July 11–15, 2022, Proceedings, Part II, pp. 720–732. Springer (2022)
18. Su, Y., Zhao, Y., Niu, C., Liu, R., Sun, W., Pei, D.: Robust anomaly detection for multivariate time series through stochastic recurrent neural network. In: Proceedings of the 25th ACM SIGKDD International Conference on Knowledge Discovery & Data Mining, KDD '19, pp. 2828–2837, New York, NY, USA, 2019. Association for Computing Machinery
19. Syarif, I., Prugel-Bennett, A., Wills, G.: Unsupervised clustering approach for network anomaly detection. In: Networked Digital Technologies: 4th International Conference, NDT 2012, Dubai, UAE, April 24–26, 2012. Proceedings, Part I 4, pp. 135–145. Springer (2012)

20. Tang, H., Cao, Z.: Machine learning-based intrusion detection algorithms. J. Comput. Inf. Syst. **5**(6), 1825–1831 (2009)
21. Toshniwal, A., Mahesh, K., Jayashree, R.: Overview of anomaly detection techniques in machine learning. In: 2020 Fourth International Conference on I-SMAC (IoT in Social, Mobile, Analytics and Cloud) (I-SMAC), pp. 808–815 (2020)
22. Wu, S.-Y., Yen, E.: Data mining-based intrusion detectors. Expert Syst. Appl. **36**(3), 5605–5612 (2009)
23. Wu, Y., Dai, H.-N., Tang, H.: Graph neural networks for anomaly detection in industrial Internet of Things. IEEE Internet Things J. **9**(12), 9214–9231 (2021)
24. Zheng, Y., Jin, M., Liu, Y., Chi, L., Phan, K.T., Chen, Y.-P.P.: Generative and contrastive self-supervised learning for graph anomaly detection. IEEE Trans. Knowl. Data Eng. (2021)

Towards Industry 5.0 by Incorporation of Trustworthy and Human-Centric Approaches

Eduardo Vyhmeister and Gabriel Gonzalez Castane

1 Introduction

The industrial sector has evolved from mechanization and steam power to mass production and automation, known as Industry 4.0. This digital revolution incorporates technologies such as Additive Manufacturing, Internet of Things (IoT), Edge and Cloud computing, Simulation, Cyber-Security, Horizontal and Vertical Integration, and Big Data analysis for efficient task execution [1].

To prioritize resilience, societal well-being, and economic growth, the European Commission (EC) introduced Industry 5.0, aiming to transform traditional factories into resilient providers of prosperity [2]. Artificial Intelligence (AI) plays a crucial role in this transition, helping in automating and getting insight in large-scale production and customization.

Building user confidence in AI products is essential, achieved through risk mitigation and facilitating AI technology adoption. Trustworthy AI (TAI) and human-centric AI integration are key challenges that need to be addressed [3]. Human-centric AI focuses on optimizing performance while prioritizing human needs, improving working conditions, and fostering a sustainable and socially responsible industrial ecosystem.

We provide an overview of the transition from Industry 4.0 to Industry 5.0, highlighting the use of AI and advanced technologies to create sustainable, human-centric, and resilient industries. It explores the challenges and opportunities of adopting TAI, offering strategies to mitigate risks and promote effective adoption.

The presented strategies for risk mitigation in TAI assets align with established approaches in the industry, particularly the Process of Risk Management (PRM)

E. Vyhmeister (✉) · G. G. Castane
The Insight SFI Research Centre of Data Analytics, University College Cork, Cork, Ireland
e-mail: eduardo.vyhmeister@insight-centre.org

J. Soldatos (ed.), *Artificial Intelligence in Manufacturing*,
https://doi.org/10.1007/978-3-031-46452-2_21

based on the ISO 31000 standard. Combining these approaches, we propose the TAI-PRM framework. The framework is designed to meet current and future regulatory requirements for AI, with specific goals that include:

- To support management units and developers in incorporating trustworthy requirements within the AI life cycle process.
- To secure the use of AI artefacts independently of legal and technical changes. The legislation heterogeneity applied in different countries on the use of AI can be varied; therefore, flexibility is key.
- To ease the combination of the frameworks that handle ethical-driven risks with other approaches commonly used for PRM. It needs to be designed as a complementary asset to these used by the industry and not a replacement to facilitate its adoption.
- To facilitate an iterative process to handle risks on AI artefacts within the framework. Many processes in software do not follow sequential development but spiral/iterative development processes. Therefore, the framework must flexibly adapt to any development cycle applied by developers for its incorporation.
- To ensure that Key Performance Indicators (KPIs) can be tracked through well-defined metrics that register the progress on the identified risks handling. Tracking KPIs is essential for their daily operations and business units. In addition, managerial levels can use these indicators to understand the impact of incorporating ethical aspects.
- To construct an architecture that addresses personal responsibilities and channels of communication. With this aim, the framework must foster communication between technical and non-technical stakeholders.
- To foster the reuse of outcomes in other research areas and market segments to avoid duplication of effort. This is translated as savings in revenue and research time on future developments. In addition, well-structured risk identification can avoid the repetition of failing conditions on AI components with similar target objectives.
- To enable a seamless path for the transition to Industry 5.0. By linking ethical considerations and risks, companies can handle TAI requirements as a PRM.
- To provide users with a tool for performing the TAI-PRM and evaluate approaches already developed for TAI.

1.1 Understanding the Transition to Industry 5.0

Industry 4.0 incorporates cyber-physical systems and digital platforms into factories to improve production processes and decision-making [4]. Different technologies have helped shape this industrial revolution. The impact of these technologies extends beyond industry, including home products, business models, clean energy, and sustainability, surpassing those of previous industrial revolutions. Industry is recognized as a driving force for sustainable transformation, necessitating consideration of societal and environmental aspects [5].

Previously, technologies focused on economic optimization, neglecting sustainable development. To address this, there is a growing demand for a shift to circular economies, prioritizing well-being, social governance, environmental efficiency, and clean energy. In 2020, the EC organized a workshop that gave rise to the concept of Industry 5.0. This vision, incorporated in industry's future roadmap, combines AI and the societal dimension as enablers [6].

Industry 5.0 entails integrating technologies to optimize workplaces, processes, and worker performance, emphasizing collaboration between humans and machines instead of replacing one another. This human-centric approach promotes the development of technologies that enhance human capabilities. To promote the adoption of Industry 5.0, the EC has established initiatives like the Skills Agenda and the Digital Education Action Plan, aiming to enhance the digital skills of European workers [7, 8]. The Commission also seeks to boost industry competitiveness by accelerating investments in research and innovation, as outlined in the Industrial Strategy [9].

Environmental sustainability is a key consideration for the EC, which promotes the use of resources efficiently and the transition to a circular economy through supporting initiatives such as the Green Deal [10]. Additionally, to promote a human-centric approach, initiatives and regulations have been developed that include the Artificial Intelligence Act, the White Paper on Artificial Intelligence, and the Trustworthy requirements [11–13].

Industry 5.0 introduces new challenges and technological enablers. These include real-time decision-making, human-centric solutions, edge computing, and transparency. Managing approaches and AI techniques can support and drive these advancements. Meanwhile, challenges from Industry 4.0, such as security, financing, talent, data analytics, integration, and procurement limitations, remain relevant [14, 15].

In terms of technological enablers, different approaches have been recognized, which includes Individualized Human–Machine Interaction, Bio-Inspired Technologies and Smart Materials, Digital Twins and simulation, Data Transmission, storage, and analysis technologies, AI, and Technologies for Energy Efficiency, Renewable, Storage, and Trustworthy Autonomy [16, 17].

1.2 AI, Trustworthy AI, and its Link to Industry 5.0

Industry 5.0 builds upon Industry 4.0, integrating ethical considerations for AI assets while prioritizing humans. It aims to promote societal well-being through the collaboration of humans and machines in smart working practices. Ethical concerns in AI vary across domains and application types, with unique considerations.

Trust development emerges as a common challenge, crucial for instilling confidence among users. The EC emphasizes TAI as a foundational ambition, acknowledging the significance of trust in advancing AI and establishing a robust framework [18]. Ensuring trustworthiness becomes imperative in AI technology, as reliable interactions between AI agents and humans are essential [19]. AI technology providers must address risks related to performance and user impact, emphasizing collaboration between AI-driven systems and humans, considering transparency, reliability, safety, and human needs, to foster user acceptance [20].

The interaction between humans and AI in manufacturing can be categorized into three types (human-in-the-loop **HITL**, human-on-the-loop **HOTL**, and human-in-command **HIC**), depending on the level of human involvement. Importantly, the decisions made by these systems can be influenced by humans and data, thus impacted by societal, legal, and physical considerations.

According to the High-Level Expert Group set up by the EC [21], TAI has three main pillars, which should be met throughout the system's entire life cycle: it should be lawful, ethical, and robust. To deem an AI Trustworthy, different requirements must be addressed, depending on the intrinsic risk of the AI assets. These requirements are related to Human Agency and Oversight, Technical Robustness and Safety, Privacy and Data Governance, Transparency, Diversity, Non-discrimination, and Fairness (DnDF), as well as Environmental and Societal Well-being, and Accountability. The Trustworthy requirements [13] aim to minimize risks associated with AI and its potential adverse outcomes throughout the AI life cycle. These requirements align with five key principles derived from 84 ethics guidelines: transparency, justice and fairness, non-maleficence, responsibility, and privacy. They ensure that AI systems respect fundamental rights, are secure and reliable, protect privacy and data, are transparent and explainable, avoid bias, promote stakeholder participation, and are subject to accountability mechanisms.

1.3   PRM and Considerations for AI Implementation

In a 2020 study, Hagendorff [22] provides guidelines and strategies for addressing critical AI issues, focusing on areas of greatest attention. However, the study lacks specific definitions for handling AI assets throughout their life cycle [23]. Moreover, the implementation of these approaches in different domains and environments can present specific challenges that need to be understood [20].

Different organizations have developed diverse methods based on multiple ethical principles to facilitate practitioners in developing AI components. These organizations include academia, trade union, business, government, and NGOs.

To ensure a trustworthy component, the actions of any agent over humans must be reliable [19]. This definition depicts the linkage between trust and risk management, in agreement with the definitions in the EC AI Act [11]. By defining a proper methodology that tracks and manages risk components derived from trustworthy requirements, the probability of producing adverse outcomes is minimized.

Developing a framework that ensures the ethical use of AI while minimizing potential harms and maximizing benefits requires a comprehensive approach. How can such a framework be implemented in the manufacturing sector, considering existing approaches?

PRM involves identifying, assessing, and controlling risks that can impact systems and organizations. Integrating ethical considerations into PRM processes requires defining the concept of ethical risks (E-risks). We propose the following definition: "E-risks are conditions and processes that can disrupt the expected behaviour of an AI asset due to the lack of consideration of TAI requirements, including values, social, legal, environmental, and other constraints."

The PRM involves intervention, communication, and involvement of various areas within the managing companies. A risk management framework consists of three key components: Risk Architecture (RA), Strategy (S), and Protocols (P), forming the RASP strategy. RA provides a formal structure for communication and reporting, S defines implementation strategies, and P includes guidelines and procedures for managing risks [24].

The risk management policy statement plays a crucial role in the framework. It outlines the organization's strategy and approach to PRM, aligned with its objectives and tailored to the specific environment. However, when establishing enterprise policies for AI management, minimal considerations may exist due to regulatory requirements, such as the Artificial Intelligence Act [11], which may influence the policy's content.

ISO 31000 is widely used as a standard for PRM in industry. It provides procedures for assessing, treating, monitoring, reviewing, recording, and reporting risks. ISO 31000 is often combined with other specific standards, such as ISO 9001 for supply chain and product quality improvement [25] and ISO/AWI23247 for the digital twin manufacturing framework [26]. However, these standards need to be reviewed and updated to align with the requirements of smart manufacturing and Industry 5.0 [27].

Although ISO 31000 does not explicitly address application security risks or provide guidance on implementing RASP controls, organizations adopting the RASP framework can benefit from considering ISO 31000 principles and guidelines. ISO 31000 promotes a systematic and proactive approach to PRM, which complements the objectives of the RASP framework.

The TAI-PRM adheres to ISO 31000 and adopts a RASP approach [28, 29]. TAI-PRM provides a supporting structure for communication and reporting of failures, implementation strategies, and guidelines for managing risks. In this chapter, we focus on developing and testing the protocols for the PRM, based on ISO 31000 with some considerations. The communication and consultation activities are part of the RASP architecture, not the PRM itself, and the Ethical PRM includes the Monitoring and Review process as a primary component after Risk Evaluation and Risk Treatment.

The mentioned works [28, 29] provide a scope contextualization for understanding the PRM and evaluation, as well as a groundwork for AI development and management, establishing connections between requirements and risk components.

2 Failure Mode and Effects Analysis

The Failure Mode and Effects Analysis (FMEA) method is commonly used for risk assessment in various industries, including the technology sector. When it comes to assessing risks derived from TAI considerations, the FMEA method can be an effective tool for several reasons:

- *Holistic approach:* The method can be applied on the entire system, rather than just focusing on individual components. This aligns with the principles of TAI, which emphasizes a comprehensive approach to ethical and technical considerations.
- *Identifying potential hazards:* TAI requires identifying potential hazards, assessing their likelihood and impact, and taking steps to mitigate them. The FMEA method provides a structured framework for this, helping also to uncover hidden hazards.
- *Systematic approach:* The method involves a systematic approach to risk assessment, helping to ensure that risks are addressed in a consistent and comprehensive manner, which is critical for TAI.
- *Continuous improvement:* The FMEA method is designed to be an iterative process. This aligns with the principles of TAI, which emphasize the need for ongoing monitoring and continuous improvement to ensure that AI systems remain ethical, transparent, and accountable over time.

The key steps of the FMEA are outlined next. These steps should be performed consecutively with the exception of the Identify Failure, Detection Methods, and Existing Risk Control; Analyse Effects; and Identify Corrective Actions:

Define the Analysis Developing TAI systems involves defining constraints, failures, and objectives while considering the system's context. Trustworthy requirements, including explainability, are outlined in the AI Act [11]. Adapting these requirements to the industrial context and AI asset goals is crucial. Methods for achieving TAI depend on functionalities, data usage, user understanding, and objectives [11].

Development of System Functional Block Diagrams Various supporting documents, including block diagrams, analyse failure modes, and their impact. Detail level should correspond to the AI asset's risk level, with worksheets and matrices as valuable tools. Additional information such as system and AI boundary descriptions, design specifications, safety measures, safeguards, and control system details are necessary for risk analysis.

Identify Failure Modes Identifying failure modes is crucial in understanding system failures. Certain failure modes align with the needs of TAI-PRM have already been identified. For instance, [30] define failure modes in IT safety. FMEA has been applied to TAI with limited success, focusing mainly on fairness [31]. To expand the scope of failure modes, we propose incorporating eleven ethical-based failure families derived from system norms and TAI requirements. These encompass failures related to robustness, safety, transparency, accountability, societal well-being, environmental well-being, human agency and oversight, privacy, data governance, bias (DnDF), and users' values. This approach facilitates detection and metric definition.

Identifying failure modes involves measuring observable conditions using supporting protocols. We propose to group the conditions based on drivers such as physical, social, data, user/system interface, and algorithms. Physical drivers include power supply, communication/data link cables, robot parts, wearables, lenses, and sensors. Internal social drivers relate to stakeholders' values and biases, including social responsibilities and ethics-in-design. Data drivers encompass sources affecting AI trustworthiness, such as biases, quality, quantity, and security. User and system interface drivers are linked to inadequate usage, absence of information display, tutorials, guidelines, and user's ill-intentioned usage. Algorithm drivers include problem-solving processes and expected functionality during AI code execution.

Identify Failure, Detection Methods, and Existing Risk Control Detecting and managing failure modes using metrics and methods to reduce risk conditions. Adequate intervention procedures are needed for high-risk AI components. Linking further actions, such as control devices and circuit breakers, is crucial to enhance system understanding and reduce risk. A lack of detection methods could affect system robustness, security, and transparency. The importance of these actions should be linked to the intrinsic risk level of the AI assets involved.

Analyse Effects This step analyses the consequences of failure modes, including the end effect and its impact on social-driven components and each HSE (Health, Safety, and Environment) element.

Identify Corrective Actions Involves identifying actions that can prevent or reduce the likelihood of failure, comply with legal requirements, ensure safe operation of the system, recover from failures, and incorporate norms based on user values and AI trustworthiness requirements.

Ranking Assigns values to each failure mode's likelihood, severity, and detectability, which can be used as KPIs for estimating the TAI state.[1] The Risk Priority Number index (RPN) with respect to a concrete failure mode enables the normalization of the AI artefacts risks. The RPN mathematical expression is $RPN_{item} = S \cdot O \cdot D$. Here S is the severity (scale of 1–10, with 10 being the most severe), O the occurrence or likelihood (scale of 1–10, with 10 being the most likely), and D corresponds to the detection ranking (scale of 1–10, with 10 being the least capable). The RPN indicates the risk level of the failure mode, with a higher RPN indicating, indirectly, a higher risk for the AI component.

---

[1] More information and tables for the ranking process are available on a GitHub page https://github.com/lebriag/TAI-PRM/tree/main/Support/RPN%20ranking%20tables.

To evaluate the overall risk of an AI component on different failure modes, the Global Risk Priority Number index ($GRPN_{item}$) can be calculated by summing the RPN of each item i to its corresponding failure mode ratio; $GRPN_{item} = \sum_{i=1}^{n} RPN_i \alpha_i$. In this equation, i represents the different Failure Modes linked to the same source; n is the number of failure modes of a specific component; and α_i is the failure mode ratio and represents the part attributed to a concrete failure mode if the failure materializes. This means the percentage of the AI asset to fail with the specific Failure Mode.

The Failure Mode Ratio can be estimated as $\alpha_i = \frac{e_i}{e_{tot}}$. e_i is the amount of failing conditions of a specific failure mode, and e_{tot} is the total amount of all failing conditions in the system. The failure mode ratio can be used to group risks with the same trustworthy requirement, by a proper weighting, thus identifying the most significant TAI risk consideration for the AI asset.

Tabulate and Report The report includes documentation and a repository for users to understand failure modes, risks, control measures, safeguards, and recommendations.

3 TAI–PRM Protocol

An overview of the TAI-RPM is shown in Fig. 1—more details at [29]. The figure shows an abstraction of the PRM with activities related to AI artefacts and TAI considerations.

The diagrams show the starting point (black circle), activities (boxes), decision points (diamonds), and endpoints (crossed circle). Some activities contain sub-flow charts, denoted by a [+] symbol, that follow the same approach explained here.

As seen in Fig. 1, the first activity is the *AI confirmation*, which involves identifying AI elements by defining and categorizing them.[2] Next, the *e-Risk identification and classification* activity focuses on identifying the AI elements' intrinsic level of risk under regulatory conditions (i.e., given the current regulatory approach in the EU, we focused on the risk levels defined by the AI Act [11]). This activity has previously been thoroughly discussed [28, 29], so it will not be covered here.

If AI classification is deemed not acceptable in the mentioned stage, the user will proceed to either the AI Scope Definition activity or the *<can be modified to acceptable?>* node—Fig. 1. This node assesses whether the AI's approach, data characteristics, and functionalities can be modified to meet TAI considerations for high-risk assets. If modification is feasible, the AI modification activity is executed to redefine the AI, followed by another round of *<e-risk identification and*

---

[2] The recommended categorization approach is based on the AI Watch definitions for European countries—https://publications.jrc.ec.europa.eu/repository/handle/JRC126426.

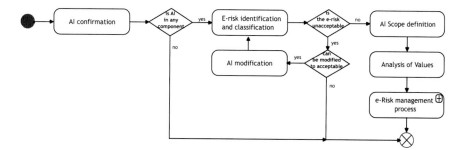

Fig. 1 Benchmark e-risk management process

classification>. If it is not possible to secure an AI asset's acceptable risk level, the design, deployment, or use process should be halted or decommissioned.

In *Analysis of Values* (see [28, 29] for more information), the potential for biases and use of personal information that could impact the requirements of DnDF are evaluated. It also considers the level of automation that is left to the AI asset and how this affects the agency of humans in decision-making processes. This activity also allows integrating users' values into the system. If there are conflicts, well-documented approaches for decision-making (e.g., ANP or AHP) can be used to resolve it. Finally, *e-Risk Management Process* encapsulates the risk assessment, treatment, monitoring, and review—described in Sect. 3.1.

3.1   e-Risk Management Process

Figure 2 shows a high-level diagram of the ethical-driven PRM.[3] Each of the activities in the figure is discussed next.

Establishing Context In this activity, a gathering and run of information is performed. This is done by: (1) Define how the AI asset interacts with other components and subsystems (i.e., software architecture documentation), (2) *Define* hierarchical extension of the components. Users must track cascade effects on risk analysis, for example, using root cause analysis. (3) Analyse the interactions between AI elements actions, UIs, and humans, defined in the activity *AI scope* (Fig. 1 in Sect. 3). (4) Analyse constraints set up as requirements to control functionalities, input behaviours, system outcomes, and component values. If relevant, these constraints were established with the physical context, enhancing the system security, especially in AI–user interaction. (5) Analyse diagrams that include Connectivity with other components and subsystems, Dependencies, and

---

[3] Full detailed diagram at: https://github.com/lebriag/TAI-PRM/tree/main/Flowcharts/e-risk%20management.

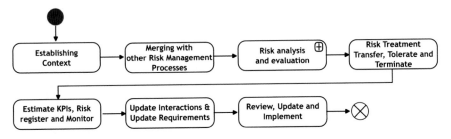

Fig. 2 High-level overview of the e-Risk Management Pipeline

Human–AI and Human–UI interactions. (6) Analyse collected information from Requirements and values.

Merging with Other Risk Management Process When several PRM instances are identified and similar approaches are used (i.e., FMEA), the analyses can be merged (e.g., merging with DFMEA and PFMEA processes; the DFMEA involves a comprehensive analysis of systems, while the PFMEA identifies and evaluates failures in processes).

Risk Analysis and Evaluation The FMEA (or Criticality Analysis (CA) if information exists—FMECA) is used as core component for risk analysis. Given the depth of the content, this activity is described in detail in Sect. 4 as it contains a detailed sub-flow—core in the management of e-risks.

Risk Treatment, Transfer, Termination, or Tolerate Depending on the risk appetite and the risk levels, this activity evaluates if an AI asset should be: **Treated:** Failing conditions can be modified, upgraded, or safeguarded. **Transferred:** External safeguards will allocate the responsibility of the failure events if materialized— Conditions can limit the option to transfer e-risks for TAI requirements. **Tolerated:** No need of AI asset modification. Periodic updates on the status must be continued according to the frequency established in the risk management policy. **Terminated:** The AI asset should not be used or developed. These activities are known as the 4T's of risk management. This process and its activities are further extended in Sect. 5.

Estimate KPIs, Risk Register, and Monitor The KPIs should be linked to each trustworthy requirement or value defined for AI and their risk level measured. The monitoring part of the activity involves analysing and evaluating stakeholder outcomes to trace contingency actions, and the implementation of PRMs. The risk register involves the process of Tabulate and Report of the FMEA.

Update Interactions and Update Requirements The outcomes of the previous process involve modifications related to AI assets and their interactions with other components, data structures managed by other AI assets, and the incorporation of additional AI assets or new functionalities that can impact the system's trustworthiness. Therefore, it is necessary to analyse these new interactions to identify potential risks. The iterative process of running the PRM must be tracked for accountability.[4]

Review, Update, and Implementation Once there are no outstanding updates, the user must implement the risk treatment strategy, which depends on the interactions between the different personas involved in the PRM as described in the Risk Architecture [29]. In addition, the status of the failure modes has to be revised, along with the mechanisms and processes to implement the 4T's—secure protocols, strategies, and control mechanisms for implementing the AI asset. Internal or external auditing processes can be used to evaluate the system status and the PRM. Finally, the risk register must be updated for accountability.

4   Risk Analysis and Evaluation Activity

Figure 3 extends the component described in the previous section. The figure depicts a flowchart that defines the instruments between the approaches FMEA, FMECA, and RCA (Root Cause Analyses) used for the risk assessment.

The first decision node, *<Protocol defined?>*, analyses if the user has defined an FMEA or alternatives (i.e., RCA—The RCA is a well-documented approach beyond the scope of this chapter, and therefore, it is not further explained.) for risk assessment. If no, the set of decision nodes drives the user to the most convenient approach to follow. The set of questions are:

- *<All failures identified?>*: Whether it is aimed to identify *all possible* failing conditions. This means that the user is interested in detecting every situation that might trigger risk outcomes.
- *<Top events limitations?>*: This question focuses on the circumstance that the number of failure events is large or could be unexpected.
- *<AI updates needed?>*: When an AI asset requires human intervention or software updates, FMEA has extensive applicability and efficiency.
- *<AI system early stage:?>* When the system is in the design or definition phase—An FMEA approach can better help to detect conditions that could lead to system failures.
- *<System modified?>*: This decision node analyses when the system will be modified considerably in future stages.
- *<Robust examination?>*: If AI assets are used in critical systems, or AI failures can have severe impacts on users and the environment, this decision node will allow the user to decide if a robust examination is required.

---

[4] Extended diagrams available at: https://github.com/lebriag/TAI-PRM/tree/main/Flowcharts/e-risk%20management/Risk%20Analysis%20and%20evaluation.

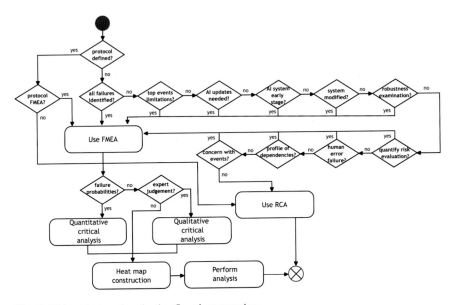

Fig. 3 Risk analysis and evaluation flowchart overview

- *<Quantity risk evaluation?>*: Since FMEA allows classification and grouping into failure modes in terms of their occurrence, severity, and detection, this decision node facilitates the definition.
- *<Human error failure?>*: This decision node addresses the systems that require considerable human intervention.
- *<Profile of dependencies?>*: If failures can trigger cascade events when they materialize, FMEA can foster a proper identification of system interdependence.
- *<Concern with events?>*: If explanation of the relations between failures that can lead to severe consequences and impacts is required.

In case the user is led to *Use FMEA* activity—detailed in Sect. 4.1—information regarding qualitative and quantitative evaluation must be gathered. Two decision nodes are used to define these: *<Failure probabilities?>* and *<Expert judgement?>*. These led to the two activities in Figure *Qualitative criticality analysis?* and *Quantitative criticality analysis?*.

Within them, a CA could be performed. In order to be performed, additional information, such as: (1) the Failure Mode Ratio (α), (2) the conditional probability β, which represents the probability that the failure effect will result in the identified severity classification, and (3) the λ, which represents the overall system failure rate due to different causes over the operating time in units of time or cycles per time, are needed. Then, the Criticality Number (C_M) provides a metric to classify a specific failure mode of an AI asset as follows $C_m = \beta\alpha\lambda$. The Overall Criticality Number (C_r) estimates how critical an AI asset is with respect to a complete system; $C_r = \sum_{i=1}^{n}(C_m)_i$.

Furthermore, a heat map and a risk matrix must be constructed to keep the quantitative information tracked and to incorporate the numerical analysis based on the probabilistic information collected about the failure modes (as described in Sect. 4.2).

4.1 Use FMEA Activity

Figure 4 shows the FMEA pipeline. The process starts with a decision node, *<design risk analysis defined?>*, where the user can check if PFMEA or DFMEA protocols are run in parallel with TAI-PRM. When this occurs, the next decision node, *<scope enable ethics?>*, checks if the scope of the framework and the ongoing approaches can be merged or extended. This means to define components, items, dependencies, and to establish similitude between policies. An activity named *Define and Merge DFMEA/PFMEA* defines the strategy to extend the functional blocks—if the process is running.

The decision node *<AI lifecycle considered?>* checks whether the user has considered analysing the complete AI asset life cycle. If that is the case, an FMEA approach must be considered for each stage of the AI life cycle and analysing the risks involved during each phase: design, development, use, and decommissioning.

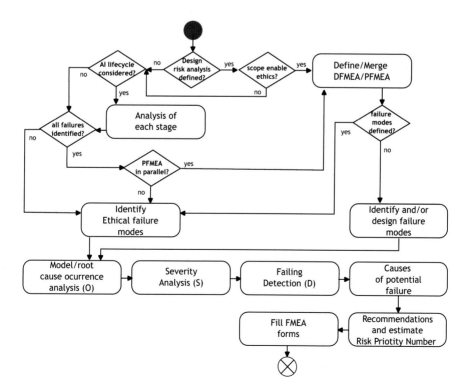

Fig. 4 FMEA pipeline

The following decision node, *<All failures identified?>*, is an internal check that evaluates that all relevant failing conditions, and their drivers, have been considered for the AI asset. After this, it is required to check if the AI asset is used within maintenance or operational processes. If it is, a PFMEA process must be set for the combination of both processes. This must be executed by the user within *<PFMEA in parallel?>*.

Independently of the path that the user takes for the FMEA flow, TAI-PRM provides two activities key to identifying failures. The first, named *Identify Ethical failure modes*, defines the scope of the analyses of ethical, trustworthy, and values to be considered by the user. The second activity is *Identifying and/or design failure modes*, where, based on the design of the AI asset, the user must analyse possible system's general failure modes and based on the analysis as overall system including DFMEA and PFMEA considerations (i.e., not ethical base only).

A **list of over 130 Failure Modes related to trustworthy considerations** have been already identified.[5] Users can access this list in the repository mentioned or through the tool described at the end of this chapter. Importantly, the possible Failure Modes are not limited to those presented, so users can expand the knowledge base for Failure Modes based on e-risks. This list is based on a literature and industrial feedback from case studies. Furthermore, the tool constructed to perform TAI-PRM (described in Sect. 6 of the document) further facilitates the identification of failure modes and adds the capability to incorporate and share Failure Modes between tool users.

After identifying the failure modes, a ranking activity of these must be executed.[6] As observed in the figure, for each failure condition—Model/Root—on each component, the likelihood/occurrence rank (O), Severity (S), and Detection (D) must be considered. It should be considered that if the detection (D) is performed by an entity external to the system, for example, throughout IoT devices, constraints of the communication channels should be implemented. This could drive the risk appetite to be set in a more stringent condition.

The final three steps of the FMEA process focus on documentation for improving the detection and accountability of future failure modes. The first, *Causes of potential failure*, focuses on the keeping control on the failure mode causes for future corrections.

In the case of *Recommendations and estimate the Risk Priority Number—RPN* activity, failure compensating previsions, functionality extensions or restrictions, or AI asset modifications should be documented to prevent/reduce the likelihood and severity, or improving the detection of the failure modes. Finally, in *Fill FMEA forms*, the user must fill the Risk Register.

---

[5] The list of failure modes is available at: https://github.com/lebriag/TAI-PRM/tree/main/Support/Failure%20modes.

[6] The ranking information is available at: https://github.com/lebriag/TAI-PRM/tree/main/Support/RPN%20ranking%20tables.

Table 1 RPN ranges in function of the intrinsic risk level

Risk level	Tolerate risk score range	Treat risk score range	Terminate risk score range
Unacceptable risk	–	–	1–1000
High risk	1–200	201–800	801–1000
Limited risk	1–400	401–1000	–
Minimal risk	1–800	801–1000	–

4.2 Heat Map Construction

The heat map—also referred to as risk matrix—provides a mechanism to identify and compare AI artefacts and failure modes associated to the risks. This component is directly linked to a process of Fig. 3 with the same name.

Authors in [32] detail how to construct a risk matrix. Nevertheless, an extension can be made depending on the intrinsic risk level of the AI asset, as defined in the AI Act.

The basis for the risk matrix is the risk definition, which is the combination of the severity of a risk when it materializes and its likelihood. To describe the risk, a classification can be used for severity and likelihood following qualitative descriptions and scales.[7] As a result, the failure modes can be allocated on a matrix constructed based on the likelihood scale, the severity scale, and the risk score, by using a direct translation of the risk appetite.[8]

4.3 Perform Analysis

To link the heat map with the AI Act, different risk appetites should be considered, depending on the intrinsic risk of the AI asset—i.e., the higher the risk of the AI asset, more stringent must be the risk appetite.

Table 1 imposes the contingency actions for the risks associated to the 4T's, described in detail in Sect. 3.1.

Further refinement is required as further knowledge is obtained from implementing the PRM in AI assets and our use cases. Importantly, users can modify the proposed ranges based on the risk policies established in their own enterprises, always considering the constraints of the AI Act.

---

[7] Tables available at: https://github.com/lebriag/TAI-PRM/tree/main/Support/RPN%20ranking %20tables.

[8] Example result available in Heat Map Construction.pdf at: https://github.com/lebriag/TAI-PRM/ tree/main/Support/.

5 Risk Treatment Transfer Terminate or Tolerate Activity

The process associated with the 4T's of risk management is part of the larger process depicted in Fig. 2. The process should consider if the PRM process is repeated or new. If it is repeated, the user should determine if new failure modes have been identified for the AI asset. If new failure modes are identified, a new evaluation of the failure modes is necessary using the 4T's analysis.

If there are no new failure modes, a confirmation over the possible modifications of the risk appetite should be performed.

The process should be run accordance to the specific needs (requirements and intrinsic risk condition) of the AI asset and thus specifying the actions of Treat, Transfer, Terminate, or Tolerate over their explainability, transparency, accountability, human agency and oversight, privacy, and robustness, among other conditions. Each of these analysis leads to specific activities that provide recommendations and options to address the corresponding risks.

A flowchart is provided in a repository[9] that also includes a decision node for other requirements and an activity called "Framework Construction" to handle new approaches and requirements. Finally, the "Risk Processing" activity summarizes the previous analyses and guides the user to determine whether to terminate, tolerate, or treat the AI assets based on the risk appetite and recommendations.

Overall, the flowchart provides a structured approach to assess and manage risks associated with AI assets, considering various factors such as failure modes, risk appetite, explainability, transparency, accountability, privacy, and robustness.

6 Validation and Real Case Scenario

The TAI-PRM validation process was executed on multiple AI assets as a cooperation between different manufacturing companies and research institutions involved in the ASSISTANT project [28, 29, 33]. The project involves five academic and seven industrial partners. The real case scenario has been used to develop and refine the framework with their inputs over multiple iterations. It is currently in use for its improvement, given the iterative process defined in its implementation.

Furthermore, a tool was developed to facilitate the implementation of the framework and evaluate the ALTAI tool and its applicability in the manufacturing sector.

The rest of this section describes the ASSISTANT scenario as a case study. The case study description starts from Sect. 3 of the TAI-PRM.

---

[9] https://github.com/lebriag/TAI-PRM/tree/main/Flowcharts/e-risk%20management.

7   TAI-PRM Tool

A tool based on TAI-PRM has been developed for the ASSISTANT use case. The tool is accessible via the **https://assistant.insight-centre.org/** page and offers comprehensive information on the PRM. It includes two main sections with user feedback. Users can access information on how to use the ALTAI tool or the TAI-PRM process. The tool is not an alternative to the ALTAI tool but rather helps to record the linkage and usefulness of Trustworthy requirements, component assessment, and AI implementation functionality. The records will be compared to provide users with trends based on their domain. Users only interested in the TAI-PRM process need not complete the ALTAI component.

To use the tool, select the My TAI-PRM tab and create as many PRMs as needed. Complete the pipeline before generating a downloadable report. The tool also allows users to create and share failure modes to extend knowledge based on the presented strategies. Users can provide feedback at the end of each TAI-PRM to improve the tool.

A total of eight steps are needed to perform the TAI-PRM including: (1) Initiation of PRM, (2) E-risk identification and classification, (3) AI Scope Definition and Analysis of Values, (4) Establishing Context, Merging with other PRM, and defining criticality analysis or FMEA, (5) FMEA or CA, (6) Ranking, (7) Risk Register, and (8) Treat, Terminate, Tolerate, and Transfer.

The tools follow a distribution similar to the ALTAI tool to familiarize users with both elements. Facilitating the acceptance of user's that have already driven ALTAI assessments.

Finally, information from the analogue of the ALTAI tool can be linked to a specific TAI-PRM, so information generated from it is also used in the final report of the TAI-PRM.

8   Conclusions

In this chapter, we explored the current paradigm in which Industry 4.0 is evolving towards Industry 5.0, where AI and other advanced technologies are being used to build services from a sustainable, human-centric, and resilient perspective.

Furthermore, we present a framework for developing and designing responsible AI artefacts in the manufacturing sector. The TAI-PRM approach combines PRM and TAI, utilizing the failure modes and effects analysis (FMEA) method. The methodology includes defining risk appetites and strategies in the industrial sector based on the AI Act. It also incorporates the 4T's of risk management.

TAI-PRM can be used to implement trustworthy considerations and manage risks in compliance with current regulations. It can also serve as a guide for developing standards in AI risk management. Feedback from various domains emphasizes the importance of considering the human factor in risk management processes. However, the lack of regulatory conditions, certification, and standards for managing AI poses obstacles to the adoption and implementation of transparent and accountable AI systems. Addressing this challenge requires substantial training and effort to transform business and development team operations.

Acknowledgments We acknowledge the support of the European Commission by funding the ASSISTANT project (no. 101000165) and, at the same time, we thank the contribution of the Science Foundation Ireland under Grant No. 12/RC/2289 for funding the Insight Centre of Data Analytics (which is co-funded under the European Regional Development Fund).

References

1. Samoili, S., Cobo, M. L., Delipetrev, B., Martinez-Plumed, F., Gomez, E., De Prato, G., et al.: AI watch. Defining Artificial Intelligence 2.0. towards an operational definition and taxonomy of AI for the AI landscape, JRC Research Reports (JRC126426) (2021)
2. Brunetti, D., Gena, C., Vernero, F.: Smart interactive technologies in the human-centric factory 5.0: a survey. Appl. Sci. **12**(16), 7965 (2022)
3. Devitt, S.K., Horne, R., Assaad, Z., Broad, E., Kurniawati, H., Cardier, B., Scott, A., Lazar, S., Gould, M., Adamson, C., Karl, C., Schrever, F., Keay, S., Tranter, K., Shellshear, E., Hunter, D., Brady, M., Putland, T.: Trust and safety Publisher: arXiv Version Number: 1. https://doi.org/10.48550/ARXIV.2104.06512
4. Zheng, T., Ardolino, M., Bacchetti, A., Perona, M.: The applications of Industry 4.0 technologies in manufacturing context: a systematic literature review. Int. J. Prod. Res. **59**(6), 1922–1954 (2021)
5. Renda, A., Schwaag Serger, S., Tataj, D., Morlet, A., Isaksson, D., Martins, F., Giovannini, E.: Industry 5.0, A transformative vision for Europe: governing systemic transformations towards a sustainable industry (2022)
6. D.-G. for Research, Innovation, Industry 5.0, a transformative vision for Europe, European Commission (2022)
7. E. C. (EC), European skills Agenda (2020)
8. Yanli, X., Danni, L., Prospect of vocational education under the background of digital age: Analysis of European Union's "Digital Education Action Plan (2021–2027)". In: 2021 International Conference on Internet, Education and Information Technology (IEIT), pp. 164–167. IEEE, Piscataway (2021)
9. Cappellin, R., Ciciotti, E., Marelli, E., Garofoli, G.: A new European industrial strategy and the European recovery program after the Covid-19 crisis. Rivista Internazionale di Scienze Sociali **128**, 265–284 (2020)
10. Fetting, C.: The European Green Deal, ESDN Report, December (2020)
11. Commission, E.: Regulation of the European parliament and of the council; laying down harmonised rurles on artificial intelligence (artificial intelligence act) and amending certain union legislative acts (2021). https://eur-lex.europa.eu/legal-content/EN/TXT/HTML/?uri=CELEX:52021PC0206&from=EN. Accessed 10 Jun 2022
12. European Parliament. Directorate General for Internal Policies of the Union. The white paper on artificial intelligence. Publications Office. https://data.europa.eu/doi/10.2861/614816
13. H.-L. E. G. on Artificial Intelligence, Ethics Guidelines for Trustworthy AI, European Commission
14. Accenture, Big success with big data - executive summary (2012). https://www.accenture.com/us-en/_acnmedia/accenture/conversion-assets/dotcom/documents/global/pdf/industries_14/accenture-big-data-pov.pdf. Accessed 18 Aug 2022
15. Finance, A.: Industry 4.0 challenges and solutions for the digital transformation and use of exponential technologies. Finance Audit Tax Consult. Corporate: Zurich, Swiss, 1–12 (2015). https://www2.deloitte.com/content/dam/Deloitte/ch/Documents/manufacturing/ch-en-manufacturing-industry-4-0-24102014.pdf
16. Xu, X., Lu, Y., Vogel-Heuser, B., Wang, L.: Industry 4.0 and Industry 5.0—inception, conception and perception. J. Manuf. Syst. **61**, 530–535 (2021). https://doi.org/10.1016/j.jmsy.2021.10.006. https://linkinghub.elsevier.com/retrieve/pii/S0278612521002119

17. P. O. o. t. E. Union: Enabling technologies for Industry 5.0: results of a workshop with Europe's technology leaders. http://op.europa.eu/en/publication-detail/-/publication/8e5de100-2a1c-11eb-9d7e-01aa75ed71a1/language-en
18. Bedué, P., Fritzsche, A.: Can we trust AI? an empirical investigation of trust requirements and guide to successful AI adoption. J. Enterprise Informat. Manag. **35**, 530–549 (2021).
19. Bartneck, C., Lütge, C., Wagner, A., Welsh, S.: An Introduction to Ethics in Robotics and AI. Springer Nature, Berlin (2021)
20. Dignum, V.: Responsible Artificial Intelligence: How to Develop and Use AI in a Responsible Way. Springer Nature, Berlin (2019)
21. European Commission. Directorate General for Communications Networks, Content and Technology., High Level Expert Group on Artificial Intelligence. Ethics guidelines for trustworthy AI. Publications Office. https://data.europa.eu/doi/10.2759/346720
22. Hagendorff, T.: The ethics of AI ethics: an evaluation of guidelines. Minds Mach. **30**(1), 99–120 (2020)
23. Eitel-Porter, R.: Beyond the promise: implementing ethical AI. AI Ethics **1**(1), 73–80 (2021)
24. Hopkin, P., Thompson, C.: Fundamentals of Risk Management: Understanding, Evaluating and Implementing Effective Enterprise Risk Management, 6th edn. Kogan Page, London (2022)
25. Palacios Guillem, M.: New methodology developed for the integration of lean manufacturing; kaizen and ISO 31000: 2009 based on ISO 9001:2015 (2019). https://www.3ciencias.com/articulos/articulo/nueva-metodologia-desarrollada-para-la-integracion-de-lean-manufacturing-kaizen-e-iso-310002009-basados-en-la-iso-90012015/. Accessed 18 July 2022. https://doi.org/10.17993/3cemp.2019.080238.12-43
26. ISO, ISO 23247-1:2021. (2021) https://www.iso.org/cms/render/live/en/sites/isoorg/contents/data/standard/07/50/75066.html. Accessed 21 July 2022
27. Lu, Y., Huang, H., Liu, C., Xu, X.: Standards for smart manufacturing: A review. In: 2019 IEEE 15th International Conference on Automation Science and Engineering (CASE), pp. 73–78, IEEE, Piscataway (2019)
28. Vyhmeister, E., Castane, G., Östberg, P.-O., Thevenin, S.: A responsible AI framework: pipeline contextualisation. AI Ethics **3**, 1–23 (2022)
29. Vyhmeister, E., Gonzalez-Castane, G., Östbergy, P.-O.: Risk as a driver for AI framework development on manufacturing. AI Ethics **3**, 1–20 (2022)
30. Kumar, R.S.S., Brien, D.O., Albert, K., Viljöen, S., Snover, J.: Failure modes in machine learning systems (2019). Preprint arXiv:1911.11034
31. Li, J., Chignell, M.: FMEA-AI: AI fairness impact assessment using failure mode and effects analysis. AI Ethics **2**, 1–14 (2022)
32. Ni, H., Chen, A., Chen, N.: Some extensions on risk matrix approach. Safety Sci. **48**(10), 1269–1278 (2010)
33. ASSISTANT: Assistant project – learning and robust decision support sytem for agile manufacturing environments. https://assistant-project.eu/. Accessed 04Jun 2022

Human in the AI Loop via xAI and Active Learning for Visual Inspection

Jože M. Rožanec, Elias Montini, Vincenzo Cutrona,
Dimitrios Papamartzivanos, Timotej Klemenčič, Blaž Fortuna,
Dunja Mladenić, Entso Veliou, Thanassis Giannetsos,
and Christos Emmanouilidis

1 Introduction

Industrial revolutions have historically disrupted manufacturing by introducing automation into the production process. Increasing automation changed worker responsibilities and roles. While past manufacturing revolutions were driven from the optimization point of view, the Industry 5.0 concepts capitalize on the technological foundations of Industry 4.0 to steer manufacturing toward human-centricity [32, 67], adding resilience and sustainability among its key targets [29]. This change is part of a holistic understanding of the industry's societal role. In particular, the

J. M. Rožanec (✉)
Jožef Stefan Institute, Ljubljana, Slovenia

Qlector d.o.o., Ljubljana, Slovenia
e-mail: joze.rozanec@ijs.si

E. Montini · V. Cutrona
University of Applied Science of Southern Switzerland, Manno, Switzerland

D. Papamartzivanos · T. Giannetsos
Ubitech Ltd., Athens, Greece

T. Klemenčič
University of Ljubljana, Ljubljana, Slovenia

B. Fortuna
Qlector d.o.o., Slovenia

D. Mladenić
Jožef Stefan Institute, Ljubljana, Slovenia

E. Veliou
University of West Attica, Aigaleo, Greece

C. Emmanouilidis
University of Groningen, Groningen, The Netherlands

J. Soldatos (ed.), *Artificial Intelligence in Manufacturing*,
https://doi.org/10.1007/978-3-031-46452-2_22

European Commission expects the industry to collaborate on achieving societal goals that transcend jobs and company growth.

Human-centric manufacturing within the Industry 5.0 aims to ensure that human well-being, needs, and values are placed at the center of the manufacturing process. Furthermore, it seeks to enable collaborative intelligence between humans and machines to enable co-innovation, co-design, and co-creation of products and services [62], thus allowing leveraging on their strengths to maximize individual and joint outcomes and their joint added value [34]. It is expected that synergies enabled within Industry 5.0 will still allow for high-speed and mass-personalized manufacturing but will shift repetitive and monotonous tasks to be more assigned to machines to capitalize more on the human propensity for critical thinking and give them to more cognitively demanding tasks [71].

The emerging shift in human roles goes beyond allowing them to move away from repetitive tasks to undertake other physical activities. As non-human actors, including artificial intelligence (AI) - enabled ones, undertake tasks that can be automated, humans are not necessarily excluded but may well play a higher added value and steering role, bringing their cognitive capabilities into the AI loop [33]. This includes active synergies between AI-enabled non-human entities and humans, resulting in novel work configurations [41]. Such configurations empower human actors in new roles rather than diminishing them [8]. As a consequence, it is increasingly recognized that involving instead of replacing the human from the AI loop not only elevates the role of humans in such work environments but also significantly enhances the machine learning process, and, therefore, the emergent capabilities of the AI-enabled actors [77]. As a result, such synergies involve humans and non-human entities who jointly contribute to shaping an emergent meta-human learning system, which in turn is more capable and powerful than human and non-human entities acting alone [70].

A possible realization of such human-machine collaboration emerges from two sub-fields of artificial intelligence: active learning and explainable artificial intelligence (XAI). Active learning is concerned with finding pieces of data that allow machine learning algorithms to learn better toward a specific goal. Human intervention is frequently required, e.g., to label selected pieces of data and enable such learning. On the other hand, XAI aims to make the machine learning models intelligible to the human person so that humans can understand the rationale behind machine learning model predictions. While active learning requires human expertise to teach machines to learn better, XAI aims to help humans learn better about how machines learn and think. This way, both paradigms play on the strengths of humans and machines to realize synergistic relationships between them.

Among the contributions of the present work are (i) a brief introduction to the state-of-the-art research on human-machine collaboration, key aspects of trustworthiness and accountability in the context of Industry 5.0, and research related to automated visual inspection; (ii) the development of a vision on how an AI-first human-centric visual inspection solution could be realized; and (iii) a description of experiments and results obtained in the field of automated visual inspection at the EU H2020 STAR project.

The rest of the work is structured as follows: Sect. 2 describes related work, providing an overview of human-machine collaboration, the industry 5.0 paradigm and human-centric manufacturing, state-of-the-art on automated quality inspection, and a vision of how human-machine collaboration can be realized in the visual inspection domain. In Sect. 3, relevant research contributions from the EU H2020 STAR project are outlined, offering concrete examples of humans and AI working in synergy. Finally, Sect. 4 provides conclusions and insights into future work.

2 Background

2.1 Overview on Human-Machine Collaboration

The advent of increasingly intelligent machines has enabled a new kind of relationship: the relationship between humans and machines. Cooperative relationships between humans and machines were envisioned back in 1960 [39, 65]. This work defines machines in a broad sense, considering intelligent systems that can make decisions autonomously and independently (e.g., automated, autonomous, or AI agents, robots, vehicles, and instruments) [39, 83, 114]. Relationships between humans and machines have been characterized through different theories, such as the Socio-Technical Systems theory (considers humans and technology shape each other while pursuing a common goal within an organization), Actor-Network theory (considers machines should be equally pondered by humans when analyzing a social system, considering the later as an association of heterogeneous elements), Cyber-Physical Social Systems theory (extends the Socio-Technical Systems theory emphasizing social dimensions where computational algorithms are used to monitor devices), the theory on social machines (considers systems that combine social participation with machine-based computation), and the Human-Machine Networks theory (considers humans and machines form interdependent networks characterized by synergistic interactions). The first three theories conceptualize humans and machines as a single unit, while the last two consider social structures mediated in human-machine networks. In particular, the Socio-Technical Systems theory considers humans and technology shape each other while pursuing a common goal within an organization. The Cyber-Physical Social Systems theory extends this vision, emphasizing social dimensions where computational algorithms are used to monitor and control devices. Moreover, the Actor-Network theory conceptualizes the social system as an association of heterogeneous elements and advocates that machines should be equally pondered to humans. The theory of social machines is interested in systems that combine social participation with machine-based computation. In contrast, the Human-Machine Networks theory considers humans and machines to form interdependent networks characterized by synergistic interactions. A thorough analysis of the abovementioned concepts can be found in [105].

Regardless of the particular theory, the goal remains the same: foster and understand mutualistic and synergistic relationships between humans and machines, where the strengths of both are optimized toward a common goal to achieve what was previously unattainable to each of them. To that end, individual roles must either be clearly defined or allow for a clear sliding of roles when a role can be shared among different types of actors. This will ensure a dynamic division of tasks, optimal use of resources, and reduced processing time. Machines are aimed at supporting, improving, and extending human capabilities. The joint outcomes of human-machine collaboration can result in systems capable of creativity and intuitive action to transcend mere automation. Communication is a critical aspect of every social system. Therefore, emphasis must be placed on the interaction interfaces between such actors. To make such interfaces effective, the concept of shared context or situation awareness between collaborating agents becomes essential and can be seen as a form of mutual understanding [33]. This shared context is enabled through interaction communication of different modalities, including direct verbal (speech, text) and non-verbal (gestures, action and intention recognition, emotions recognition). On the other hand, means must be designed so that humans can understand the machine's goals and rationale for acting to reach such goals in a human-like form. In this regard, human-machine interfaces to support multi-modal interaction play a crucial role. These aspects were also identified by Jwo et al. [53], who described the 3I (Intellect, Interaction, and Interface) aspects that must be considered for achieving human-in-the-loop smart manufacturing.

Beyond shared context, human-machine cooperation requires adequate communication and shared or sliding control [101]. To realize an effective bidirectional information exchange, theory and methods must address how data and machine reasoning can be presented intuitively to humans. Frameworks and models abstracting human cognitive capabilities [60] are key to achieving this. Aligning the design of interactive interfaces and support tools for human-machine interactions with such concepts can be critically important for making effective human–machine interfaces. Enhancements in the interactivity, multisensitivity, and autonomy of feedback functions implemented on such interfaces allow for deeper integration between humans and machines. Shared control can be articulated at operational, tactical, and strategic levels, affecting information-gathering, information-analysis, decision-making, and action implementation.

Human-machine interactions can be viewed from multiple perspectives, necessitating a thorough consideration of several factors influencing such collaborations. These factors encompass emotional and social responses, task design and assignment, trust, acceptance, decision-making, and accountability [23]. Notably, research indicates that machines in collaborative settings impact human behavior, resulting in a diminished emotional response toward them. Consequently, this reduced emotional response can foster more rational interactions. Moreover, studies reveal that humans perceive a team more favorably when machines acknowledge and admit their errors. Additionally, the absence of social pressure from humans can detrimentally affect overall human productivity. Furthermore, concerning accountability for decision-making, humans tend to shift responsibility onto machines.

Trust, a critical aspect to consider, has been explored extensively. Studies demonstrate that trust in machines is closely linked to perceived aptness [23]. Instances of machine errors often lead to a loss of trust, particularly when machines act autonomously. However, if machines operate in an advisory capacity, trust can be amended over time. Additionally, research reveals that while humans value machine advice, they hesitate to relinquish decision-making authority entirely. Nevertheless, relying excessively on machines can result in sub-optimal outcomes, as humans may fail to identify specific scenarios that necessitate their attention and judgment. For further details about the abovementioned experiments and additional insights the reader may be interested on the works by Chugunova et al. [23].

2.2 Industry 5.0 and Human-Centric Manufacturing

2.2.1 New Technological Opportunities to Reshape the Human Workforce

Digital transformation in production environments demands new digital skills and radically reshapes the roles of plant and machine operators [22, 104]. While Industry 4.0 emphasizes the use of technologies to interconnect different stages of the value chain and the use of data analytics to increase productivity, Industry 5.0 emphasizes the role of humans in the manufacturing context [16, 54]. Furthermore, it aims to develop means that enable humans to work alongside advanced technologies to enhance industry-related processes [68]. An extensive review of this paradigm and its components was written by Leng et al. [62]. Nevertheless, two components are relevant to this work: collaborative intelligence and multi-objective interweaving. Collaborative Intelligence is the fusion of human and AI [110]. In the context of Industry 5.0, the fusion of both types of intelligence entails the cognitive coordination between humans and AI in machines, enabling them to collaborate in the innovation, design, and creation of tailored products and services. Complementarity between humans and AI (see Table 1) leads to the more effective execution of such tasks than would be possible if relegated to humans or machines only [18, 51, 75, 86].

When analyzing complementarities, humans have the knowledge and skills to develop and train machines by framing the problems to be solved and providing feedback regarding their actions or outputs [50, 62, 82, 112]. Furthermore, humans can enrich machine outcomes by interpreting results and insights and deciding how to act upon them [113]. Machines amplify workers' cognitive abilities: they can track many data sources and decide what information is potentially relevant to humans. Furthermore, machines can excel at repetitive tasks and free humans from such a burden. Such complementary is considered within the multi-objective interweaving nature of Industry 5.0, which enables optimizing multiple goals beyond process performance and social and environmental sustainability [13].

Table 1 Overview on humans and AI complementarity (adapted from [50, 62, 82], and complemented with our observations). **x**: capability completely fulfilled; **o**: capability partially fulfilled

Capability		Humans	Machines
Strengths and capabilities	Leadership	x	
	Teamwork	x	
	Creativity	x	o
	Problem-solving	x	x
	Risk assessment	x	o
	Intuition	x	
	Interpretation	x	
	Empathy	x	
	Adapt behavior	x	o
	Learn from experience	x	o
	Speed		x
	Scalability		x
	Endurance		x
	Quantitative accuracy		x
	Process large amounts of data		x
	Process different kinds of data in parallel	x	
	Perform continuous operations		x
	Consistent decision-making		x
	Physical and cognitive abilities	x	o
Weaknesses	Prone to biases and errors	x	
	Affected by emotions	x	
	Affected by distractions	x	
	Prone to frauds and adversarial attacks	o	x
	Affected by fatigue	x	
	Limited to certain scope and goals		x
	Lack of emotional intelligence		x
	Lack of social skills		x

Moreover, research suggests that leading companies are beginning to recognize the benefit of using machines and automation systems to supplement human labor rather than replacing the human workforce entirely [5, 28]. While AI was already able to tackle certain tasks with super-human capability [24], it has recently shown progress in areas such as creativity (e.g., through generative models such as DALL·E 2 [84]) or problem-solving [20], opening new frontiers of human-machine collaboration, such as co-creativity [9, 64].

In addition to the direct human involvement described above, digital twins [74] are another way to incorporate human insights into the AI processes. By creating virtual models of human behavior and mental processes, more profound insights into how humans interact with the world and use this information to improve AI systems. Digital twins can also support explainability and transparency in AI systems, making explaining how they arrive at their decisions easier [12]. Moreover, digital representations can be used to consider users' preferences in the AI system behaviors, e.g., type of support [48, 106].

2.2.2 Trustworthiness and Implications for AI-Driven Industrial Systems

Trustworthiness for systems and their associated services and characteristics is defined according to the International Organization for Standardization (ISO) as *"the ability to meet stakeholders' expectations in a verifiable way"* [3]. It follows that trustworthiness can refer to products, services, technology, and data and, ultimately, to organizations. Therefore, the concept of trustworthiness is directly applicable to AI-driven systems, particularly to human-centric AI-enabled solutions. However, it should be understood that trustworthiness is a multifaceted concept, incorporating distinct characteristics such as accountability, accuracy, authenticity, availability, controllability, integrity, privacy, quality, reliability, resilience, robustness, safety, security, transparency, and usability [3].

Some of these characteristics should be seen as emerging characteristics of AI-enabled systems, which are not solely determined by the AI's contribution to an overall solution. Focusing specifically on the AI components of such solutions, ethics guidelines published by the European Commission (EC) identifies seven key requirements for trustworthiness characteristics that must be addressed [35]. These include (i) human agency and oversight; (ii) technical robustness and safety; (iii) privacy and data governance; (iv) transparency; (v) diversity, non-discrimination, and fairness; (vi) societal and environmental well-being; and (vii) accountability. Regarding some of these characteristics, there is a direct correspondence between broader trustworthiness as documented according to ISO and the EC guidelines. Technical robustness, safety, privacy, transparency, and accountability are identified in both sources. Human agency and oversight are directly linked to controllability, and so is governance, which is also the prime focus of ISO recommendations [2]. Given the societal impacts that AI-induced outcomes can have, the EC has also highlighted diversity, non-discrimination, fairness, and societal and environmental well-being as key characteristics of trusted AI solutions. However, these aspects are also partly addressed as part of the broader concept of "freedom from risk", which can be defined as the extent to which a system avoids or mitigates risks to economic status, human life, health, and well-being and or the environment [1].

The trustworthiness of an AI system can be affected by multiple factors. Some of them relate to cybersecurity. In particular, machine learning algorithms are vulnerable to poison and evasion attacks. During poisoning attacks, the adversary aims to tamper with the training data used to create the machine learning models and distort the AI model on its foundation [42, 96]. Evasion attacks are performed during inference, where the attacker crafts adversarial inputs that may seem normal to humans but drive the models to classify the inputs wrongly [49, 72]. Such an adversarial landscape poses significant challenges and requires a collaborative approach between humans and machines to build defenses that can lead to more robust and trustworthy AI solutions. While human intelligence can be used for the human-in-the-loop adversarial generation, where humans are guided to break models [108], AI solutions can be trained to detect adversarial inputs and uncover potentially malicious instances that try to evade the AI models [10]. Furthermore, human-machine collaboration can be fostered to detect such attacks promptly.

Accountability refers to the state of being accountable and relates to allocated responsibility [3]. At the system level, accountability is a property that ensures that the actions of an entity can be traced uniquely to the entity [4]. However, when considering governance, accountability is the obligation of an individual or organization to account for its activities, accept responsibility for them, and disclose the results in a transparent manner [2]. Therefore, accountability is closely linked to transparency for AI-enabled systems, which is served via XAI and interpretable AI. XAI and interpretable AI ensure that AI systems can be trusted when analyzing model outcomes that impact costs and investments or whenever their outputs provide information to guide human decision-making. Accuracy generally refers to the closeness of results and estimates to true values but, in the context of AI, further attains the meaning appropriate for specific machine learning tasks. Any entity that is what it claims to be is said to be characterized by authenticity, with relevant connotations for what AI-enabled systems claim to deliver. Such systems may furthermore be characterized by enhanced availability to the extent that they are usable on demand. Other characteristics such as integrity, privacy, and security attain additional meaning and importance in AI-driven systems and are further discussed in the next section. They can contribute to and affect the overall quality, reliability, resilience, robustness, and safety, whether the unit of interest is a component, a product, a production asset, or a service, with implications for individual workers all the way to the organization as a whole. When considering accountability for AI systems from the legal perspective, the EU AI Act [36] in its current form considers developers and manufacturers responsible for AI failures or unexpected outcomes. Nevertheless, the concept of accountability will evolve based on the issues found in practice and the corresponding jurisprudence that will shape the learning on how different risks, contexts, and outcomes must be considered in the industry context [46].

2.3 Automated Quality Inspection

2.3.1 The Role of Robotics

The increasing prevalence of human-robot collaboration in diverse industries show-cases the efforts to enhance workplace productivity, efficiency, and safety through the symbiotic interaction of robots and humans [44]. In manufacturing, robots are employed for repetitive and physically demanding tasks, enabling human workers to allocate their skills toward more intricate and creative endeavors. This collaborative partnership allows for the fusion of human and robot capabilities, maximizing the overall outcomes.

The successful implementation of human-robot interaction owes credit to collaborative robots, commonly called cobots [56]. These advanced robots have sophisticated sensors and programming that facilitate safe and intuitive human interaction. This collaboration improves productivity and fosters a work environment where humans and robots can coexist harmoniously. This approach harmoniously merges robots' precision and accuracy with human workers' adaptability and dexterity.

Robotic integration in product quality control has become widespread across diverse industries and production sectors. Robots offer exceptional advantages within quality inspection processes, including precise repeatability and accurate movements [17]. They possess the capability to analyze various product aspects such as dimensions, surface defects, color, texture, and alignment, ensuring adherence to predefined standards. Robots' superior accuracy and efficiency make them an ideal choice for quality control applications.

To facilitate quality testing, robots are equipped with a range of sensors. These sensors enable precise measurement, detection, and sorting operations. Robots with cameras utilize advanced machine vision techniques to analyze image and video streams and identify anomalies like cracks, scratches, and other imperfections [107]. Subsequently, defective items are segregated from conforming ones, elevating overall production quality. The industry is witnessing an increasing adoption of 3D vision systems, particularly in applications requiring object grasping and precise information about object position and orientation.

Specially designed robots, such as coordinate measuring machines, are employed for dimensional and precision measurements. These robots feature high-precision axis encoders and accurate touch probes, enabling them to detect part measurements and consistently evaluate adherence to quality standards [61].

The active learning paradigm can be applied to enable efficient and flexible learning in robots. This can be particularly useful in resource-constrained industrial environments, where data scarcity and limited human knowledge prevail, acquiring essential data through unsupervised discovery becomes imperative [26]. Active learning demonstrates extensive applicability in robotics, encompassing prioritized decision-making, inspection, object recognition, and classification. Within quality control, active learning algorithms optimize machine learning models' defect detection and quality assessment training process. By actively selecting informative samples for labeling, active learning minimizes labeling efforts, augments model training efficiency, and ultimately enhances the accuracy and performance of quality control systems.

An intriguing domain of investigation pertains to the advancement of intuitive and natural interfaces that foster seamless communication and interaction between humans and robots. This entails the exploration of innovative interaction modalities, encompassing speech, gestures, and facial expressions, or even using augmented reality to customize the robots' appearance and foster better interaction with humans [59]. Other key research areas involve developing adaptive and flexible robotic systems that dynamically adapt their behavior and actions to the prevailing context and the human collaborator's preferences, achieving low processing times [78]. These could be critical to enable real-time human intent recognition, situational awareness, and decision-making, all aimed at augmenting the adaptability and responsiveness of robots during collaborative tasks.

2.3.2 Artificial Intelligence—Enabled Visual Inspection

Visual inspection is frequently used to assess whether the manufactured product complies with quality standards and allows for the detection of functional and cosmetic defects [21]. It has historically involved human inspectors in determining whether the manufactured pieces are defective. Nevertheless, the human visual system excels in a world of variety and change, while the visual inspection process requires repeatedly observing the same product type. Furthermore, human visual inspection suffers from poor scalability and the fact that it is subjective, creating an inherent inspector-to-inspector inconsistency. The quality of visual inspection can be affected by many factors. See [94] classified them into five categories, whether they are related to the (i) task, (ii) individual, (iii) environment, (iv) organization, or (v) social aspects.

To solve the issues described above, much effort has been invested in automated visual inspection by creating software capable of inspecting manufactured products and determining whether they are defective. Cameras are used to provide visual input. Different approaches have been developed to determine whether a defect exists or not.

Automated optical quality control may target visual features as simple as colors, but more complex ones are involved in crack detection, the orientation of threads, defects in bolts [52] and metallic nuts [14]. Through automated optical inspection, it is also possible to detect defects on product surfaces of wide-ranging sizes [19, 103, 117]. Furthermore, it is also possible to target the actual manufacturing process, for example, welding [102], injection molding [66], or assembly of manufactured components [37]. Additionally, automated visual inspection applies to remanufacturing products at the end of their useful life [91].

State-of-the-art (SOTA) automated visual inspection techniques are dominated by deep learning approaches, achieving high-performance levels [6]. Among the many types of learning from data for visual inspection, unsupervised, weakly supervised, and supervised methods can be named. Unsupervised methods aim to discriminate defective manufactured pieces without labeled data. The weakly supervised approach assumes that data has an inherent cluster structure (instances of the same class are close to each other) and that the data lies in a manifold (nearby data instances have similar predictions). Therefore, it leverages a small amount of annotated data and unlabeled data to learn and issue predictions. Finally, supervised methods require annotated data and usually perform best among the three approaches. Often, labeled data are unavailable in sufficient range and numbers to enable fully supervised learning and additional exemplar images can be produced through data augmentation [55]. In addition, multiple strategies have been developed to reduce the labeled data required to train and enhance a given classifier. Among them are active learning, generative AI, and few-shot learning. In the context of visual inspection, active learning studies how to select data instances that can be presented to a human annotator to maximize the models' learning. Generative AI aims to learn how to create data instances that resemble a particular class. Finally, few-shot learning aims to develop means by which the learner can acquire experience to solve a specific task with only a few labeled examples. To compensate for the lack of labeled data, it can either augment the dataset with samples from other datasets or use unlabeled data, acquire knowledge on another dataset, or algorithm (e.g., by adapting hyperparameters based on prior meta-learned knowledge) [109].

Regardless of the progress made in automated visual inspection, many challenges remain. First, there is no universal solution for automated visual inspection: solutions and approaches have been developed to target a specific product. Flexibility to address the inspection of multiple manufactured products with a single visual inspection system is a complex challenge and remains an open issue [21, 25, 80]. Second, unsupervised machine learning models do not require annotating data and may provide a certain level of defect detection when associating data clusters to categories (e.g., types of defects or no defects). Furthermore, given that no prior annotation of expected defects is required, they are suitable when various defects exist. Nevertheless, their detection rates are lower than those obtained by supervised machine learning models. Therefore, it should be examined use case by use case whether the unsupervised machine learning models are a suitable solution. Third, data collection and annotation are expensive. While data collection affects unsupervised machine learning models, data collection and annotations directly impact supervised machine learning approaches. While multiple strategies have been envisioned to overcome this issue (e.g., generative models, active learning, and few-shot learning), data collection and annotation remain an open challenge. Finally, better explainability techniques and intuitive ways to convey information to humans must be developed to understand whether the models learn and predict properly.

2.4 Realizing Human-Machine Collaboration in Visual Inspection

While much progress has been made in automated visual inspection, the authors recognize that most solutions are custom and developed for a particular product type. Developing systems that could adapt to a broad set of products and requirements remain a challenge. In human-centered manufacturing, it is critical to rethink and redesign the role of humans in the visual inspection process. The role of humans in automated visual inspection is shifting away from repetitive and manual tasks to roles with more cognitive involvement, which can still not be replicated by machines and AI. In the simplest case, this involves humans labeling acquired image samples to guide the machine learning process [98]. However, the role of humans extends beyond data labeling and may involve interaction loops between humans and AI as part of the machine learning process [79].

In this regard, two machine learning paradigms are particularly important: active learning and XAI. On the one side, active learning is an AI paradigm that seeks the intervention of an oracle (usually a human person) to help the machine learning model learn better toward an objective. XAI, on the other side, aims to explain the rationale behind a machine learning model action or prediction. Doing so enables a fruitful dialogue between humans and machines by providing insights into the machines' rationale and decision-making process.

Active learning for classification is based on the premises that unlabeled data (either collected or generated) is abundant, the data labeling is expensive, and the models' generalization error can be minimized by carefully selecting new input instances with which the model is trained [95, 99]. Active learning for classification has traditionally focused on the data (selecting or generating the data without further consideration for the model at hand) and the model's learning (e.g., considering the uncertainty at the predicted scores). Nevertheless, approaches have been developed to consider both dimensions and provide a holistic solution. One of them is the Robust Zero-Sum Game (RZSG) framework [119], which attempts to optimize both objectives at once, framing the data selection as a robust optimization problem to find the best weights for unlabeled data to minimize the actual risk, reduce the average loss (to achieve greater robustness to outliers) and minimize the maximal loss (increasing the robustness to imbalanced data distributions). Another perspective has been considered by Zajec et al. [118] and Križnar et al. [57], who aim to select data based on insights provided by XAI methods and, therefore, benefit from direct insights into the model's learning dynamics. Regardless of the approach, Wu et al. [111] propose that three aspects must be considered when searching for the most valuable samples: informativeness (contains rich information that would benefit the objective function), representativeness (how many other samples are similar to it), and diversity (the samples do not concentrate in a particular region but rather are scattered across the whole space). Strategies will be conditioned by particular requirements (e.g., whether the data instances are drawn from a pool of samples or a data stream). For a detailed review of active learning, the reader may be interested in some high-quality surveys of this domain. In particular, the works by Settles [95] and Rožanec et al. [87] can serve as an introduction to this topic. Furthermore, the surveys by Fu et al. [38] and Kumar et al. [58] provide an overview of querying strategies in a batch setting; the survey by Lughofer [69] give an overview of active learning in online settings, and the study by Ren et al. [85] describes active learning approaches related to deep learning models.

While AI models have the potential to automate many tasks and achieve super-human performance levels, in most cases, such models are opaque to humans: their predictions are mostly accurate, but no intuition regarding their reasoning process is conveyed to humans. Understanding the rationale behind a model's prediction is of utmost importance, given it provides a means to assess whether the predictions are based on accurate facts and intuitions. Furthermore, it is crucial to develop means to understand the model's reasoning process given the impact such techniques have on the real world, either in fully automated settings or when decision-making is delegated to humans. Such insights enable responsible decision-making and accountability. The subfield of AI research developing techniques and mechanisms to elucidate the models' rationale and how to present them to humans is known as XAI. While the field can be traced back to the 1970s [93], it has recently flourished with the advent of modern deep learning [115]. When dealing with XAI, it is important to understand what makes a good explanation. A good explanation must consider at least three elements [11]: (a) reasons for a given model output (e.g., features and their values, how strongly do features influence a forecast, whether the features at which the model looks at make sense w.r.t. the forecast, how did training data influence the model's learning), (b) context (e.g., the data on which the machine learning model was trained, the context on which inference is performed), and (c) how is the abovementioned information conveyed to the users (e.g., target audience, the terminology used by such an audience, what information can be disclosed to it). XAI can be valuable in enhancing human understanding with new (machine-based) perspectives. It can also help to understand whether the model is optimizing for one or few of all required goals and, therefore, identify an appropriate compromise between the different goals that must be satisfied for the problem at hand [30]. To assess the goodness of an explanation, aspects such as user satisfaction, the explanation persuasiveness, the improvement of human judgment, the improvement of human-AI system performance, the automation capability, and the novelty of explanation must be considered [92]. For a detailed review of XAI, the reader may consider the works of Arrieta et al. [11], Doshi-Velez et al. [30], and Schwalbe et al. [92]. The work of Bodria et al. [15] provides a comprehensive introduction to XAI black box methods, and the works of Doshi-Velez et al. [30], Hoffman et al. [45], and Das et al. [27] focus on insights about how to measure the quality of explanations.

Active learning and XAI can complement each other. Understanding the rationale behind a model prediction provides valuable insight to humans and can also be leveraged in an active learning setting. In the particular case of defect inspection, insights obtained by XAI techniques are usually presented in anomaly maps. Such anomaly maps highlight regions of the image the machine learning models consider to issue a prediction. The more perfect the learning of a machine learning model, the better those anomaly maps should annotate a given image indicating defective regions. Therefore, the insights obtained from those anomaly maps can be used in at least two ways. First, the anomaly maps can be handed to the oracle (human inspector), who, aided by the anomaly map and the image of the product, may realize better where the manufacturing errors are, if any. Second, anomaly maps can be used to develop novel models and active learning policies that allow for data selection, considering what was learned by the model and how the model perceives unlabeled data. This approach is detailed in Fig. 1, which depicts how an initial dataset is used to train machine learning models for defect classification or data generation. In the model training process, XAI can be used to debug and iterate the

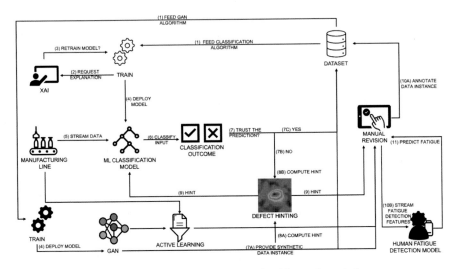

Fig. 1 Envisioned setup for an AI-first human-centric visual inspection solution

model until getting satisfactory results. The classification model is then deployed to perform inference on incoming product images from the manufacturing line. If the classification scores for certain classes are high enough, the product can be classified as good or defective. When the uncertainty around the predicted scores is not low enough, the case can be sent for manual revision. Insights obtained through XAI and unsupervised classification models can be used to hint to the human inspector where the defects may be located. Alternative data sources for the manual revision or data labeling process can be generative models (e.g., generative adversarial networks), which can be used to generate labeled synthetic data and validate the level of attention of a human inspector. When collecting data, active learning techniques can be used to select the most promising data instances from either generative models or incoming images from the manufacturing line, reducing the labeling effort. Finally, a separate model can monitor human inspectors to predict fatigue and performance. Such models can be a valuable tool to ensure workplace well-being and enhance work quality. Some of the results obtained within the STAR project are presented in Sect. 3.1.

In recent years, researchers have made significant progress in understanding and quantifying fatigue and recognizing its impact on human performance and overall well-being. Through AI techniques, new approaches have emerged to accurately estimate the fatigue levels of individuals during different tasks and in different contexts [7, 47]. One notable area of inquiry concerns the assessment of fatigue in the workplace. Understanding and managing worker fatigue has become essential given the increasing demands and pressures of modern work environments. AI models can consider various factors and features to assess employee fatigue levels accurately. These models can provide valuable insights for organizations looking to implement strategies and interventions to optimize productivity and ensure employee well-being or to support workflows, including quality controls, such as identifying when operators need a break. Although laboratory experiments have been conducted in this area [63], industrial applications remain relatively restricted compared to other fields, such as driving [97].

3 Industrial Applications

This section briefly describes how some ideas presented in the previous sections have been realized within the EU H2020 STAR project. Three domains are considered: artificial intelligence for visual inspection, digital twins, and cybersecurity.

3.1 Machine Learning and Visual Inspection

In the domain of visual inspection, multiple use cases were considered. The datasets were provided by two industrial partners: *Philips Consumer Lifestyle BV* (Drachten, The Netherlands) and *Iber-Oleff - Componentes Tecnicos Em Plástico, S.A.* (Portugal). The *Philips Consumer Lifestyle BV* manufacturing plant is considered one of Europe's most important Philips development centers and is devoted to producing household appliances. They provided us with three datasets corresponding to different products. The first one corresponded to logo prints on shavers. The visual inspection task required understanding whether the logo was correctly printed or had some printing defect (e.g., double printing or interrupted printing). The second one corresponded to decorative caps covering the shaving head's center, and it required identifying whether the caps were correctly manufactured or if some flow lines or marks existed. Finally, the third dataset was about toothbrush shafts transferring motion from the handle to the brush. It required identifying whether the handles were manufactured without defects or if big dents, small dents, or some stripes could be appreciated. *Iber-Oleff - Componentes Tecnicos Em Plástico, S.A.* provided us with another dataset about automobile air vents they manufacture. The air vents have three components of interest: housing, lamellas (used to direct the air), and plastic links (which keep the lamellas tied together). The visual inspection task required us to determine whether (a) the fork is leaning against the support and correctly positioned, (b) the plastic link is present, (c) the lamella 1 is present, and the link is correctly assembled, and (d) the lamella 3 is present, and the link is correctly assembled.

Through the research, the researchers aimed to develop a comprehensive AI-first and human-centric approach to automated visual inspection. In particular, they (i) developed machine learning models to detect defects, (ii) used active learning to enhance the models' learning process while alleviating the need to label data, (iii) used XAI to enhance the labeling process, (iv) analyzed how data augmentation techniques at embeddings and image level, along with anomaly maps can enhance the machine learning discriminative capabilities, (v) how human fatigue can be detected and predicted in humans, and (vi) how to calibrate and measure models' calibration quality to provide probabilistic predictive scores.

Research at the EU H2020 STAR project confirmed that active learning could alleviate the need for data labeling and help machine learning models learn better based on fewer data instances [90]. Nevertheless, the effort saved depends on the pool of unlabeled images, the use case, and the active learning strategy. Data augmentation techniques at an image or embedding level have increased the models' discriminative performance [89]. Furthermore, complementing images with anomaly maps as input to supervised classification models has substantially improved discriminative capabilities [88]. The data labeling experiments showed decreased labeling accuracy by humans over time [86], which was attributed to human fatigue. While the future labeling quality can be predicted, it requires ground truth data. This can be acquired by showing synthetically generated images. Nevertheless, more research is required to devise new models that would consider other cues and predict human fatigue in data labeling without the requirement of annotated data. Finally, predictive scores alone provide little information to the decision-maker: predictive score distributions differ across different models. Therefore, performing probability calibration is paramount to ensure probability scores have the same semantics across the models. The research compared some of the existing probability calibration techniques and developed metrics to measure and assess calibration quality regardless of ground truth availability [90].

3.2 Human-Digital Twins in Quality Control

In the context of STAR, significant advancements have been made in developing human-digital twins (HDTs). In particular, the project has developed an infrastructure (Clawdite Platform [74]) that allows the effortless creation of replicas of human workers through instantiating their digital counterparts. These HDTs have diverse features, encompassing static characteristics, dynamic data, and behavioral and functional models [73].

To ensure a comprehensive representation of the human worker, STAR's HDT incorporates two crucial data types. Firstly, it assimilates physiological data collected from wearable devices. Secondly, it utilizes quasi-static data, which encapsulates characteristic attributes of the human, offering a holistic perspective on their traits. Central to STAR's HDT is an AI model designed to detect mental stress and physical fatigue. By leveraging physiological and quasi-static data, this AI model effectively gauges the stress and fatigue levels experienced by the human worker. This breakthrough in automated quality control holds remarkable significance, manifesting in two distinct ways:

- During user manual inspection, the HDT continuously monitors the quality control process, actively identifying instances where the worker may be under significant mental or physical stress. In such cases, the system promptly suggests the worker take a break, ensuring their well-being and preventing any potential decline in performance.

- During the training of automatic quality assessment models, as the worker evaluates and labels pictures during the dataset creation, the system periodically assigns a confidence score to each label provided by the user. This confidence score is computed based on evaluating the worker's mental and physical stress levels estimated through the HDT's AI model. By considering these stress levels as an integral part of the quality evaluation process, the HDT provides valuable insights into the worker's state of mind and physical condition, allowing one to consider these features during the training of AI models for quality assessment and control.

The integration of the HDTs, supported by the Clawdite Platform, in STAR's operations signifies a significant step forward in human-AI collaboration. This innovative approach prioritizes human workers' well-being and empowers automated quality control systems, ensuring optimal productivity and efficiency in various industrial settings.

3.3   Making AI Visual Inspection Robust Against Adversarial Attacks

In the context of the STAR project, an AI architecture was created for evaluating adversarial tactics and defense algorithms intended to safeguard, secure, and make the environments of manufacturing AI systems more reliable. More specifically, it was focused on AI-based visual inspection and tackled multiple use cases provided by two industrial partners: *Philips Consumer Lifestyle BV* (Drachten, The Netherlands) and *Iber-Oleff - Componentes Tecnicos Em Plástico, S.A.* (Portugal). Current production lines are often tailored for the mass production of one product or product series in the most efficient way. Given its many advantages, AI is being increasingly adopted for quality inspection. Such models are usually trained considering some convolutional neural network (CNN), which then classifies whether a product is defective through inference upon receiving images captured by the inspection cameras. Nevertheless, such models can be attacked through adversarial data, leading AI models to wrongly classify the products (e.g., not detecting defects). For instance, the adversary may exploit a vulnerability in the visual inspection camera and compromise the integrity of the captured data by manipulating the operational behavior of this business resource.

Among the various experimental testbeds built in the context of the STAR project, the ones created with soother cherries provided by *Philips Consumer Lifestyle BV* were the most challenging. The cherry is the upper part of the soother. The high quality of the cherry must be guaranteed to avoid any harm to the babies. Therefore, detecting any adversarial attack is of primary importance, given the

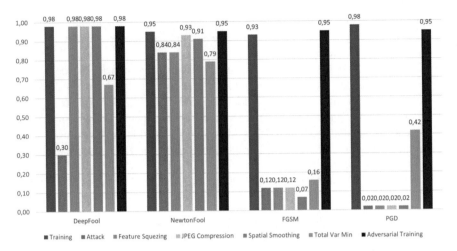

Fig. 2 Evaluation results of pairwise comparison of adversarial attacks and defenses

consequences of the attack can directly impact children's health. The goal of the testbed was to quantify the impact of adversarial attacks on classification models performing a visual inspection and evaluate how effective the defenses against such attacks were. To build the testbed, the Adversarial Robustness Toolbox [81] was used. In the experiments, the following adversarial methods were used: Fast Gradient Sign Attack (FGSM) [40], DeepFool [76], NewtonFool [49], and Projected Gradient Descent (PGD) [72]. The aim was to utilize these well-documented adversarial methods to derive crafted instances that can be used to attack the baseline classification model. Experiments were performed with defense strategies, namely feature squeezing [116], JPEG compression [31], spatial smoothing [116], TotalVarMin [43], and adversarial training [100].

To gather insights regarding adversarial tactics and defenses, they were evaluated pairwise. This enabled us to identify adversarial training as the best defense strategy to enhance the robustness of the CNN models. The basic idea behind adversarial training is to create examples that will be used later in the training process, creating a model aware of adversarial vectors launched against the quality control system. The results of the pairwise evaluation of the attacks and defenses are summarized in Fig. 2. The results are grouped into four sets based on the attack strategy. A baseline classifier was initially trained for each of the four experiments (see tag "Training") to get the perception of the accuracy level that the quality inspection algorithm can achieve. The baseline model achieved an accuracy between 93% and 98%. The "Attack" bar indicates the accuracy of the classifier when posed against the adversarial attack. The DeepFool, FGSM, and PGD attacks strongly affected the classifier, causing the model's accuracy to drop below 30%. This was not the case for the NewtonFool attack, where the classifier's accuracy dropped to 84%. When considering defense strategies, feature squeezing, JPEG compression, and spatial smoothing can defend against the DeepFool attack: for the given dataset, they led to an accuracy of 98%. However, TotalVarMin failed to defend the model. All the defenses failed against the FGSM and the PGD attacks. Based on the acquired results of the pairwise evaluations, it became clear that no clear mapping exists between types of attacks and defenses. Therefore, it can be challenging for defenders to plan a strategy to cope against any attack successfully. This outcome advocates the criticality and the challenge of defending against adversarial AI attacks. While the off-the-shelf and state-of-the-art defenses cannot perform in a stable manner under different adversarial methods, the adversarial training approach seems robust. The results agree with the literature, advocating that Adversarial Training can be a robust solution that can cope with adversaries despite its simplicity. A more detailed description of the abovementioned work can be found in [10].

4 Conclusion

This work has briefly introduced state-of-the-art research on human-machine collaboration, perspectives on human-centric manufacturing, and the key aspects of trustworthiness and accountability in the context of Industry 5.0. It described research on automated quality inspection, considering the role of robotics, AI approaches, and solutions to visual inspection and how a fruitful human-machine collaboration can be developed in the visual inspection domain. Finally, it described the experience and results obtained through research performed in the EU H2020 STAR project.

The converging view from the literature analysis is that human-machine cooperation requires adequate communication and control realized through effective bidirectional information exchange. Studies have been performed to understand peoples' emotional and social responses in human-machine interactions, understand task design, and how humans' trust, acceptance, decision-making, and accountability are developed or impacted in the presence of machines. In the field of visual inspection, much research was invested in automating the task of visual inspection by developing machine learning models to detect product defects. Furthermore, many research efforts targeted the development of techniques for XAI related to machine vision. Visual aids and hints derived by XAI are conveyed to humans through heat maps. Similarly, insights obtained from unsupervised machine learning models are conveyed to humans as anomaly maps. While such approaches solve particular problems, little research describes how a human-in-the-loop approach could be developed for visual inspection in manufacturing settings. This research aims to bridge the gap by implementing existing and researching novel active learning techniques for data selection to enhance the learning of machine learning algorithms. It also explores how labeling requirements could be reduced by employing few-shot learning and active learning techniques. Furthermore, research was conducted to understand how XAI and unsupervised classification methods can be used to generate heat maps and anomaly maps to facilitate data labeling in the context of manual revision or data annotation tasks. Moreover, predictive models were developed to predict how heat maps and anomaly maps should be adapted over time to bridge the gap between the information conveyed by machine learning algorithms and explainability techniques and human perception. In addition, experiments were performed to gain insights related to human fatigue monitoring in the context of visual inspection. The present work described a complete and modular infrastructure developed to instantiate HDT, and different AI models for perceived fatigue exertion and mental stress have been trained to derive relevant features for human-centered production systems. Finally, it describes some research on adversarial attacks and defenses to enhance the understanding of protecting visual inspection setups in manufacturing environments.

While the research presented above advances the understanding of developing a human-in-the-loop approach for visual inspection in manufacturing, many open issues remain to be solved. Further research is required to understand how adaptive humans perceive hinting and how the many solutions described above contribute to building trust between humans and machines. Furthermore, effort must be invested to quantify the benefits such solutions bring to a manufacturing plant when implemented. Future research will encompass the integration of these solutions, aiming to achieve a comprehensive and synergistic implementation. The research will aim to develop new approaches that interleave active learning and XAI. Furthermore, novel few-shot learning solutions will be considered to allow for greater flexibility of the visual inspection while reducing data labeling requirements to a minimum. Finally, integrating AI visual inspection models and the HDT is expected to significantly augment the efficacy of quality inspection processes during user manual assessment and AI model training.

Acknowledgments This work was supported by the Slovenian Research Agency and the European Union's Horizon 2020 program project STAR under grant agreement number H2020-956573.

References

1. 25022:2015 I: System and software quality requirements and evaluation (square) - measurement of quality in use (2015)
2. 38500:2015 I: Information technology — governance of it for the organization (2015)
3. 5723 IT: Trustworthiness - vocabulary (2022)
4. 7498-2:1989 I: Information processing systems - security architecture (1989)
5. Accenture: Reworking the revolution (2018). https://www.accenture.com/_acnmedia/pdf-69/accenture-reworking-the-revolution-jan-2018-pov.pdf
6. Aggour, K.S., Gupta, V.K., Ruscitto, D., et al.: Artificial intelligence/machine learning in manufacturing and inspection: A GE perspective. MRS Bull. **44**(7), 545–558 (2019)
7. Aguirre, A., Pinto, M.J., Cifuentes, C.A., et al.: Machine learning approach for fatigue estimation in sit-to-stand exercise. Sensors **21**(15), 5006 (2021)
8. Amershi, S., Cakmak, M., Bradley Knox, W., Kulesza, T.: Power to the people: the role of humans in interactive machine learning. AI Mag. **35**, 105–120 (2014). https://doi.org/10.1609/aimag.v35i4.2513

9. Anantrasirichai, N., Bull, D.: Artificial intelligence in the creative industries: a review. Artif. Intell. Rev., 1–68 (2022)
10. Anastasiou, T., Karagiorgou, S., Petrou, P., et al.: Towards robustifying image classifiers against the perils of adversarial attacks on artificial intelligence systems. Sensors 22(18), (2022). https://doi.org/10.3390/s22186905, https://www.mdpi.com/1424-8220/22/18/6905
11. Arrieta, A.B., Díaz-Rodríguez, N., Del Ser, J., et al.: Explainable artificial intelligence (XAI): Concepts, taxonomies, opportunities and challenges toward responsible AI. Inf. Fusion 58, 82–115 (2020)
12. Bansal, G., Nushi, B., Kamar, E., et al.: Beyond accuracy: The role of mental models in human-AI team performance. In: Proceedings of the AAAI Conference on Human Computation and Crowdsourcing, pp. 2–11 (2019)
13. Bettoni, A., Montini, E., Righi, M., et al.: Mutualistic and adaptive human-machine collaboration based on machine learning in an injection moulding manufacturing line. Procedia CIRP 93, 395–400 (2020)
14. Bharti, S., McGibney, A., O'Gorman, T.: Edge-enabled federated learning for vision based product quality inspection. In: 2022 33rd Irish Signals and Systems Conference (ISSC), pp. 1–6 (2022). https://doi.org/10.1109/ISSC55427.2022.9826185
15. Bodria, F., Giannotti, F., Guidotti, R., et al.: Benchmarking and survey of explanation methods for black box models. Data Mining Knowl. Discovery, 1–60 (2023)
16. Breque, M., De Nul, L., Petridis, A.: Industry 5.0: Towards a sustainable, human-centric and resilient European industry (2021)
17. Brito, T., Queiroz, J., Piardi, L., et al.: A machine learning approach for collaborative robot smart manufacturing inspection for quality control systems. Procedia Manuf. 51, 11–18 (2020)
18. Cai, C.J., Reif, E., Hegde, N., et al.: Human-centered tools for coping with imperfect algorithms during medical decision-making. In: Proceedings of the 2019 Chi Conference on Human Factors in Computing Systems, pp. 1–14 (2019)
19. Cao, G., Ruan, S., Peng, Y., et al.: Large-complex-surface defect detection by hybrid gradient threshold segmentation and image registration. IEEE Access 6, 36235–36246 (2018). https://doi.org/10.1109/ACCESS.2018.2842028
20. Cao, L.: A new age of AI: Features and futures. IEEE Intell. Syst. 37(1), 25–37 (2022)
21. Chin, R.T., Harlow, C.A.: Automated visual inspection: A survey. IEEE Trans. Pattern Anal. Mach. Intell. 4(6), 557–573 (1982)
22. Chuang, S.: Indispensable skills for human employees in the age of robots and AI. Eur. J. Train. Dev. (ahead-of-print) (2022)
23. Chugunova, M., Sele, D.: We and it: An interdisciplinary review of the experimental evidence on human-machine interaction. Center for law & economics working paper series 12 (2020)
24. Ciregan, D., Meier, U., Schmidhuber, J.: Multi-column deep neural networks for image classification. In: 2012 IEEE Conference on Computer Vision and Pattern Recognition, pp 3642–3649. IEEE (2012)
25. Czimmermann, T., Ciuti, G., Milazzo, M., et al.: Visual-based defect detection and classification approaches for industrial applications—a survey. Sensors 20(5), 1459 (2020)
26. Daniel, C., Viering, M., Metz, J., et al.: Active reward learning. In: Robotics: Science and Systems (2014)
27. Das, A., Rad, P.: Opportunities and challenges in explainable artificial intelligence (XAI): A survey. Preprint (2020). arXiv:200611371
28. Deloitte: The rise of the social enterprise. 2018 Deloitte global human capital trends (2018). https://www2.deloitte.com/content/dam/insights/us/articles/HCTrends2018/2018-HCtrends_Rise-of-the-social-enterprise.pdf
29. Directorate-General for Research and Innovation, European Commission, Breque, M., De Nul, L., Petridis, A.: Industry 5.0: towards a sustainable, human centric and resilient European industry. Publications Office of the European Union (2021). https://op.europa.eu/en/publication-detail/-/publication/468a892a-5097-11eb-b59f-01aa75ed71a1/language-en

30. Doshi-Velez, F., Kim, B.: Towards a rigorous science of interpretable machine learning. Preprint (2017). arXiv:170208608
31. Dziugaite, G.K., Ghahramani, Z., Roy, D.M.: A study of the effect of jpg compression on adversarial images. Preprint (2016). arXiv:160800853
32. EESC: Industry 5.0 (2018). https://www.eesc.europa.eu/en/agenda/our-events/events/industry-50, Accessed: 24 May 2023
33. Emmanouilidis, C., Pistofidis, P., Bertoncelj, L., et al.: Enabling the human in the loop: Linked data and knowledge in industrial cyber-physical systems. Annu. Rev. Control **47**, 249–265 (2019). https://doi.org/10.1016/j.arcontrol.2019.03.004
34. Emmanouilidis, C., Waschull, S., Bokhorst, J.A., et al.: Human in the AI Loop in Production Environments, vol. 633 IFIP. Springer Science and Business Media Deutschland GmbH, pp. 331–342 (2021). https://doi.org/10.1007/978-3-030-85910-7_35
35. European Commission: Ethics guidelines for trustworthy AI (2019)
36. European Commision: Laying down harmonised rules on artificial intelligence (artificial intelligence act) and amending certain union legislative acts (2021)
37. Frustaci, F., Spagnolo, F., Perri, S., et al.: Robust and high-performance machine vision system for automatic quality inspection in assembly processes. Sensors **22**, 2839 (2022). https://doi.org/10.3390/s22082839, https://www.mdpi.com/1424-8220/22/8/2839
38. Fu, Y., Zhu, X., Li, B.: A survey on instance selection for active learning. Knowl. Inf. Syst. **35**(2), 249–283 (2013)
39. Gerber, A., Derckx, P., Döppner, D.A., et al.: Conceptualization of the human-machine symbiosis—a literature review. In: Proceedings of the 53rd Hawaii International Conference on System Sciences (2020)
40. Goodfellow, I.J., Shlens, J., Szegedy, C.: Explaining and harnessing adversarial examples. In: Bengio Y, LeCun Y (eds) 3rd International Conference on Learning Representations, ICLR 2015, San Diego, CA, USA, May 7–9, 2015, Conference Track Proceedings (2015)
41. Grønsund, T., Aanestad, M.: Augmenting the algorithm: Emerging human-in-the-loop work configurations. J. Strat. Inf. Syst. **29**, 101614 (2020). https://doi.org/10.1016/j.jsis.2020.101614
42. Gu, T., Dolan-Gavitt, B., Garg, S.: Badnets: Identifying vulnerabilities in the machine learning model supply chain. Preprint (2017). arXiv:170806733
43. Guo, C., Rana, M., Cisse, M., et al.: Countering adversarial images using input transformations. In: International Conference on Learning Representations (2018). https://openreview.net/forum?id=SyJ7ClWCb
44. Heyer, C.: Human-robot interaction and future industrial robotics applications. In: 2010 IEEE/RSJ International Conference on Intelligent Robots and Systems, pp 4749–4754. IEEE (2010)
45. Hoffman, R.R., Mueller, S.T., Klein, G., et al.: Metrics for explainable ai: Challenges and prospects. Tech. rep., DARPA Explainable AI Program (2018)
46. Hohma, E., Boch, A., Trauth, R., et al.: Investigating accountability for artificial intelligence through risk governance: A workshop-based exploratory study. Front. Psychol. **14**, 86 (2023)
47. Hooda, R., Joshi, V., Shah, M.: A comprehensive review of approaches to detect fatigue using machine learning techniques. Chronic Dis. Transl. Med. **8**(1), 26–35 (2022)
48. Hu, Z., Lou, S., Xing, Y., et al.: Review and perspectives on driver digital twin and its enabling technologies for intelligent vehicles. IEEE Trans. Intell. Veh. (2022)
49. Jang, U., Wu, X., Jha, S.: Objective metrics and gradient descent algorithms for adversarial examples in machine learning. In: Proceedings of the 33rd Annual Computer Security Applications Conference. Association for Computing Machinery, New York, NY, USA, ACSAC '17, pp. 262–277 (2017). https://doi.org/10.1145/3134600.3134635
50. Jarrahi, M.H.: Artificial intelligence and the future of work: Human-AI symbiosis in organizational decision making. Bus. Horizons **61**(4), 577–586 (2018)
51. Jarrahi, M.H., Davoudi, V., Haeri, M.: The key to an effective AI-powered digital pathology: Establishing a symbiotic workflow between pathologists and machine. J. Pathol. Inf. **13**, 100156 (2022)

52. John Rajan, A., Jayakrishna, K., Vignesh, T., et al.: Development of computer vision for inspection of bolt using convolutional neural network. Mater. Today Proc. **45**, 6931–6935 (2021). https://doi.org/10.1016/j.matpr.2021.01.372, https://www.sciencedirect.com/science/article/pii/S2214785321004636. International Conference on Mechanical, Electronics and Computer Engineering 2020: Materials Science

53. Jwo, J.S., Lin, C.S., Lee, C.H.: Smart technology–driven aspects for human-in-the-loop smart manufacturing. Int. J. Adv. Manuf. Technol. **114**, 1741–1752 (2021)

54. Kaasinen, E., Anttila, A.H., Heikkilä, P., et al.: Smooth and resilient human–machine teamwork as an industry 5.0 design challenge. Sustainability **14**(5), 2773 (2022)

55. Kim, T.H., Kim, H.R., Cho, Y.J.: Product inspection methodology via deep learning: An overview. Sensors **21**(15), 5039 (2021). https://doi.org/10.3390/s21155039

56. Kosuge, K., Hirata, Y.: Human-robot interaction. In: 2004 IEEE International Conference on Robotics and Biomimetics, pp. 8–11. IEEE (2004)

57. Križnar, K., Rožanec, J.M., Fortuna, B., et al.: Explainable artificial intelligence meets active learning: A novel gradcam-based active learning strategy, submitted (2023)

58. Kumar, P., Gupta, A.: Active learning query strategies for classification, regression, and clustering: A survey. J. Comput. Sci. Technol. **35**(4), 913–945 (2020)

59. Lambert, A., Norouzi, N., Bruder, G., et al.: A systematic review of ten years of research on human interaction with social robots. Int. J. Human Comput. Interact. **36**(19), 1804–1817 (2020)

60. Langley, P.: Interactive cognitive systems and social intelligence. IEEE Intell. Syst. **32**, 22–30 (2017). https://doi.org/10.1109/MIS.2017.3121556

61. Leach, R., Bourell, D., Carmignato, S., et al.: Geometrical metrology for metal additive manufacturing. CIRP Ann. **68**(2), 677–700 (2019)

62. Leng, J., Sha, W., Wang, B., et al.: Industry 5.0: Prospect and retrospect. J. Manuf. Syst. **65**, 279–295 (2022)

63. Leone, A., Rescio, G., Siciliano, P., et al.: Multi sensors platform for stress monitoring of workers in smart manufacturing context. In: 2020 IEEE International Instrumentation and Measurement Technology Conference (I2MTC), pp 1–5. IEEE (2020)

64. Liapis, A., Yannakakis, G.N., Alexopoulos, C., et al.: Can computers foster human users' creativity? theory and praxis of mixed-initiative co-creativity. Digit. Cult. Educ. **8**, (2016)

65. Licklider, J.C.R.: Man-computer symbiosis. IRE Trans. Human Fact. Electron. **HFE-1**(1), 4–11 (1960). https://doi.org/10.1109/THFE2.1960.4503259

66. Liu, J., Guo, F., Gao, H., et al.: Defect detection of injection molding products on small datasets using transfer learning. J. Manuf. Process. **70**, 400–413 (2021). https://doi.org/10.1016/j.jmapro.2021.08.034

67. Longo, F., Padovano, A., Umbrello, S.: Value-oriented and ethical technology engineering in industry 5.0: A human-centric perspective for the design of the factory of the future. Appl. Sci. **10**(12), 4182 (2020)

68. Lu, Y.: The current status and developing trends of industry 4.0: A review. Inf. Syst. Front., 1–20 (2021)

69. Lughofer, E.: On-line active learning: A new paradigm to improve practical useability of data stream modeling methods. Inf. Sci. **415**, 356–376 (2017)

70. Lyytinen, K., Nickerson, J.V., King, J.L.: Metahuman systems = humans + machines that learn. J. Inf. Technol. (2020). https://doi.org/10.1177/0268396220915917

71. Maddikunta, P.K.R., Pham, Q.V., Prabadevi, B., et al.: Industry 5.0: A survey on enabling technologies and potential applications. J. Ind. Inf. Integr. **26**, 100257 (2022)

72. Madry, A., Makelov, A., Schmidt, L., et al.: Towards deep learning models resistant to adversarial attacks. In: 6th International Conference on Learning Representations, ICLR 2018, Vancouver, BC, Canada, April 30–May 3, 2018, Conference Track Proceedings (2018). OpenReview.net. https://openreview.net/forum?id=rJzIBfZAb

73. Montini, E., Bettoni, A., Ciavotta, M., et al.: A meta-model for modular composition of tailored human digital twins in production. Procedia CIRP **104**, 689–695 (2021)

74. Montini, E., Cutrona, V., Bonomi, N., et al.: An iiot platform for human-aware factory digital twins. Procedia CIRP **107**, 661–667 (2022)
75. Montini, E., Cutrona, V., Dell'Oca, S., et al.: A framework for human-aware collaborative robotics systems development. Procedia CIRP (2023)
76. Moosavi-Dezfooli, S.M., Fawzi, A., Frossard, P.: DeepFool: a simple and accurate method to fool deep neural networks. In: Proceedings of the IEEE Conference on Computer Vision and Pattern Recognition, pp. 2574–2582 (2016)
77. Mosqueira-Rey, E., Hernández-Pereira, E., Alonso-Ríos, D., et al.: Human-in-the-loop machine learning: a state of the art. Artif. Intell. Rev. (2022). https://doi.org/10.1007/s10462-022-10246-w
78. Mukherjee, D., Gupta, K., Chang, L.H., et al.: A survey of robot learning strategies for human-robot collaboration in industrial settings. Robot. Comput. Integr. Manuf. **73**, 102231 (2022)
79. Müller, D., März, M., Scheele, S., et al.: An interactive explanatory AI system for industrial quality control. In: Thirty-Sixth AAAI Conference on Artificial Intelligence, AAAI 2022, Thirty-Fourth Conference on Innovative Applications of Artificial Intelligence, IAAI 2022, The Twelfth Symposium on Educational Advances in Artificial Intelligence, EAAI 2022 Virtual Event, February 22–March 1, 2022, pp. 12580–12586. AAAI Press (2022)
80. Newman, T.S., Jain, A.K.: A survey of automated visual inspection. Comput. Vis. Image Understand. **61**(2), 231–262 (1995)
81. Nicolae, M.I., Sinn, M., Tran, M.N., et al.: Adversarial robustness toolbox v1.0.0. Preprint (2018). arXiv:180701069
82. Paul, S., Yuan, L., Jain, H.K., et al.: Intelligence augmentation: Human factors in ai and future of work. AIS Trans. Human Comput. Interact. **14**(3), 426–445 (2022)
83. Rahwan, I., Cebrian, M., Obradovich, N., et al.: Machine behaviour. Machine Learning and the City: Applications in Architecture and Urban Design, pp. 143–166 (2022)
84. Ramesh, A., Dhariwal, P., Nichol, A., et al.: Hierarchical text-conditional image generation with CLIP latents. Preprint (2022). arXiv:220406125
85. Ren, P., Xiao, Y., Chang, X., et al.: A survey of deep active learning. ACM Comput. Surv. (CSUR) **54**(9), 1–40 (2021)
86. Rožanec, J.M., Karel, K., Montini, E., et al.: Predicting operators' fatigue in a human in the artificial intelligence loop for defect detection in manufacturing. In: Proceedings of the 2023 IFAC World Congress (2023)
87. Rožanec, J.M., Fortuna, B., Mladenić, D.: The future of data mining. chapter 6: Active learning (2022). https://doi.org/10.52305/KCIN5931
88. Rožanec, J.M., Zajec, P., Theodoropoulos, S., et al.: Robust anomaly map assisted multiple defect detection with supervised classification techniques. Preprint (2022). arXiv:221209352
89. Rožanec, J.M., Zajec, P., Theodoropoulos, S., et al.: Synthetic data augmentation using GAN for improved automated visual inspection. Preprint (2022). arXiv:221209317
90. Rožanec, J.M., Bizjak, L., Trajkova, E., et al.: Active learning and novel model calibration measurements for automated visual inspection in manufacturing. J. Intell. Manuf., 1–22 (2023)
91. Saiz, F.A., Alfaro, G., Barandiaran, I.: An inspection and classification system for automotive component remanufacturing industry based on ensemble learning. Information **12**(12), (2021)
92. Schwalbe, G., Finzel, B.: A comprehensive taxonomy for explainable artificial intelligence: a systematic survey of surveys on methods and concepts. Data Mining Knowl. Disc., 1–59 (2023)
93. Scott, A.C., Clancey, W.J., Davis, R., et al.: Explanation capabilities of production-based consultation systems. Tech. rep., Stanford Univ CA Dept Of Computer Science (1977)
94. See, J.E.: Visual inspection: a review of the literature. Sandia Report SAND2012-8590, Sandia National Laboratories, Albuquerque, New Mexico (2012)
95. Settles, B.: Active learning literature survey. Tech. rep., University of Wisconsin-Madison Department of Computer Sciences (2009)
96. Shokri, R., et al.: Bypassing backdoor detection algorithms in deep learning. In: 2020 IEEE European Symposium on Security and Privacy (EuroS&P), pp 175–183. IEEE (2020)

97. Sikander, G., Anwar, S.: Driver fatigue detection systems: A review. IEEE Trans. Intell. Transp. Syst. **20**(6), 2339–2352 (2018)
98. Silva, B., Marques, R., Faustino, D., et al.: Enhance the injection molding quality prediction with artificial intelligence to reach zero-defect manufacturing. Processes **11**, (2023). https://doi.org/10.3390/pr11010062
99. Sugiyama, M., Kawanabe, M.: Active Learning, pp 183–214. MIT Press (2012)
100. Szegedy, C., Zaremba, W., Sutskever, I., et al.: Intriguing properties of neural networks. In: Bengio, Y., LeCun, Y. (eds.) 2nd International Conference on Learning Representations, ICLR 2014, Banff, AB, Canada, April 14–16, 2014, Conference Track Proceedings (2014)
101. Tang, F., Mohammed, M., Longazo, J.: Experiments of human-robot teaming under sliding autonomy. In: 2016 IEEE International Conference on Advanced Intelligent Mechatronics (AIM), pp 113–118 (2016). https://doi.org/10.1109/AIM.2016.7576752
102. Tripicchio, P., Camacho-Gonzalez, G., D'Avella, S.: Welding defect detection: coping with artifacts in the production line. Int. J. Adv. Manuf. Technol. **111**, 1659–1669 (2020). https://doi.org/10.1007/s00170-020-06146-4
103. Tsai, D.M., Jen, P.H.: Autoencoder-based anomaly detection for surface defect inspection. Adv. Eng. Inf. **48**, (2021). https://doi.org/10.1016/j.aei.2021.101272
104. Tschang, F.T., Almirall, E.: Artificial intelligence as augmenting automation: Implications for employment. Acad. Manag. Perspect. **35**(4), 642–659 (2021)
105. Tsvetkova, M., Yasseri, T., Meyer, E.T., et al.: Understanding human-machine networks: a cross-disciplinary survey. ACM Comput. Surv. (CSUR) **50**(1), 1–35 (2017)
106. van Berkel, N., Skov, M.B., Kjeldskov, J.: Human-AI interaction: intermittent, continuous, and proactive. Interactions **28**(6), 67–71 (2021)
107. Villalba-Diez, J., Schmidt, D., Gevers, R., et al.: Deep learning for industrial computer vision quality control in the printing industry 4.0. Sensors **19**(18), 3987 (2019)
108. Wallace, E., Rodriguez, P., Feng, S., et al.: Trick me if you can: Human-in-the-loop generation of adversarial examples for question answering. Trans. Assoc. Comput. Linguist. **7**, 387–401 (2019)
109. Wang, Y., Yao, Q., Kwok, J.T., et al.: Generalizing from a few examples: A survey on few-shot learning. ACM Comput. Surv. (CSUR) **53**(3), 1–34 (2020)
110. Wilson, H.J., Daugherty, P.R.: Collaborative intelligence: Humans and AI are joining forces. Harv. Bus. Rev. **96**(4), 114–123 (2018)
111. Wu, D.: Pool-based sequential active learning for regression. IEEE Trans. Neural Networks Learn. Syst. **30**(5), 1348–1359 (2018)
112. Wu, J., Huang, Z., Hu, Z., et al.: Toward human-in-the-loop AI: Enhancing deep reinforcement learning via real-time human guidance for autonomous driving. Engineering **21**, 75–91 (2023). https://doi.org/10.1016/j.eng.2022.05.017, https://www.sciencedirect.com/science/article/pii/S2095809922004878
113. Wu, X., Xiao, L., Sun, Y., et al.: A survey of human-in-the-loop for machine learning. Fut. Gener. Comput. Syst. **135**, 364–381 (2022). https://doi.org/10.1016/j.future.2022.05.014, https://www.sciencedirect.com/science/article/pii/S0167739X22001790
114. Xiong, W., Fan, H., Ma, L., et al.: Challenges of human—machine collaboration in risky decision-making. Front. Eng. Manag. **9**(1), 89–103 (2022)
115. Xu, F., Uszkoreit, H., Du, Y., et al.: Explainable AI: A brief survey on history, research areas, approaches and challenges. In: CCF International Conference on Natural Language Processing and Chinese Computing, pp 563–574. Springer (2019)
116. Xu, W., Evans, D., Qi, Y.: Feature squeezing: Detecting adversarial examples in deep neural networks. Preprint (2017). arXiv:170401155
117. Yun, J.P., Shin, W.C., Koo, G., et al.: Automated defect inspection system for metal surfaces based on deep learning and data augmentation. J. Manuf. Syst. **55**, 317–324 (2020)
118. Zajec, P., Rožanec, J.M., Theodoropoulos, S., et al.: Few-shot learning for defect detection in manufacturing, submitted (2023)
119. Zhu, D., Li, Z., Wang, X., et al.: A robust zero-sum game framework for pool-based active learning. In: The 22nd International Conference on Artificial Intelligence and Statistics, pp 517–526. PMLR (2019)

Multi-Stakeholder Perspective on Human-AI Collaboration in Industry 5.0

Thomas Hoch, Jorge Martinez-Gil, Mario Pichler, Agastya Silvina, Bernhard Heinzl, Bernhard Moser, Dimitris Eleftheriou, Hector Diego Estrada-Lugo, and Maria Chiara Leva

1 Introduction

The potential applications of AI in smart manufacturing are numerous, ranging from improving the efficiency of machinery maintenance to detecting defects in the machine or the product to preventing worker injury. AI-based systems can identify bottlenecks, optimize production schedules, and adjust settings to maximize efficiency by analyzing large amounts of data from sensors and other sources in real time.

Furthermore, AI-based software systems can provide context-specific support to machine operators. By monitoring machine performance in real time, these systems can detect potential issues, give the operators recommended actions to solve the problem, and even automate the resolution, if necessary. This support can reduce operator errors, improve machine up-time, and increase productivity.

In general, collaborative processes in smart manufacturing are characterized by alternating phases of reactive and proactive elements, with each actor supporting the other alternately [1]. AI-enabled smart manufacturing systems can be self-sensing, self-adapting, self-organizing, and self-decision [2, 3], enabling them to respond to physical changes in the production environment in a variety of ways.

T. Hoch · J. Martinez-Gil (✉) · M. Pichler · A. Silvina · B. Heinzl · B. Moser
Software Competence Center Hagenberg GmbH, Hagenberg, Austria
e-mail: thomas.hoch@scch.at; Jorge.Martinez-Gil@scch.at; Mario.Pichler@scch.at;
Agastya.Silvina@scch.at; Bernhard.Heinzl@scch.at; Bernhard.Moser@scch.at

D. Eleftheriou
CORE Innovation, Athens, Greece
e-mail: d.eleftheriou@protonmail.com

H. D. Estrada-Lugo · M. C. Leva
Technological University Dublin, School of Environmental Health, Dublin, Ireland
e-mail: hectordiego.estradalugo@tudublin.ie; mariachiara.leva@TUDublin.ie

© The Author(s) 2024
J. Soldatos (ed.), *Artificial Intelligence in Manufacturing*,
https://doi.org/10.1007/978-3-031-46452-2_23

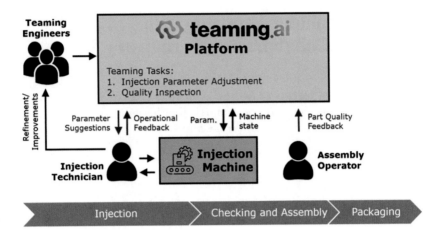

Fig. 1 Teaming.AI project overview

AI-guided interactions in the manufacturing process include stopping machines, adapting production tasks, or suggesting a change in production parameters. However, achieving effective teaming between machine operators and AI-enabled manufacturing systems requires mutual trust based primarily on the self-sensing and self-adaptation of each actor [4].

The increased situational awareness through an improved Human-AI collaboration enables operators to make informed decisions about optimizing machine settings and adjusting production schedules. This collaboration can improve product quality, reduce waste, and increase efficiency [5]. As AI continues to evolve, we can expect to see even more significant advances in smart manufacturing in the years to come.

In the frame of the international research project Teaming.AI,[1] we develop a software platform to facilitate human-AI teaming in smart manufacturing as shown if Fig. 1. We already presented reference architecture in [6]. However, in this work, we elaborate on different stakeholders' requirements regarding the quality characteristics of AI software platforms. For this purpose, we conducted 14 structured interviews with various stakeholders of the prospective platform. They rated a set of 11 different quality characteristics and provided vital success factors that can evaluate the fulfillment of these quality characteristics during the development and operation of the platform.

The results of our study provide valuable insights into the different stakeholders' expectations and remark on the importance of addressing their specific needs in the platform's design and development. Considering these quality characteristics and critical success factors, we can ensure effective collaboration between human operators and AI systems.

---

[1] https://www.teamingai-project.eu/.

The remainder of this work is structured as follows: Sect. 2 presents the related work regarding stakeholder interaction in AI-related projects. Section 3 addresses the three use cases we have faced in the context of the Teaming.AI project. Section 4 details stakeholders' different roles in projects of this type. Section 5 discusses the pains identified when implementing such a solution. Section 6 discusses the expectations toward the technical realization. Section 7 discusses the characteristics of the high-level teaming concept. Finally, we point out the lessons we have learned in this work and some lines of future work.

2   Related Work

The field of human-AI collaboration has gained significant attention in recent years, driven by the emphasis on integrating AI technologies into collaborative work settings in Industry 5.0 [7–12]. This growing interest revolves around the exploration of how AI systems can complement human abilities rather than replace them. Numerous studies have delved into different aspects of human-AI collaboration, including the design of intelligent systems [13], the development of new interaction paradigms [14], and the evaluation of the usefulness of these approaches in real-world scenarios [15].

One area of study in human-AI collaboration strives to design AI systems that can work effectively with human counterparts. Researchers have examined various strategies for designing intelligent systems to communicate and collaborate with human users, including natural language processing, machine learning, and cognitive modeling [14]. Additionally, some studies have focused on designing new interaction paradigms that enable seamless collaboration between humans and AI systems. For example, researchers have investigated using augmented and virtual reality to create immersive environments that improve human-AI interaction.

Another focus is evaluating their usage in real-world scenarios. Several studies analyze the impact of AI systems on the performance of human workers, as well as their acceptance and adoption of these systems. These studies have explored different elements that influence the success of human-AI collaboration, such as trust, transparency, and the nature of the tasks being performed [16]. Additionally, some researchers have investigated the ethical implications of human-AI collaboration, such as the potential for bias in decision-making processes.

Within this context, Knowledge graphs (KGs) have also become a powerful tool for making production lines more efficient and flexible in manufacturing and production [17]. They provide a means of organizing and processing vast amounts of data about devices, equipment, machine models, location, usage, and other related data [18]. KGs can also help make manufacturing smarter by providing insights into the complex and competitive landscape [19]. This can enable manufacturers to identify patterns, trends, and correlations that were previously hidden, leading to more informed decision-making and improved operational efficiency. The potential benefits of KGs in manufacturing and production make them an essential technology for the future of industrial operations and highlight the importance of continued research and development in this field [20].

Especially interesting and relevant in this context are recent AI developments like ChatGPT by OpenAI[2] and Luminous by Aleph Alpha.[3] Both of them are providing natural language interfaces for their human users, so that they are able to express their problems, information needs etc. through their most natural communication means. This seems beneficial especially in situations of mental pressure or other forms of stress that workers might have to cope with in their daily routines. The TEAMING.AI sister project COALA[4] performs research on such kinds of voice-enabled digital intelligent assistants. With Luminous-Explore, Aleph Alpha points out the importance of semantic representations,[5] so that humans are no more forced to represent their thoughts and intentions in machine representations but are enabled to expressing them in a more natural way. With those kind of developments, Aleph Alpha is also focusing on industrial use cases of their technology.[6]

3   Manufacturing Context

The following use cases (UC) describe concrete applications where an AI-based smart manufacturing solution could support a Human and AI collaboration in a manufacturing context. UC1 and UC2 derive from automotive suppliers and cover the process of plastic injection molding. In UC3, we investigate the ergonomic risk assessment during large-part manufacturing. Optimization focuses on the interplay between AI-controlled machine tasks and manual human labor.

3.1   UC1: Quality Inspection

The main objective of UC1 is to support the machine operator during the visual quality inspection of plastic parts produced by injection molding. The software platform shall classify products as OK or not-OK (including the type of defect), with the machine operator double-checking the latter. The software system interacts with the machine operator during the quality inspection and provides context-specific information for fault analysis and adjusting parameters to mitigate product defects.

---

[2] https://openai.com/blog/chatgpt.

[3] https://www.aleph-alpha.com/.

[4] https://www.coala-h2020.eu/.

[5] https://www.aleph-alpha.com/luminous-explore-a-model-\for-world-class-semantic-representation.

[6] https://de.nachrichten.yahoo.com/aleph-alpha-weg-halben-einhorn-125545116.html.

The main focus is on integrating human feedback: The machine operator should have the chance to overrule and correct the suggestions of the AI system, e.g., by manually marking defective regions if they were classified wrong.

This collaborative interaction between the machine operator and the software platform reinforces the notion of human-AI partnership, with each contributing their unique strengths to achieve the best possible outcomes. It empowers the machine operator with the authority to validate and correct AI decisions while ensuring continuous learning and improvement of the AI system by incorporating the operator's feedback [21].

3.2   UC2: Parameter Optimization

UC2 is also concerned with injection molding. However, the produced plastic parts are bigger, and cycle times are longer. Therefore, the software platform should provide a more proactive way to reduce and prevent non-OK parts effectively (zero waste production). The software platform should predict possible process deviations and identify the likely root failure causes before they materialize in faulty parts. It should be able to explain its findings (e.g., likelihoods), present recommendations (e.g., on parameter changes), and give the machine operator the ability to provide feedback to the software platform.

To achieve this proactive approach, the software platform leverages its analytical capabilities to predict possible process deviations. It analyzes real-time data from various sensors and monitors the production parameters to identify patterns or anomalies that might indicate an impending issue. The software platform can provide the machine operator with early warnings and proactive recommendations by continuously monitoring and analyzing the production process.

3.3   UC3: Ergonomic Risk Assessment

UC3 focuses on high-precision manufacturing of significant components (e.g., gear cases for wind turbines). This manufacturing process is typically time-consuming, physically strenuous, and involves a combination of automated and manual labor.

The objective is to analyze the ergonomic risks of human workers in terms of static loads and repetitive strains, especially during workpiece setup (involving manual part positioning, clamping, unloading, etc.). Using a camera-based tracking system, the software platform can determine the location of the machine operators on the shop floor, analyze their pose for ergonomic suitability and give feedback to the operator, e.g., via alerts. In addition, the software platform should also identify manual tasks associated with milling operations (e.g., taking measurements) and collect information about the tooling used during these tasks. This way, the software platform mediates between the milling machine and the operator by combining this information with context information, such as machine data. Overall, the software system should

1. Improve communication between the operator and the machine
2. Perform a continuous ergonomic risk assessment
3. Allow rescheduling of similar assembly tasks to reduce repetitive strains

By incorporating these functionalities, the software platform empowers both the machine operator and the milling machine to work in harmony, prioritizing the well-being and safety of the operator. It is an intelligent assistant providing real-time insights, guidance, and risk assessments to optimize ergonomics and prevent work-related injuries. This holistic approach promotes a healthier and more productive working environment, ensuring the efficient manufacturing of large parts while prioritizing the workers' welfare.

4 Stakeholder Roles

In the following, we describe the different stakeholder roles with their exemplary activities identified during requirements engineering.

- **Data Protection Officer (DPO)** enforces the laws protecting the company and individuals' data (e.g., the GDPR) by controlling the processing of data and properly auditing the system.
- **Software Scientist (SS)** queries runtime data of the software components of the software platform, such as logging information, for evaluating and optimizing the system's base code and behavior.
- **Data Scientist (DS)** applies statistical methods to the data processed by a software platform.
- **Machine Operator (MO)** performs a visual inspection of the produced parts, clamping, adjusting the workpieces, and performing manual tasks on the machine, such as obtaining measurements and making parameter adjustments.
- **Production Line Manager (PLM)** monitors and optimizes the processes for producing and assembling the product or its parts on the shop floor.

The involvement of these stakeholders, each with their unique roles and activities, highlights the multi-dimensional nature of the software platform and its impact on various aspects of the manufacturing process. By incorporating these stakeholders' expertise and responsibilities, the software platform's development and operation can benefit from a well-rounded perspective, ensuring compliance, optimization, data analysis, production efficiency, and quality assurance.

5   Identified Pains

According to [22], pains are *"bad outcomes, risks, and obstacles related to customer jobs."* In a collaborative project like Teaming.AI, the end users' participation enables the technical side to address real market needs. To this end, two questionnaires were circulated among all use case partners to identify the existing issues and pains related to their processes and understand the potential benefits they would expect from the technologies developed. Each questionnaire is directed toward two categories of employees: (1) Managers, who can present more managerial challenges of the organization, and (2) Operators, who can more effectively depict their day-to-day challenges and are the active users of the machines that will be retrofitted.

Specifically, the questionnaire was circulated among 18 individuals who participated in the study and distributed evenly for each use case through the EUSurvey platform.[7] The personas analyzed included:

- Injection Technician
- Production Shift Coordinator
- Operator
- Engineering Director
- Process Managers
- R&D Manager
- Innovation Manager
- Production Manager
- Head of Automotive Digital Transformation
- Data Scientist

The first section of the questionnaire aimed to analyze the profile of each end user. Overall, all end users are more results-driven organizations. When asked about the top 2 priorities in selecting third-party collaboration for Industry 4.0 initiatives, 78% preferred parties with proven pilot cases. The next most selected priority involved the ability to ensure an easier integration of solutions by 45% of the respondents. The third and fourth most selected priorities were results-oriented and involved the capacity to promise short-term value and the market participant's brand acknowledgment with 34% and 23%, respectively. Less prominent options also included the proximity of the technology provider and the sustainability improvement.

5.1   UC1: Quality Inspection

This particular use case involved the participation of the following roles:

---

[7] https://ec.europa.eu/eusurvey/auth/login.

- Data scientist
- Head of Mobile and Digital Transformation
- R&D Manager
- Operator

Overall, the results indicate that the end user faces pains primarily within the production department, which also affects coordinating activities. Specifically:

1. **Setup parametrization:** The most significant bottleneck in the particular production system, according to the respondents. As the manufacturing context section demonstrates, adjusting the parameters is necessary to mitigate product defects.
2. **Scrap Generation:** Causes for scrap generation can emerge from lack of quality raw materials, setup mistakes, machine issues, etc. Although scrap generation is considered financially sustainable, it impacts planning and financing.
3. **Unexpected downtime/equipment failures:** Mechanical and electrical failures accompanied by non-optimal maintenance are primary factors leading to failures.
4. **Loss of time:** Issues related to time loss refer to production delays, waste generation, lack of raw materials, non-productive processes, etc.
5. **Not optimal production quality:** The capacity to be prone to errors is affected by the level of control in the existing production system.
6. **Increased inventory:** Supply chain disruptions and unexpected failures lead to material unavailability, which leads planners to over-order to ensure that production does not remain stagnant.
7. **Lack of Flexibility in tasks and product design:** Flexibility is not limited to how operators are liberated to move between activities but also to the ability to switch between orders and in the adaptability to produce different types of products.

5.2 UC2: Parameter Optimization

The second use case of the project involved the following six roles:

- Engineering Director
- Injection Technician
- Operator
- Process Manager
- Production shift technician
- R&D Manager

As mentioned above, the first two use cases encounter similarities between each other contextually and, by extension, similar pains. Specifically:

1. **Setup parameterization:** The level of operator expertise influences the possibility for a defect to occur.

2. **Unexpected downtimes/equipment failures:** The main consequence of unexpected downtimes leads to production pauses and redirection of employees to other places.
3. **Lack of human-machine interaction:** Similarly with the first pain, the level of expertise has a dominant impact on production and handling impending issues.
4. **Loss of time:** The primary concern is from a managerial perspective. Identified loss of time lies in sales, dispatch, unexpected failures, and scrap generation for the organization.
5. **Scrap generation:** At a similar level with UC1. However, areas such as cost management and logistics would be the first areas to be improved by reducing scrap in production.
6. **Increased inventory:** Even though scrap generation is considered sustainable, it directly influences the purchasing and warehouse departments, which need to add to their risk management and planning activities.
7. **Increasing Costs:** Although all UCs experience increased costs due to other pains, UC2 respondents have highlighted the challenges in their cost management activities.

5.3 UC3: Ergonomic Risk Assessment

The third and final use case focused on the following roles:

- Injection Technician
- Innovation Manager
- Machine Operator
- Operator
- Production Manager
- R&D Manager

Following the questionnaire results, the ergonomic risk assessment use case is characterized by the following pains:

1. **Setup parameters:** In UC3, operators indicate that setup difficulties make them feel there is a lack of time.
2. **Unexpected downtimes/equipment failures:** Similarly to other use cases, it leads to high rework costs.
3. **Waste generation:** Classified as below average, there is an excess of production material waste in the current form of processes.
4. **The system does not help meet scheduling demands:** Bureaucracy leads to a lack of control over increasing equipment productivity.
5. **Delivery delays:** Inability to meet scheduling demands on time lead to delivery delays and profit reductions.
6. **Increased inventory**: Like all the use cases mentioned above, over-ordering leads to an increased inventory and profit reduction.

7. **Not optimal planning:** The current production system impedes stakeholders from conducting optimal planning activities.
8. **Loss of time:** A consequence of those mentioned above and other relevant factors reduces the production system's time and productivity.

5.4   Total Results: Pains

To summarize, although the project involves three different use cases, which may signify different needs from different stakeholders, some common themes offer a common ground to build upon. Particularly,

1. Setup parameters
2. Unexpected downtimes/equipment failures
3. Loss of time
4. Waste/scrap generation
5. Increased inventory

The five pains are the primary market needs that are the groundwork to construct compelling value propositions, which is one of the building blocks of a business model.

6   Expectations Toward the Technical Realization

In previous work [23], we conducted 14 interviews with stakeholders from three industry partners and three specialized SMEs for software development of AI-based systems.

We defined candidate scenarios [24] that describe the context and the anticipated functionality from the stakeholders' perspectives when interacting with the prospective software platform. In an interview-based case study, we assessed each of the 11 quality characteristics in terms of their importance to the overall platform from the stakeholders' perspective. We elicited the critical success criteria related to the software platform. The quality characteristics comprised the 11 characteristics of the ISO 25010:2011 standard for software quality (SQuaRE) [25] and 3 AI-specific quality characteristics, such as trustworthiness and explicability.

At the beginning of the interviews, we explained the research context of our study (i.e., human-AI teaming in smart manufacturing) to the interviewees. Each interviewee thoroughly understood the research context since they had participated in the project for over 1 year. For the relevance assessment, we adapted the *Quality Attribute Workshop* format [26] and asked the interviewees to assign, in total, 100 points to the different quality characteristics according to their subjective relevance for human-AI teaming in smart manufacturing.

The interviewees rated trustworthiness, functional suitability, reliability, and security as the most important quality characteristics. In contrast, portability, compatibility, and maintainability are rated as the least important. Furthermore, the results indicate consensus regarding the relevance of the quality characteristics among interviewees with the same role. However, we also recognized that the relevance of the quality characteristics varies according to the concrete use case for the prospective software platform. In addition, we asked interviewees to discuss critical success factors related to the prospective software platform. According to the interviewees, critical success criteria for human-AI teaming in smart manufacturing are improved production cycle efficiency, fewer faulty parts and scrap, and a shorter period for detecting deviations (product or process quality). This response was unsurprising since similar pains had already been expressed earlier (see Sect. 5).

7   Team Effectiveness

As described in [27], well-designed coordination mechanisms can improve team effectiveness to ensure that relevant information is distributed throughout the team. These coordination mechanisms, which have first been described by Salas, Sims, and Burke [4] as part of their *big five* framework for team effectiveness, are:

- **Shared Mental Models:** Shared mental models facilitate a common understanding of the environment by creating knowledge structures that promote the information exchange about state changes and team member needs. The knowledge structures need to be designed to be comprehensible by humans and AI.
- **Mutual Trust:** Trust in the team setting has been defined by Webber [28] as *"the shared perception ... that individuals in the team will perform particular actions important to its members and ... will recognize and protect the rights and interests of all the team members engaged in their joint endeavor."* A culture of mutual trust is essential in supporting the core components of teamwork, especially since, as [29] shows, trust critically influences how individuals within a team will interpret others' behaviors.
- **Closed-Loop Communication:** Communication between humans and AI may suffer from similar issues as communication between humans. Communication may be hindered because of misinterpretation of messages due to their perspectives and biases or because team members have become focused on their tasks rather than on how those tasks affect other team members' tasks.

Although the original *big five* framework focused purely on teaming between humans, it nonetheless builds a solid foundation for human-AI teaming digitalization. By prioritizing team effectiveness as a goal rather than just performance output, the emphasis remains on human team members instead of AI, acknowledging that the interactions among team members are equally vital.

Effective communication is, therefore, essential for teams to function correctly. In the context of human-AI teams, communication can help to ensure that AI systems are correctly interpreting human input and that humans are correctly interpreting the output of AI systems. This can be particularly important in high-stakes environments where errors can have serious consequences.

8 Conclusions and Future Work

In this work, we have seen how the development of AI-based software platforms that facilitate collaboration between human operators and AI services needs the integration of the different stakeholder perspectives into a common framework. In this regard, it is vital to identify the individual relevance of different quality characteristics per stakeholder and propose key success factors related to human-AI teaming to measure fulfillment. This can help ensure that the software platform is user-friendly and practical, meeting the expectations and needs of all stakeholders involved in the collaboration. Furthermore, it can mitigate conflicts arising from differing stakeholder perspectives during the projects.

Our research has thoroughly analyzed the critical issues, challenges, and opportunities of integrating AI technologies into collaborative work environments. To do that, we have adopted a multi-stakeholder perspective, considering the perspectives of different actors involved in the human-AI collaboration process. We aim to provide insights and recommendations for designing effective human-AI collaboration systems that enhance productivity, innovation, and social welfare.

We have observed that human-AI collaboration in Industry 5.0 requires careful consideration of various factors, such as the design of intelligent systems, the development of new interaction paradigms, the evaluation of the effectiveness of these systems in real-world scenarios, and the ethical implications of human-AI collaboration [24, 30]. Moreover, we have highlighted the importance of adopting a human-centric approach to AI system design, prioritizing human users' needs, preferences, and capabilities. Other elements (e.g., establishing trust and transparency in human-AI collaboration systems and ensuring fairness, accountability, and transparency in decision-making processes) are also essential in this manufacturing context.

In conclusion, integrating AI technologies into collaborative work environments offers immense potential for enhancing productivity, innovation, and social welfare. However, it also presents numerous challenges that require careful consideration and proactive measures. By adopting a multi-stakeholder perspective, prioritizing human-centric design, fostering interdisciplinary collaborations, and implementing responsible governance, we can pave the way for practical and ethical human-AI collaboration systems that maximize the benefits while minimizing the risks associated with this transformative technology.

As future lines of research, it is necessary to remark that as AI technologies continue to advance, it becomes increasingly essential to handle the issue of AI bias in collaborative work environments. Bias in AI systems can perpetuate existing social imbalances, support discriminatory practices, and limit opportunities for specific groups. Therefore, it is crucial to develop mechanisms that detect and mitigate bias in AI algorithms and data sets used in human-AI collaboration. In addition, integrating AI technologies into collaborative work environments necessitates ongoing training and upskilling programs for people. These programs aim to introduce individuals to AI capabilities, promote digital literacy, and provide them with the necessary skills to collaborate with intelligent systems effectively.

Acknowledgments We would like to thank the anonymous reviewers for their constructive comments to improve this work. SCCH co-authors has been partially funded by the Federal Ministry for Climate Action, Environment, Energy, Mobility, Innovation, and Technology (BMK), the Federal Ministry for Digital and Economic Affairs (BMDW), and the State of Upper Austria in the frame of SCCH, a center in the COMET—Competence Centers for Excellent Technologies Programme managed by Austrian Research Promotion Agency FFG. All co-authors involved in this study have also received funding from Teaming.AI, a project supported by the European Union's Horizon 2020 research and innovation program, under grant agreement No. 957402.

References

1. Johnson, M., Vera, A.: No AI is an island: the case for teaming intelligence. AI Mag. **40**(1), 16–28 (2019)
2. Qu, Y.J., Ming, X.G., Liu, Z.W., Zhang, X.Y., Hou, Z.T.: Smart manufacturing systems: state of the art and future trends. Int. J. Adv. Manuf. Technol. **103**, 3751–3768 (2019)
3. Phuyal, S., Bista, D., Bista, R.: Challenges, opportunities and future directions of smart manufacturing: a state of art review. Sustainable Futures **2**, 100023 (2020)
4. Salas, E., Sims, D.E., Burke, C.S.: Is there a "big five" in teamwork? Small Group Res. **36**(5), 555–599 (2005)
5. Daugherty, P.R., Wilson, H.J.: Human + Machine: Reimagining Work in the Age of AI. Harvard Business Press, Cambridge (2018)
6. Haindl, P., Buchgeher, G., Khan, M., Moser, B.: Towards a reference software architecture for human-ai teaming in smart manufacturing. In: Proceedings of the ACM/IEEE 44th International Conference on Software Engineering: New Ideas and Emerging Results (2022), pp. 96–100
7. Bauer, A., Wollherr, D., Buss, M.: Human–robot collaboration: a survey. Int. J. Humanoid Rob. **05**(01), 47–66 (2008)
8. Mingyue Ma, L., Fong, T., Micire, M., Kim, Y., Feigh, K.M.: Human-robot teaming: Concepts and components for design. In: Field and Service Robotics: Results of the 11th International Conference (FSR) (2017)
9. Chella, A., Lanza, F., Pipitone, A., Seidita, V.: Human-robot teaming: Perspective on analysis and implementation issues. In: AIRO@AI*IA (2018)
10. Krämer, N.C., Rosenthal von der Pütten, A.M., Eimler, S.C.: Human-agent and human-robot interaction theory: similarities to and differences from human-human interaction. In: Human-Computer Interaction: The Agency Perspective (2012)
11. Nikolaidis, S., Shah, J.A.: Human-robot cross-training: Computational formulation, modeling and evaluation of a human team training strategy. In: 2013 8th ACM/IEEE International Conference on Human-Robot Interaction (HRI) (2013), pp. 33–40
12. Chen, J.Y.C., Barnes, M.J.: Human–agent teaming for multirobot control: a review of human factors issues. IEEE Trans. Hum.-Mach. Syst. **44**(1), 13–29 (2014)

13. Freudenthaler, B., Martinez-Gil, J., Fensel, A., Höfig, K., Huber, S., Jacob, D.: Ki-net: Ai-based optimization in industrial manufacturing—A project overview. In: Computer Aided Systems Theory—EUROCAST 2022—18th International Conference, Las Palmas de Gran Canaria, Spain, February 20–25, 2022, Revised Selected Papers. Lecture Notes in Computer Science, vol. 13789, pp. 554–561. Springer, Berlin (2022)
14. Şahinel, D., Akpolat, C., Görür, O.C., Sivrikaya, F., Albayrak, S.: Human modeling and interaction in cyber-physical systems: a reference framework. J. Manuf. Syst. **59**, 367–385 (2021)
15. Johnson, M., Vignatti, M., Duran, D.: Understanding human-machine teaming through interdependence analysis. In: Contemporary Research, pp. 209–233. CRC Press, New York (2020)
16. Chen, M., Nikolaidis, S., Soh, H., Hsu, D., Srinivasa, S.: Planning with trust for human-robot collaboration. In: Proceedings of the 2018 ACM/IEEE International Conference on Human-Robot Interaction, HRI '18, pp. 307–315. Association for Computing Machinery, New York (2018)
17. Buchgeher, G., Gabauer, D., Martinez-Gil, J., Ehrlinger, L.: Knowledge graphs in manufacturing and production: a systematic literature review. IEEE Access **9**, 55537–55554 (2021)
18. Martinez-Gil, J., Buchgeher, G., Gabauer, D., Freudenthaler, B., Filipiak, D., Fensel, A.: Root cause analysis in the industrial domain using knowledge graphs: a case study on power transformers. In: Longo, F., Affenzeller, M., Padovano, A. (eds.) Proceedings of the 3rd International Conference on Industry 4.0 and Smart Manufacturing (ISM 2022), Virtual Event/Upper Austria University of Applied Sciences—Hagenberg Campus—Linz, Austria, 17–19 November 2021. Procedia Computer Science, vol. 200, pp. 944–953. Elsevier, Amsterdam (2021)
19. Noy, N.F., Gao, Y., Jain, A., Narayanan, A., Patterson, A., Taylor, J.: Industry-scale knowledge graphs: lessons and challenges. Commun. ACM **62**(8), 36–43 (2019)
20. Hogan, A., Blomqvist, E., Cochez, M., d'Amato, C., de Melo, G., Gutierrez, C., Kirrane, S., Gayo, J.E.L., Navigli, R., Neumaier, S., et al.: Knowledge graphs. Synthesis Lectures on Data, Semantics, and Knowledge **12**(2), 1–257 (2021)
21. Hoi, S.C.H., Sahoo, D., Lu, J., Zhao, P.: Online learning: a comprehensive survey. Neurocomputing **459**, 249–289 (2021)
22. Osterwalder, A., Pigneur, Y., Bernarda, G., Smith, A.: Value Proposition Design. Wiley, New York (2014)
23. Haindl, P., Hoch, T., Dominguez, J., Aperribai, J., Ure, N.K., Tunçel, M.: Quality characteristics of a software platform for human-ai teaming in smart manufacturing. In: Quality of Information and Communications Technology: 15th International Conference, QUATIC 2022, Talavera de la Reina, Spain, September 12–14, 2022, Proceedings, pp. 3–17. Springer, Berlin (2022)
24. Sutcliffe, A.: Scenario-based requirements engineering. In: Proceedings. 11th IEEE International Requirements Engineering Conference, 2003, pp. 320–329 (2003). ISSN: 1090-705X
25. ISO/IEC 25010. ISO/IEC 25010:2011, Systems and Software Engineering—Systems and Software Quality Requirements and Evaluation (SQuaRE)—System and Software Quality Models (2011)
26. Barbacci, M.R., Ellison, R., Lattanze, A.J., Stafford, J.A., Weinstock, C.B.: Quality attribute workshops (QAWS). Technical report, Carnegie Mellon University, Pittsburgh PA (2003)
27. Hoch, T., Heinzl, B., Czech, G., Khan, M., Waibel, P., Bachhofner, S., Kiesling, E., Moser, B.: Teaming.AI: enabling human-AI teaming intelligence in manufacturing. In: Proceedings http://ceur-ws.org ISSN, 1613:0073 (2022)
28. Webber, S.S.: Leadership and trust facilitating cross-functional team success. J. Manag. Dev. **21**(3), 201–214 (2002)
29. Simons, T.L., Peterson, R.S.: Task conflict and relationship conflict in top management teams: the pivotal role of intragroup trust. J. Appl. Psychol. **85**(1), 102 (2000)
30. Weyns, D.: Software engineering of self-adaptive systems: an organised tour and future challenges. In: Chapter in Handbook of Software Engineering (2017)

Holistic Production Overview: Using XAI for Production Optimization

Sergi Perez-Castanos, Ausias Prieto-Roig, David Monzo, and Javier Colomer-Barbera

1 Use Case Context

The eXplainable MANufacturing Artificial Intelligence (XMANAI) Ford use case focuses on managing the complexity of manufacturing in large production lines. These lines are composed of multiple work stations that mostly work sequentially, creating a direct dependence among them. Moreover, each of the stations is also made up of diverse machines and assets, with interconnected processes that range from fully automated to manual labor. All this considered, in a manufacturing line there are numerous complexities and challenges that can arise, making it difficult to anticipate and address potential production problems effectively. For example, a minor problem undetected, and thus not solved, at an asset at the beginning of the line may propagate affecting with great impact to the subsequent stations at the end of the line, and creating bottlenecks that hamper to reach the production goals. These kind of difficulties emphasize the crucial need for developing intelligent systems to ensure that the objective quality and quantity of produced items reach each industry goal, while keeping the production times within a profitable margin; such systems should not only anticipate unwanted situations during production, but in order to help the line operators to make proper decisions to manage them, it is also of great relevance for them to know the root causes that may cause production deviations.

The Ford use case has been designed in order to tackle these inherent complexities applying AI-based optimization systems, complemented with an explainability layer, applied on a real engine production line to monitor the overall standards that

S. Perez-Castanos (✉) · A. Prieto-Roig · D. Monzo
Tyris AI, Valencia, Spain
e-mail: sergi.perez@tyris.ai; ausias.prieto@tyris.ai; david.monzo@tyris.ai

J. Colomer-Barbera
Ford, Valencia, Spain
e-mail: jcolome5@ford.com

© The Author(s) 2024
J. Soldatos (ed.), *Artificial Intelligence in Manufacturing*,
https://doi.org/10.1007/978-3-031-46452-2_24

are needed, and therefore minimizing deviations on the expected production. One of the main challenges addressed in this case is the high variability in engine types and their corresponding components. In an engine manufacturing facility, different engine derivatives are produced to meet the requirements of various vehicle models. Each engine type may require specific components, such as the engine crankcase, fuel pump, oil pump, clutch, and more. Managing the diverse range of components and ensuring their availability and correct installation on the assembly line can be a daunting task. The sheer number of engine types and components increases the likelihood of errors, delays, and production bottlenecks.

Another complexity lies in the planning and scheduling of production batches. Currently, manual processes driven by the expertise of the MP&L (Material Planning & Logistics) team and production staff are employed to manage weekly production batches. However, accurately determining the optimal batch size, sequencing, and allocation of resources is a complex task. The planning engineer relies on customer demand and their own experience to make decisions, which can result in suboptimal production plans and resource allocation. This can lead to inefficiencies, increased downtime, and compromised production capacity.

Moreover, unforeseen issues and disruptions during the manufacturing process can significantly impact production efficiency. Shift foremen must make decisions on the fly to minimize planned stoppages and address unexpected failures on the assembly line. Without a comprehensive understanding of the root causes and potential solutions for such issues, decision-making becomes challenging, and it becomes difficult to maintain consistent line availability and performance.

By developing AI models specifically designed for optimizing production on the engines line, these challenges can be effectively addressed. AI systems can analyze real-time and batch data simultaneously acquired from various systems to identify patterns, detect anomalies, and provide recommendations for line optimization. With advanced Machine Learning techniques, AI models can simulate different scenarios, predict the impact of changes, and suggest the best course of action to maximize line performance. These AI models can assist operators and engineers in making data-driven decisions, reducing errors, improving resource allocation, and minimizing downtime.

Overall, the complexities and uncertainties inherent in engine manufacturing highlight the critical need for AI models to optimize production. By leveraging AI technologies, manufacturers can enhance their ability to anticipate and tackle potential problems, leading to increased efficiency, reduced costs, and improved overall performance on the assembly line.

The current situation at Ford Engine Plant does not allow the power of quasi-real-time data to be harnessed for decision-making. There are records on the status of the different operations on the production line, the quantity of engines produced and their parts, quality reports, and production plans. Despite having this information, there is not a centralized database and all the information is disaggregated in different corporate databases. This lack of centralized information is the first problem that needs to be solved in order to optimize the different processes that occur on the production line. This problem implies another one, which is the lack of artificial intelligence applied to the different decision-making processes due to the impossibility of taking advantage of all the available data. The proposed application aims to mitigate these problems by means of a set of functionalities that will be explained in the following sections.

This use case consists of a set of actions related to the current status of the line within a shift. By means of the information provided by the different disaggregated data sources, it is possible to analyze this information jointly to establish trends and to make predictions about anomalous situations in the line or the total amount of produced engines at the end of the shift. Thus, this use case is focused on the estimation of the production at the end of the shift, detection of unwanted scenarios, and simulations of new hypothetical situations, while giving insights on the assets that may cause potential deviations regarding the expected production goals.

Ford internal databases have different information about the status of operations (whether an operation is cycling a new component, waiting for a new part, blocked or in another possible state), operation failures, cycle times (both actual and design time), number of parts produced in a shift and data related to the quality of the parts produced. In this use case, different data sources related to production data will be joined to represent the historical status of the production line and to make predictions about the number of engines produced at the end of the shift following the current trend of the line. Both information will help the business experts to understand the significant deviations that may occur between the predicted (planned) production and the actual engines produced at the end of the shift.

2 XAI Approach

The application of explainability techniques is not a merely technical process, since explainability is precisely the bridge that connects intelligent systems with the users who are using them. Therefore, it is important that the end-users of XAI solutions are involved in the design process of explainable systems, looking for a human-in-the-loop approach to be followed. Therefore, in order to provide a solid explainability layer to the AI system developed in relation to the Ford use case, two key activities were carried out prior to the development of the XAI models.

The initial activity involved identifying the specific XAI needs of the end-users, i.e. the operators of the engine production line. This step is crucial in understanding the requirements and preferences of the stakeholders who will be interacting with the AI system. By closely collaborating with the end-users, their expectations and concerns regarding the interpretability of the AI models are effectively captured. This process ensures that the subsequent selection of Machine Learning models and explainability tools is aligned with the identified needs of the end-users.

Consequently, the second activity focused on selecting the appropriate methods that fulfills the identified XAI needs. Drawing from a range of available techniques, the selection process took into consideration the specific requirements and constraints of the manufacturing problem. The chosen methods were evaluated based on their capability to provide interpretable insights into the decision-making process of the AI models. Through this meticulous selection process, the chosen methods effectively address the XAI needs identified during the initial activity.

By conducting these two activities, the system can ensure that the AI and explainability aspects of the demonstrator are tailored to the requirements of the end-users at the plant. This approach fosters a collaborative and user-centric approach, guaranteeing that the selected methods provide meaningful and actionable explanations. Ultimately, the identification of explainability needs and the subsequent selection of appropriate methods enable the development of explainable AI models that empowers plant personnel to understand and trust the decisions made by the AI models, facilitating effective decision-making and driving the optimization of manufacturing processes.

2.1 Identification of XAI Needs

Prior to the development of XAI models, several tasks need to be addressed to effectively solve the intended problem. The general workflow follows the next steps.

The initial step is identifying the relevant data sources and determining the technical requirements necessary for data collection, storage, and processing. This involves understanding the data ecosystem within the manufacturing environment, including sources such as corporate systems, maintenance records, tooling systems, and real-time data acquisition. Concretely, for this problem, the data employed are related to the status of the production of the line and quality data.

Next, it is essential to assess the AI needs specific to the manufacturing problem at hand. This includes identifying the key challenges and objectives, such as optimizing production, minimizing downtime, and improving resource allocation. Understanding the desired outcomes helps in defining the scope and purpose of the AI models to be developed. The objective for this problem consists on finding deviations from the expected productions and preventing line bottlenecks.

Simultaneously, it is important to recognize the explainability needs of the stakeholders involved. This entails considering the requirements for transparency, interpretability, and trust in the decision-making process. Different stakeholders may have varying levels of expertise and understanding of AI systems, so it is crucial to determine the appropriate level of explanation needed to ensure effective collaboration and decision-making. For this problem, it is essential to understand which elements of the line and to what degree they have influenced predicted deviations and analyze the best action to fit it.

By applying XAI models, several advantages can be achieved in the manufacturing context. The foremost advantage is that the end-user involved in the production lines will be able to make better decisions based on results of the analysis of the available data and understand why these results are originated, in terms of knowing the specific assets that have the most influence on them.

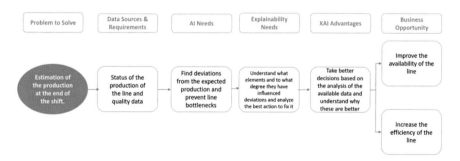

Fig. 1 Identification of AI and explainability needs for the use case

The combination of XAI models and their advantages opens up significant business opportunities in this use case. Specifically, the application of XAI models will affect positively the line in two aspects: firstly, the line will increase its availability, preventing line bottlenecks by quick fixing issues that can affect the whole production, and secondly, the efficiency of the production will also be increased, by reducing unnecessary maintenance stops.

Overall, the tasks preceding the development of XAI models involve identifying data sources, technical requirements, AI needs, explainability needs, recognizing the advantages of using XAI models, and uncovering the resulting business opportunities. In Fig. 1, a diagram of the whole identification process is presented.

By addressing these aspects, it is possible to identify which Machine Learning Models and explainability tools meet the XAI needs of the demonstrator as it is explained in the next section.

2.2 Hybrid Models

In the context of the XMANAI project, a hybrid model [2] refers to the combination of two components: a Machine Learning model and an explainability tool used to interpret the results produced by the Machine Learning model.

The component with the Machine Learning model is responsible for learning patterns and making predictions based on input data. It leverages algorithms and techniques to extract information, generalize from training examples, and generate predictions for new data instances. The Machine Learning model may include various approaches such as decision trees [3] or random forests [4].

On the other hand, the component with the explainability tool is employed to provide insights into the decision-making process of the Machine Learning model. It helps to uncover the underlying factors, features, or patterns that influence the model's predictions. The explainability tool enhances transparency and interpretability by providing explanations, visualizations, or metrics that shed light on how the Machine Learning model arrives at its results.

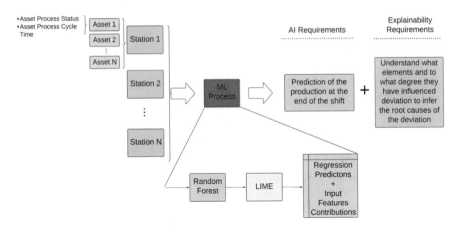

Fig. 2 Ford experiment diagram

By combining these two components, the hybrid model aims to address the "black box" nature of some Machine Learning algorithms, providing added transparency. The explainability tool provides insights into the most relevant features on a specific prediction, the relative contributions of those features, and how they have impacted on the final model decision.

Since the goal of this approach is the development and training of a model for the estimation of the production at the end of a shift, a regression model has been determined to be a good fit to solve the problem, and thus it has been selected for that purpose. As this estimation is a regression task, this model can be employed to that end and the output prediction of the system will be the number of engines produced at the end of one production shift. Based on the comparison of this prediction with regard to the expected production, the plant manager and the operators will be able to make decisions to correct potential deviations.

In order to solve this use case, an experiment based on the architecture shown in Fig. 2 has been built in order to configure the use case assets and train the AI model. From left to right, the use case data asset will be generated retrieving operation data directly from the different production line stations of the engine assembly line. Those assets will feed the Machine Learning pipeline performing a regression prediction on the number of engines that will be produced at the end of the shift. Different ML predictive models have been evaluated to analyze which of them provide more accurate results. Finally, Random Forest was selected as the most suitable AI base-model for this goal.

Therefore, the explainability requirement of this problem is to understand what elements and how much have influenced the deviation so as to infer the root causes of the deviation as mentioned above. Considering the explainability requirement, LIME [8] has been selected as the optimal XAI Tool due to its feature contribution explanation, which gives much value to the model explanation in this use case scenario. Local Interpretable Model-Agnostic Explanations (LIME) is an explainability tool that provides interpretable explanations for individual predictions made by Machine Learning models. It creates local surrogate models to approximate the behavior of the original model, generating feature importance weights that indicate the relative influence of each input feature. LIME is model-agnostic, making it applicable to various models without requiring access to their internal workings. Its explanations enhance transparency and understanding, fostering trust in complex Machine Learning systems.

Taking into account the selected Machine Learning model and the explainability tool, the expected output of the ML pipeline is a prediction which indicates the expected performance of the production at the end of the shift, which ultimately will be compared with the actual production goals to prevent potential deviations. Thus, the explanations given should help to understand what the key elements are and how much they have influenced on production deviations, and supporting the end-users to infer root causes that could be tackled to minimize the impact.

2.2.1 Interpretation of XAI Outputs

Using a Random Forest as a model and LIME as an explainability tool, an example of explainability is shown in Fig. 3 where an entire shift of production is analyzed to estimate the production of the current shift through the regression of the input data.

On the left side, it can be seen that the model predicts a value of 19.16. This value has to be interpreted as the average value of engines produced by the line in time intervals of 10 minutes. For this use case, values in the range between 15 and 20 are

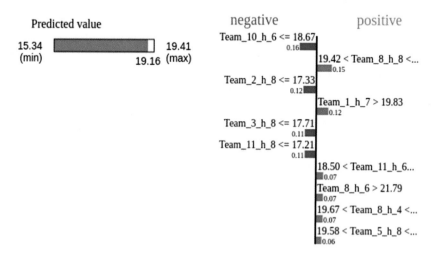

Fig. 3 Lime explainability result example. On the left side, it is shown the ML prediction information. On the right side, it can be see, the contributions (both positive and negative) of the different input features

considered as proper values, representing a nominal behavior of the line. When the value is lower than 15, a deviation from the expected value has to be considered.

On the right side, the information provided by the LIME explainability tool is presented. As this technique consists on a feature relevance representation, the different rows in the diagram represents the input features, coming from the input data used for this concrete prediction, that are contributing the most to this concrete output and in what degree they are contributing. The negative or positive side represents how much they are affecting to an upper o lower value from the average output value of the training data employed to train the model.

2.3 Graph Machine Learning Models

Graph Machine Learning (Graph ML) [7] models are a class of Machine Learning algorithms designed to operate on structured data represented as graphs. In contrast to traditional Machine Learning approaches that work with tabular data or sequential data, graph ML models leverage the inherent relational information present in graph structures to make predictions or gain insights.

Graph ML models are particularly suited for tasks that involve complex relationships and dependencies among data points. They find applications in various domains, including social network analysis, recommendation systems, bioinformatics, fraud detection, knowledge graph completion, and many more [10].

This topology of models can be correlated with the complex structure of interconnected processes present at production plants of manufacturing industries. Thus, a research line within XMANAI has been dedicated to determine the fit of this type of approach to provide IA-based predictive models adapted to the specifics of manufacturing considering the addition of an explainability layer. Specifically, among our experiments with the data, Heterogeneous Graph ML models were selected as they are focused on Explainability. After a research among different beyond-state-of-the-art options, we were inclined to use in this case Tensorflow GNN (TFGNN) [5]. This library is capable of handling Heterogeneous Graphs to fit a dynamic number of nodes and edges, with different node classes and edge classes, and their own set of data. This also leverages Tensorflow's [1] graph execution, which speeds up the training and inference process, allowing for faster iteration and less waiting time.

2.3.1 Graph Models

Graph models offer distinct advantages over regular models in the context of predicting the output of a manufacturing line. By leveraging the inherent relational structure present in the data, graph models can incorporate contextual information and capture the interdependencies between different components. Manufacturing lines often exhibit complex relationships, such as cascading effects or feedback loops, which graph models excel at representing and exploiting. Additionally, graph models can effectively utilize both the graph topology and the node-level features associated with each component, enabling them to learn embeddings that encode intrinsic properties and interactions.

The transferability capabilities of graph models allow them to generalize knowledge across different manufacturing line states with shared characteristics. Furthermore, graph models provide interpretability and explainability by analyzing learned embeddings and the influence of components on the output.

GNNs have gained significant attention in recent years due to their ability to capture both the local and global structural information of the graph. They leverage a combination of node-level features, graph topology, and message passing mechanisms to update the representations of nodes throughout the network layers. GNNs can learn expressive node embeddings that encode both the inherent features of the nodes and their relationships with neighboring nodes.

During the training stage of the graph ML models, they are often optimized using gradient-based methods, where the gradients are computed through backpropagation. Graph ML models can be trained in a supervised manner, where labeled data is available, or in an unsupervised or semi-supervised manner, where only a subset of the data is labeled or no labels are available.

Graph ML models can provide various types of outputs that capture different aspects of the graph data. For example, in node classification tasks, the model may assign a label or class to each node in the graph, indicating the predicted category or behavior of the corresponding entity. Graph ML models can also generate outputs related to link prediction, where they estimate the likelihood or presence of connections between nodes in the graph. Additionally, graph ML models may produce node embeddings or representations that capture the learned features and relationships of each node, enabling downstream tasks or further analysis.

In the scenario analyzed, we work on temporal slices of the manufacturing line and we will be predicting the output at the final slice of the manufacturing line, encoded in the graph training as an attribute of the nodes. This has been proven to be possible, per example by Google in their paper [6], where they improved the state of the art by reducing the loss by 6% (using RMSLE).

2.3.2 Explainability Techniques

For the explainability, we used Graph ATtention (GAT) [9] layers to collect the weight that each input is given by the model. Graph Machine Learning not only allows us to adapt both the inference and explainability to the layout of the manufacturing process when operators are added or removed, but it also allows the model to have more contextual information about what operators are connected or what kinds of data there is, which in the end, this improves the accuracy of the model and explanations

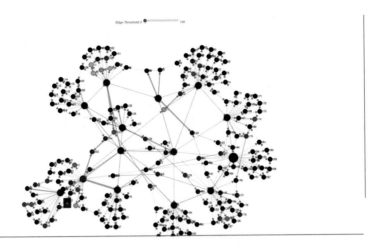

Fig. 4 Graph ML GAT contribution weights full graph. Brighter colors and wider edges means greater contribution

Upon completion of training and explanation, we obtain a graph that depicts the manufacturing line, including all the relationships between its components and their respective importance.

In Fig. 4, we present a visualization that represents the connections between operators (green dots) and teams (red dots) in our project. The relevance of each node is indicated by the darkness of its color, with darker colors indicating less relevance and brighter colors indicating more relevance. The size of the edges connecting the nodes also represents the strength of the connection, with larger edges denoting more relevant connections.

Specifically, the green dots represent operators, and a connection between two green dots implies that those operators are connected in line, with the next operator receiving the output of the previous one. Similarly, the red dots represent teams, and connections between red dots indicate that one team receives the output of the previous team's work. Additionally, there are connections between the red and green dots, indicating which operators belong to which team.

Figure 5 is a modified version of the first graph, where we have removed some of the less relevant connections. This filtering process results in retaining only groups of highly influential operators, represented by the remaining connections.

The utility of presenting this information to a user is that it provides a clear understanding of the relationships and dependencies between operators and teams in the project. By visualizing the graph, the user can identify the most relevant operators and teams based on their brightness and the size of their connections. This information helps in understanding the flow of work, identifying key contributors, and potentially optimizing the project by focusing on the most influential aspects.

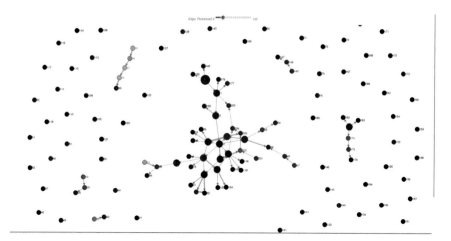

Fig. 5 Graph ML GAT contribution weights trimmed graph to only highest contributors. Brighter colors and wider edges means greater contribution

By removing less relevant connections in the second graph, the user can quickly grasp the core components and key dependencies, simplifying the visualization and highlighting the most critical elements. This focused view is simpler and straighter.

Finally, if this were to be too hard to understand for users, this has the potential to be added in a map of the line to better represent the positions of each team and operator.

3 XMANAI Platform Usage

All training of Hybrid models and the exploration of the available data is performed under the XMANAI platform. For this, Ford uploads their datasets to the platform, ensuring the data's security and privacy by sharing it under a contractual agreement while retaining ownership.

The platform provides tools for visualizing the data, allowing us to gain insights and understand its characteristics. After visualizing the data, we can further explore the problem and develop solutions using notebooks available on the platform, leveraging its computational capabilities and pre-built libraries.

To create a hybrid model and explainable artificial intelligence (XAI) tool, we can upload their model and associated artifacts to the platform. We can configure the model with its hyperparameters with a viable explainability tool, and fine-tuning them to achieve optimal performance. The output is a session which will be used for training, and that will ensure that the model can always be used with the selected explainability tool.

After creating the session, we can use the platform's pipeline functionality to establish a streamlined process for training both the model and the explainer component.

Once the pipeline is set up, Ford can run it whenever they wish to make inferences or predictions based on the trained model. The pipeline ensures consistent and automated execution, saving time and effort.

Finally, we can visualize the explanations provided by the XAI tool through the platform's visualization capabilities. This allows them to gain interpretability and insights into how the model arrived at its predictions, aiding in decision-making and model validation.

4   Achievements, Conclusions, and Open work lines

In the following paragraphs, we will delve into the accomplishments we have made in the areas of data veracity, data ingestion automation, data processing and analytics, hybrid and graph model development, explainability methods, manufacturing app creation, and infrastructure enhancement.

In terms of data veracity, we have allocated a team of programmers to thoroughly review and ensure that all machines adhere to the standardized manufacturing data reporting. This serves as the starting point for all other aspects, as the availability of accurate data is paramount. Without reliable data, we cannot construct models that accurately represent the reality of our manufacturing processes.

We have also made significant progress in automating the data ingestion process. Previously done manually, the data ingestion is now automated, creating dumps of data periodically and producing new batches of data that can be used to retrain existing models.

Our focus on data processing and data analytics has involved extracting features from different datasets and determining the best approach for ingesting the input data to achieve relevant results.

Furthermore, we have been actively working on developing hybrid models that combine machine learning algorithms with explainability tools. Specifically, we have utilized Local Interpretable Model-Agnostic Explanations to create explanations that help identify the parts of the production line that are affecting specific predictions. Alongside this, we have also been working on graph models that provide explanations in a similar manner to our hybrid models.

Additionally, we have utilized the eXplainable MANufacturing Artificial Intelligence platform to develop a manufacturing app, as described in Sect. 3. This app has been designed for use by demonstrators in their respective use cases.

To accommodate the on-premise segment of the platform and effectively host the entire backend infrastructure for our final applications, we have recently acquired a dedicated physical server. This acquisition ensures efficient and reliable hosting for our platform.

In conclusion, our achievements in data veracity have ensured a solid foundation for our manufacturing processes by guaranteeing accurate and reliable data. This has enabled us to construct models that accurately represent the reality of our operations. The automation of data ingestion has not only saved valuable time and resources but also provided us with fresh batches of data for continuous model improvement. Our focus on data processing and analytics has allowed us to extract meaningful insights and achieve relevant results. The development of hybrid and graph models, coupled with explainability tools, has enhanced our understanding of predictions and identified the specific factors influencing them. The implementation of explainability methods has increased transparency and trust in our models' outputs, empowering us to make informed decisions. Furthermore, the creation of the manufacturing app has facilitated streamlined processes and improved collaboration among demonstrators in their respective use cases. Lastly, our investment in infrastructure, including the acquisition of a dedicated physical server, ensures a robust and efficient hosting environment for our final applications.

As we look toward the future, one of our key objectives is to bridge the gap between scientific explanations and end-user accessibility. While our current focus on developing explanations has been rooted in scientific rigor, we recognize the need to make these explanations more comprehensible and user-friendly for a non-technical audience. Our next step involves refining and adapting the explanations derived from our models, transforming them into a format that is easily understandable and meaningful to end-users. By translating complex technical concepts into accessible language and visualizations, we aim to empower the end-users to make informed decisions and derive maximum value from our manufacturing processes. This user-centric approach will enable us to deliver explanations that not only provide scientific insights but also serve as practical tools for enhancing productivity, efficiency, and overall business profitability.

References

1. Abadi, M., et al.: TensorFlow, Large-scale Machine Learning on Heterogeneous Systems (2015). https://doi.org/10.5281/zenodo.4724125
2. Amini, M., et al.: Discovering injury severity risk factors in automobile crashes: a hybrid explainable AI framework for decision support. Reliab. Eng. Syst. Saf. **226**, 108720 (2022). https://doi.org/10.1016/j.ress.2022.108720. https://www.sciencedirect.com/science/article/pii/S0951832022003441
3. Breiman, L.: Classification and regression trees. In: The Wadsworth & Brooks/Cole (1984)
4. Brieman, L.: Random Forests Machine Learning (2001)
5. Ferludin, O., et al.: TF-GNN: graph neural networks in TensorFlow. CoRR abs/2207.03522 (2022). http://arxiv.org/abs/2207.03522
6. Kapoor, A., et al.: Examining covid-19 forecasting using spatio-temporal graph neural networks. In: MLG workshop @ KDD'2020, epiDAMIK workshop @ KDD'2020 (2020). https://arxiv.org/abs/2007.03113
7. Liu, Z., et al.: Heterogeneous graph neural networks for malicious account detection. In: Proceedings of the 27th ACM International Conference on Information and Knowledge

Management (CIKM '18). Association for Computing Machinery, New York (2018), pp. 2077–2085. https://doi.org/10.1145/3269206.3272010

8. Ribeiro, M.T., et al.: Model-agnostic interpretability of machine learning (2016). arXiv preprint arXiv:160605386

9. Wang, X., et al.: Heterogeneous graph attention network. In: The World Wide Web Conference (WWW '19). Association for Computing Machinery, New York (2019), pp. 2022–2032. https://doi.org/10.1145/3308558.3313562

10. Xia, F., et al.: Graph learning: a survey. IEEE Trans. Artif. Intell. **2**(2), 109–127 (2021). https://doi.org/10.1109/TAI.2021.3076021

XAI for Product Demand Planning: Models, Experiences, and Lessons Learnt

Fenareti Lampathaki, Enrica Bosani, Evmorfia Biliri, Erifili Ichtiaroglou, Andreas Louca, Dimitris Syrrafos, Mattia Calabresi, Michele Sesana, Veronica Antonello, and Andrea Capaccioli

1 Introduction

The H2020 XMANAI project represents a unique experience in Whirlpool's Operations Excellence, within the I4.0 technology research work stream, as it really faces one of the most commonly experienced obstacles to the successful introduction of Artificial Intelligence (AI) technologies. Human engagement in the AI Loop is commonly addressed through a structured change management which acts on communication and training to achieve the readiness level that enables adoption by real users. XMANAI has developed dedicated tools that serve the customized need of the user to *understand* and *trust* the AI system.

Besides, the project's strength lies not only on the implemented solution but also in the journey to arrive there. The latter unveils the complexity of the *explainability* requirements for a user, even when one is aware of what to look for. The conversion of desired information into a format which is really usable by the business experts,

F. Lampathaki (✉) · E. Biliri · E. Ichtiaroglou · A. Louca · D. Syrrafos
Suite5 Data Intelligence Solutions, Limassol, Cyprus
e-mail: fenareti@suite5.eu; evmorfia@suite5.eu; erifili@suite5.eu; andreas@suite5.eu;
dimitris.syrrafos@suite5.eu

E. Bosani
Whirlpool Management EMEA, Milan, Italy
e-mail: enrica_bosani@whirlpool.com

M. Calabresi · M. Sesana · V. Antonello
TXT e-solutions SpA, Milan, Italy
e-mail: mattia.calabresi@txtgroup.com; michele.sesana@txtgroup.com;
veronica.antonello@txtgroup.com

A. Capaccioli
Deep Blue, Rome, Italy
e-mail: andrea.capaccioli@dblue.it

J. Soldatos (ed.), *Artificial Intelligence in Manufacturing*,
https://doi.org/10.1007/978-3-031-46452-2_25

and sometimes also by the technical experts, has been demonstrated to be a long path. The successful navigation of this path requires the adoption of structured mapping techniques and methods, along with an agile approach to the development of the solution that ensures a smooth, step-by-step, progress.

The continuous and strict collaboration between technical experts and business experts has also been a key success factor for achieving the final goal in the manufacturing environment. This is a success factor not only for AI adoption but also for the successful deployment of most of the innovative technologies offered by I4.0.

This chapter explores and documents the experience gained by Whirlpool in the XMANAI project [1], where the explainability of AI technology is applied in a sales demand forecasting scenario. In this scenario, the level of adoption of the AI tool is highly dependent on the trust that can be generated in the users. In the Whirlpool's case, the possibility to explain AI results to users has been identified as one of the key enablers to gain users' trust and to fully engage human stakeholders within the AI loop. XAI is the door to open toward full awareness on why a specific result has been generated. The possibility to achieve a deeper understanding of the processes simulated by the AI rises the awareness and the belief in XAI: It is the spark to achieve better decisions and results in daily business management.

In this context, this chapter starts with a detailed description of the Whirlpool's use case in the H2020 XMANAI project in Sect. 2. It presents the motivation driving the project and the business context of the use cases. Then, the chapter describes the current state and the identified business requirements driving the XMANAI solution design, providing an outline of the "to-be" scenario and the key objectives to be achieved in Sect. 3.

Accordingly, Sect. 4 presents the technical implementation journey, followed by the explainability value presentation. The process used to detail the explainability requirements with the users is outlined and the meaningfulness of the explanation mode, deployed into the system, is justified in Sect. 5.

Moreover, Sect. 6 includes the description of the XMANAI platform, showing details of the users' journey in the XAI platform and in the XAI Manufacturing application of Whirlpool's use case. Finally, Sect. 7 presents the evaluation results of the demonstration sessions held with the users, along with a summary of the lessons learnt so far in Sect. 8.

2   Whirlpool as XMANAI Demonstrator

Whirlpool is the biggest player in white goods business at global level and one of the most important players in Europe, where it counts 9 industrial sites in 5 countries and more than 50 OEM (Original Equipment Manufacturers) producing and delivering in the 35 European markets and more than 100 other destinations in the world.

Overall, the white goods business is characterized by high levels of competition in the European market. Several very aggressive global competitors play a big role in the traditional B2B (Business to Business) market, where currently the key success factors are price and brand reputation. In this context, all the players started, some years ago, to enrich their product offers with services to final customers. Most of these services were partially driven by product IoT (Internet of Things) functionalities and partially leveraged on post-sales organization. Nevertheless, most of these offerings had no real and disruptive effect on business setup and market share footprint.

Today, Whirlpool's products are mainly considered a commodity. Hence, due to the maturity of the market, the overall business setup had been quite stable until 2020. However, the pandemic event of COVID19 has significantly modified such a market setup through the competitive advantage of companies, like Whirlpool, which early addressed the safety constraints for contagion avoidance and quickly restarted the production flow after the lockdowns. In addition, the pandemic lockdown has provided a unique opportunity, for white goods producers, to finally exploit the direct sales to final customers (D2C), leveraging the functionalities of the web and by-passing the B2B selling constraints. This element is opening up new opportunities for market share acquisition, and the successful competition on this new channel is expected to be crucial in the future. This new business channel is characterized by a set of specific and challenging requirements, which are mainly related to the speed of the buying experience and the hard competition on the web based on product offers and brand reputation. Specifically, key success factors include the range of the available offerings, the speed of order-to-delivery, the pricing policy with focused and personalized promotions, as well as additional services that can be offered (e.g., free installation, guarantee extension, home delivery, old product scrapping, and waste disposal).

In this context, the complexity of the overall process and the speed required in decision making can be barely addressed by humans, even if experts, with traditional analytic methods. Thus, AI becomes the key enabler for a significant improvement in decision-making process, ensuring a reliable forecasting service which may support people in "acting."

It needs to be noted that Whirlpool's experience in applying AI technology in the manufacturing environment started more than 10 years ago, mostly with applications to quality control (like vision systems or product testing) and in predictive maintenance. After every successful project implementation, the company is faced with the challenge of sustaining the implemented solutions in the mid/long term and of exploiting the successful pilot application in other areas of the factories.

Some key factors have been identified as root cause of this effect: firstly, the users' difficulty to understand AI technology fully and deeply. The awareness of AI's potential and limitations prevents a dangerous misalignment between expectations and results. This weak awareness was mainly due to the poor control and understanding of the data fed to the AI, implying a progressive and unexpected derail in the quality of AI results, which often end in systems shutdowns. In these cases, the original trust of the users in AI system was progressively consumed, sometimes even resulting in rejection or boycott. Moreover, in the case of predictive analytics applications, the user's trust was negatively affected by a weak understanding of the AI results, mainly when they were far from their usual experience or expertise. In these cases, the users' approach was to tentatively reject the recommendation, i.e., to bypass the AI system. In both cases, the capability to create and sustain the trust in AI technology has been demonstrated as a key factor to really make the most out of this approach.

In this context, eXplainable AI (XAI) may represent the key to unlock the achievement of all the AI solution objectives, as it may enable the users to deep dive into AI results with a language that makes them understandable (i.e., in terms of *how* and *why* such results are generated) and, therefore, actionable and sustainable in the long term. Thus, Whirlpool has joined the XMANAI project experience with a use case that is quite far from the traditional manufacturing domain, yet which can be used to demonstrate, in terms of business impact, the full potential of this technology.

3   White Appliances Use Case Description

Whirlpool, as other competitors in these years, has launched the D2C (Direct To Consumer) channels in the biggest European markets ensuring a wide product offer among its several important European brands (e.g., Whirlpool, KitchenAid, Hotpoint, Ariston, Ignis, Indesit) with a promised order-to-delivery time of 3 working days.

With a highly complex manufacturing footprint, composed by mono-product factories which produce one single product platform distributed in all the destination markets, the supply network imposes a transportation time, and consequent order-to-delivery time, that cannot fulfill the request of the D2C business. The traditional approach is based on inventory strategy, focused on ensuring a certain safety level of stock to be able to serve any customer, at any place, for any product, and at any moment. Due to the need to extend as much as possible the product range offer to the customer, the risk of obsolescence and the high blocked working capital must be balanced with the customer.

A reliable demand forecasting may minimize the protected inventory, maximizing the possibility for customers to find exactly what they are looking for. Thus, a reliable forecasting functionality represents the first business requirement posted to XMANAI. As explained earlier, this is not enough to ensure the full and conscious adoption of an AI solution. Here is where the explainability functionality has to enable the "actionability" of the AI results by answering users' questions about how and why an outcome is generated.

The second, perhaps more important, business requirement which XMANAI project needs to address is a deep understanding of the business process. This is based on the development of a tool capable of supporting users with a clear and user-friendly visualization of how the forecast is generated and why. This enables them to understand what they can do to change (when possible) the forecasting results toward achieving their business goals.

In summary, the Whirlpool use case in the H2020 XMANAI project is focused on the creation of a reliable, explainable, and actionable demand forecasting tool capable of providing the following:

1. Demand forecast reliability, including:

   - Getting a reliable demand forecasting
   - Minimizing inventory for D2C
   - Maximizing product availability on request
   - Maximizing customer satisfaction

2. Business dynamics understanding, including:

   - Understanding of demand evolution
   - Understanding of customers' behavior
   - Understanding of buying patterns
   - Supporting of promotional initiatives
   - Supporting simulations (i.e., execution of "what-if" scenarios)

Currently, the demand forecasting of D2C market is embedded into the full ODP process (Operational Demand Plan), which funnels all the expected sales demand of all the markets in a unique master production plan for all the factories and OEM sources. This process is coordinated by a central Demand Planning team, which enriches the unconstrained forecast generated by statistical analytics on historical data, with adjustments driven by factory capacity planning, supply base constraints, inventory and transportation strategy guidelines. They work starting from each single market demand profile to generate a total demand, which is then split to supply each single market. At this moment, the marketing and sales team of each market analyzes the forecast to take decisions about pricing strategy, promotional actions, and product range offer. The result of this second enrichment is the input for the manufacturing production plan.

In the "to-be" situation, as depicted in Fig. 1, the XMANAI platform will be used to support the decisions made by these two groups of users to achieve better results, using the reliability of AI functionality and the business dynamic understanding of XAI.

It needs to be noted that the sustainable usage of AI and XAI pipelines generates the need to include an IT role with the specific responsibility of managing the XMANAI platform daily. Data scientists and data engineers become key organizational roles to ensure that the XMANAI platform will perform according to high standard levels, granting the optimal forecast accuracy and ensuring the models' and pipelines' maintenance during the whole system lifecycle. This element is new in the organizational structure of a traditional industrial company, which, in most cases, does not have internally the competence to effectively support the business users. In this case, the answer provided by XMANAI platform highlights this organizational gap and the strategy that must be adopted to fill it.

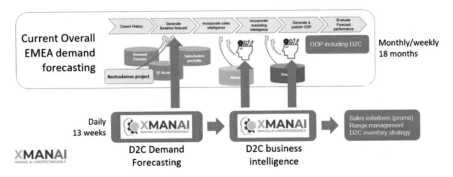

Fig. 1 XMANAI and Whirlpool ODP management process

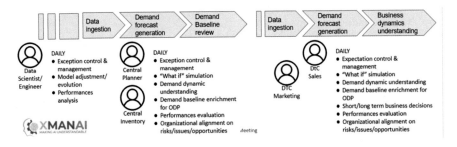

Fig. 2 Users' scenarios in Whirlpool's use case

The so-defined scenarios presented in Fig. 2 have facilitated the identification of the user's stories, which generated the specifications for the explainability requirements, as described in the following chapters, and set the foundation for demonstrator validation.

Last but not least, taking into consideration the fact that the XMANAI platform has the credentials to eventually become a very strategic business tool that can set Whirlpool apart from its competitors, it must guarantee the security and protection of the strictly confidential and sensitive data that are used. Thus, a security-by-design approach and a full GDPR (General Data Protection Regulation) compliance must be embedded into the solution development since the beginning.

4 Explainable AI Approach

The technical activities performed and the achievements reached during the preparation and execution of Whirlpool demonstrator span across different work areas, including data exploration, predictive model development and performance improvement, development of explainability methods, adaptation of the results for the needs of business users, as well as development and validation of the initial release of the corresponding XMANAI manufacturing app (i.e., Demand Forecasting Manufacturing App).

The main achievements across these axes are as follows: (a) use case detailed analysis and elaboration presented in Sect. 4.1; (b) data acquisition and exploration as explained in Sect. 4.2; (c) development and validation of hybrid XAI models outlined in Sect. 4.3; and (d) design and development of the XMANAI manufacturing app (i.e., the Demand Forecasting Manufacturing App) as elaborated in Sect. 4.4.

4.1 In-Depth Analysis

A close collaboration and discussions between the technical partners and the business users needed to occur in order to ensure that the business problem, the "as-is" situation, and the business requirements were explained in depth and that all partners acquired a common understanding in order to properly address the challenges. Various user stories were formulated based on the business needs, including the generation of an accurate demand forecast per each single product in specified time horizons and steps, in an explainable way for different target audiences. The AI needs and the data requirements were identified and discussed, along with the different technical and business aspects that needed to be aligned. The main identified AI and explainability needs were the following:

- AI requirements, including the extraction of D2C profiles and the improvement of forecasting reliability.
- General explainability requirements, including the extraction of insights about demand trends to optimize supply flow, as well as the identification of influential factors for demand and of critical situations that need to be handled to avoid stock breakage.

4.2 Data Acquisition and Exploration

The data employed for the forecasting task were acquired mainly by Google Analytics datasets, which track information from the Whirlpool's website. The data include information regarding product sales and customer-related data and are accessed through frequently updated database tables that are extracted as CSV (Comma Separated Values) files. Google Analytics generates data related to various fields that are not necessarily of use for the task at hand. Hence, there was a need to identify and isolate only a valuable subset of information that could aid the sales forecasting task and provide useful insights for the business users. Based on the Google Analytics documentation, some initial data investigation of the provided fields and various statistics, it was possible to select a subset of approximately 30 fields that were considered the most relevant. Accordingly, data from other database tables, containing Whirlpools internal product information, were also considered and added as needed to the previous extracted subset.

The next phase of data manipulation included data cleaning, identification of irregularities, and exploratory data analysis to detect interesting properties and correlations among the different data fields. The extracted results were discussed among the XMANAI partners, in order to get feedback from the business users regarding some of the findings, the assumptions, and the insights made by the technical users. Based on the findings of this procedure, the business partners agreed to provide additional data, such as product hierarchies/categories, product price, and campaigns.

Next, more features were extracted based on the fields of the selected subset including sales, visits, price, calendar information. Sales' lag features, average number of unique visits, future product price, day of the week, and month are some of the features that were created. An exploratory data analysis focused on the extracted features was performed to conclude on which of them might be the most promising for the forecasting model.

Finally, the datasets that were to be used for the next steps of the analysis were anonymized and uploaded (as CSV files) to the XMANAI platform. Note that certain extensions to the data model were required to ensure that the data uploaded are accompanied by proper semantics (i.e., explanations at the data level).

4.3 Development and Validation of Hybrid XAI Models

Once data exploration concluded and the Whirlpool data were available in the XMANAI platform, the implementation activities for an appropriate predictive model for the sales forecasting use case, along with a suitable explainability model, started.

During the initial experimentation, the focus was on the development of various forecasting models for the most sold individual products and product categories. Whirlpool's D2C channel is new and, therefore, during the first experimentation, Google Analytics contained sufficient information only for a subset of the company's products that permitted the implementation of high-performing predictive models. The examined time horizons were (a) 1 week, (b) 1 month, and (c) 3 months ahead, and the models were implemented both for weekly and daily predictions, whenever this was possible, since daily predictions for 3 months ahead were not feasible. The models were evaluated using appropriate evaluation metrics and the most promising results were obtained by boosting models and more specifically XGBoost.

As the implementation phase progressed over time, the focus moved toward improving the performance of the forecasting models. The selected horizons were fixed to 1 week and 1 month ahead with weekly steps. More data were available through Google Analytics, and the walk forward approach for training the XGBoost models was examined and found to improve the results. Additionally, different product hierarchies and categories were considered based on the input from the business users, and the hierarchical approach was employed to provide more coherent forecasts for the individual levels of the products' hierarchies.

Regarding the explainability aspect of the task, descriptive visualizations depicting correlations among the features and their influence on the target value were employed to provide initial explainability insights. After developing the predictive models, a range of explainability tools [2, 3] has been used, including SHAP, permutation importance, counterfactuals, and what-if scenarios, to generate additional explanation results. This allowed the provision of clear and meaningful insights to business users, while also offering them the flexibility to explore hypothetical scenarios as needed. The work performed included "offline" experimentation, as well as experimentation and configuration of the hybrid XAI models within XAI pipelines in the XMANAI platform.

4.4 Delivery of the XMANAI Demand Forecasting Manufacturing App

Through dedicated brainstorming sessions among the business users and the technical partners of the XMANAI project, the detailed design of mockups for the user interface of the manufacturing app has been defined. The identification of the right visualization tools has been supported by the QFD (Quality Function Deployment) methodology and led to the definition of the type of diagrams and the exact information to be displayed.

The subsequent development activities of the manufacturing app fulfilled the specified visualization requirements, through front-end and back-end development. In addition, they involved integration activities with the XMANAI Platform to retrieve necessary data, results, and explanations from pre-configured XAI pipelines (as described in Sect. 4.3) and display them in the dedicated dashboard presented in Fig. 3. Finally, they insured that only authorized partners could utilize the app through single sign-on functionalities with the XMANAI Platform.

5   Explainable AI Implications and Added Value

For the Whirlpool's use case, the purpose of the explainability activities is to address the needs of the business users who want to comprehend how product sales are influenced by multiple parameters. The concerned business users are the central demand planner, the D2C marketing and sales specialist, and the data scientist/engineer. Different explainability needs have been identified for each of the users. For example, the central demand planner is interested in understanding the causes of critical situations in forecasting process and the root causes that affect

Fig. 3 XMANAI Demand Forecasting Manufacturing App

the forecast accuracy the most, visualizing forecasting plots and getting the trends and the directions based on the past, understanding customer behaviors and the impact of marketing strategies (campaigns, promotions), and finally being able to examine how the output would change based on different company's decisions. The D2C marketing and sales specialist is interested in understanding demand evolution, customers' buying patterns, and the effect of the marketing strategies.

The user's stories developed in the aforementioned activities and indicatively presented in Fig. 4 have been the key drivers to guide the finalization of the visualization tools for XAI deployment. The design methodology was extracted based on user's questionnaires, personas identification, and user's stories definition. All these were integrated into the user's journey description, which embeds the full user's experience within XMANAI platform and Manufacturing App.

The main explainability functionalities that have been identified for business users are the following:

- Demand forecasting visibility
- Demand root cause analysis
- Feature relations visualization
- Feature impact visualization
- Target scenario simulation
- Demand trends identification

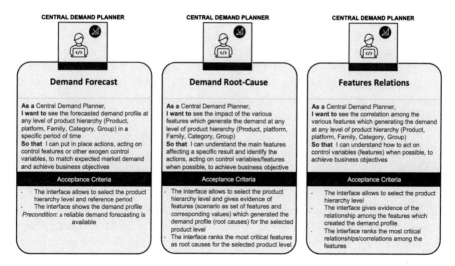

Fig. 4 User's stories example

- Demand trends root cause analysis
- Demand anomalies visualization
- Demand anomalies root cause analysis
- Forecast accuracy visualization
- Buying patterns identification
- Buying patterns root cause analysis
- Customer's behaviors visualization
- Customer's behavior root cause

In order to meet the above-mentioned business needs, the implementations of the explainability approaches were divided into two categories:

(a) Explanation at data level
(b) Explanations at instance and model levels

The primary focus was on the generation of appropriate explanations, followed by the selection and development of suitable visualizations that are more user-friendly and easily interpreted by the business users.

5.1 Explanations at Data Level

In the case of explanations at the data level, the aim has been to extract descriptive visualizations that provide insights in a comprehensive way for the business needs. At this stage, line plots and histograms were employed to explain demand evolution, historical and forecasting patterns, trend lines, and seasonality. Pie charts were employed to explain fluctuations in the forecasts' accuracy based on the inclusion/exclusion of individual features. Partial dependency plots were also examined to illustrate the influence of the features on the output. Finally, heat maps were used to depict the relationships among the features, the sales, and the buying patterns of the customers, as well as the sales with respect to different calendar information. Upon discussion with the business users, the technical partners included only the visualizations that were most appropriate for their needs.

5.2 *Explanations at Instance and Model Levels*

In the case of explanations at instance and model levels, the aim has been to carry out a post-hoc processing of the already trained models using SHAP and permutation importance explainability techniques. In addition, a tool for the creation of what-if scenarios has been provided to inspect the effect of features on hypothetical test cases. The instance-level explanations provide insights into the direction and the contribution of individual features to the model's output. The explanations at this level are significant to the business users as they are capable of understanding individual extreme situations and the factors that were the most influential. The model-level explanations provide a more holistic view of how the features affect the outputs, showing their relevance. Instance-level explanations are typically provided by SHAP force plots. The forecasted value of the model is explained by showing in a graph the feature contributions and the impact direction. Shapley values (average marginal contribution of a feature value over all possible coalitions) are calculated locally for specific instances and shown in a graph provided by the SHAP library (Fig. 5).

After discussing with the business users, a more self-explanatory graph than the default SHAP force plots was requested and designed, with a less data scientist–oriented approach: alluvial plots, as depicted in Fig. 6.

The possibility to generate and explore what-if scenarios is also provided to the business users, who can examine how the predicted value of an instance would change in response to a modification of the input features. Such scenarios (as indicatively presented in Fig. 7) allow the business users to understand how strong the correlation of the input features to the models output is, potentially even suggesting changes that could increase the company's sales.

The model-level explanations are indicatively provided by SHAP and permutation importance. The SHAP library provides and visualizes global explanations by creating a bees-warm plot and aggregating all local feature contributions, while permutation importance generates global feature contributions by computing the change in the forecasting error, as it randomly shuffles each input feature. The results are presented in a table with the most influential features and their weights. In the context of the first phase of the manufacturing app, global explanation results were decided to be depicted through bar plots, upon discussions with the business users (as indicatively shown in Fig. 8).

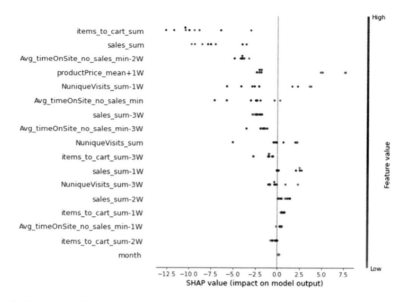

Fig. 5 Instance-level explanations through a SHAP force plots – data scientist–oriented

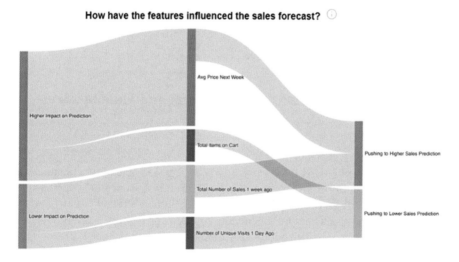

Fig. 6 Alluvial plot example presenting the feature contributions on the output at instance level – business user–oriented

Fig. 7 What-if scenario
example. Impact of the
product price to the products'
sales

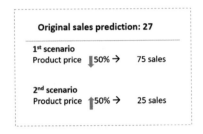

Original sales prediction: 27

1st scenario
Product price ↓50% → 75 sales

2nd scenario
Product price ↑50% → 25 sales

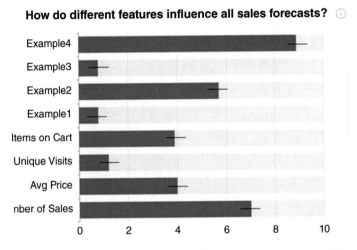

How do different features influence all sales forecasts? ⓘ

Fig. 8 Bar plot example presenting the feature contributions on the output at model level

6 Application of the XMANAI Platform and Manufacturing App

In order to deliver explainable AI results to the target end-users, Whirlpool has leveraged (a) the XMANAI Platform that provides a wide range of functionalities to address explainability from a data, model, and result perspective [4] (as presented in Sect. 6.1) and (b) a dedicated Manufacturing App that offers the user interface with appropriate visualizations specifically selected for demand analysis, what-if simulation, and customer behavior visualization (as outlined in Sect. 6.2).

6.1 XMANAI Platform

In brief, the main aim of the XMANAI XAI platform is to enable the efficient XAI pipelines lifecycle management and to ensure the capability of the user to maintain the best performance level. It's mainly dedicated to IT roles (data scientists and data

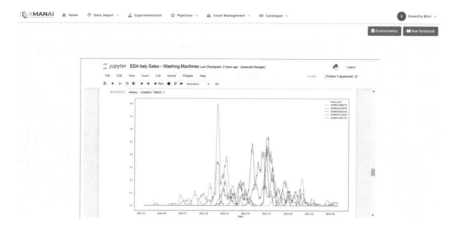

Fig. 9 XMANAI Platform – data exploration over a selected Whirlpool dataset

engineers) and business roles (demonstrator's business/technical users) and follows the specific users' journey identified during the previous design phases [5].

In particular, for the Whirlpool demonstrator, the relevant CSV files that have been extracted from the Whirlpool databases and systems (as explained in Sect. 4.2) have been uploaded in the XAI Platform and then mapped to their equivalent fields in XMANAI Data Model. Once the dataset has been ingested, the metadata have been provided by Whirlpool data scientists. As a last step, an appropriate access policy has been defined for the uploaded datasets by Whirlpool users, in order to ensure full data protection: a dedicated sharing contract signature management function has been created to ensure a secure access to allowed partners. Once the available datasets are available and access rights have been granted, the data exploration phase may start in the Interactive Exploration & Experimentation menu depicted in Fig. 9.

Based on the data exploration outcomes, different models have been trained in the XMANAI Platform and different XAI pipelines have been configured, utilizing both the data preparation functionalities and the ML/XAI functionalities. Once the pipelines were configured, the different execution logs were leveraged by the data scientists of the technical partner to check the progresses made (Fig. 10).

6.2 XMANAI Manufacturing App

The main aim of the XMANAI manufacturing app of Whirlpool is to offer business users the possibility to get access to the explainability dashboard, which is targeted to the "demand forecasting" problem and consists in 3 different web pages offering the functionalities of Demand profile, What-If Scenario and Customer's behavior that map to corresponding menu items.

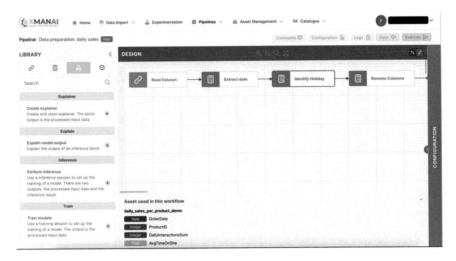

Fig. 10 XMANAI Platform – XAI Pipeline utilizing the Whirlpool data

The "Demand Profile" is dedicated to the demand forecast analysis results visualization for both central planning users and market users. It includes information regarding the sales predictions and the features' influence on the predictions. The information may be presented with a weekly scale in different relevant time horizons in the future (1 week, 1 month) and in the past. Here, the user may get important information not only on predicted sales value for specific products or product families but may also see which features have higher impact on the prediction value and prediction accuracy. This is relevant, as it allows them to understand which levers to use to drive the values. Also, they may get visibility of the correlations among features and awareness about the final result of a potential action. An extract of the "Demand Profile" page is presented in Fig. 11.

The "What-If Scenario" page provides insights into how the sales predictions change depending on the values of the input features, as well as on which features are to be changed (and how) in order to reach a predefined sales prediction. The user can also create new scenarios using a wizard, edit the configuration of the existing ones, view an existing scenario of their interest (as depicted in Fig. 12), or delete those which are no longer useful. All the created scenarios provide the possibility to see the "Demand Profile" diagrams. This function, dedicated to central planning users and market users, allows to see the effect on the output by modifying one or more features. Conversely, it also provides information about the specific values of the modifiable features required to achieve the desired forecast.

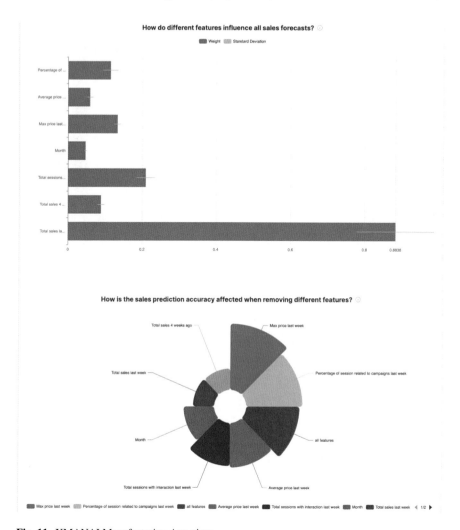

Fig. 11 XMANAI Manufacturing App view

7 Evaluation of the XMANAI Solution

The validity of the XMANAI solution and the status of achievement of the project objectives have been captured, after the completion of the demonstrator sessions held by the users, through questionnaires that identified key results from a user's perspective on different dimensions (in a scale 1–5, where 1 = not at all and 5 = completely).

As the main strategic objective of the XMANAI platform is to gain the users' trust through the explainability associated with artificial intelligence, the users'

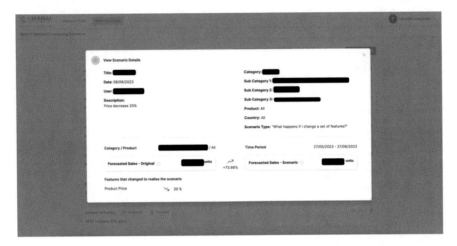

Fig. 12 View of a configured what-if scenario

Fig. 13 Users' XAI
Manufacturing App
evaluation questionnaire for
Whirlpool's use case

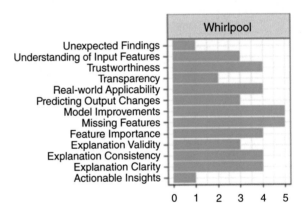

feedback has been particularly meaningful, even in a first release of the XMANAI
results.

As depicted in detail in Fig. 13, the questionnaire results highlighted a strong
improvement on the business impact driven by XAI solution, even if the alpha
release that was assessed still needs to be significantly improved to achieve the
expected excellence. The effort spent on explainability and the refinement path
(which started from visualization diagrams fit for data scientists to a dashboard fit
for business users) has been appreciated even if, at the moment, the expectations are
not fully satisfied. The questionnaire focused on what is still missing and what can
be done to fully reach the goals.

The business objectives that have been identified during the preliminary project
phases and are mainly related to the reliability of the forecast and the explainability
of the results have been summarized as follows:

Fig. 14 Whirlpool's use case
KPIs

Demonstrator KPI	Meaning
DFE Monthly target (Lag1, Lag2)	Demand Forecast Error at SKU level 1/2 months before
DFE Weekly target (Lag 2,3,4)	Demand Forecast Error at SKU/week/market level 2/3/4 weeks before
ATP	Availability To Promise
Sales trend	Revenue's monthly variation
5STARS	Customer appreciation

- Optimization of the **order-to-delivery process**, maximizing the **customer satisfaction** (revenues, margins) with minimal **required resources** (inventory minimization, supply management just-in-time)
- Business **dynamics knowledge** acquisition for people empowerment in driving process

As a consequence, the corresponding business KPIs (Key Performance Indicators) set has been identified in Fig. 14 in order to capture the business impact of XMANAI usage.

In this initial demonstration phase, a detailed and solid measurement of the KPIs set has not been feasible, but the data gathering related to weekly DFE (Demand Forecast Error) generated by XMANAI platform vs the demand generated in the actual management process showed a significant improvement as shown in Fig. 15, notwithstanding an overall fluctuation effect.

To conclude, the users' feedback has been very positive, demonstrating a high confidence on the possibility of further improvement in reliability of prediction, completeness, and effectiveness of the visualization tools.

Finally, an evaluation tool based on 6P methodology [6] has been applied to the demonstrator session after the sessions with the end users. The gathered results have been compared with the same questionnaire submitted at the beginning of the project, in order to capture the "as-is" situation. The assessment has been focused on various aspects of AI and XAI adoption to measure the progress in the development journey during the project lifecycle. The specificity of Whirlpool's use case excluded the possibility to measure some of the dimensions, specific for manufacturing environment. However, the final result has provided clear evidence of the main gap in the dimensions of people readiness for AI and explainability technologies (as depicted in Fig. 16). The strategy to address this gap coverage

KPI	Base Level value (Without XMANAI)*	Measured value with first alpha release*	Expected Value
DFE weekly (Lag1) (reference at total IT market level)	w9: 67% w10: 61.2% w11: 68.2% w12: 81.9% w13: 82% w14: 83.4% w15: 71.1% w16: 76.3% w17: 74.8% w18: 66.3% w19: 71.6%	W9: 0% w10: -6% w11: 1.1% w12: 45.7% w13: 68.5% w14: 100% w15: -8.2% w16: -32% w17: -22.8% w18: 21.7% w19: 0%	<50%

Fig. 15 Whirlpool's use case – preliminary KPIs results for example product range

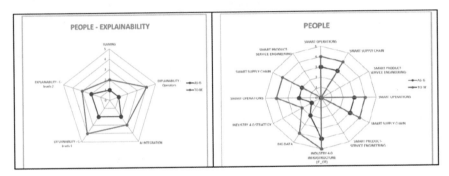

Fig. 16 Whirlpool's use case – 6P assessment result for the PEOPLE dimension after the initial demonstrator phase

represents one of the most interesting indirect results expected out of the XMANAI project experience.

8 Conclusions and Lessons Learnt

Following the completion of the first demonstration phase with the end users of the XMANAI solution, some key lessons learnt have been captured and are to be used for a more effective continuation both of the next project phases and of any XAI initiative:

- As with any AI initiative, the availability of appropriate data is instrumental. In the case of Whirlpool, the need to enrich the underlying data pipelines has been further highlighted (after the initial results) to improve accuracy and explainability of the demand forecasts.

- The end users have significant difficulty in vocalizing and describing their explainability requirements: a preliminary preparation of the target users is recommended in terms of change management to enable awareness of XAI earlier and then competence to get the most out of it.
- Predictions at multiple product hierarchy levels are not trivial and need to be reconciled with appropriate techniques to ensure homogeneous results. Such an activity requires additional work on the XAI pipelines to ensure that all predictions across all product hierarchy levels (e.g., category, item, product function) are consistent, while always in consultation with the business experts to understand the business dynamics.
- The product demand forecasting problem has significant complexity since each product has its own peculiarities and one-size-fits-all XAI models do not work. Therefore, the selection of a single product hierarchy that brings maximum impact and development of targeted XAI models to improve accuracy is a most recommended approach (instead of trying to solve the full product demand forecasting problem at once).

Overall, from the business perspective, the support of a structured change management, which includes communication and focused training actions, seems to be the key success factor for the introduction of XMANAI platform into a business organization. Besides, from a technical point of view, the communication between data scientists and business experts represents some of the real challenges to be addressed and to which XMANAI has tried to contribute.

Acknowledgments The research leading to this work has received funding from the European Union's Horizon 2020 research and innovation programme under Grant Agreement No: 957362.

References

1. Lampathaki, F., Agostinho, C., Glikman, Y., Sesana, M.: Moving from 'black box' to 'glass box' artificial intelligence in manufacturing with XMANAI. In: 2021 IEEE International Conference on Engineering, Technology and Innovation (ICE/ITMC), Cardiff, United Kingdom, vol. 2021, pp. 1–6 (2021). https://doi.org/10.1109/ICE/ITMC52061.2021.9570236
2. Lundberg, S.M., Lee, S.-I.: A unified approach to interpreting model predictions. In: Proceedings of the 31st International Conference on Neural Information Processing Systems (NIPS'17), pp. 4768–4777. Curran Associates Inc., Red Hook (2017)
3. Arrieta, A.B., Díaz-Rodríguez, N., Del Ser, J., Bennetot, A., Tabik, S., Barbado, A., García, S., Gil-López, S., Molina, D., Benjamins, R., Chatila, R.: Explainable Artificial Intelligence (XAI): concepts, taxonomies, opportunities and challenges toward responsible AI. Inf. Fusion. **58**, 82–115 (2020)
4. Miltiadou, D., Perakis, K., Sesana, M., Calabresi, M., Lampathaki, F., Biliri, E.: A novel explainable artificial intelligence and secure artificial intelligence asset sharing platform for the manufacturing industry. In: Proceedings of 2023 IEEE International Conference on Engineering, Technology and Innovation (ICE/ITMC) (2023)

5. Branco, R., Agostinho, C., Gusmeroli, S., Lavasa, E., Dikopoulou, Z., Monzo, D., Lampathaki, F.: Explainable AI in manufacturing: an analysis of transparency and interpretability methods for the XMANAI platform. In: Proceedings of 2023 IEEE International Conference on Engineering, Technology and Innovation (ICE/ITMC) (2023)
6. Spaltini, M., Acerbi, F., Pinzone, M., Gusmeroli, S., Taisch, M.: Defining the roadmap towards industry 4.0: the 6Ps maturity model for manufacturing SMEs. Procedia CIRP. **105**, 631–636 (2022)

Process and Product Quality Optimization with Explainable Artificial Intelligence

Michele Sesana, Sara Cavallaro, Mattia Calabresi, Andrea Capaccioli, Linda Napoletano, Veronica Antonello, and Fabio Grandi

1 Introduction

XAI holds great potential for enhancing operators in modern factories, revolutionizing the way they interact with AI systems and improving their decision-making capabilities. By providing understandable and transparent explanations for the outputs of AI models, XAI empowers operators to gain deeper insights into the underlying processes and logic behind AI-driven recommendations or predictions. This enables operators to make informed judgments, validate AI-generated suggestions, and identify potential errors or biases in the system's output. With XAI, operators can trust AI systems as reliable assistants, leveraging their expertise while retaining control and accountability. The ability to understand and interpret AI's reasoning fosters collaboration between operators and AI systems, leading to more effective problem-solving, optimized processes, and improved overall factory performance. XAI serves as a valuable tool in bridging the gap between human

M. Sesana (✉) · M. Calabresi · V. Antonello
TXT e-tech, Milan, Italy
e-mail: michele.sesana@txtgroup.com; mattia.calabresi@txtgroup.com;
veronica.antonello@txtgroup.com

S. Cavallaro
CNH Industrial, Modena, Italy
e-mail: sara.cavallaro@cnhind.com

A. Capaccioli · L. Napoletano
Deep Blue, Rome, Italy
e-mail: andrea.capaccioli@dblue.it; linda.napoletano@dblue.it

F. Grandi
Università di Modena e Reggio Emilia, Modena, Italy
e-mail: fabio.grandi@unimore.it

J. Soldatos (ed.), *Artificial Intelligence in Manufacturing*,
https://doi.org/10.1007/978-3-031-46452-2_26

operators and intelligent machines, driving a symbiotic relationship that maximizes efficiency and productivity in modern factory settings.

Potential applications of XAI in product optimization are as follows [1]:

1. Feature Importance Analysis: XAI techniques can identify the most influential features or factors affecting product performance, allowing businesses to prioritize and optimize those aspects for enhanced product quality.
2. Failure Analysis and Predictive Maintenance: XAI techniques aid in identifying potential failure modes, understanding failure causes, and predicting maintenance needs. This enables operators to proactively address maintenance issues, reducing downtime and optimizing product performance.
3. Continuous Improvement and Iterative Optimization: XAI enables operators to continuously monitor and evaluate product performance, identifying areas for improvement and iterative optimization. By understanding the factors contributing to product success or failure, operators can drive ongoing enhancements and innovation.

Potential applications of XAI in process optimization are as follows [2]:

1. Anomaly Detection and Root Cause Analysis: XAI techniques can identify anomalies in process data and provide interpretable explanations for their occurrence. This helps operators understand the underlying causes of process deviations and take corrective actions to optimize process performance.
2. Process Monitoring and Control: XAI enables operators to monitor and control complex processes by providing transparent explanations for system outputs and recommendations. This enhances operators' understanding of the process dynamics, facilitating real-time decision-making and adjustments to optimize process efficiency.
3. Process Bottleneck Identification: XAI helps identify bottlenecks in complex process workflows by providing interpretable insights into the factors limiting throughput or efficiency. Operators can then focus on optimizing these bottlenecks to improve overall process performance and productivity.
4. Quality Control and Defect Prevention: XAI techniques can explain the relationship between process parameters and product quality. By understanding the key factors affecting quality, operators can adjust process settings and reduce the likelihood of defects, ensuring consistent and high-quality output.

These applications highlight the valuable role of XAI in process and product optimization, enabling operators to gain insights, make informed decisions, and drive continuous improvement in complex manufacturing processes. By enhancing transparency, trust, and collaboration between operators and AI systems, XAI paves the way for more efficient, cost-effective, and high-quality process operations.

This chapter begins describing CNH Industrial, which is a partner of H2020 XMANAI European research project [3]. CNH explores the use of XAI in a practical manufacturing context. This chapter provides a comprehensive overview of the use case and its application in today's competitive business environment. It delves into the challenges faced in developing XAI within the current manufacturing scenario considering the maintenance operator's needs during the XAI implementation.

Subsequently, the focus shifts to the XAI platform, briefly describing the scope of each component useful to avoid conventional "black boxes" AI approaches, due to their lack of transparency, introducing XAI techniques that brought a new level of interpretability to the field traducing it into "glass box."

Furthermore, this chapter explores the explainability value and the methods to design XAI web-app considering user needs applied to a real-world case study where organizations have successfully employed XAI in their quality optimization processes. This case study highlights the tangible benefits derived from the utilization of explainable AI, such as improved anomaly detection and enhanced process efficiency.

Finally, the chapter concludes by discussing the XAI evaluation within the application on a real manufacturing environment for quality optimization and the Key Performance Indicator selected to measure the global benefits of the explainability. By the end of this chapter, readers will have gained some take aways of the role of explainable artificial intelligence in process and product quality optimization.

2   The CNH Industrial XMANAI Demonstrator

CNH Industrial is a world leader in the design and manufacture of machinery and services for agriculture and construction. It employs more than 40,000 people in 43 manufacturing plants and 40 research and development centers all over the world. The CNH Industrial use case for XMANAI European project focused on San Matteo plant, located in Modena, in Italy, where there is one of the most important research centers in the tractor field in Europe, using the most advanced technologies for design and engineering purposes. In addition to the San Matteo research site, Modena is also home to one of Italy's manufacturing plants where the medium tractor transmissions are produced. They are used to assemble tractors in CNHi factories around the world. It is within this plant that the application cases of the European-funded XMANAI project originate.

The use cases focus on Modena Plant which is currently manufacturing 60,000 APL (all purpose low) and APH (all purpose high) tractors drivelines used to equip all tractors assembled in CNHi plants worldwide. The case study stems from a real problem that the Modena production plant frequently encounters nowadays. In fact, today's production lines are mainly affected by unexpected failures of the production machine that stops the line for undefined periods of time.

Within CNHi's production plant, the needs of the operators who face downtime issues every day lie in restoring the machinery, the beating heart of the plant, in such a quick timeframe that it does not cause major production losses. Downtimes are usually due to replacement of defective parts or for maintenance. XAI could act as a bridge between the machines and the operators, enabling them to understand in a quicker way the status of the machines and improving their productivity through a fast-responsive intervention.

The primary objectives CNHi aims to address through XAI within the XMANAI project, are ingest, manage and analyze the real-time and batch data acquired by CNHi systems, and the faults history data related to the maintenance and tooling systems to give user simplified suggestion thanks to explainability to restore the machine. The goal is to implement XAI models that should help the operator with recommendations, to optimize the production line avoiding waste of time and cost for the company.

In the next sections are described the use case and the current challenges the operators face during the use of machineries and how the proposed platform could enhance their work through the use of XAI.

2.1 Use Case Description

In the current state, within the Modena CNHi production plant, when a machine stops, maintenance operators must exclude different parts of the machine step by step to get to the faulty component and understand where the fault occurred and which anomaly caused the stoppages. As a result, operators waste significant amounts of time troubleshooting the faulty component and for replacing the component. Moreover, if the operator is not able to restore the machine, it is necessary to call for external maintenance operators, which slows down the process even more.

By implementing the XAI platform in the production process, the maintenance operators receive assistance in diagnosing machine errors. This support involves utilizing XAI suggestions derived from the sensor values installed on the machine. Thus, a twofold benefit is targeted:

1. Detect where the fault occurs so that maintenance operators will know which component is responsible and why, which allows to replace immediately the faulty part (for faster recovery).
2. Identifying the specific anomaly responsible for the failure assists operators in tracking the occurrence, minimizing troubleshooting duration, and enabling them to concentrate on pinpointing the root cause, thereby reducing recovery time.

To implement XAI within CNHi plant, the developed platform takes data from the current systems, carrying information about the status of the Heller 400, the CNC machine that is selected for the use cases (Fig. 1). A data pipeline will be created to ingest data and to train the XAI models and give suggestions from AI to the operator in an explainable form. In such a way, the user will test how to create a more organized data management and sharing and how to generate knowledge graphs with clear relationships between data and defined actions. Finally, the trained

Fig. 1 The CNC machine selected for the use case

algorithm provides scheduling suggestions to the production manager to improve the scheduling in the Plant Management System (PMS) and visualization tools employed to aid operators by displaying and elucidating the AI results.

For the CNHi case study, explainability is indispensable to provide the worker with the necessary explanations and clarity to understand which part of the machine equipped with appropriate sensors may have caused the blockage. The worker must therefore be put in a position to decide and understand whether the artificial intelligence algorithm has correctly processed and considered the various possibilities of machine error and why it has arrived at that suggestion, according to which correlation between the various data extrapolated from the sensors. Several sensors were considered to equip the machine to be monitored by AI, but it could also occur that some machine malfunction could not be correlated with the currently installed sensors. Hence, it is crucial that the algorithm through explainability provides all the logical connections used by machine learning to make the operator aware of the machine's operating/malfunctioning state.

2.2 XAI Technical Implementation

The data sources, collected by the sensors installed in the selected plant machine, a CNC work center machine named Heller 400, will interface with the advanced XAI platform developed specifically for the manufacturing industries, called XMANAI platform. The CNC work center is managed by machine conductor and shopfloor people in CNH. The machine is equipped with different sensors that collect data on the operating status of the machine and they are conned to the network. In detail, a dedicated management system is set up to collect sensor data (SmartObserver).

This use case focuses on anomaly detection on sensor data and explanation of machinery faults. The activities carried out for the development of the case study are discussed below, grouped in these different steps:

- Data processing and data analytics: In this task, the complete history of sensor measurements was collected from the Heller 400 CNC machine. Data have been cleaned, sorted, and completed with the missing information, in order to obtain a usable dataset to feed to the selected ML models. Finally, the features selected have been identified as the complete set of 76 sensors data and a collection of recorded anomalies throughout the years.
- Development of intelligent analytical model: In this task, a suitable set of ML models (Isolation Forest) was selected among others, taking into account the type and amount of data at disposal and the end goal of the detection task to be performed. For a more detailed explanation of the selected model, please refer to the upcoming section.
- Training of the selected ML models: In this task, the selected Isolation Forest models were trained on a portion of the available dataset. The goal of the training was to correctly categorize a set of sensor configurations as anomalies or regular data. The set of models was then tested on the remaining portion of the dataset to generate insights about the model accuracy and quality of the predictions.
- Definition of the explainability requirements and visualization tools: As part of this task, the explainability requirements were collected and analyzed to find the best output format in which the model explanation could be displayed, considering the target audience, their knowledge in term of explainability charts, and their explainability needs.
- Production of visual explanations: The trained ML models are exploited to generate the needed explanation using a combination of custom tools and standard graphs, taken from the SHAP [4] library, and taking into account the requirements stated above. In this phase, the models produced explainability charts targeting individual sensor configurations (the one classified as anomalies) with local explanations, grouped sets of sensors (for multivariate explanations), as well as global explanations concerning the overall correlation between anomalies and the sensors having the higher contribution impact on these.

A high-level overview on the motivations behind the selection of an Isolation Forest approach as the basis for the models is provided in following paragraphs, along with the general idea behind the implementation of such technique in both univariate and multivariate cases. To this end, a crucial aspect that has been carefully considered for the current use case was the selection of the model to be used to accomplish the described tasks with the support of the XMANAI platform. The model needed to perform well with various irregular time series of varying lengths. It also needed to keep into account the possibility to add new sensors in the future without compromising accuracy or forcing to homogenize data every time. For all these reasons, an Isolation Forest approach was identified as the most suitable to use. This algorithm selects a feature and then randomly selects a split between minimum and maximum values of the selected parameter. The core idea is that many splits are required to isolate a *normal* point while a small number of splits are required to isolate an *anomaly*. The sequence of splits that brings to an individual data point is called *path*.

Depending on the path length, an anomaly score is computed and interpreted as follows:

- A score close to 0.5 indicates a *normal* point.
- A score close to -0.5 indicates an *anomaly*.

With this basic concept in mind, two approaches were explored:

1. Initially, the method involved fitting an Isolation Forest for each accessible sensor, followed by evaluating the models' outcomes on the day when the failure occurred. Finally build plots to visualize and interpret the results. This method allows each individual model to be fitted on the specific sensor, increasing classification accuracy and making each prediction independent to the number of samples available for other sensors. This approach, however, prevents the user to catch correlation between sensors (since each model only refers to an individual sensor). This aspect is crucial in identifying possible points of failures in the machine, as multiple sensors can contribute to the same anomaly together. For this reason, a second approach was proposed to account for this scenario.

2. The second approach relies in fitting an Isolation Forest for each group of sensors and checking the results of the models in the day of occurrence of the failure. Groups were identified by domain experts considering the placement of each machine component with respect to the others and their reciprocal influence. Finally, some summary plots have been built exploiting the visualization capabilities of the SHAP library to interpret the results. The crucial aspect of this method is the selection of the groups; these should be large enough to allow the model to correlate together as many sensors as possible, while avoid grouping sensors with too different time series shape, so that only a few samples need to be discarded/extended to harmonize all sensors in the group.

The combination of the two approaches described allows the overall process to be precise enough, thanks to the individual models tailored for each sensor, while providing good correlation information, thanks to the categorization of the sensors. The visualizations produced by each of the two approaches are presented in the section below.

2.3   Explainability Value

Methods like questionnaires, user stories, user journey, and personas play a crucial role in exploiting XAI for product and process optimization. These methods facilitate a deeper understanding of user needs, preferences, and experiences, which are essential for designing and implementing effective XAI systems. The importance of these methods is expressed in more detail below:

	Title: **Maintenance operator** Tasks: He is an operator with different level of expertise, specialised in machine maintenance and he is authorised to call the (maintenance) team leader depending on the type and severity of the fault. There are internal procedures that he can carry out based on his experience/training. Devices: Doesn't own a PC or company smartphone → Needs a dedicated workstation to use XMANAI's web-app
	Title: **Maintenance team leader** or maintenance manager (Specialised Technician/Engineer) Tasks: He is responsible for general maintenance of the plant and he has general overview/experience in machine maintenance, explanation of troubleshooting procedures and replacement of components to support operators and maintenance staff. Devices: Owns PC and company mobile phone → Can control XMANAI's web-app.
	Title: **Maintenance Engineer** or maintenance technician (Specialised Technician/Engineer) Tasks: He is responsible for general maintenance of the plant and he has general overview/experience in machine maintenance, explanation of troubleshooting procedures and replacement of components to support operators and maintenance staff. Devices: Owns PC and company mobile phone → Can control XMANAI's cloud platform.

Fig. 2 Personas developed for the CNHi use case

1. Questionnaires: valuable tools for gathering quantitative and qualitative data from users. They help in capturing user perspectives, expectations, and feedback related to product or process optimization, collecting valuable insights that inform the development of XAI models.

2. Personas: represent fictional archetypes of target users, based on real user research. They provide a human-centered perspective, helping businesses empathize with and understand the needs and motivations of different user groups. Personas facilitate the creation of XAI systems that meet various user requirements, ensuring, in the case of CNHi, the optimization of the XAI web app interface for different maintenance profiles (Fig. 2).

3. User Stories: they provide a narrative description of users' interactions with a product or process. They capture users' goals, motivations, and pain points, highlighting key aspects that need to be considered for optimization. User stories also help in identifying specific areas where explainability is crucial to enhance user trust and decision-making. User story descriptions typically follow a simple template as a Card: As a < role>, I want <goal> so that <Benefit>. Finally, the key output from user story is a series of <Acceptance Criteria> preparatory to the interface design of the XAI web app (Fig. 3).

User Journey: maps to visualize the end-to-end experience of users throughout their interaction with a product or process. These maps illustrate touchpoints, pain points, and opportunities for optimization. Understanding the user journey helps in designing XAI systems that provide relevant explanations at the right moments, ensuring a seamless and trusted user experience.

Incorporating user perspectives through the use of these methods enhances the user-centric design of XAI, ensuring that explanations provided by AI systems align with user expectations, facilitate informed decision-making, and build trust. Ultimately, these methods contribute to the successful adoption and exploitation of XAI in optimizing products and processes to meet user needs and improve overall performance.

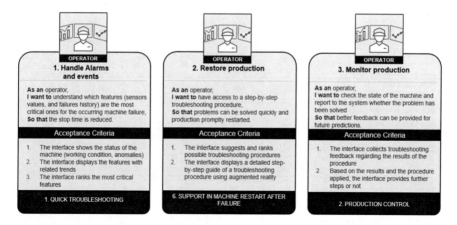

Fig. 3 User stories developed for the CNHi use case

2.4 XMANAI Manufacturing App Experience

This section introduces the XMANAI platform key features and capabilities and discusses their usage in the CNH pilot. Subsequently, the chapter describes the manufacturing Web App developed to connect the generic platform functionalities to the specific final user to solve real manufacturing problems like troubleshooting.

2.4.1 XMANAI Platform and Components Usage

The XMANAI platform was developed to enable XAI takeup specifically for the manufacturing industry. It bridges the gap between the complex nature of AI algorithms and the demand for transparency, interpretability, and trust in decision-making processes within manufacturing operations.

Fig. 4 shows an overview of the components constituting the XMANAI cloud platform. In particular, the components surrounded in a red box are being employed in the context of the CNH demonstrator.

Starting with the *XAI Insight Services* and *Data Manipulation Services*, those are used at an early stage to perform tests on the models and the training data, in order to understand the challenges related to the use case realization (Fig. 5).

The *XAI Secure Sharing Services* allows secure sharing of the input data between the CNH demonstrator and the technical supporting partners, granting CNH full ownership of their data in the platform and the ability to setup sharing policies according to company internal policies.

The *XAI Execution Services* allow to execute the XAI algorithms experimented earlier in a robust production-ready environment, once the models have been finalized.

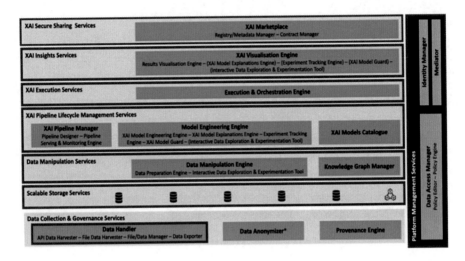

Fig. 4 XMANAI platform components usage

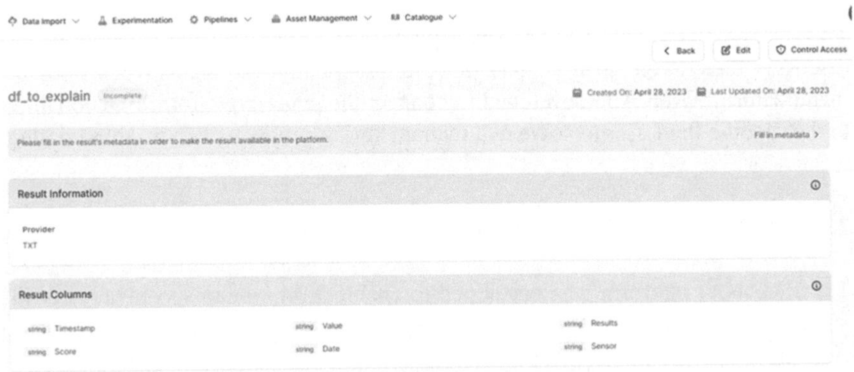

Fig. 5 Visualization of the explanations through the XAI Pipeline Designer

Moving on to the *XAI Pipeline Lifecycle Management Services*, those are being used in all aspects of the process, in order to securely create, schedule, and monitor the execution of XAI pipelines (Figs. 6, 7 and 8).

The *Scalable Storage Services* are used both as the source of input data to be processed and to store the refined data undergoing the process of data ingestion. Those are further utilized to store the trained models and the associated prediction.

Moving to the Data Collection & Governance Services and, in particular, to the *Data Handler* category, the *File Data Harvester* component finds particular use to ingest the raw machinery data in the proper format, following the CNH data model and ensuring type consistency (Fig. 9).

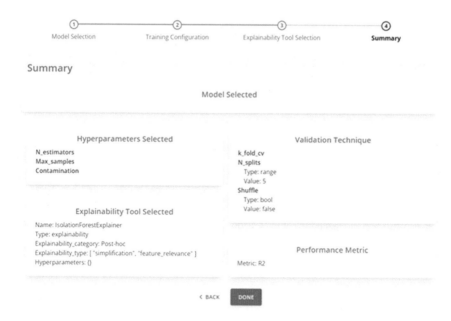

Fig. 6 Configuration of the model through the XAI Model Engineering Engine

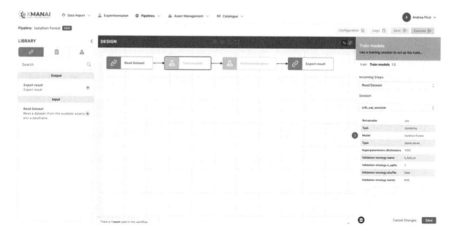

Fig. 7 Design of the Isolation Forest Univariate pipeline through the XAI Pipeline Designer

Finally, the *Platform Management Services* are used to guarantee security and authorization of the users in the platform, allowing access to data and pipelines only to CNH and the supportive partners of choice.

It is important to highlight that, although some of the components haven't been marked as used, they are exploited indirectly through other components of the XMANAI platform. The above description only refers to the components directly used in the realization of the use case.

Fig. 8 Execution and monitoring of the Isolation Forest Univariate pipeline through the XMANAI Platform

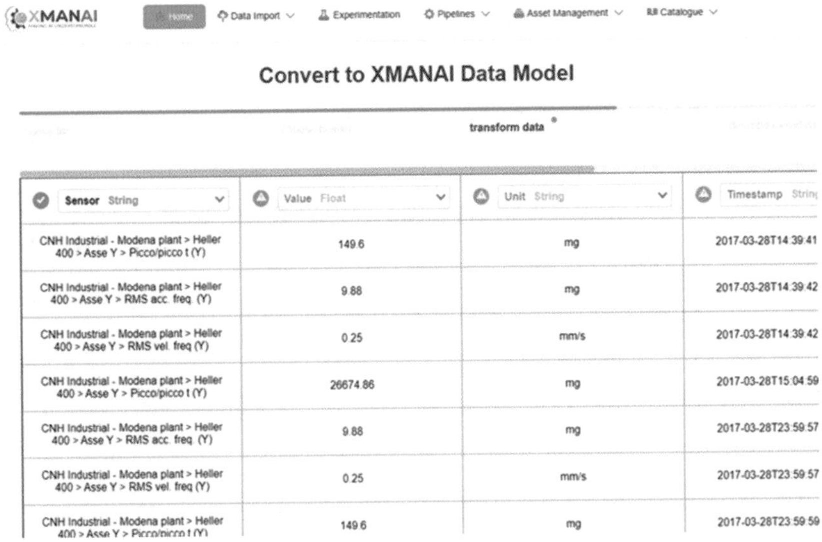

Fig. 9 Dataset is imported into the platform through the File Data Harvester

2.4.2 XAI Powered Manufacturing Web App

The overall explainability design process resulted in the development of an interface for the manufacturing Web App that encompasses the explainability requirements and the acceptance criteria defined in the user stories. The manufacturing Web App, as an explainability tool, refers to a web-based application specifically designed to provide transparent and interpretable explanations for the outputs and decision-making processes of AI systems. It serves as a user interface that enables users to

Fig. 10 (**a**) Home screen; (**b**) sensor values screen; (**c, d**) sensor details

interact with AI models and gain insights into the reasoning behind AI-generated outcomes, coming from the XMANAI platform. The purpose of the Web App is to bridge the gap between the complexity of AI algorithms and the need for human understanding and trust. It provides users, such as domain experts, decision-makers, or end-users, with a user-friendly interface to access and interpret the explanations generated by AI models.

Through the Web App prototype, showed in Fig. 10, end-users can monitor information about the operating condition of the monitored machine. Looking closely at the different interfaces designed, in the Home screen there are information regarding the number of active sensors, the number of alarms and anomalies detected, together with the historical data of failures and a specific feature to compare trends of various sensors. The Sensor Values page displays the hierarchy of the most crucial sensors, utilizing a chromatic scale of red to indicate their level of criticality based on the algorithm confidence value. In the specific Sensor page, there is possibility to view its trend over time and compare different suggested and ranked anomalies. Finally, there is also a tab where the user can provide feedback on whether the platform's suggestion was helpful in solving the fault or not.

The proposed manufacturing Web App prototype design was changed during the development phase due to technical issues related to the availability of data. The final developed manufacturing Web App has a simplified Home screen with just two widgets: Sensors values and Anomalies (Fig. 11a). The Sensor values menu allows the user to check the value of all sensors connected to the faults on different days, while the Anomalies menu includes the entire fault history of the machine. By clicking on Sensor values page (Fig. 11b), the application will show all the sensors in the selected time range, ranked according to the algorithm confidence value for the specific fault, highlighted by an increasingly bright red color showing the increasing criticality of the fault.

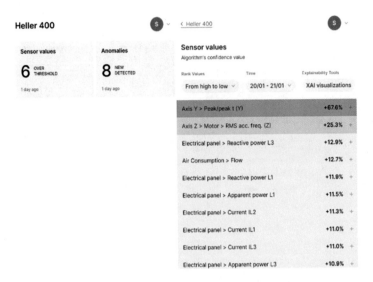

Fig. 11 Web App pages. (**a**) Home screen; (**b**) Sensor values screen

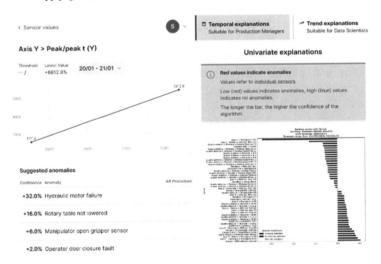

Fig. 12 Web App pages for specific sensor. (**a**) Sensor's trend; (**b**) XAI Visualization button

By clicking on the specific sensor (Fig. 12a), the Web App opens a page with the sensor's trend where the user can check the value for each day and the potential anomalies (i.e., the top ten potential anomalies ranked by the algorithm) associated with that specific sensor. Displayed above the graph are the sensor threshold (if applicable) and the most recent measured value. In Fig. 12b, the graph shows suggested anomalies that can be associated with that specific sensor according to the value. The anomalies are ranked from the most probable according to the algorithm.

On the contrary, by clicking on the XAI Visualization button (see Fig. 12b), the Web App will open a page with all the XAI standard SHAP plots mainly designed for advanced users. Moreover, it is possible to select the type of explanation according to user expertise and role.

In addition to the Web App, an Augmented Reality (AR) application will be available to aid workers in executing maintenance procedures. It will provide step-by-step guidance for machine recovery through a suggested wizard interface. The AR app will assist the operator in solving maintenance problems and help the explainability to take place with the XAI algorithms.

2.4.3 AR App Design

Augmented reality (AR) combined with an XAI approach holds immense potential in supporting maintenance operators in manufacturing context. By overlaying digital information onto the real-world environment, AR provides operators with contextualized and visual guidance, enabling them to perform complex maintenance tasks more effectively. When coupled with XAI, AR becomes even more powerful by offering transparent explanations for AI-generated recommendations or insights, empowering operators to understand and trust the AI's assistance. In the considered use case, the integration of AR and XAI benefits maintenance operators from different points of view:

- Real-Time Visualization and Guidance: AR overlays digital information, such as step-by-step instructions, diagrams, or annotations, onto the physical equipment or machinery being maintained. This visual guidance helps operators locate components, identify potential issues, and follow the correct procedures. By combining XAI, operators can understand the reasoning behind AI-generated instructions, enhancing their confidence and ensuring accurate execution of maintenance tasks.
- Predictive Maintenance and Anomaly Detection: XAI techniques can analyze real-time sensor data from the equipment to detect anomalies or potential failures. AR can then visualize this information, highlighting critical areas that require attention. By providing transparent explanations for the AI's predictions or alerts, operators can understand the factors contributing to potential equipment failures, enabling them to take preventive actions or plan maintenance activities effectively.
- Historical Data Analysis and Process Optimization: XAI techniques can analyze historical maintenance data, identifying patterns and correlations that are not easily noticeable by humans alone. AR can present this analysis visually, enabling operators to understand how past maintenance actions have affected equipment performance and reliability. By comprehending the insights provided by XAI, operators can make data-driven decisions, optimize maintenance processes, and improve overall equipment effectiveness.

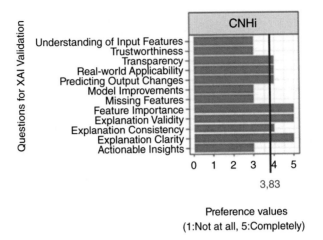

Fig. 13 Questionnaire for XAI evaluation

The integration of AR and XAI in maintenance operations not only improves efficiency, accuracy, and safety but also empowers operators with explainable AI support. By visualizing information and providing transparent explanations, this approach enhances operators' understanding, trust, and collaboration with AI systems, ultimately leading to optimized maintenance processes and increased equipment reliability.

From a practical point of view, the operator launches the AR application from the list of anomalies highlighted in the Web App. The AR application is running on a hand-held device like a smartphone or tablet, so the operator is able to intervene on the machine to restore the production through some digital instruction. Once the procedure is finished, the operator can give a feedback on the proposed workflow, in order to understand if it helped in restoring the machine.

2.4.4 Evaluation of XAI Platform

Definitely, the added value of XAI platform is the explainability associated with artificial intelligence. The evaluation of the platform through the questionnaire shows a positive impact of the XAI implemented in industrial environment. In fact, the ratings recorded during the questionnaire for the validation of the XAI Web App show that the explainability plays a key role (Fig. 13). Each parameter was rated by experts using a scale ranging from 1 to 5 (in which 1 is not at all and 5 is completely), obtaining a mean value of 3.83 (std dev = 0.83). For instance, when considering feature importance, a high value signifies that users can comprehend the significance of each feature in the final algorithm suggestion through the confidence score, expressed as a percentage in the output. And another example is the explanation clarity obtained by adding standard explainability graphs.

Table 1 KPIs selected to measure the benefits of the implementation of XAI

Demonstrator Key Performance Indicators (KPI)	Measured value	Expected value	Means of verification
Trust in XAI predictions for production managers	Medium (60%)	High (at least 70%)	Number of decisions following XAI suggestions vs number of suggestions to solve downtime
Accuracy of XAI assistance in providing predictions relieving the production manager from onerous tasks with low "human value" is low, while maintaining situational awareness and control of the task	Medium (45%)	Medium–high (65%)	Number of preventive maintenance executed/number of unplanned stoppages
Relevant information sharing between the XAI and blue-collar worker in collaborative troubleshooting	High (75%)	High (85%)	Heuristic evaluation/user observation

The chosen Key Performance Indicators (KPIs) used to evaluate the advantages of integrating XAI into the production line primarily center around accuracy, reliability, and human-machine collaboration. These are outlined in detail in Table 1. Despite the current medium-high values displayed by the Key Performance Indicators (KPIs), there remains a necessity for additional optimization. This requirement arises from the fault grouping process, which was conducted solely by data scientists without input or support from maintenance experts. This optimization requires additional support from the maintenance staff to ensure greater accuracy and achieve the expected values.

Furthermore, samples of sensors' data currently available are only based on few months of data coming from the new sensors (recently installed), but it is planned to expand the sample to optimize the predictions of the algorithm and achieve the expected levels of KPIs.

3   Conclusion and Lessons Learnt

During the development and integration of XAI modules for manufacturing tasks in the scope of the XMANAI, several issues have been addressed and overcome, which is summarized in Table 2.

Table 2 Lesson learnt table

Category	Problem/success	Impact	Recommendation
Technology	Difficulties in extracting data from the internal server	No real-time updated data in the final web app	Store data in an internal environment that can communicate easily with the XAI web app
Coordination	Difficulty in coordinating some people from plant (maintenance technicians) available and dedicated for a long time for innovation project	More time to develop certain part of the project that directly involves the maintenance people	Be more flexible when considering the availability of a maintenance worker who is constantly busy within a production plant
Implementation planning	Considering short time for implementing web app in production	More time to develop certain part of the project that directly involves the maintenance people	Planning by paying attention to the availability of operators within the production facilities considering the frequent unplanned stoppage

At the technological level, an initial challenge was encountered in retrieving sensor data stored within the internal server to ensure real-time updates on the XAI web app. There should be an on-premises solution with a dedicated internal server with which the XAI Web App can easily communicate to update data in real time.

At pilot coordination level, there was a difficulty in collaborating constantly and effectively with the plant maintenance staff, who were often busy with emergencies to be solved on the production lines. Greater flexibility should be observed when assessing the availability of maintenance personnel, especially during the planning phase of implementation. It is essential to meticulously plan activities, considering the schedules and availability of maintenance workers.

The expected benefits of using the XMANAI platform will be the beginning of a human-machine collaboration in which the XAI will actively assist the maintenance technician during fault diagnosis. This collaboration will be achieved with a certain level of accuracy and trustworthiness of the XAI that will significantly increase the operators' trust in XAI suggestions.

Acknowledgments The research leading to this work has been carried out within the context of the XMANAI project. The XMANAI project is being funded by the European Commission under the Horizon 2020 Programme (Grant Agreement No 957362).

References

1. Chen, T.C.T., et al.: Explainable artificial intelligence (XAI) in manufacturing. In: Explainable Artificial Intelligence (XAI) in Manufacturing: Methodology, Tools, and Applications, pp. 1–11. Springer International Publishing, Cham (2023)
2. Sidahmed Alamin, K.S., et al.: SMART-IC: smart monitoring and production optimization for zero-waste semiconductor manufacturing. In: 2022 IEEE 23rd Latin American Test Symposium (LATS), Montevideo, Uruguay, pp. 1–6 (2022). https://doi.org/10.1109/LATS57337.2022.9937011
3. XMANAI. https://cordis.europa.eu/project/id/957362, https://ai4manufacturing.eu/
4. Lundberg, S.M., Lee, S.-I.: A unified approach to interpreting model predictions. In: Proceedings of the 31st International Conference on Neural Information Processing Systems (NIPS'17), pp. 4768–4777. Curran Associates Inc., Red Hook (2017)

Toward Explainable Metrology 4.0: Utilizing Explainable AI to Predict the Pointwise Accuracy of Laser Scanning Devices in Industrial Manufacturing

Eleni Lavasa, Christos Chadoulos, Athanasios Siouras, Ainhoa Etxabarri Llana, Silvia Rodríguez Del Rey, Theodore Dalamagas, and Serafeim Moustakidis

1 Introduction

Quality control for manufactured parts in the automotive, aeronautics, and energy sectors is performed by certified metrological laboratories, often by using 3D laser scanners. The part under study is captured in a set of points constituting a point cloud that is the virtual representation of the actual object. The point cloud is filtered to reduce noise and then processed to achieve the final goal of the study, which is dimensional and/or positional measurement and tolerancing. Based on the geometry of the features of the object, the metrologist initially constructs a measurement plan. However, this preliminary study may need to be repeated for less-experienced operators who are less likely to define an appropriate measurement plan from the beginning. This obstacle may lead to both time-consuming processes and inconsistent results between operators.

E. Lavasa · T. Dalamagas
Athena Research Center, Marousi, Greece
e-mail: elavasa@athenarc.gr; dalamag@athenarc.gr

C. Chadoulos · A. Siouras · S. Moustakidis (✉)
AIDEAS OU, Tallinn, Estonia
e-mail: s.moustakidis@aideas.eu

A. Etxabarri Llana
UNIMETRIK S.A., Legutiano, Álava, Spain
e-mail: aetxabarri@unimetrik.es

S. Rodríguez Del Rey
Asociación de Empresas Tecnológicas Innovalia, Calle Rodrígez Arias, Bilbao, Spain
e-mail: srodriguez@innovalia.org

© The Author(s) 2024
J. Soldatos (ed.), *Artificial Intelligence in Manufacturing*,
https://doi.org/10.1007/978-3-031-46452-2_27

479

The goal of this study is to model the differential accuracy of the laser scanning device in response to different geometries and under various scanning configurations. Capture errors are driven by several sources, such as the incident angle, measurement distance, and surface texture. Measurement error is introduced by the instrument itself, which has an anisotropic response by design: accuracy is maximum in the axial direction (axis of movement of the rotary head) and diminishes laterally. Another important source of errors can be the surface orientations that the object's geometry encompasses with respect to the laser source orientation and the viewing direction of the CMOS sensor. In essence, very small incidence angles of the laser light on the surface (i.e., when surface and laser source orientations are quasi-parallel) result in severe backscattering of the light toward the CMOS camera, so large capture errors are introduced. On the other hand, very large incidence angles (close to 90°) also introduce large errors, but for a different reason: most of the laser light is scattered away from the CMOS sensor rather than being collected to capture an accurate representation of the surface.

To address these issues, we propose a novel approach that leverages the power of artificial intelligence (AI) and a specific subfield known as Explainable AI (XAI). XAI is an emerging area of AI that aims at making the decision-making process of AI models transparent and understandable. Unlike traditional AI, where decision-making processes can be black boxes, XAI allows users to understand, trust, and manage AI outcomes. By breaking down complex AI decisions into digestible, human-interpretable components, XAI tools help enhance trust and improve adoption of AI systems. In this context, we propose a novel approach to assist metrologists in defining the optimal scanning setup for the measurement. This approach focuses on the geometrical properties of the objects and enables manufacturers to save time and costs while providing consistent results. We develop an AI-based decision support system to predict the point-wise accuracy of the laser scanner across the surface of the part under analysis. We further apply an explainability tool to provide a comprehensive view of the most important parameters affecting the model's predictions. These user-centered explanations are the key ingredients in building trust and allowing inexperienced operators to have better understanding about the different responses of the scanning device and make informed decisions.

The rest of this chapter is organized as follows: the necessary background on optical industrial metrology as well as XAI applications in industry 4.0 is provided in Sect. 2; the formulation of the problem and the available data sources are outlined in Sect. 3; The proposed methodology and experimental setting are explained in detail in Sect. 4; experimental results and associated explanations are presented in Sect. 5; and conclusions and practical implications are presented in Sect. 6.

2 Background and Related Work

2.1 Optical Metrology

Optical measurement in digital manufacturing has been extensively researched, with a growing emphasis on integrating metrology into the production flow to optimize processes and enable fully automated manufacturing cells [1]. Laser-based instruments have emerged as the most common technology for industrial applications, although on-machine inspection still faces challenges due to high processing temperatures and random process variations. Key trends in this field include the shift toward zero-defect manufacturing strategies, facilitated by in-line measurement and in-process monitoring. However, several challenges must be addressed to ensure effective and efficient integrated metrology, including limitations in measurement and data processing speed, part complexity (size, shape, and surface texture), user-dependent constraints, and measurements in challenging environments [2].

Measurement speed is crucial, as faster measurements without sacrificing accuracy and precision lead to increased throughput and reduced production costs [3]. Addressing part complexity necessitates adaptable and versatile measurement solutions to accommodate various manufacturing scenarios [4]. Challenging environments present additional difficulties, requiring alternative design strategies and robust sensors capable of reliable in situ operation [5]. Furthermore, multi-sensor integration and data fusion are essential to enhance overall measurement quality and provide comprehensive information about parts and processes [6]. This demands robust and efficient methods for sensor compatibility, data synchronization, and data fusion techniques. Addressing these challenges is vital for advancing integrated metrology and improving its effectiveness in modern manufacturing.

Optical inspection methods, particularly laser triangulation measurements, have developed significantly in recent years, providing fast data acquisition, contactless measurements, and high sampling rates. However, challenges remain with the accuracy and reliability of these measurements. The random measurement error is an important aspect to consider, as it quantifies the variation of the actual measurement from its expected value [7]. Factors that affect the measurement error of scanners include the angle of inclination [8], sensor distance from the surface, color [9], texture, and surface reflectiveness [10]. Various studies have investigated the impact of different factors on measurement performance and proposed methods to improve the quality of this process. These include the use of least-squares methods [11], color-error compensation methods [12], and mathematical frameworks for statistical modeling of uncertainties [13].

Some machine learning (ML) methods have also been applied to determine the measurement capacity of scanning devices and improve processes related to non-contact 3D scanning [14–16]. These ML applications include simplifying information about geometry obtained from point clouds and identifying distinct geometrical features of objects. With the ongoing advancement of optical inspection methods, researchers are investigating various techniques to improve the accuracy and reliability of measurements across various contexts [17]. Despite the numerous advantages and improvements offered by ML methods, there remains a notable gap in the current literature. A major concern is that most ML models are considered "black boxes," meaning their inner workings and decision-making processes are not easily interpretable or transparent. Transparency is of crucial importance in various industries, including metrology, since it ensures that stakeholders can trust the results generated by these models. When implementing ML techniques, it is vital to have a clear understanding of how the models arrive at their conclusions, particularly when making critical decisions that directly impact product quality, safety, or compliance with industry regulations. The lack of transparency in current ML models raises questions about their reliability and suitability for widespread adoption in the field of metrology.

To address this gap, our work focuses on developing more transparent and interpretable ML models that can be easily understood by practitioners and stakeholders in the metrology industry. By providing transparency for ML models, researchers can foster greater confidence in their adoption, leading to more accurate and reliable measurements and, ultimately, enhancing the overall quality of optical inspection methods.

2.2 Explainable Artificial Intelligence

The research and application field of XAI has been very active, following the widespread use of AI systems in many sectors related to everyday life and the rising need for interpreting the predictions of these complex systems. Interpretable by design models such as Decision Trees and linear models are outperformed by more complex methods such as Neural Networks, Support Vector Machines (SVMs), or ensemble models. The term XAI mostly refers to post-hoc explainability being applied to interpret-trained ML/DL models with complex internal mechanics that are otherwise opaque. Many different approaches have been proposed to this end (see for example [18] for a comprehensive review), and although taxonomies in the literature are not identical, a first-level distinction of XAI methods mainly pertains to the following:

The applicability of the method: Model-agnostic methods are methods applicable to any kind of model, whereas model-specific methods are designed to explain a certain type of model, considering their particular properties and internal design.

The scale of explanations: global methods explain the overall behavior of the model (at the global level), whereas local explanations refer to individual predictions (at the instance level).

XAI methods are further classified by their scope, with those related to our study being feature attribution methods that aim to quantify the effect of individual input features on the model's output. Global techniques estimate feature importance by the overall change in model's performance in the absence of an input feature or under shuffling of its values, such as permutation importance [19]. Local feature attributions, on the other hand, explain how individual predictions change with each input feature. Shapley Additive eXplanations, or SHAP [20], apply methods from cooperative game theory to machine learning. Like Shapley values measure a player's impact in a game, SHAP values measure each input feature's impact on a model's prediction, compared to the average prediction. Local explanations are then aggregated to provide the ranking of global feature importances.

SHAP is a well-established, model-agnostic method with robust theoretical background that has been widely employed to interpret the predictions of ML models across various application domains, including the industrial sector. In the field of machine prognosis and health monitoring, the authors of [21] designed a deep-stacked convolutional bidirectional Long Short-Term Memory (Bi-LSTM) network. This network predicts a turbofan engine's Remaining Useful Life (RUL) using time-series data measured from 21 sensors. They apply SHAP values to identify the sensors that are mostly affecting the prediction and produce rich visualizations to meticulously study the explanations. In a similar line of work, [22] used SVM and k-Nearest Neighbor models on vibration signals to achieve fault diagnosis of industrial bearings. They depict the most important features influencing fault prediction with SHAP. An XAI approach for process quality optimization is proposed by [23]. They train a gradient-boosting tree-based ensemble on transistor chip production data of Hitachi ABB and identify the most significant production parameters affecting the industry's yield using SHAP. A detailed study on how these parameters affect the production allowed for the prioritization of processes and the selection of improvement actions. Experimental validation of the method on a new production batch showed a remarkable improvement in yield losses, leading the company to adopt the method for further use. These indicative studies show the added value brought about by the SHAP method explaining complex models' predictions, allowing thus otherwise "black-box" models to be transparent to stakeholders.

Despite the active research and applications of XAI in various sectors, its application within the realm of metrology remains relatively unexplored. Specifically, there is a notable gap in the literature regarding work focusing on developing an explainable pipeline for predicting point-wise accuracy of laser scanning devices in the metrology domain. This chapter aims to bridge this gap. Our work contributes to the nascent field of "Explainable Metrology 4.0" by designing a robust, interpretable model that not only predicts the accuracy of laser scanning devices but also provides valuable insights into the features that drive these predictions. By incorporating XAI techniques, particularly the model-agnostic SHAP, into metrology, we aim to transform "black-box" models into transparent tools that empower stakeholders to understand and trust the AI-driven measurement process. This novel approach is envisioned to bring about a significant positive shift in the field of industrial metrology.

3 Methodology

The proposed methodology is described step by step in this section, from data preparation and feature extraction to the established pipelines for model training, optimization, evaluation, and explanation.

3.1 Setting Up a Supervised Learning Task

Our study aims to model the behavior of the scanning instrument, taking into account both the scanning configuration and surface orientation. Specifically, our objective is to predict point-wise measurement errors at individual points along the three axes (X, Y, and Z). This prediction is a function of the scanning conditions, surface orientation, laser orientation, and the viewing direction of the CMOS sensor. Therefore, we can formulate the problem as:

$$\textbf{measurment error} = f \begin{pmatrix} \textbf{scanning conditions, surface orientation,} \\ \textbf{laser orientation, CMOS viewing direction} \end{pmatrix}$$

Having defined the problem to be solved, we seek to establish a supervised learning setting to address its solution. Supervised learning requires access to a certain ground truth that the models are trained to predict. This is achieved with the use of calibrated object measurements. Calibrated artifacts have properties known to be a very high level of precision (of the order of 1 μm = 10^{-6} m). Characteristic properties include dimension/position of features, while shape perfection is also certified. For these objects, we can safely consider that the actual properties, such as dimension, are identical to the nominal ones (provided by the manufacturer). Assuming that the center of the calibrated object is perfectly positioned at the center of the scanning instrument's reference frame, the problem can be formulated as a supervised regression task, where the ground truth is found in the location of points (in X, Y, Z coordinates) on the nominal surface of the calibrated object.

3.2 Data Sources

Our database so far comprises measurements of three calibrated spheres (three diameters), under different scanning configurations.

Scanning configurations are obtained by variation between three levels (low–medium–high) of the following scanning conditions:

- *Lateral density*: Point density in the lateral direction with respect to the movement axis of the rotary head.

- *Direction density*: the velocity of the rotary head, inversely proportional to point density along the axial direction (axis of movement).
- *Exposure time*: the duration of each laser pulse.

As a result, we have an initial database comprising 108 raw point cloud files (3 objects × 3^3 scanning configurations). The number of points in each point cloud varies (min: 16 k, max:180 k) depending on the size of the object and scanning conditions. The total number of points adds up to ~6.5 M.

3.3 Data Pre-processing

Preliminary processing of raw Point Clouds is performed with the Open3D open-source library [24]. Open3D offers a rich collection of data structures and geometry processing algorithms to support the analysis of 3D data. With the use of Open3D functions, two key processing steps are performed:

- *Statistical outlier removal*: Points that are, on average, further away from their neighbors are described as outliers. Configurable parameters are the number of nearest neighbors and the standard deviation ratio to be considered in the calculations.
- *Estimation of surface normal*: The normal vectors are estimated for points in the point cloud based on the number of nearest neighbors within a given radius (these are configurable parameters). Once normal vectors are calculated, a second function is applied to ensure that they are consistently aligned. The alignment of surface normals is verified via visual inspection.

3.4 Feature Extraction

Based on the estimated normal vectors that stand for the surface orientation at each point, a set of informative features is extracted to describe the geometric setup between (a) surface orientation $\vec{N}$, (b) the laser source orientation $\vec{L}$, and (c) the CMOS sensor viewing direction $\vec{V}$. Geometric features include cosine similarities between vectors and vector differences. In addition, the vertical distance between each point and the laser source is calculated, as well as the four-quadrant angle between the laser and the surface orientation (Fig. 1).

Overall, the **input features** to the ML algorithms are listed below:

- Scanning conditions: lateral density, direction density, and exposure time
- Components of surface normal:

Fig. 1 The vectors contributing to the measurement geometric setup are surface orientation (N), laser orientation (L), and CMOS viewing direction (V). Each measured point P is projected along surface orientation to retrieve the closest point P' on the nominal surface

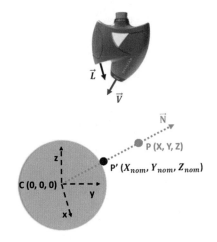

$$\vec{N} = \left[N_x, N_y, N_z\right] \tag{1}$$

- Components of the difference vector $\overrightarrow{ori}$:

$$\overrightarrow{ori} = \vec{N} - \vec{L} = [ori_X, ori_Y, ori_Z] \tag{2}$$

- Components of the difference vector $\overrightarrow{oriCMOS}$ (we use only the Y-component, because X & Z components are equal to those of $\overrightarrow{ori}$ above by design):

$$\overrightarrow{oriCMOS} = \vec{N} - \vec{V} = [oriCMOS_X, oriCMOS_Y, oriCMOS_Z] \tag{3}$$

- Cosine similarity Inc between $\left(\vec{N}, \vec{L}\right)$, that is, the cosine of the incidence angle of light on the surface:

$$Inc = \cos\left(\frac{\vec{N} \cdot \vec{L}}{\left\|\vec{N}\right\|\left\|\vec{L}\right\|}\right) \tag{4}$$

- Cosine similarity *ViewAng* between $\left(\vec{N}, \vec{V}\right)$, that is, the cosine of the viewing angle of the surface from the CMOS camera:

$$ViewAng = \cos \left(\frac{\vec{N} \bullet \vec{V}}{\left\| \vec{N} \right\| \left\| \vec{V} \right\|} \right) \tag{5}$$

- The vertical distance R_s from each point to the laser source, which is always at $Z_{\text{laser}} \approx 90$ mm:

$$R_s = 90 - Z \tag{6}$$

- The four-quadrant angle ang between laser orientation and surface orientation:

$$ang = \arctan2d \left(\left\| \vec{N} \times \vec{L} \right\|, \vec{N} \bullet \vec{L} \right) \tag{7}$$

The target variables, i.e., point capture errors in the x, y, z axes, are defined by projecting captured points on the nominal surface of the calibrated object. The projection is radial in the case of the sphere and axial for the cylinder (the cylinder's axis lies along the Y-axis of the instrument's reference frame). In practice, by projecting each measured point $P(X, Y, Z)$, we find the closest point $P'(X_{\text{nom}}, Y_{\text{nom}}, Z_{\text{nom}})$ on the nominal surface of the calibrated object. For this "reference" point we can assume that:

- If it belongs to a sphere, its actual radial distance to the center of the sphere $C(0, 0, 0)$ is equal to the nominal radius of the sphere R_{nom}. This can be compared to the radial distance R of the measured point P, where:

$$R = \sqrt{X^2 + Y^2 + Z^2} \tag{8}$$

- If it belongs to the cylinder, its axial distance (distance to the Y-axis) is equal to the nominal radius of the cylinder, R_{nom}. This can be compared to the axial distance R of the measured point P, where:

$$R = \sqrt{X^2 + Z^2} \tag{9}$$

This analysis is based on the underlying assumption that the center of the sphere is perfectly positioned at the center of the scanning instrument's reference frame and that the cylinder's axis is perfectly aligned to the Y-axis of the reference frame.

The coordinates of the reference/projected point P' for the sphere are calculated as follows:

$$X_{nom} = X \frac{R_{nom}}{R} \tag{10}$$

$$Y_{nom} = Y \frac{R_{nom}}{R} \tag{11}$$

$$Z_{nom} = Z \frac{R_{nom}}{R} \tag{12}$$

The same formulas as above also apply to X_{nom} and Z_{nom} for the cylinder, whereas $Y_{nom} = Y$, since the projection is axial in this case.

It follows that the target variables for our models, that is, point-wise measurement errors in three axes, can be computed as the difference between projected (nominal) and measured point coordinates:

$$Xerr = X_{nom} - X \tag{13}$$

$$Yerr = Y_{nom} - Y \tag{14}$$

$$Zerr = Z_{nom} - Z \tag{15}$$

Obviously, $Ydev$ is applicable to spherical geometry, while for the cylinder $Ydev = 0$.

Equations 13, 14 and 15 imply that, in the positive X, Y, Z coordinate range, negative errors correspond to points captured outward to the surface, while points with positive errors are captured inward to the surface. The situation is reversed in the negative coordinate range.

Models will be trained and optimized to predict capture errors in each axis, with the same input features, while the target variable will be the measurement error in the respective axis.

3.5 Model Training, Optimization, and Validation

The original PointNet architecture [25] was proposed to deal with tasks pertaining to the segmentation and classification of point clouds. Taking n points as input (each point represented by its spatial coordinates), it initially feeds the input points in a Spatial Transformer Network (STN), with the aim of endowing the overall model with invariability with respect to affine transformations, specifically rotations and translations. The learned features are then passed through a Max Pooling layer, producing a global feature vector. In classification-oriented tasks, this global feature vector is fed to a standard Multi-layer Perceptron (MLP) which outputs scores for each one of the classes under consideration; in segmentation-oriented tasks, an additional step takes place, whereby the global features produced by the Max Pooling layer are concatenated with the output of the last layer of the model, resulting in a feature vector comprising both local and global features. A final pass through a second MLP takes place, and the model outputs a label for each point, indicating the object it belongs to.

In order to utilize PointNet for the prediction of deviations for each point, certain modifications must first take place:

1. Since the problem we are dealing with is neither classification nor segmentation, but regression, the output layer of PointNet must be modified accordingly to reflect the difference in objective. Therefore, the final layer of the model is replaced by a Linear Layer, and the evaluation metric which drives the updates of the model's parameters is changed to the MAE function. The model outputs the deviation of each point in a specific axis; therefore, three distinct models must be trained in order to obtain the overall deviation in 3D space.

2. The employment of the Spatial Transformer Network is redundant in our case, since spherical objects are by default invariant to rotations, and the objects in consideration are not subject to potential translations. Having to learn additional parameters for the STN submodule would only lead to overfitting of the overall model. Therefore, the first step of passing the inputs through an STN submodule is skipped (Fig. 2).

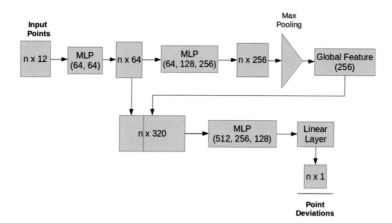

Fig. 2 Modified PointNet Architecture. The network takes n points as input, passes the data through two series of MLPs, and extracts a global feature vector through a MaxPooling layer. The global feature vector is concatenated with the output of the first MLP sub-module and a final MLP is used to perform regression on point deviations

In training the model, grid search and cross validation were performed for the final MLP submodule. Specifically, a range of different numbers of layers and the respective units comprising each layer were tested to identify the optimal combination. The Rectified Linear Unit (ReLU) activation function is used for all MLP layers, and the learning rate is set to adaptive in order to avoid training stagnation.

3.6 Comparative Analysis

An extended search over competing solutions was carried out, with the employment of diverse ML algorithms to address the formulated regression tasks. These include Decision Tree (DT), Support Vector Machine (SVM), and Multi-layer Perceptron (MLP) regressors implemented by using the *Scikit-Learn* Python library [26], as well as the CatBoost regressor from the *Catboost* Python library [27]. MLP is a feed-forward Neural Network, while CatBoost is an ensemble model that combines multiple DT base learners.

Next, we define the strategy that is applied to train, optimize, and validate the performance of the proposed and the competing ML algorithms, which comprised the steps listed below.

1. **Train – test data partitions**

Our training and testing data partitions were built using a random selection of points from each of the available Point Clouds. Specifically, we selected 100 points per Point Cloud randomly and without replacement for the training dataset. To comprehensively test the model's performance, we generated five separate testing datasets, each containing 50 points per Point Cloud, also selected randomly and without replacement. The rationale behind this approach was to obtain a more unbiased assessment of model performance and gain an understanding of the variance in performance scores across different testing datasets.

2. **Performance assessment**

In assessing the performance of our models for the given regression tasks, we used Mean Absolute Error (MAE) and R-Squared (R^2) as our chosen metrics. Importantly, the model optimization process was primarily driven by the MAE score. Performance was separately assessed for each axis, thereby allowing us to individually tune and optimize performance per axis.

3. **Model training & fine-tuning**

Each regression algorithm was provided with a broad hyper-parameter grid for model training and fine-tuning. This was followed by a Randomized Grid Search combined with fivefold cross-validation on the training data partitions. This process allowed us to identify the optimal model of each type based on the mean validation scores, primarily using the MAE score. This randomized grid search helped us explore a wide range of possible parameters and their combinations to find the most effective ones.

4. **Model deployment**

All models were then deployed on the testing datasets. Importantly, these datasets were unseen by the models during the training and optimization phases, safeguarding against overfitting and information leakage. The final performance scores were calculated as the mean scores plus or minus the standard deviation (mean $\pm$ std) derived from the 5 testing data partitions. This strategy gave us a robust and reliable measurement of the model's ability to generalize to new, unseen data.

3.7   Generation of Explanations

Post-hoc explainability techniques refer to methods used to interpret the predictions or decisions of machine learning models after they have been trained. These techniques aim to provide insights and explanations for the decisions made by the model to users. Specifically, we used SHAP (SHapley Additive exPlanations) [20], a method specifically designed to measure the impact of input features on the predictions of a machine learning model. The SHAP method is rooted in cooperative game theory, providing a unified measure of feature importance that allocates the "contribution" of each feature to the prediction for each instance in a principled, theoretically sound manner. SHAP values quantify the change in the expected model prediction when conditioning on that feature, reflecting the feature's contribution to the prediction. They can handle interactions between features and provide fairness attributes because they sum up to the difference between the prediction for the instance and the average prediction. For individual predictions or instances (local interpretability), SHAP values indicate how much each feature in the data set contributes, either positively or negatively, to the target variable. In essence, the SHAP method provides a more detailed view of how specific values of features, whether low or high, impact a prediction by nudging it above or below the average prediction value. This localized insight is aggregated to provide a global understanding of the model, i.e., which features are most important across all instances.

Regarding the integration of XAI into our work, we primarily used the SHAP method to interpret our model's predictions and decisions. After the model was trained and optimized, we computed SHAP values for all features over the entire dataset. This gave us an overall picture of how each feature contributes to the model's decision-making process. In addition, we generated SHAP plots to visually illustrate the impact of individual features on the model's predictions, making it easy to understand the relationships between feature values and their effects on the model's output. This way, the SHAP method provided us with an interpretability tool that helped in understanding the model's complex decision-making process, promoting transparency and trust in the model's outcomes. This utilization of XAI ensured that our model was not just a black box producing predictions, but a comprehensible system where both its decisions and processes are understandable and explainable.

4 Results

4.1 Model Performance – Quantitative Analysis

Tables 1, 2 and 3 showcase the performance of the proposed PointNet-based approach compared to the other machine learning models. PointNet outperformed all other models, achieving the best MAE across all three axes (X, Y, Z) and the highest R^2 score on the Y axis. The second-best performance was demonstrated by the MLP model, which achieved the highest R^2 scores on the X and Z axes. However, the remaining models performed poorly in comparison. These results can likely be attributed to PointNet's inherent design to manage point clouds, which renders it superior to other models that do not consider the adjacency of data points.

Table 1 Model performance for X-axis errors

Model	MAE (mm)	R^2
Decision tree	0.0179 ± 0.0002	0.54 ± 0.02
SVM	0.0203 ± 0.0002	0.52 ± 0.02
MLP	0.0153 ± 0.0004	**0.64 ± 0.02**
CatBoost	0.0181 ± 0.0001	0.56 ± 0.02
PointNet+	**0.0084 (0.0003)**	0.61 (0.013)

Table 2 Model performance for Y-axis errors

Model	MAE (mm)	R^2
Decision tree	0.0191 ± 0.0003	0.69 ± 0.01
SVM	0.0263 ± 0.0003	0.49 ± 0.02
MLP	0.0216 ± 0.0004	0.61 ± 0.03
CatBoost	0.0178 ± 0.0002	0.73 ± 0.02
PointNet+	**0.0018 (0.0003)**	**0.76 (0.018)**

Table 3 Model performance for Z-axis errors

Model	MAE (mm)	R^2
Decision tree	0.0256 ± 0.0005	0.37 ± 0.02
SVM	0.0288 ± 0.0004	0.26 ± 0.01
MLP	0.0268 ± 0.0005	**0.57 ± 0.03**
CatBoost	0.0249 ± 0.0004	0.40 ± 0.01
PointNet+	**0.011 (0.0002)**	0.41 (0.009)

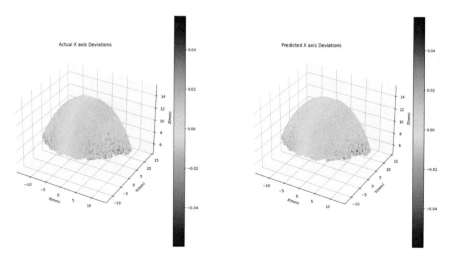

Fig. 3 Predicted versus actual deviation maps for the X-axis

4.2 Qualitative Analysis – Error Maps

Figures 3, 4 and 5 illustrate the predicted deviations of the data points, presenting a distinct 3D deviation map for each axis. Both predicted and actual deviations are represented in each figure for all three axes. These figures effectively highlight the proficiency of the proposed model in predicting the measurement accuracy (deviation) of the scanning device, with the predicted deviations closely mirroring the actual ones. However, a noticeable discrepancy between the predicted and actual deviations is only discernible in the case of the Z axis, corroborating the findings from Tables 1, 2 and 3. Overall, the consistency between the predicted and actual deviations underscores the efficacy of the proposed model, demonstrating its potential to be a reliable tool for estimating the accuracy of scanning devices. Even in the case of the Z axis, where slight deviations occur, the model provides insightful data that could lead to further improvements and enhanced predictability in the next iterations.

4.3 Explanations

Figures 6, 7 and 8 feature the SHAP summary plots of the trained PointNet model, presented individually for each of the three axes. Features appear on the y-axis, ordered by their global importance in the model's output. SHAP values are measured on the x-axis, so that each point in the summary plot corresponds to a SHAP value for a given sample and feature. The magnitude of feature values is color-coded from blue to magenta, corresponding to low and high values,

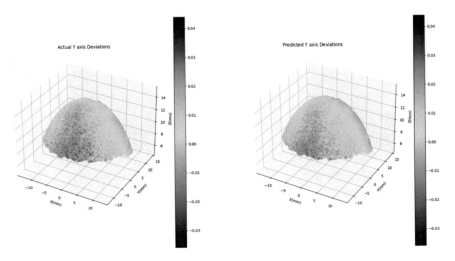

Fig. 4 Predicted versus actual deviation maps for the Y-axis

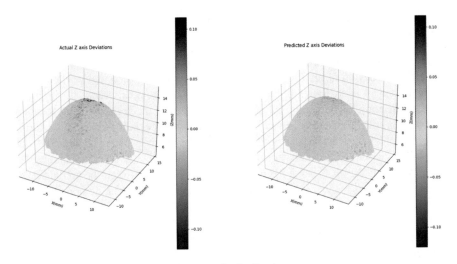

Fig. 5 Predicted versus actual deviation maps for the Z-axis

respectively, so that the effect of different feature values on the model's output can be deduced by the summary plot. More specifically, the summary plot is indicative of how feature values (high/low) push individual predictions upward or downward with respect to the average prediction, which is specified by the vertical line at SHAP value zero.

The most important features affecting the measurement errors of the predicted point on the X-axis are found to be the x-component of surface orientation (Nx) and the x-component of the vector difference between laser and surface orientations

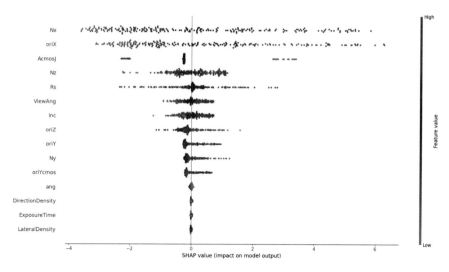

Fig. 6 SHAP summary plots of the trained PointNet model for the X-axis

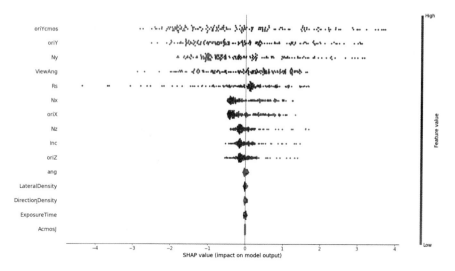

Fig. 7 SHAP summary plots of the trained PointNet model for the Y-axis

(oriX). For both of these features, low values correspond to outputs lower than the average prediction (i.e., toward negative measurement errors), while outputs larger than average (i.e., toward positive errors) are induced by high feature values. Points with low Nx values are those facing the negative side of the X-axis, while large Nx corresponds to points facing the positive side. Taking into account the calculation formula (Eq. 13), it follows that points on both sides are captured inward to the surface, with the larger measurement errors corresponding to parts of the object

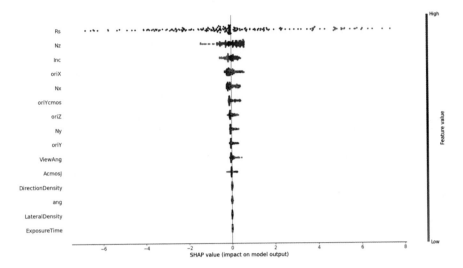

Fig. 8 SHAP summary plots of the trained PointNet model for the Z-axis

where the surface is oriented toward the x-axis (lateral direction), scattering light away from the CMOS sensor.

Regarding predicted point measurement errors in the Y-axis, these are mostly affected by the Y-components of vector differences between surface and CMOS orientation (oriYcmos) and surface and laser orientation (oriY). Large values for these features correspond to points mostly facing the positive side of the Y-axis, while low values indicate points mostly facing the negative side of the axis. Considering the calculation formula for measurement errors (Eq. 14), these correspond to points being misplaced inward to the surface. The effect is most visible in areas of the surface facing toward (positive or negative) the Y-axis (scan direction), where most of the incident light is scattered away from the CMOS sensor due to large reflecting angles.

Predicted measurement errors in the Z-axis are highly affected by the measurement distance (Rs) between each point and the rotary head of the scanner, where the laser source and CMOS camera are located. Considering the calculation formula (Eq. 15) and the fact that the objects under study lie in the positive Z-coordinate range, we find that points at large measurement distances are misplaced outward to the surface (negative errors), while points closer to the laser and CMOS (low measurement distance) tend to be captured inward to the surface (positive errors).

Overall, we observe that the model's response in the prediction of errors along each axis is affected by relevant properties of the geometric setup that match optics intuition.

5   Discussion

5.1   Discussion on Experimental Results

This study presents an innovative approach using a PointNet-based model to handle point cloud data with a focus on predicting measurement accuracy in scanning devices. Unlike traditional machine learning models, the design of the proposed approach is inherently equipped to manage point clouds, accounting for the adjacency of data points, making it ideal for this specific task. Leveraging the unique capabilities of the PointNet-based model, our research aimed to develop a more precise and reliable tool for estimating the accuracy of point cloud scanning devices, an area of increasing importance in numerous fields such as robotics, geospatial sciences, and architecture.

The results of our study demonstrate the superior performance of the PointNet-based approach compared to traditional machine learning models. Our quantitative analysis revealed that the PointNet-based model achieved the lowest MAE on all three axes (X, Y, Z) and attained the highest R^2 score on the Y axis, reflecting its enhanced predictive accuracy. The qualitative analysis, in the form of 3D deviation maps, visually underscored the proficiency of the model in predicting deviations. These maps revealed a significant consistency between the predicted and actual deviations, validating the model's effectiveness. However, a slight discrepancy observed on the Z axis highlights an area for potential improvement. Despite this, even the deviation on the Z-axis offers valuable insights that can be used to refine future iterations of the model, ultimately improving its predictability and reliability.

The SHAP summary plots, represented in Figs. 6, 7 and 8, play a crucial role in explaining how the PointNet-based model operates, highlighting that the model's predictions are largely guided by key geometric properties consistent with principles of optics. Specifically, measurement errors on the X-axis are influenced by the surface's orientation and its vector difference with the laser, while Y-axis errors are significantly impacted by the surface's vector differences with the CMOS and laser orientation. For the Z-axis, the model recognizes that errors are largely dependent on the measurement distance between the point and the scanner's rotary head. This ability of the proposed model to incorporate and leverage these geometric and optical principles when making predictions not only underscores its robustness but also provides a high degree of explainability that enhances its practical utility in the field of metrology.

5.2   Limitations of the Analysis

This study presents some limitations, mainly regarding the data utilized. We opted to demonstrate the models' efficiency using spherical geometrical objects, which embody diverse conditions in terms of angles between the laser orientation, surface orientation, and CMOS viewing orientation. However, to assess the generalizability of the proposed approach, the analysis should be extended to include non-spherical geometrical objects. The key point here is that this experimentation requires calibrated objects, rendering the data collection process both costly and time-consuming. Looking ahead, we plan to expand our research to incorporate additional geometries, including both convex and concave structures, to further challenge and refine our model's capability. By considering diverse geometrical shapes, we aim to enhance the model's robustness and ensure its applicability across a wider array of scenarios. This extension, though demanding in terms of resources, would enable a more comprehensive understanding of the model's performance, contributing significantly to the development of more accurate and versatile scanning devices in the future.

5.3 Practical Implications and Future Perspectives

The application of AI in metrology, particularly with the advent of XAI, holds the potential to dramatically enhance both the accuracy and efficiency of measurements, a revolution with profound implications for sectors such as aeronautics and automotive. The integration of ML algorithms can improve precision by identifying patterns and correcting errors within large datasets, consequently fostering automation and promoting real-time data analysis. This approach can result in advanced decision-making, where metrologists are empowered with data-driven insights and correlations, potentially reducing manual interventions significantly. XAI is poised to bring a new layer of transparency and interpretability to these intricate AI systems, providing users with a clearer understanding of the decision-making processes of these models, thus encouraging trust and validation in AI-driven measurements. The use of XAI can also allow for the identification of biases or errors within AI models, promoting a deeper understanding of the measurement process as a whole. This transparency will enhance the credibility, accountability, and reliability of AI in metrology, reinforcing confidence in its application.

Looking forward, we envisage that AI will facilitate adaptive calibration and compensation by dynamically adjusting measurement systems to cater to environmental changes and complex geometries. XAI, in particular, is expected to provide significant advancements in the metrology sector by enabling professionals to gain deeper insights into the inner workings of AI models. This understanding will facilitate data-driven decision-making and allow for potential error sources to be identified and mitigated, promoting continuous improvement in the metrological process. Furthermore, XAI's role in knowledge transfer, specifically in assisting junior metrologists to understand and validate results, will be instrumental in achieving higher accuracy, repeatability, and reliability in measurements, ultimately leading to zero-defect manufacturing. Companies focused on metrology products and services, like Unimetrik and Trimek, could significantly benefit from XAI's ability to support decision-making during measurement plan definition, enabling time-saving and error reduction, particularly for less experienced professionals. The adoption of XAI in metrology is thus poised not only to improve process efficiency but also to enhance understanding of the metrological process by providing professionals insights into AI models and the impact of key configuration parameters on measurement accuracy.

6 Conclusions

This chapter effectively demonstrated the superiority of a PointNet-based model in predicting 3D scanning device accuracy, significantly outperforming traditional machine learning models. The unique 3D deviation maps contributed greatly to the model's explainability, offering clear insight into its decision-making process and resultant errors. Despite a minor discrepancy noted on the Z axis, this study represents a promising advancement in metrology, demonstrating the potential of machine learning models to improve the accuracy, reliability, and explainability of measurements in industries that rely on 3D scanning technology. The consequent benefits have broad implications, spanning from precision manufacturing to preserving archaeological artifacts, bringing about a new era of improved accuracy and accountability in metrology.

Acknowledgments This research was supported by the European Union's Horizon 2020 research and innovation program under grant agreement No 957362, project XMANAI (eXplainable MANufacturing Artificial Intelligence).

References

1. Gao, W., Haitjema, H., Fang, F.Z., Leach, R.K., Cheung, C.F., Savio, E., Linares, J.M.: On-machine and in-process surface metrology for precision manufacturing. Ann. CIRP. **68**, 843–866 (2019)
2. Catalucci, S., et al.: Optical metrology for digital manufacturing: a review. Int. J. Adv. Manuf. Technol. **120**, 4271–4290 (2022). https://doi.org/10.1007/s00170-022-09084-5
3. Caggiano, A.: Cloud-based manufacturing process monitoring for smart diagnosis services. Int. J. Comput. Integr. Manuf. **31**, 612–623 (2018)
4. Leach, R.K., Bourell, D., Carmignato, S., Donmez, A., Senin, N., Dewulf, W.: Geometrical metrology for metal additive manufacturing. Ann. CIRP. **68**, 677–700 (2019)
5. French, P., Krijnen, G., Roozeboom, F.: Precision in harsh environments. Microsyst. Nanoeng. **2**, 1–12 (2016)
6. Remani, A., Williams, R., Thompson, A., Dardis, J., Jones, N., Hooper, P., Leach, R.: Design of a multi-sensor measurement system for in-situ defect identification in metal additive manufacturing. In: Proceedings ASPE/Euspen Advancing Precision in Additive. Manufacturing (2021)
7. Joint Committee for Guides in Metrology: Evaluation of measurement data—the role of measurement uncertainty in conformity assessment. JCGM. **106**, 2012 (2012)

8. Pathak, V.K., Singh, A.K.: Optimization of morphological process parameters in contactless laser scanning system using modified particle swarm algorithm. Measurement. **109**, 27–35 (2017)
9. Vukašinović, N., Bracun, D., Mozina, J., Duhovnik, J.: The influence of incident angle, object colour and distance on CNC laser scanning. Int. J. Adv. Manuf. Technol. **50**, 265–274 (2010). https://doi.org/10.1007/s00170-009-2493-x
10. Mueller, T., Poesch, A., Reithmeier, E.: Measurement uncertainty of microscopic laser triangulation on technical surfaces. Microsc. Microanal. **21**, 1443–1454 (2015)
11. Isa, M.A., Lazoglu, I.: Design and analysis of a 3D laser scanner. Measurement. **111**, 122–133 (2017)
12. Li, S., Jia, X., Chen, M., Yang, Y.: Error analysis and correction for color in laser triangulation measurement. Optik. **168**, 165–173 (2018)
13. Mohammadikaji, M., Bergmann, S., Irgenfried, S., Beyerer, J., Dachsbacher, C., Wörn, H.: A framework for uncertainty propagation in 3D shape measurement using laser triangulation. In: Proceedings, IEEE International Instrumentation and Measurement Technology Conference, pp. 1–6 (2016)
14. Wissel, T., Wagner, B., Stüber, P., Schweikard, A., Ernst, F.: Data-driven learning for calibrating galvanometric laser scanners. IEEE Sens. J. **15**, 5709–5717 (2015)
15. Bos, A., Bos, M., van der Linden, W.E.: Artificial neural networks as a multivariate calibration tool: modeling the Fe–Cr–Ni system in x-ray fluorescence spectroscopy. Theor. Chim. Acta. **277**, 289–295 (1993)
16. Urbas, U., Vlah, D., Vukašinović, N.: Machine learning method for predicting the influence of scanning parameters on random measurement error. Meas. Sci. Technol. **32**(6), 065201 (2021). https://doi.org/10.1088/1361-6501/abd57a
17. Vallejo, M., de la Espriella, C., Gómez-Santamaría, J., Ramírez-Barrera, A.F., Delgado-Trejos, E.: Soft metrology based on machine learning: a review. Meas. Sci.Technol. **31**, 032001 (2019)
18. Barredo Arrieta, A., Díaz-Rodríguez, N., Del Ser, J., Bennetot, A., Tabik, S., Barbado González, A., Garcia, S., Gil-Lopez, S., Molina, D., Benjamins, V.R., Chatila, R., Herrera, F.: Explainable Artificial Intelligence (XAI): concepts, taxonomies, opportunities and challenges toward responsible AI. Inf. Fusion. **58** (2019). https://doi.org/10.1016/j.inffus.2019.12.012
19. Breiman, L.: Random forests. Mach. Learn. **45**(1), 5–32 (2001)
20. Lundberg, S., Lee, S.-I.: A unified approach to interpreting model predictions. In: Advances in Neural Information Processing Systems, vol. 30. Curran Associates, Inc. (2017)
21. Hong, C.W., Lee, C., Lee, K., Ko, M.-S., Kim, D.E., Hur, K.: Remaining useful life prognosis for turbofan engine using explainable deep neural networks with dimensionality reduction. Sensors. **20**(22), 6626 (2020)
22. Brusa, E., Cibrario, L., Delprete, C., Di Maggio, L.G.: Explainable AI for machine fault diagnosis: understanding features' contribution in machine learning models for industrial condition monitoring. Appl. Sci. **13**(4), 2038 (2023)
23. Senoner, J., Netland, T., Feuerriegel, S.: Using explainable artificial intelligence to improve process quality: evidence from semiconductor manufacturing. Manag. Sci. **68**(8), 5704–5723 (2021)
24. Zhou, Q.-Y., Park, J., Koltun, V.: Open3D: a modern library for 3D data processing. arXiv preprint arXiv, 1801.09847 (2018)
25. Qi, C.R., Hao, S., Mo, K., Guibas, L.J.: Pointnet: Deep learning on point sets for 3d classification and segmentation. In: Proceedings of the IEEE Conference on Computer Vision and Pattern Recognition, pp. 652–660 (2017)
26. Pedregosa, F., et al.: Scikit-learn: machine learning in python. JMLR. **12**, 2825–2830 (2011)
27. Prokhorenkova, L., Gusev, G., Vorobev, A., Dorogush, A.V., Gulin, A.: CatBoost: unbiased boosting with categorical features. In: Advances in Neural Information Processing Systems (2018)

Index

Printed in the United States
by Baker & Taylor Publisher Services